Lecture Notes in Earth Sciences

Edited by Gerald M. Friedman

Rochester, 20. II. 87

1

Sedimentary and Evolutionary Cycles

Edited by Ulf Bayer and Adolf Seilacher

Springer-Verlag
Berlin Heidelberg New York Tokyo

Editors

Dr. Ulf Bayer
Prof. Adolf Seilacher
Institut für Geologie und Paläontologie der Universität Tübingen
Sigwartstr. 10, D-7400 Tübingen, F.R.G.

ISBN 3-540-13982-6 Springer-Verlag Berlin Heidelberg New York Tokyo
ISBN 0-387-13982-6 Springer-Verlag New York Heidelberg Berlin Tokyo

Printed in Germany

Printing and binding: Beltz Offsetdruck, Hemsbach/Bergstr.
2132/3140-543210

Preface

This volume is the outcome of a Symposium held at Tübingen Sept. 15.-17. 1983 which was sponsored by the DFG and the 'Sonderforschungsbereich 53'. It provides a final report of the project B20: "Bankungsrhythmen in sedimentologischer, ökologischer und diagenetischer Sicht" (directed by U. Bayer) although research continues until the end of 1984 and then will be briefly summarized elsewhere.

We are indebted to G.R. McGhee and D. Bayer for help in preparing this volume, to W. Wetzel for photographic work, R. Stephani and H. Wörner for typewriting, and D. Wiesner for organizing the symposium. We are greatful to the Springer Verlag for supporting the production of this volume and, of course, to the authors for their contributions.

Tübingen, December 1984

U. Bayer   A. Seilacher

CONTENT

INTRODUCTION

In the context of evolutionary studies, it is the privilege of paleontologists to trace the actual course of evolutionary change over time spans that are adequate for such a slow process. At the same time it is their crux that they can not always hope to do this with the resolution necessary to reveal the causal relationships involved.

The Tübingen Sonderforschungsbereich 53, "Palökologie", was primarily geared to study the interrelationships between organisms and environments in the fossil record. As is pointed out in this volume, such an approach will necessarily emphasize the static aspect of this relationship, all the more since this is what we need for the practical purposes of facies recognition. This was done during a time interval of thirteen years at the level of individual species and taxonomic groups ("Konstruktions-Morphologie"), of characteristic facies complexes ("Fossil-Lagerstätten") and of assemblages ("Fossil--Vergesellschaftungen") with the aim to recognize general patterns that persist in spite of the historical and evolutionary changes in the biosphere.

But as our project came closer to its end, the possible causal relationships between physical and evolutionary changes became more tangible. This trend is expressed by symposia devoted to the biological effects of long term tectonic changes (KULLMANN & SCHÖNENBERG, eds., 1983) and of short term physical events (EINSELE & SEILACHER, eds., 1982). But in retrospect it appears that the time scales of the environmental changes chosen were either too large or too small to reveal the mechanisms of evolutionary response.

The present volume is the outcome of a symposium of the projects B 20 ("Bankungsrhythmen in sedimentologischer, ökologischer und diagenetischer Sicht", directed by U. BAYER), D 40 ("Analoge Gehäuse-Aberrationen bei Ammonoideen", directed by J. WIEDMANN) and D 60 ("Substratwechsel im marinen Benthos", directed by A. SEILACHER) in September, 1983. It addresses environmental changes at time scales large enough to produce more than a local ecological response and short enough to observe evolutionary and/or migratory changes at the species and genus levels. It also focusses on basins which by various degrees of isolation provided suitable sites for "evolutionary experiments", such as lakes and marginal epicontinental basins.

In a way, this book is a successor of the previous one on "Cyclic and event stratification" (EINSELE & SEILACHER, eds., 1982). Small scale cycles and events are the 'primitives' of a sedimentary sequence, the lowermost scale from which it can be deciphered. However, medium and long term physical cycles commonly impress sedimentological and lithological trends on the stratigraphic column which are accompanied by faunal replacements and cycles. But since sedimentation is controlled both by physical and biological processes, which are intercorrelated in complicated ways, we also need to decode the stratigraphic text. In this effort, paleontological and sedimentological interpretation must go hand in hand. On the 'megascale' of global sea-level changes faunal and species evolution is triggered by opening and closing of migration pathways, sometimes providing us with major biostratigraphic boundaries.

As it turns out, however, integrated research and the choice of suitable scales do not free us from problems of resolution. Thus our inability to distinguish local speciation from ecophenotypic modification and from immigration in the fossil record excludes definite evolutionary answers even in well studied cases. Nevertheless we hope that this approach opens a fruitful discussion, in which stratigraphy, systematic paleontology and paleoecology will be reconciled in a concerted effort to eventually understand the evolutionary mechanisms of our biosphere.

U. Bayer   A. Seilacher

References

Einsele, G. & Seilacher, A., eds. 1982: Cyclic and event stratification. (Springer) Berlin, 536 pp.

Kullmann, J., Schönenberg, R., Wiedmann J., eds. 1982: Subsidenz-Entwicklung im kantabrischen Variszikum und an passiven Kontinentalrändern der Kreide. Teil 1, Variszikum: N. Jb. Geol. Paläont. Abh. 163(2), 137-300; Teil 2, Kreide: N.Jb. Geol. Paläont. Abh. 165(1), 1-183.

PART 1

SEA LEVEL CHANGES:

General Consequences

Concerning biological evolution, there are two main viewpoints:

> *(1) Organisms depend on other organisms, as predator and prey, and they are always in competition with individuals of their own species and with other species. As they move and shift their ecological niches -- which they define by their existence -- they change the biological environment. Therefore, other species have to react in the struggle of the fittest, and thus a small move may initiate a snowball effect comming back to its origin much later and forcing it to move again.*

If this biological 'Red Queen Hypothesis' is correct, the only task left for paleontology would be to register the succession of form because any causality is lost in the biological system, and evolution should appear random. This contrasts with paleontological experience and the adaptionist's program which forms the paleontological leg of this volume.

> *(2) Species are adapted to certain environments, and they move as these environments change. The external forces are manifold, such as gradual plate tectonic changes, or astronomic events that cause short but intensive bottle neck effects etc.. Between such long term changes and events the earth provides rather stable environments which allow the biosphere to adapt and to reach a kind of equilibrium state.*

Perfect quietness, however, is also not characteristic for environments. As major events vanish smaller scale changes and events take over, still altering environmental conditions. Major forces on this level are sea level changes which over and over have been recognized in the earth's history. Besides climatic changes, which are likely to accompany them, major sea-level changes may trigger faunal evolution in three ways:

- ** *the opening and closing of migration pathways -- geographical isolation on intercontinental levels -- are addressed by the contributions of A. Hallam and G.E.G. Westermann*
- ** *the change of evolutionary rates due to variable extensions of environments is discussed by J. Kullmann*
- ** *the diversity of faunas controlled by migration and evolutionary rates is addressed by D.T. Donovan.*

These factors are the topic of the first part. However, the possible reactions of the biosphere to such changes depend on the flexibility of species. Besides this, Tintant & Kabamba remind us that the biological potential 'preadaptation' of a taxonomic group is another important factor and that our knowledge of adaptive pathways depends on the available taxonomy.

The always present external perturbations are important in two ways for marine environments: They dampen the 'Red Queen' oscillations by changing ecological relationships between faunas in geological times, and they repeatedly provide isolated areas allowing for adaptation under slow selection followed by spreading of the few 'revolutionary organisms' during transgressive times.

JURASSIC MOLLUSCAN MIGRATION AND EVOLUTION IN RELATION TO SEA LEVEL CHANGES

A. Hallam
Birmingham

Jurassic sea level exhibited a secular trend, on which were superimposed short-term oscillations (HALLAM, 1978b), from a low stand at the beginning of the period to a high stand in the Oxfordian-Kimmeridgian, followed by a reversal to a low stand at the end. Analysis of the global distribution of marine bivalve genera indicates broadly speaking an inverse correlation between sea-level stand and endemicity. High endemicity correlates with times of comparative regression and low endemicity (or high cosmopolitanism) with times of comparative transgression of epicontinental seas. This broad relationship, which can be matched with marine invertebrate data from other periods, is readily explained in terms of the comparative freedom of migration of larvae between epicontinental seas across the globe, which is obviously facilitated when sea level is high, but there remains some dispute about the migration pathways.

For most of the Jurassic Pangaea remained a coherent supercontinent. Therefore the number of possible routes was confined, most obviously to the periphery, with the northern route being the one utilised by organisms belonging to the Boreal Realm. Of three possible "Atlantic" routes, that between the southern Andean region and East Africa was probably not created before the end of the period, while that between Greenland and Scandinavia excluded Tethyan faunas. Evidence from Jurassic bivalves, gastropods, ammonites and belemnites suggests that the central Atlantic sector between Africa and North America, the so-called Hispanic Corridor, operated as an intermittent shallow epicontinental seaway permitting only restricted intermigration, during the Pliensbachian-Callovian time interval, with intermigration being relatively free only during the Toarcian and early Bajocian (HALLAM, 1977, 1983). Not until Oxfordian times, when the central sector of the Atlantic was opening as rapid sea-floor spreading commenced, did a distinction between European and East Pacific faunal provinces finally break down. This was evidently the result of the creation of a true oceanic strait between the western Tethys and the proto-Pacific.

With regard to speciation, a general survey of Jurassic bivalve species in Europe has led to the conclusion that the overwhelmingly predominant mode is one of punctuated equilibria, with only phyletic size increase, which is widespread, being more or less gradualistic (HALLAM, 1978a). Detailed biometric analysis of species of *Gryphaea* provides evidence for punctuated equilibria (HALLAM, 1982), phyletic size increase and morphological trends involving paedomorphosis, a kind of "punctuated gradualism". In critical cases migration betwen provinces can be ruled out and direct one-to-one ancestor-descendent relationships between species traced, rendering a strict cladistic approach to taxonomic distinctions inoperable. A review of the distribution of *Gryphaea* species across the world supports migration from centres of origin rather than a vicariance model. Thus the ancestor of the European *G. arcuata* originated in the late Triassic of the Arctic, and the European mid Jurassic bilobate lineage originated in South America.

Among the ammonites, it has long been recognised that the genera of Phylloceratina, which occupied relatively deep water habitats, had longer stratigraphic ranges than the shallower-water Ammonitina. It appears now that this distinction is even recognisable at species level within the Ammonitina. Thus such well known Liassic ammonites as *Arnioceras* and Liparoceras had longer -ranging species than their respective contemporaries *Coroniceras* and *Androgynoceras*, and facies analysis suggests that they lived in deeper neritic habitats.

On both a large and small scale, marine regressions, related to regional or global sea level falls, correlate with episodes of increased extinction rate among both bivalves and ammonites, and marine transgressions correlate with episodes of radiation of new groups. Further correlations support a speciation model whereby times of regression cause a deterioration of the environment, increasing the stress on organisms and promoting the evolutionary strategy known by ecologists as r selection, while times of transgression promote k selection. Thus new species evolved at times of low sea-level stand are often smaller than their ancestors, and increased in size phyletically during the subsequent, environmentally less stressful times of high sea-level stand.

REFERENCES

Hallam, A. 1977: Jurassic bivalve biogeography. Paleobiol. 3, 58-73.

--- 1978a: How rare is phyletic gradualism? Evidence from Jurassic bivalves. Paleobiol. 4, 16-25.

--- 1978b: Eustatic cycles in the Jurassic. Palaeogeogr., Palaeoclimatol., Palaeoecol. 23, 1-32.

--- 1982: Patterns of speciation in Jurassic Gryphaea. Paleobiol. 8, 354-366.

--- 1983: Early and mid-Jurassic molluscan biogeography and the establishment of the central Atlantic seaway. Palaeogeogr., Palaeoclimatol., Palaeoecol. 43, 181-193.

MIDDLE JURASSIC AMMONITE EVOLUTION IN THE ANDEAN PROVINCE AND EMIGRATION TO TETHYS *)

Gerd E. G. Westermann and Alberto C. Riccardi

Hamilton, Canada and La Plata, Argentina

Abstract: The Neuquen Basin represents the southeastern embayment of the Jurassic marginal sea. The rather complete Middle Jurassic ammonite-bearing sequence interfingers with continental deposits toward basin margin. It includes mainly the Cuyan Sedimentary Cycle, with Pliensbachian -- Upper Bajocian Lower Subcycle and Upper Bathonian -- Middle Callovian Upper Subcycle; followed by Upper Callovian -- Upper Jurassic Lotenian-Chacayan Sedimentary Cycle.

Middle Jurassic Ammonitina of the Neuquen Basin are composed of endemic, Andean (province), East-Pacific, Tethyan (realms) and cosmopolitan genera. Northward, endemic genera become rare while cosmopolitan and Tethyan genera predominate, probably reflecting paleolatitude and corresponding with lithofacies change from terrigenous to carbonate-dominated. The entire Andean Province (or Region) belonged to the Tethyan Realm up to Early Bajocian, to the East-Pacific Realm up to Early Callovian, and then again to the Tethyan Realm.

Middle Jurassic phyletic clades and grades are documented, with major radiations in late Aalenian - earliest Bajocian and late Bathonian, coincident with high sea levels regionally and globally. Several major clades probably originated in the Andes, including the Neuquen Embayment, and spread to western Tethys via the Hispanic Corridor ("proto-Atlantic"). The most important are: Sonniniidae with *Euhoploceras* derived from Andean *Puchenquia(Gerthiceras)*, Sphaeroceratidae derived with *Chondroceras* from East-Pacific and mostly Andean *Emileia* (*Chondromileia*), and ? Reineckeiinae (?) perhaps derived with *Reineckeia* (*Rehmannia*) from *Neuqueniceras* (*Frickites*). The eastward passage through the Hispanic Corridor is supported by the ecotonic overlap of Andean/East-Pacific with west Tethyan faunas in the Mixteca terrane, south Mexico. The Macrocephalitinae derived from the Eurycephalitinae in the south Pacific and spread westward into Tethys. Major migrations through the Hispanic Corridor coincided with high eustatic sea levels which presumably removed barriers in the cratonic seaway.

*) A Contribution to Project #171: Circum-Pacific Jurassic

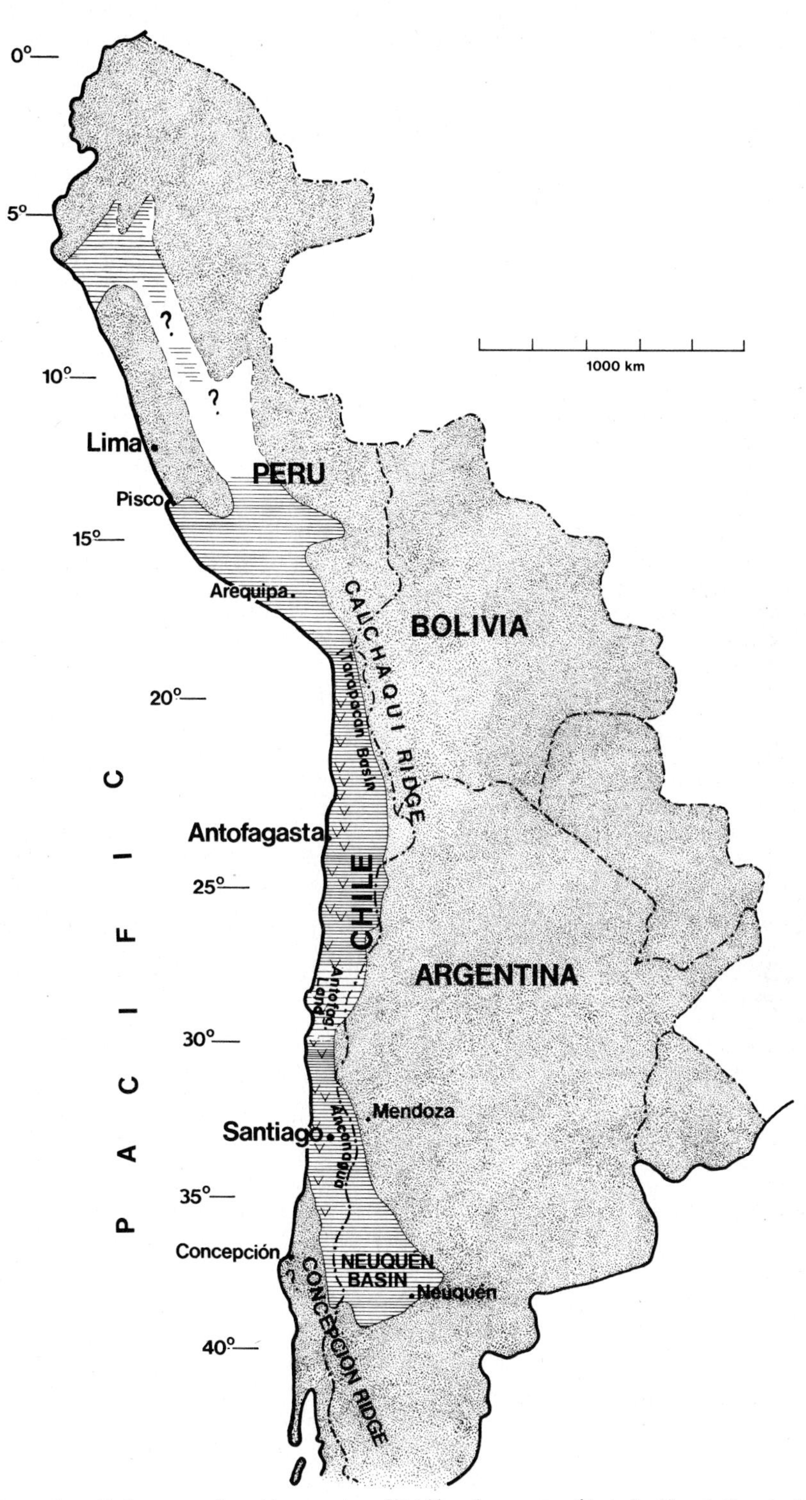

Fig. 1: Index map for the marine Middle Jurassic (hatched) of South America.

INTRODUCTION

Established opinion holds that pre-Callovian Jurassic ammonite faunas of the eastern Pacific margin were either entirely isolated from western Tethys by the western Pangea land barrier, or that in case restricted faunal exchange through a cratonic proto-Atlantic (Hispanic Corridor) occurred, the migrations were only westward; accordingly, the major clades evolved in western Tethys and adjoining epicontinental seas, i.e. Europe, and spread into the eastern Pacific with the equatorial current (e.g. ENAY, 1980; THIERRY, 1982). Other authors generalized two-way migratory paths only, because of the overwhelming ambivalence in this respect of the paleontologic record (HALLAM 1977; WESTERMANN, 1981; HILLEBRANDT, 1981). In contrast, eastward migration has recently been documented for Early Jurassic bivalves and gastropods (DAMBORENEA & MENCEÑIDO, 1979; HILLEBRANDT, 1981; HALLAM, 1983). The evidence presented here appears therefore the first for Jurassic ammonites with eastward path through the Hispanic Corridor and, hence, for Pacific origin of several major pan-Tethyan taxa for which the occurrence in Europe is cryptogenic.

GEOLOGY

The Early Jurassic sea invaded what are now the Andes of southern Peru, Chile and western Argentina from the west and northwest and formed at the southern end an embayment with a large eastward extension, i.e. the Neuquen Embayment (Figs. 1-3). In the Neuquen Basin, open-marine deposits up to 2000 m in thickness are extensively represented by the fine-grained sediments of the Los Molles Formation. This unit, which is Sinemurian-Pliensbachian at its base, changes upwards and laterally into sandstones representing progradation in a deltaic environment, i.e. the Lajas Formation. Progradation ended with development of a fluviatile system, which in central areas of the basin intergrades laterally into evaporites, i.e. the Tabanos Formation (DELAPE et al., 1979; RICCARDI, 1983).

The Lajas Formation reaches its maximum thickness near the southeastern margin of the embayment, while it thins towards the central and northern areas. Thus, in Chacay Melehue the Lajas Formation is absent and the Los Molles Formation continues upwards into Bathonian-Callovian levels (i.e. the Chacay Melehue Formation). In that region the marine shales of the Los Molles and Chacay Melehue Formations are directly overlain by the evaporites of the Tabanos Formation.

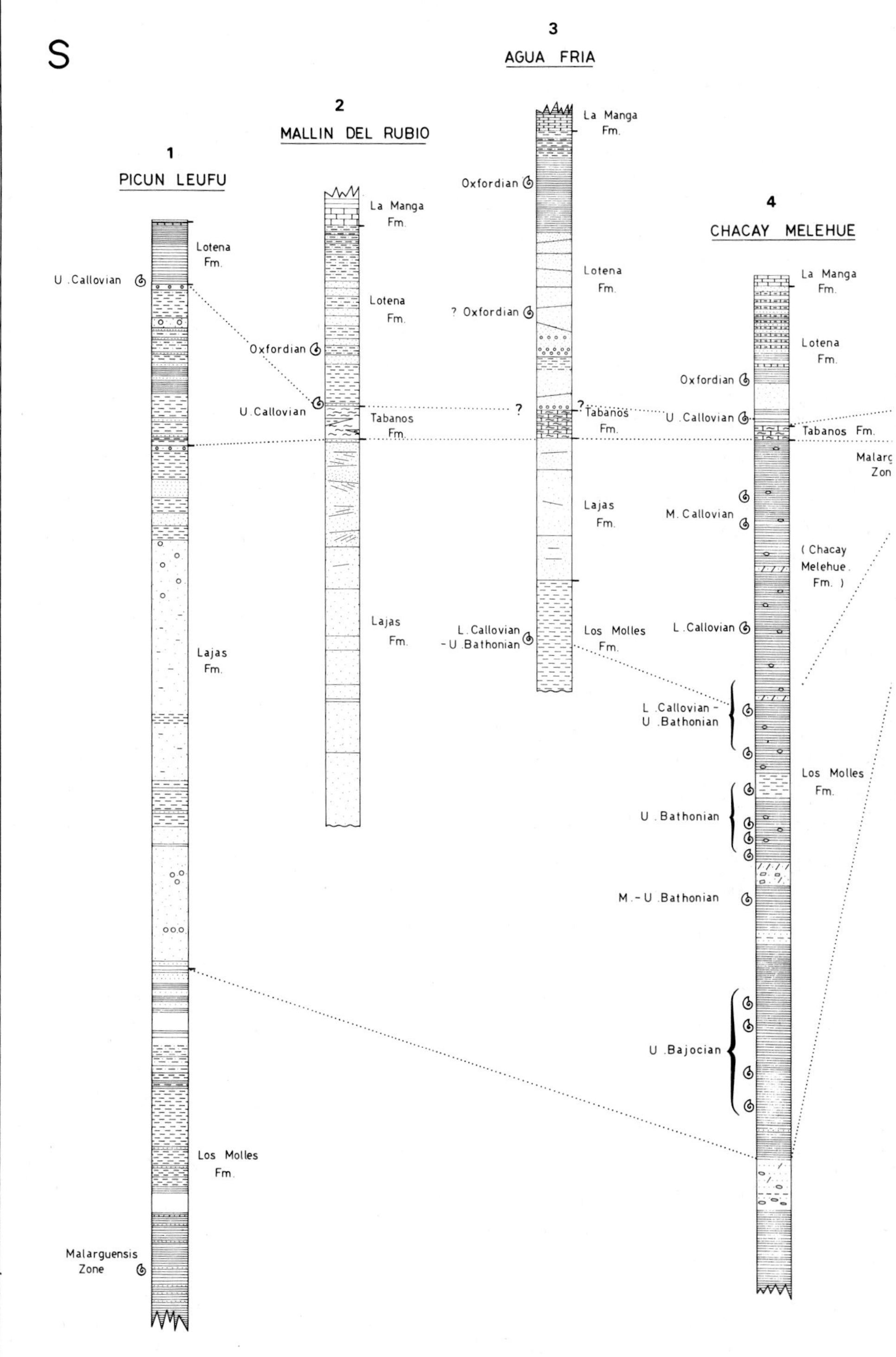
S
1
PICUN LEUFU
2
MALLIN DEL RUBIO
3
AGUA FRIA
4
CHACAY MELEHUE
La Manga Fm.
Lotena Fm.
Tabanos Fm.
Lajas Fm.
Los Molles Fm.
(Chacay Melehue Fm.)
U. Callovian
Oxfordian
? Oxfordian
M. Callovian
L. Callovian
L. Callovian - U. Bathonian
U. Bathonian
M.- U. Bathonian
U. Bajocian
Malarguensis Zone
Dib. C.R. Tremouilles

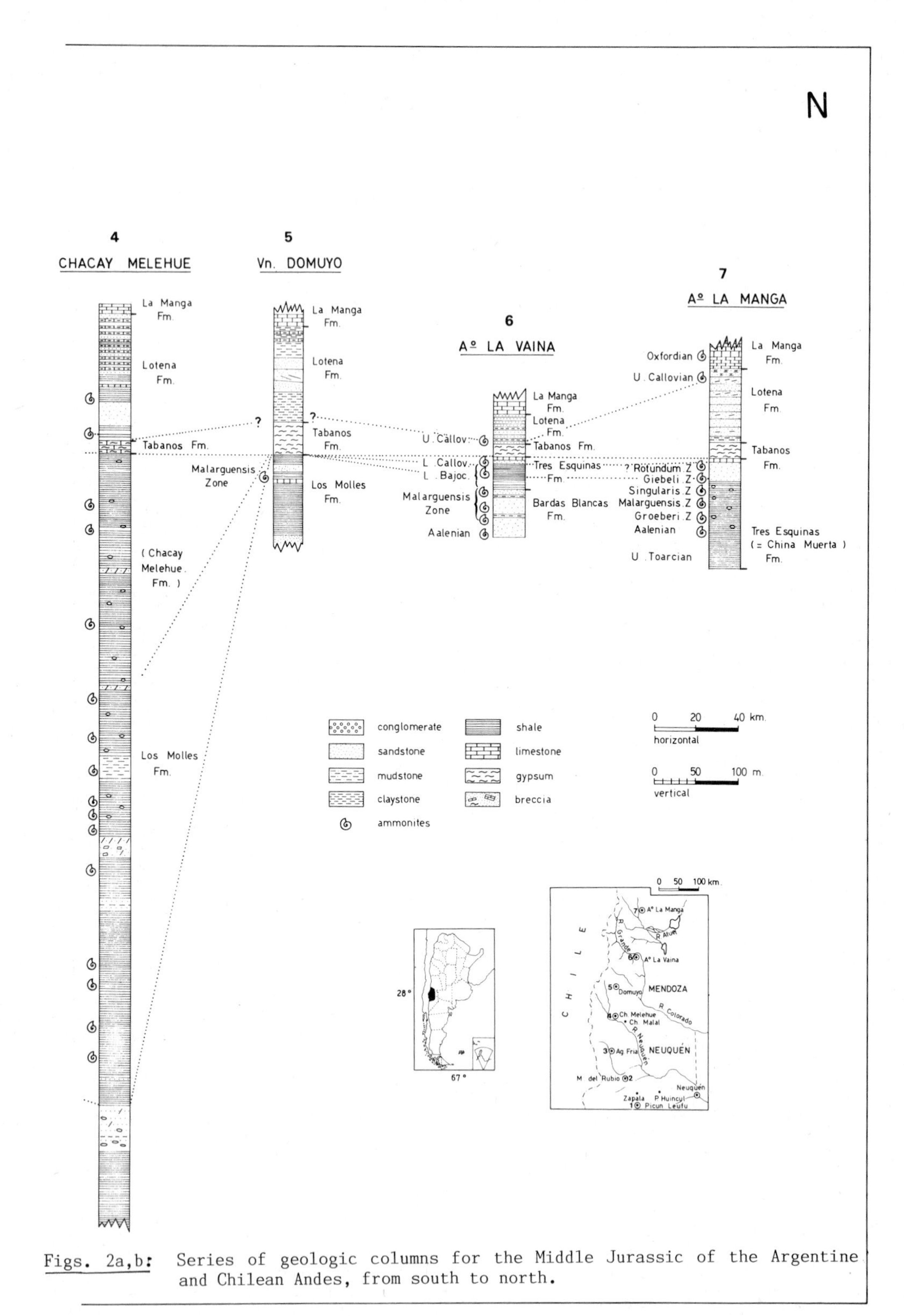

Figs. 2a,b: Series of geologic columns for the Middle Jurassic of the Argentine and Chilean Andes, from south to north.

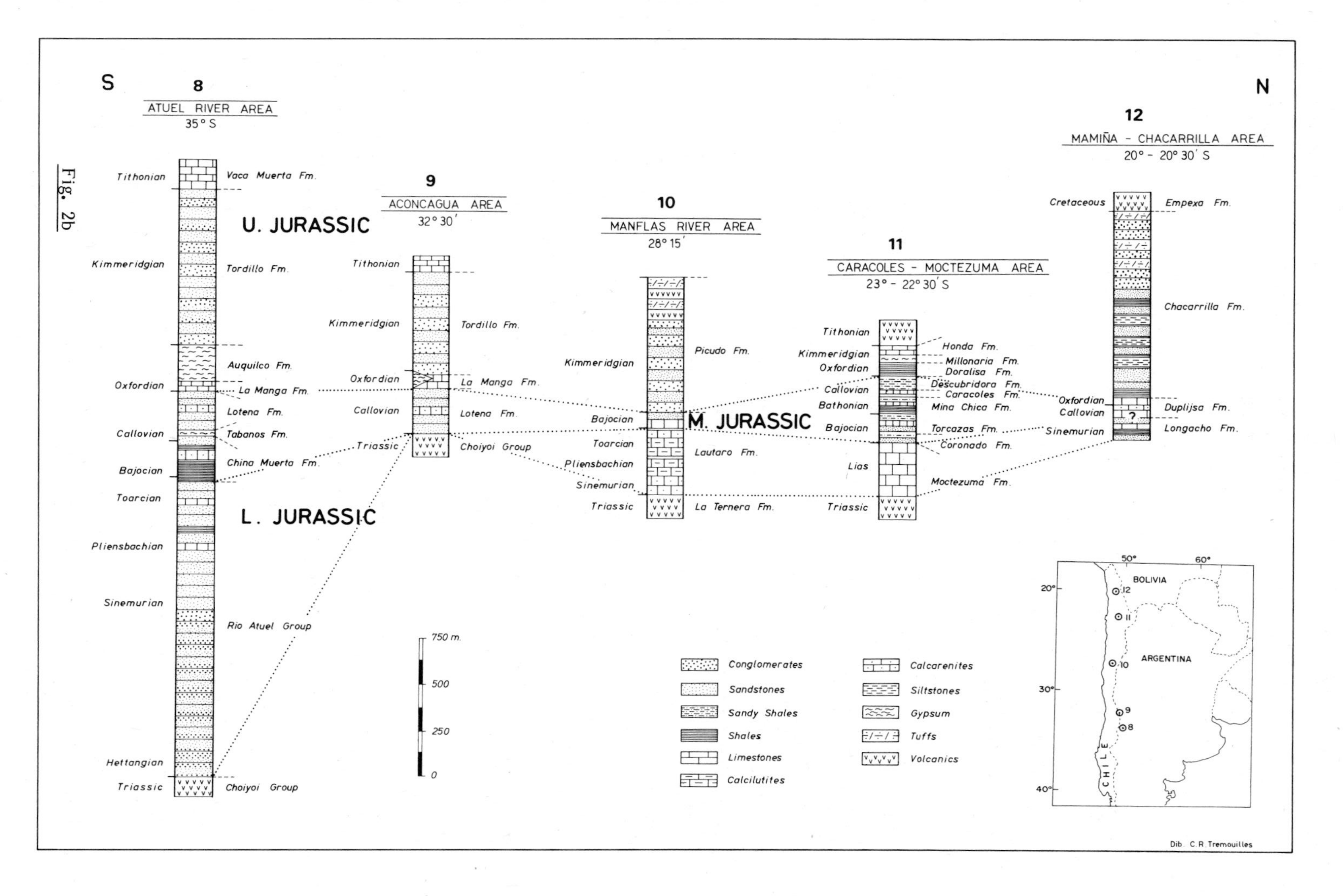
S
N
8
ATUEL RIVER AREA
35° S
9
ACONCAGUA AREA
32° 30'
10
MANFLAS RIVER AREA
28° 15'
11
CARACOLES - MOCTEZUMA AREA
23° - 22° 30' S
12
MAMIÑA - CHACARRILLA AREA
20° - 20° 30' S
U. JURASSIC
M. JURASSIC
L. JURASSIC
Tithonian
Kimmeridgian
Oxfordian
Callovian
Bajocian
Toarcian
Pliensbachian
Sinemurian
Hettangian
Triassic
Vaca Muerta Fm.
Tordillo Fm.
Auquilco Fm.
La Manga Fm.
Lotena Fm.
Tabanos Fm.
China Muerta Fm.
Rio Atuel Group
Choiyoi Group
Picudo Fm.
Lautaro Fm.
La Ternera Fm.
Lias
Bathonian
Honda Fm.
Millonaria Fm.
Doralisa Fm.
Descubridora Fm.
Caracoles Fm.
Mina Chica Fm.
Torcazas Fm.
Coronado Fm.
Moctezuma Fm.
Cretaceous
Empexa Fm.
Chacarrilla Fm.
Duplijsa Fm.
Longacho Fm.
750 m.
500
250
0
Conglomerates
Sandstones
Sandy Shales
Shales
Limestones
Calcilutites
Calcarenites
Siltstones
Gypsum
Tuffs
Volcanics
BOLIVIA
ARGENTINA
CHILE
20°
30°
40°
50°
60°
Dib. C.R.Tremouilles

Fig. 2b

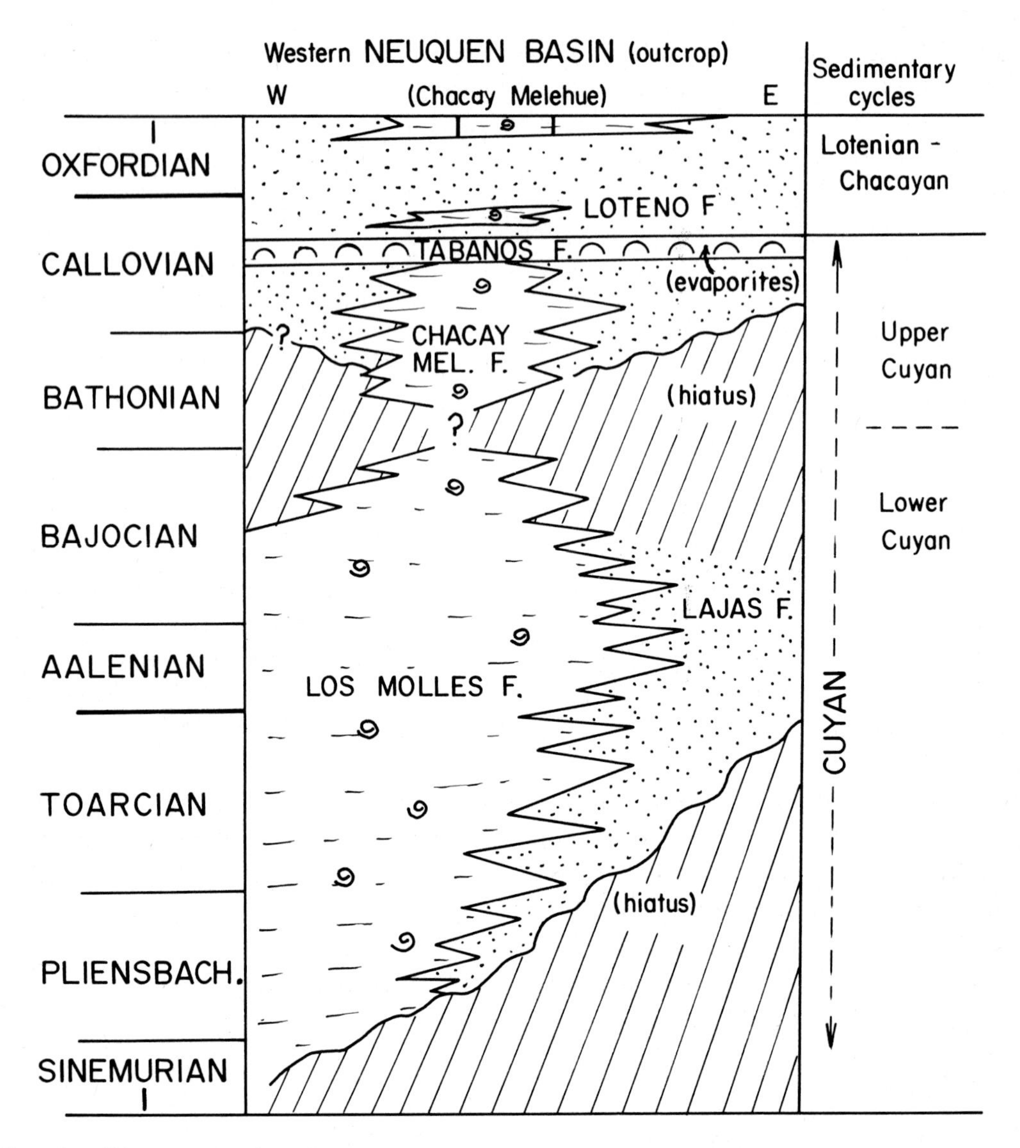

Fig. 3: Biostratigraphy, facies and sedimentary cycles of the Middle Jurassic of western Neuquen Basin (outcrop). For transgressive-regressive cycle see Fig. 8.

The Los Molles, Lajas and Tabanos formations constitute the Cuyan Sedimentary Cycle. Paleontological evidences indicate that the initial transgression reached its maximum by the Aalenian-earliest Bajocian, while the regressive phase began in the late Early Bajocian and culminated in the Early Bathonian. A more restricted and short-lived transgression occurred from Late Bathonian to Early Callovian and is terminated by the Middle Callovian regressive phase as documented by the evaporites of the Tabanos Formation. These two transgressive-regressive episodes have been named the Lower and Upper Cuyan Subcycles.

In the Neuquen Basin the Tabanos Formation and its equivalents are overlain by the littoral to continental sandstones and conglomerates of the Lotena Formation. Within this unit there are, however, marine intercalations with Callovian ammonites. This unit belongs to the lower part of the Lotenian-Chacayan transgressive-regressive sedimentary cycle (cf. RICCARDI, 1983).

In Chilean territory, a similar pattern of marine sedimentation for the Lias-Bajocian is represented north of 28° S, although between 28° and 31° S an early northward regression beginning with the Early Toarcian onwards is indicated. The Bathonian regression is clearly evident between 26° and 37° S where marine Callovian (and Oxfordian) sediments rest directly on Bajocian and even older strata. This regression began an irreversible trend that resulted in the delineation of two basins, the Tarapacan to the north and the Aconcaguan-Neuquenian to the south, separated by an emergent land, i.e. the Antofagasta Land.

Late Bathonian sediments, restricted to the more centrally located areas of both basins, i.e. Chacay Melehue in northern Neuquen and El Profeta-Caracoles in Chile, followed by more widespread Early to Middle Callovian marine sediments, attest to relatively minor fluctuations in a regressive cycle which culminated in the late Early to Middle Callovian.

A renewed and more widespread transgression occurred in the Middle to Late Callovian over areas that during the Bathonian had been above sea level. But the sea did not extend over the whole region previously covered by the Bajocian sea, i.e. no record exists of marine Callovian between 28° S and 30-31° S.

In the Neuquen Basin only the western part is exposed and thus capable of yielding ammonites, while the eastern part is known from numerous oil wells.

PHYLOGENY AND BIOGEOGRAPHY

The Aalenian-Bajocian Ammonitina of the Andes of Chile, Argentina and Peru have now largely been described or revised in recent years (HILLEBRANDT, 1970; WESTERMANN & RICCARDI, 1972, 1979, 1980 a, b, 1982; HILLEBRANDT & WESTERMANN, 1984; WESTERMANN et al., 1980), while the Bathonian-Callovian taxa have also been investigated by us (manuscripts ready for press). Middle Jurassic ammonite phylogeny and ammonite biogeography of the Andes have been discussed previously by WESTERMANN & RICCARDI (1976, 1979; WESTERMANN, 1981); see Fig. 4.

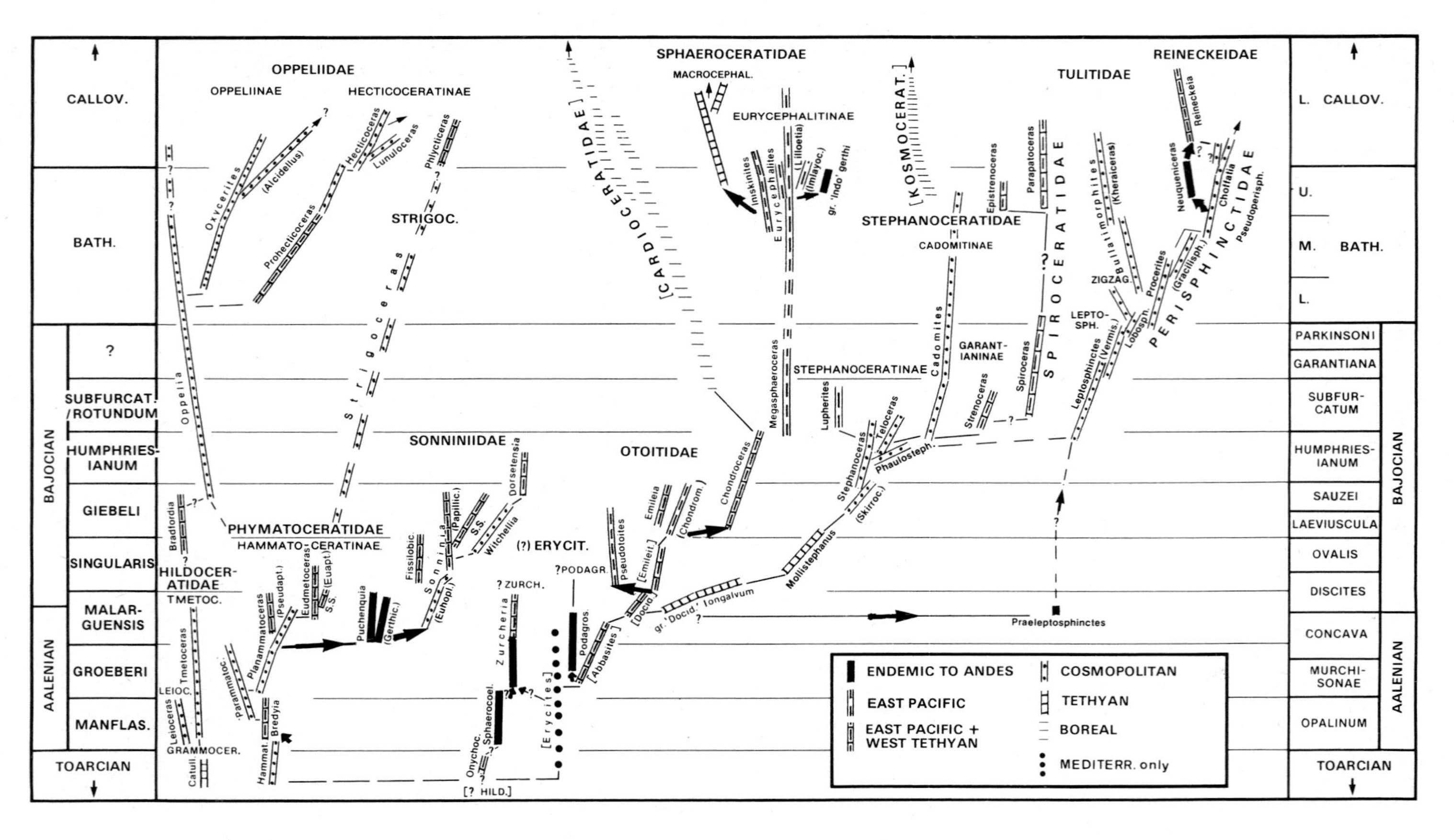

Fig. 4: Phylogeny and biogeographic affinity of the Middle Jurassic Ammonitina with Andean representatives. Heavy black arrows indicate the Andean origin postulated for the major clades, i.e. Sonniniidae, Sphaeroceratidae, Macrocephalitinae, and (?)Reineckeiidae; all except Macrocephalitinae expanded into Tethys via the Hispanic Corridor.

Superfamily HILDOCERATACEAE

1) Family HILDOCERATIDAE

In the Aalenian, the cosmopolitan *Leioceras* is now known to occur scarcely in northern Chile and west-central Argentina. The Tethyan *Tmetoceras* is an important guide fossil along the eastern Pacific for the entire Aalenian (HILLEBRANDT, 1970, 1973; WESTERMANN, 1964, 1981). The Upper Toarcian ancestor, the Mediterranean *Catulloceras*, is missing in the Pacific, so that *Tmetoceras* probably originated in western Tethys. In contrast to Tethys, however, *Tmetoceras* survived abundantly and with little change and diversity to the Aalenian/Bajocian boundary in the East-Pacific, e.g. the circumglobal *T. scissum* (Benecke) in South Alaska, Oregon and the Andes (Fig. 4).

As in the North-Cordilleran Province, the ubiquitous Tethyan family Graphoceratidae is unknown from the Andes (possibly excepting the single ? *Staufenia* reported by WESTERMANN & RICCARDI, 1972), contributing to the difficulties of global correlation.

The late Aalenian-early Bajocian "*Fontannesia*" *austroamericana* Jaworski is now tentatively placed in *Puchenquia* of the Hammatoceratidae.

2) Family PHYMATOCERATIDAE

The almost cosmopolitan *Hammatoceras* of the Upper Toarcian evolved by palingenesis ("recapitulation") into the early Aalenian *Bredyia*, known from westernmost Tethys and the Andean Province only. The inner whorls of the abundant North Chilean form (HILLEBRANDT & WESTERMANN, 1984) are ogival without sulci, resembling the inflated late Toarcian "*Pachammatoceras*" Buckman, considered synonymous with *Hammatoceras*, while the outer whorls resemble typical *Bredyia* from the *scissum* Zone of southern England and Morocco (SENIOR, 1977), where it is rare in both regions. The incipient ventral costae alternation with obsolete keel as well as the suture with short ventral lobe are reminiscent of *Erycites*. A central Andean origin of *Bredyia* is possible.

In the upper *Bredyia manflasensis* Zone (*scissum/murchisonae* Zone boundary) of northwern Chile, *Bredyia* appears to be the root of *Parammatoceras*, continuing the palingenetic series. Only the nucleus resembles *Hammatoceras* while the intermediate and outer whorls have extended primaries and a tendency toward falcoid costae. These newly described forms (HILLEBRAND & WESTERMANN, 1984) thrive in the *Zucheria groeberi* Zone (upper *murchisonae*-lower *concavum* Standard Zone or Chron).

In the Tethyan *murchisonae* Zone and the Andean *Z. groeberi* Zone occurs the circum-global (cosmopolitan except Arctic) *Planammatoceras*. This genus is now understood to include all Aalenian hammatoceratids with sharp umbilical margin, ogival whorl section, high floored keel, and sinuous ribbing with well developed primaries (WESTERMANN & RICCARDI, 1982). It evolved perhaps from *Parammatoceras* in either the Tethys or the Andean Province. The mostly early, evolute species, e.g. *P. gerthi* Jaworski which are placed in *Planammatoceras* s.s., range in the Mendoza Andes exceptionally high, i.e. *Puchenquia malarguensis* Zone. The involute, late Aalenian to basal Bajocian subgenus *Pseudaptetoceras* Geczy is represented by the eastern Pacific (south Alaska + Andes) and Mediterranean *P. klimakomphalum* (Vacek). In Mendoza province and Northern Chile occurs a morphologically intermediate form, *P. tricolore* Westermann and Riccardi, which is perhaps the phyletic link between *Planammatoceras* s.s. and *Pseudaptetoceras* s.s.. An apparently endemic Andean species is *P. (Pseudaptetoceras) moerickei* (Jaworski).

The lectotye of *Planammatoceras gerthi* (Jaworski), one of the latest known species of the restricted genus, was placed previously in *Eudmetoceras* s.s., while the paralectotypes were considered affiliated with " *Eudmetoceras* " *(Planammatoceras) insignoides* (Quenstedt). The entire type set is probably conspecific and has the strongly flexed ribbing typical of *Planammatoceras*. *P. gerthi* occurs together with *Eudmetoceras (Eudmetoceras) eudmetum jaworskii* Westermann and the involute *E.* (*Euaptetoceras*) *amplectens* Buckman. The latter two species occur also in Europe and South Alaska, making this an important guide assemblage of the Aalenian-Bajocian boundary. *Eudmetoceras* could therefore have evolved in either the eastern Pacific or western Tethys.

Of great interest is the quasi-endemic *Puchenquia*, from the *P. malarguensis* Zone of Neuquen to Peru. A single specimen has now been found by R. MOUTERDE (Lyon, pers. comm. 1982, confirmed by G.E.G. WESTERMANN 1983 and J. SANDOVAL, 1984) in an upper *concavum - discites* Zone assemblage from Portugal. The ancestor was probably *Planammatoceras* and the origin presumably Andean. Two subgenera are associated in the Andean Province, (1) the small *Puchenquia (Puchenquia) malarguensis*, and (2) the large *P. (Gerthiceras)* West. and Ricc., characterized by inner whorls as in *Puchenquia* s.s. and outer whorls as in *Pseudammatoceras*, with prominent primaries sometimes bearing bullate spines. Both occur in Mendoza (Cerro Tricolor area) and the Sierra Domeyko, northern Chile (HILLEBRANDT & WESTERMANN, 1984). *"Fontannesia" austroamericana* Jaworski from the *P. malarguensis* Zone of the entire Andean province, is tentatively transferred to the endemic *Puchenquia* s.s. and considered to be an extremely evolute microconch.

3) Family OPPELIIDAE

This truly cosmopolitan family ranges in the Andes from the *Pseudotoites singularis* Zone through the Upper Jurassic. As in the Mediterranean, the earliest genus is the rare and cryptogenic *Bradfordia* which evolved in either the western Tethys or the eastern Pacific from lissoceratids or late hammatoceratids (WESTERMANN, 1969). Of the latter, the Andean *Planammatoceras (Pseudaptetoceras) moerickei* Jaworski one of the possible oppeliid ancestors. Curiously, however, Bajocian oppeliids are rare in South America. In the Bathonian, *Oxycerites* evolved from *Oppelia*, again by palingenesis (a supposedly rare mode of evolution!), the tertiary ribs on the ventral shoulder becoming progressively more restricted to the inner whorls. An interesting trend occurs in the late Bathonian when the tricarinate subgenus *Paroxycerites (Alcidellus, Paralcidia* auct). becomes prominent, again with progressive reduction of ornament and with size increase. These phyletic developments occur both in the western Tethys and the eastern Pacific, without trace of centre of origin.

4) Family HECTICOCERATIDAE

The Late Bathonian *Prohecticoceras* includes important guide fossils in northern Chile, e.g. *P. retrocostatum* (Grossouvre) and *P.* cf. *ochraceum* Elmi, previously known only from western Tethys (HILLEBRANDT, 1970 and unpublished). The Callovian descendants, *Hecticoceras* s.l., are known in the eastern Pacific in single specimens only from southern Mexico (WESTERMANN et al., 1984) and the Neuquen Basin.

5) Family STRIGOCERATIDAE

A collection of G. Chong from northern Chile contained an interesting, highly involute form related to the almost cosmopolitan late hammatoceratid *Planammatoceras klimakomphalum* (Vacek), but with features intermediate to *Strigoceras* s.l.. Similar transient forms are also present in the Upper Aalenian of Oregon and western Europe, however, so that the centre of origin remains equivocal.

Of particular interest is the rare presence in northern Chile of the West-Tethyan *Phlycticeras* (Covacevich pers. comm., and unpubl. observ. University of Chile).

6) Family SONNINIIDAE

In the Neuquen Basin, the hammatoceratoid *Puchenquia* is accompanied by *Euhoploceras amosi* Westermann and Riccardi. Its age is at least as old as the first sonniniid in western Europe (and South Alaska?), i.e. upper *concavum* Zone.

Hitherto, the almost cosmopolitan Sonniniidae were believed to have originated with *Euhoploceras* from *Eudmetoceras*, a hammatoceratid of the same or slightly younger age. *Eudmetoceras* and the several species now transferred to *Planammatoceras* s.s. as discussed above, differ significantly from *Euhoploceras* and other sonniniids in the complicated and strongly retracted septal suture, and from the first sonniniid, *Euhoploceras*, also in the high keel. *Puchenquia*, on the other hand, has inner whorls with low keel and a rather simple, weakly retracted suture, closely resembling the non-spinose early *Euhoploceras* ex gr. *amosi* West. and Ricc. et *modesta* Buckman. The quasi-cosmopolitan Sonniniidae therefore probably evolved in the Andean Province from *Puchenquia* and, most likely, from the subgenus *Gerthiceras*. The slightly later *Fissilobiceras* (?subgenus of *Sonninia*), occurring abundantly also in the Andes, i.e. *F. zitteli* (Gottsche), matches in the inner whorls the non-tuberculate *Euhoploceras* and *Puchenquia*.

Most sonniniid genera, subgenera, and several species are in common between the Andes (WESTERMANN & RICCARDI, 1972) and northwest Europe. Thus, the early *Dorsetensia* species *D. mendozana* and *P. blancoensis* West. and Ricc. from the upper *E. giebeli* Zone of Mendoza, may be conspecific with the earliest *Dorsetensia* from Europe, i.e. *D. hebridica* MORTON (1972)from Scotland; *D. romani* (Oppel) and *D. liostraca* Buckman occur in the Humphriesianum Zone of Europe and the Andes. While *Witchellia* likely originated in northwest Europe where it is most diverse and abundant, *Dorsetensia* may have originated in either province.

7) Family ERYCITIDAE

?Podagrosiceratinae.

The late Toarcian and Aalenian Erycitinae are unknown in the Andean Province, but an offshoot of the long-ranging *Erycites*, near the root of *Abbasites*, probably gave rise to the endemic Andean *Podagrosiceras* of the *Z. groeberi* and *P.malarguensis* Zones (WESTERMANN & RICCARDI, 1975). The genus ranges from Mendoza to northern Chile and Peru (HILLEBRANDT & WESTERMANN, 1984; WESTERMANN et al., 1980). *Podagrosiceras* perhaps gave rise to the mainly Boreal *Arkelloceras* (*P. crassicostatus* Zone (*sauzei* Zone of South Alaska and the North Slope) which could not previously be placed satisfactorily, and perhaps even to the cryptogenic *Ermoceras* of the Mediterranean Upper Bajocian. The Erycitinae affinity of *Podagrosiceras* has become evident in the macroconch becoming smooth and round, and in the septal suture which is unlike that of *Hammatoceras*. The Boreal *Erycitoides* Westermann, erroneously considered synonymous with *Podagrosiceras* by HOWARTH (in DONOVAN et al., 1981), is superficially similar in the microconch only (WESTERMANN & RICCARDI, 1979).

Another possible descendant or relative of *Erycites* is *Zurcheria* (with family-group name by HYATT 1900 available), hitherto placed in the Sonniniidae, but bearing the short E lobe and the broad ventral interruption of *Erycites*. *Zurcheria* occurs much earlier in the Andes than in Europe, i.e. middle vs. latest Aalenian, and is probably a descendant of *Sphaerocoeloceras* (HILLEBRANDT & WESTERMANN, 1984).

Superfamily S T E P H A N O C E R A T A C E A E

8) Family OTOITIDAE

Typical sphaeroconic *Docidoceras* of the basal Bajocian have a North-Cordilleran and West-Tethys distribution, resembling that of their ancestor *Abbasites* which is still closely linked with *Erycites*. The North (and ? South) American *Docidoceras (Pseudocidoceras)* can be considered ancestral to the ambi-Pacific *Pseudotoites* of the *P. singularis* and lower *E. giebeli* Zones (*ovalis-laeviuscula* Zones). *Pseudotoites* occurs rarely in South Alaska and Oregon, abundantly in the Southern Andes, moderately common in Western Australia, and rarely in Indonesia. The genus resembles the northwest European *Emileites* of similar age and the early development of its probable descendant *Emileia* (incl. *"Otoites"*). *Emileia* is abundant in the Americas and most of Tethys (WESTERMANN, 1969; WESTERMANN & RICCARDI, 1979). The subgenus *Chondromileia*, based on the Andean *"Sphaeroceras"/"Chondroceras" giebeli* and *"S./C." submicrostoma* , Gottsche spp., occurs rarely in Oregon.

9) Family SPHAEROCERATIDAE

Sphaeroceratinae.

Emileia (Chondromileia), morphologically transitional from *Emileia* to *Chondroceras* but still with a lappet-bearing microconch, was probably at the root of the almost cosmopolitan *Chondroceras*. Its oldest representative, *C. recticostatum* West. and Ricc. is now known from the upper *P. singularis*-lower *E. giebeli* Zone (*laeviuscula* Zone) of the Neuquen Basin WESTERMANN & RICCARDI (1979). The Andean *Chondroceras* thus makes its first appearance together with *Emileia*, while in Europe it is unknown before the *sauzei* Zone: In Europe and North America alike, the genus flourishes in the *humphriesianum* Zone. The other morphologically intermediate form between Otoitidae and Sphaeroceratidae, i.e. *Labyrinthoceras* with its microconch *Frogdenites* bearing reduced lappets, is known only from the *sauzei* Zone of Europe and North America and is thus probably another parallel development, rather than the phyletic link between Otoitidae and Sphaeroceratidae as assumed previously (WESTERMANN 1964;

CALLOMON in DONOVAN et al. 1981).

The East Pacific *"Defonticeras"*, a subgenus or synonym of *Chondroceras* which is particularly common in North America, gave origin in the Upper Bajocian to the important Boreal family Cardioceratidae, beginning with the subfamily Arctocephalitinae (CALLOMON, 1984). *Megasphaeroceras*, the ancestor assumed by THIERRY (1978) and ENAY (1980), is already a typical eurycephalitine.

Eurycephalitinae

Of particular interest are the basal Bajocian to ?Early Oxfordian Eurycephalitinae, a mainly East-Pacific subfamily which has previously been consistently mistaken for Macrocephalitinae. (Note: we follow CALLOMON in DONAVAN et al., 1981 in uniting the Sphaeroceratidae and Macrocephalitidae into one family). The earliest member, *Megasphaeroceras*, occurs in the *subfurcatum* Zone from South Alaska to the Southern Andes (and ?Antarctica, WESTERMANN, 1981), together with the latest stephanoceratids and *Spiroceras*, Thus the eurycephalitines evolved likewise from late *Chondroceras*, somewhere along the eastern Pacific. The exceptional conservatism (stasis) of this homogeneous group is remarkable, from the Late Bajocian to the Callovian or even Oxfordian, with sphaeroconic, often almost smooth, *Phylloceras*-like macroconchs and much smaller, ornate microconchs, e.g. *Xenocephalites*. Even the early Oxfordian *Araucanites* WESTERMANN & RICCARDI (in STIPANICICet al., 1975) from Mendoza, formerly placed in *Mayaites*, is also very close to *Eurycephalites (Lilloettia)*. In contrast to the wide geographic range of *Megasphaeroceras* and *E. (Lilloettia)*, extreme endemism is represented by the group *"Indocephalites"* (n. gen.) *gerthi* Spath, occurring, in profusion, only in the vicinity of Chacay Melehue, Neuquen province (STEHN, 1923; WESTERMANN & RICCARDI, unpublished). The North American *Iniskinites* (and ?*Imlayoceras*) have now also been found in the Southern Andes, stratigraphically between *Megasphaeroceras* and *Eurycephalites (Lilloettia)*. One or two finds of *Eurycephalites* s.l. have also been made in the western Pacific, i.e. New Guinea and ?Japan (WESTERMANN, 1981). Perhaps the presence of these rare specimens is best explained by post-mortem drift or rare migrants outside the "proper" biogeographic limits (ENAY, 1980). These occurrences, nevertheless, indicate the possibility of genetic exchange between these provinces. (The Indonesian eurycephalitines recorded by CALLOMON in DONOVAN et al., 1981, p. 147, are now placed in a new endemic genus; WESTERMANN & CALLOMON, unpublished).

Macrocephalitinae.

Of great interest is also the origin of the Macrocephalitinae which were customarily derived from the Bathonian Tulitidae (e.g. ARKELL et al., 1957), but later from Sphaeroceratinae (e.g. WESTERMANN, 1956). Throughout Tethys they are assumed to mark the base of the Callovian (e.g. THIERRY, 1976), with significant exceptions in the southeastern Tethys (WESTERMANN & CALLOMON, unpublished) and rare occurrences in southern Germany (DIETL, 1981) where they appear to occur already in the late Bathonian. The probably oldest macrocephalitines have particularly involute, compressed and finely ribbed macroconchs, resembling the east Pacific *Eurycephalites (Lilloettia)*. Accordingly, the south Andean (to Antarctic Peninsula?) Middle Bathonian eurycephalitines were probably the root of the macrocephalitines which spread rapidly from the southeastern Tethys into the Mediterranean and Submediterranean Tethys. Their dispersal along the north shore of Gondwana (WESTERMANN, 1981; CALLOMAN in DONOVAN et al., 1981) is more plausible than through a narrow cratonic "Austral Seaway" between southern Africa and Antarctica (THIERRY, 1976; HALLAM, 1977; ENAY, 1980) which at this time was still entirely hypothetical.

10) Family STEPHANOCERATIDAE

Stephanoceratinae

In the upper *E. giebeli* Zone = *sauzei* Zone , the family appears in the Andes with rare representatives of the early *Stephanoceras* subgenus *Skirroceras* which probably migrated here from Tethys. In the superjacent *humphrisianum* Zone *Stephanoceras* s.l. becomes abundant and diverse, just as in the North-Cordilleran Region, displaying close specific ties with western Tethys. *Stephanoceras* s.kl. and *Teloceras* continue into the basal Upper Bajocian, as in the Mediterranean Province. The peculiar Late Bajocian offshoot *Lupherites ("Domeykoceras")* is in common with Oregon and, perhaps, Oaxaca, Mexico (WESTERMANN, 1983, 1984; WESTERMANN *et al.*, 1984).

Cadomitinae

In northern Chile and Neuquen, typical *Cadomites* occur together with the latest Stephanoceratinae, again as in western Tethys. The Mediterranean origin of *Cadomites* in the Stephanoceratinae remains, therefore, somewhat uncertain (note that SCHINDEWOLF 1965 derived *Cadomites* from Otoitidae, based on their septal suture). In the Neuquen Basin, typically Tethyan species, e.g. *C. rectelobatus* (Hauer), extend also into at least the Middle(?) Bathonian (unpublished).

(Superfamily S P I R O C E R A T A C E A E)

11) Family SPIROCERATIDAE

The late Bajocian *Spiroceras*, an open-coiled heteromorph with cosmopolitan (circum-global) distribution, and the very similar Late Bathonian to Early Callovian *Parapatoceras* occur in northern Chile and Neuquen (HILLEBRANDT, 1970; STEHN, 1923; unpublished). Their origins remain cryptic, taxonomically and geographically, with the best choice of an ancestor remaining the normally coiled *Strenoceras* which also occurs in northern Chile (WESTERMANN, 1956; WESTERMANN & RICCARDI, 1980). The coeval *Strenoceras* and *Spiroceras* differ essentially only in coiling and associated sutural deviation. The micromorph *Strenoceras* is now placed in the Garantianinae of the (ancestral) Stephanoceratidae (CALLOMON in DONOVAN *et al.*, 1981). Significantly the occurrence of *Strenoceras* in southern Mexico appears to predate that in Europe by a subzone or so (WESTERMANN, 1984). (Note: *Orthogarantinana* has also been found recently in northern Chile; v. HILLEBRANDT and DIETL, unpublished). Callomon's (in DONOVAN et al, 1981) supposition of the spiroceratid derivation directly from the Aalenian grammoceratid *Tmetoceras*, is barely a remote possibility. The root of the Late Bathonian (recoiled) *Epistrenoceras* may be in *Parapatoceras* rather than in the coiled *Strenoceras* (DIETL, 1978).

Superfamily P E R I S P H I N C T A C E A E

12) Family PERISPHINCTIDAE

Incertae sedis and Leptosphinctinae

Praeleptosphinctes jaworskii Westermann from the basal Bajocian of Mendoza is the earliest known apparent perisphinctacean, known only in the holotype. A thorough search for topotypes yielded no additional specimens, but the age of Gerth and Jaworski's type horizon, i.e. latest Aalenian - earliest Bajocian, has been confirmed by RICCARDI (in WESTERMANN & RICCARDI, 1982). Neither direct ancestors nor descendants are known so that *Praeleptosphinctes* could have been an early, unsuccessful perisphinctid offshoot from hammatoceratids, by heterochronous iteration. Although typical constrictions are present, the primary ribs of perisphinctids are missing. Additional material is obviously needed before the hypothesis of a Mediterranean origin of the Perisphinctaceae in *Phaulostephanus* of the Stephanoceratinae (PAVIA, 1971, 1983) can be challenged. As in most of Europe and North America, the stratigraphic record of the true Leptosphinctinae (and the entire superfamily) in the Andean Province begins abruptly near the Lower/Upper Bajocian boundary.

Zigzagiceratinae? and Pseudoperisphinctinae

While *Procerites* probably occurs locally in the Neuquen Basin, the pseudoperisphinctine *Choffatia* s.l. is ubiquitously throughout the Andean province.

13) Family REINECKEIIDAE

In the Late Bathonian, the cosmopolitan *Choffatia* s.l. gave origin to a genus with an apparently unique distribution, i.e. *Neuqueniceras* , known with several species only from the Bathonian-Callovian boundary beds of the southern Andes, southern Mexico and ?Japan (WESTERMANN, 1981; WESTERMANN *et al.*, 1984). The mode of evolution, furthermore, is palingenetic (as in the Hammatoceratinae and the Sonniniidae discussed above), with the inner whorls resembling *Choffatia* and the outer whorls displaying the descendant features: coronate mature whorls with lateral spines. The problem of the apparently ambi-Pacific distribution of *Neuqueniceras*, while seemingly absent from the northeastern and southwestern Pacific margins, remains unresolved. (If the Japanese form is indeed the same as in the southeastern Pacific, was then this taxon dispersed across the Pacific Ocean; or did the Japanese locality belong to an exotic tectonic terrane?).

Ancestor and "centre of evolution" of the typically Tethyan reineckeiids remains equivocal, i.e. whether they evolved during the latest Bathonian-earliest Callovian in the southeastern Pacific (southern Mexico or Andes) from *Neuqueniceras* s.s. via *N. (Frickites)* and *Rehmannia* (WESTERMANN *et al.*, 1984), or in Tethys directly from *Choffatia (Homeoplanulites)* as advocated by CARIOU (1980). True *Reineckeia* occurs well above *Neuqueniceras* throughout the Andes as well as Carribean and Pacific Mexico, i.e. *Reineckeia stehni* Zeiss ; *R. patagoniensis* Weaver and ? *R. leufuensis* Weaver (DELLAPE *et al.*, 1979; WESTERMANN *et al.*, 1980, 1984; but see CARIOU, 1980).

14) Family TULITIDAE

Only the circum-global *Bullatimorphites* s. l. is known from the Andean Province, providing much needed time-correlations with the European standard sequence. The early forms of the restricted genus, e.g. the Tethyan *B. (B.)* cf. *sofanum* Boehm, occur rarely in the Neuquen Basin, indicating the Middle Bathonian (SANDOVAL, 1983), while the Late Bathonian-Early Callovian, pandemic dimorphic pair *B. (Kheraiceras) bullatum* (d'Orbigny) ♀ -- "*Bomburites* cf. *microstoma* (d'Orbigny)" ♂ becomes somewhat more abundant in northern Chile. The rare *B. (K.) v-costatus* (Burckhardt) has previously been known only from southern Mexico (WESTERMANN et al., 1984).

BIOGEOGRAPHIC SUMMARY

Intra-Provincial Differentiation and Development

The biogeographic differences of the Middle Jurassic Ammonitina from Peru to the Neuquen Basin in the south are relatively minor and probably largely due to increasing latitude (Table 1; Figs. 1, 5, 8). The entire Andean Middle Jurassic has 4 known endemic genera (11%). The mostly shallow onshore seas of (the present) Peru and northern Chile -- the Tarapacan Basin -- deposited mainly limestones and contained highly diverse ammonite faunas including typically Tethyan elements. Further south in the northern Aconcagua-Neuquen Basin s.l. (San Juan and Mendoza provinces) limestones become less prominent and several Tethyan taxa disappear; in the Neuquen Basin s.s. terrigenous deposits predominate, Tethyan and also cosmopolitan taxa were further reduced and strictly endemic elements became locally abundant (belonging to a single genus).

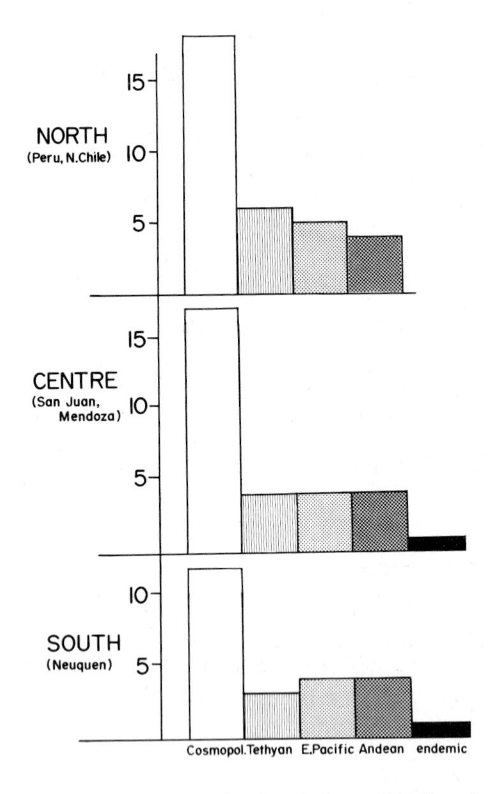

Fig. 5: Generic compositions of the Andean Middle Jurassic ammonite fauna, from Peru to Neuquen Basin. Note the southward decrease in the number of Tethyan and cosmopolitan genera and the occurrence of strict endemics in Neuquen Basin.

Genus	North (Peru, N. Chile	Centre (S. Juan, Mendoza)	South (Neuquen)
Leioceras	o	o?	
Tmetoceras	o	o	o
Sphaerocoeloceras	+	+	+?
Bredyia	●	●	
Parammatoceras	●		
Eudmetoceras	o	o	
Planammatoceras	o	o	o?
Puchenquia	+	+	+
Podagrosiceras	+	+	+
Oxycerites	o	o	o
Phlycticeras	(●)		
Euhoploceras		o	o
Sonninia	o	o	o
Dorsetensia	o	o	
Zurcheria	●	●	●
Pseudotoites	+	+	+
Emileia	o	o	o
Chondroceras	o	o	o
Sphaeroceras	o	o?	
Megasphaeroceras	x	x	x
Eurycephalites	x	x	x
Xenocephalites	x	x	x
"Indoc." gerthi gen.			*
Stephanoceras	o	o	o
Teloceras	o	o	o
Lupherites	x		
Cadomites	o/●	o/●	o/●
Spiroceras	o	(o)	(o)
Parapatoceras	o	o?	o
Epistrenoceras	o		
Praeleptosphinctes		(*)	
Leptosphinctes	o	o	o
Choffatia	o	o	o
Neuqueniceras	+	+	+
Reineckeia	●	●	●
Bullatimorphites	o/●	o/●	o/●

Table 1 - Biogeographic affinities of the Middle Jurassic ammonites of the Andes. Symbols: o cosmopolitan, ● Tethyan, x (East) Pacific, + Andean, * endemic, () rare, ? questionable

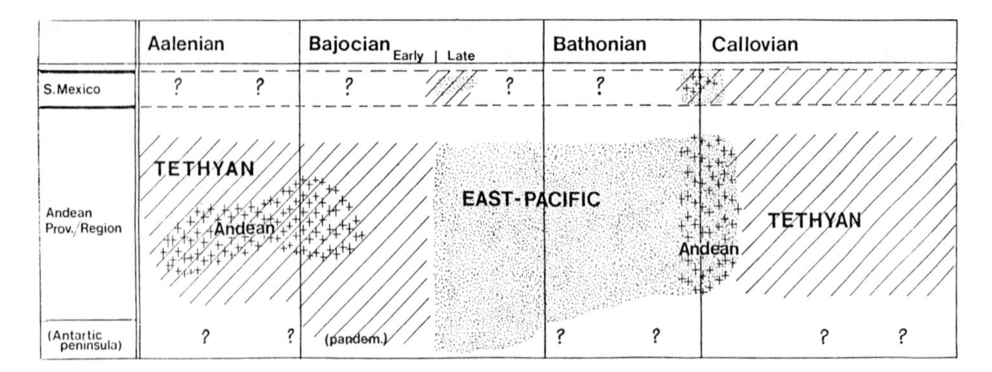

Fig. 6: Evolution of Middle Jurassic ammonite biogeography along the southeastern Pacific margin, from south Mexico (Mixteca terrane) to Antarctic Peninsula. The Tethyan Realm was circum-global except for the Late Bajocian-Early Callovian interval when the East-Pacific Realm was developed. An Andean Ammonite Province or Region developed repeatedly as part of either realm. An ecotonal overlap between Tethyan and East-Pacific realms is present on the Mixteca terrane located at the entrance to the Hispanic Corridor.

The entire Andean Ammonite Province changed biogeographically through Middle Jurassic time (Fig. 6), much more markedly than between basins. The province was part of the circum-global Tethyan Realm up to the end of the Early Bajocian, when it became part of the East-Pacific Realm which extended from the present south Alaska to the Antarctic Peninsula. While an East-Pacific Ammonite Province can be distinguished from time to time during the Early Jurassic, the more distinct development as a realm began almost simultaneously with the Boreal Realm. The Late Bajocian - Early Callovian East-Pacific Realm was probably largely caused by barrier development within the Hispanic Corridor (proto-Atlantic), the cratonic seaway connecting the eastern part of the Pacific with western Tethys and associated epicontinental seas. In the Middle Callovian, the Andean Province became rapidly again part of the expanded, circum-global Tethyan Realm, significantly coincident with the beginning of middle Atlantic seafloor spreading, the birth of the true southern North Atlantic ocean. Four of the 5 recorded genera with eastern Pacific distribution belong to this interval (Table 1, Figure 5). But only 1 genus, *Neuqueniceras*, has Andean Province distribution during the existence of the East-Pacific Realm, while one other, "*Indocephalites*" gr. *gerthi* (n. gen.), is at this time strictly endemic to the Neuquen Basin.

Extension of the Andean Province

Far north of the Andes Jurassic occurs a small isolated area with Andean Province ammonite faunas, i.e. the Cualac area of Guerrero state in southern Mexico which has yielded a typical *Neuqueniceras* assemblage (WESTERMANN *et al.*, 1984). Not far away, at Mixtepec in Oaxaca state, occurs the late Bajocian *Duashnoceras* assemblage which contains several Andean species (WESTERMANN, 1983, 1984). The two Mexican faunas belong to the ecotonal overlap between the East-Pacific and Tethyan realms, and both are located on the Mixteca tectonostratigraphic terrane which was then presumably located at the entrance of the Hispanic Corridor.

Inter-Realm Relations and Migrations

Before the birth of the true middle Atlantic ocean (southern North Atlantic) in the Callovian, the Hispanic Corridor provided sporadic faunal exchange between the eastern Pacific and western Tethys since the Pliensbachian (DAMBORENEA & MANCEÑIDO, 1979; HILLEBRANDT, 1981; HALLAM, 1983). Before the Callovian, Middle Jurassic faunal exchange occurred particularly during the Early Bajocian culminating in the Humpriesianum Chron (corresponding to that pan-Tethyan standard Zone). Some brief and restricted migration through the Corridor happened presumably also at other times, at least once every one or two million years during the Aalenian and even during the existence of the Late Bajocian-Early Callovian East-Pacific Realm (WESTERMANN, 1981). The passage of this faunal interchange through the Hispanic Corridor (not around the extremes of Pangea) is attested by the ecotonic faunas of the Mixteca terrane, south Mexico, and corroborated by the recent documentation of Middle Jurassic ammonites from Venezuela (BARTOK *et al.*, 1984). Significant questions remaining include the migratory direction through the Hispanic Corridor and the origin of the major clades (center of evolution in either eastern Pacific or western Tethys) (Figs. 6, 7).

For many clades the evidence for migratory direction and center of evolution is ambivalent, because presumed ancestral and descendant clades occur in both Pacific and Tethys contemporaneously -- inter-continental correlations permitting. The evidence required for directional discrimination is either the presence in one area of suitable phyletic intermediates or of missing links, and/or the evidence of earlier appearance of the clade in question. Since the established theory holds that most clades originated in Europe (coincident with the higher density of researchers) and dispersed from there to the Pacific, the point is made here for the contrary biogeographic evolution.

The Hammatoceratidae-Sonniniidae clade is more completely represented in the Andean Aalenian-Early Bajocian than in Eurafrica (Europe + North Africa). In the Phymatoceratinae, successive grades connect *Hammatoceras*, *Bredyia* and *Parammatoceras*; *Bredyia* migrated to Eurafrica. A related (perhaps connected) clade is *Planammatoceras* - *P. (Pseudaptetoceras)*, branching off into the endemic Andean clade *Puchenquia* s. s. (with rare emigrants to Portugal) and *P. (Gerthiceras)*, and continuing into the quasi-

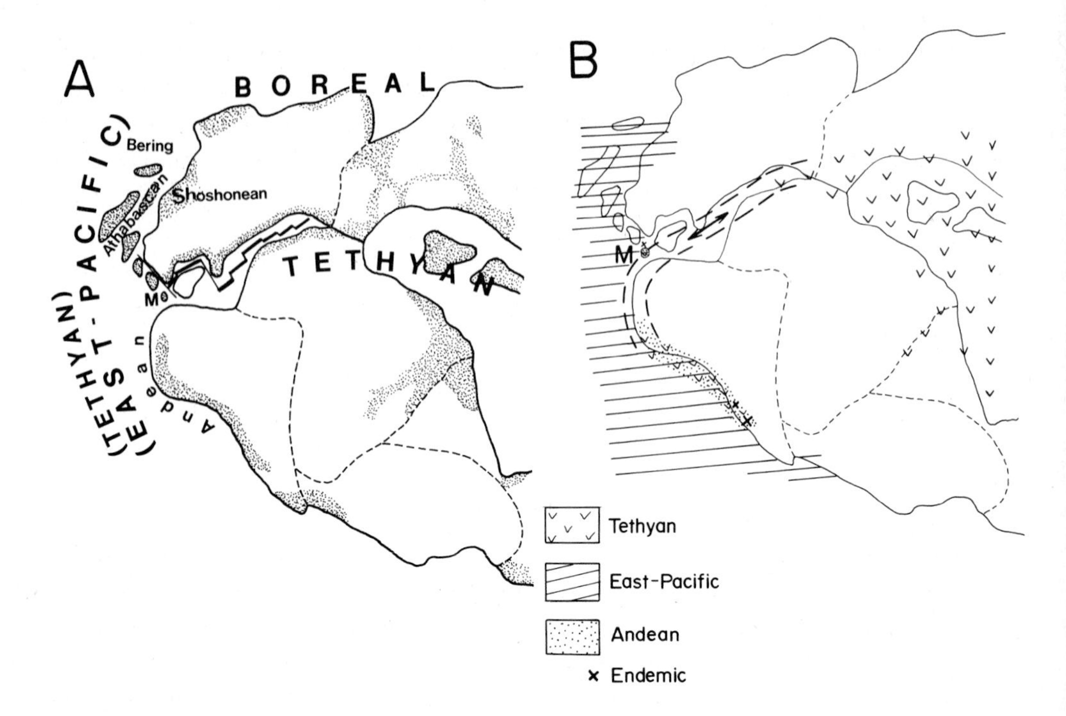

Fig. 7: Summary account of biogeography, taxonomic affinity and migration routes for the Middle Jurassic Ammonitina of the southeastern Pacific margin. Pacific (Panthalassa) and Tethys were sporadically connected via the Hispanic Corridor which began near the Mixteca terrane, Mexico as indicated by its mixed faunas (Fig. 6). Note that Tethyan influence decreases southward toward the Neuquen Basin while endemism increases.

cosmopolitan first sonniniid *Euhoploceras*. *"Fontannesia" austroamericana*, a probable *Puchenquia*, could indeed be ancestral to the Tethyan *Fontannesia* ; while *Zurcheria* occurs much earlier in the Andes than in Europe and is probably closer to erycitids than to sonniniids.

The Andean *Planammatoceras (Pseudaptetoceras) moerickei* is among the possible ancestors of the Oppeliidae, and the Andean representatives of *P. (Ps.) klimakomphalum* to the Strigoceratidae, although the evidence remains unconvincing in both cases.

The Otoitidae and Sphaeroceratidae are another good example of Andean evolution. While the centre of evolution of the Otoitidae remains ambivalent, the important phyletic link between the two families is present in the eastern Pacific, particularly in the Andean Province, i.e. *Emileia (Chondromileia)* leads to the earliest known sphaeroceratid, *Chondroceras recticostatum* of Laeviuscula Chron. It was in fact only the lappet-bearing, associated microconch *"Otoites"* that led us to transfer the Andean *"Sphaeroceras/Chondroceras" submicrostoma* and *giebeli* to *Emileia*.

The northeast Pacific *Chondroceras (Defonticeras)* is believed to have given rise to the Boreal Arctocephalitinae. In the southeast or south Pacific, Middle Bathonian Andean Eurycephalitinae presumably branched off into the Indo-West-Pacific and Tethyan Macrocephalitinae. The first rare immigrants reached central Europe in the Late Bathonian while the main host arrived during about the first million years of the Callovian.

The southeastern Pacific (including south Mexico) may even hold the solution to the Spiroceratidae dilemma, whither they came from and if and how these heteromorphs are related to the coiled *Strenoceras* and *Epistrenoceras*.

Among the Perisphinctidae, the very early Bajocian *Praeleptosphinctes* from Mendoza is either ancestral to all Perisphinctaceae or, more probably, an early "unsuccessful" offshoot.

The Reineckeiidae occur earlier in the Andes and south Mexico than in Europe. The cosmopolitan *Choffatia* s.l. either gave rise iteratively and heterochronously to the Andean *Neuqueniceras* (?subfam. Neuqueniceratinae) and the Tethyan *Reineckeia* s.l. (?subfam. Reineckeiinae) as believed by CARIOU (1980), or the Tethyan *Reineckeia* s.l. evolved from the Andean *Neuqueniceras*. The south Mexican clade *Neuqueniceras (Frickites) - Reineckeia (Rehmannia) - Reineckeia* s.s. indicates the second possibility (WESTERMANN et al., 1984), so that *R. (Rehmannia)* would have immigrated into Tethys. Middle to Late Callovian typical *Reineckeia* probably returned from Tethys to the eastern Pacific.

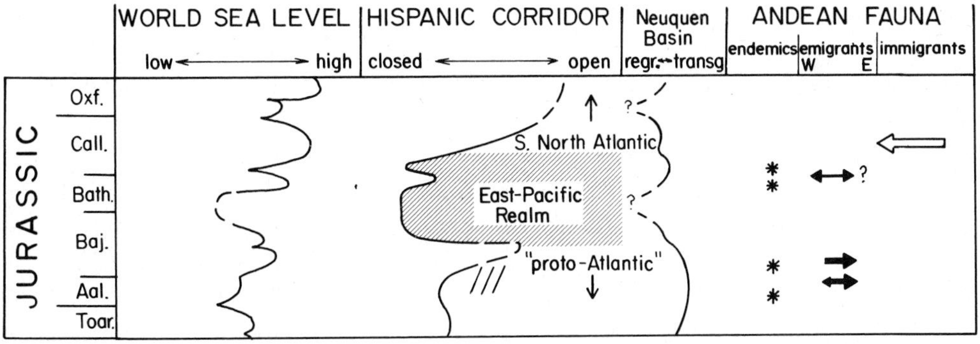

Fig. 8: Comparison of the Middle Jurassic sea level curve of HALLAM (1978) with ammonite biogeography. The evolution of the Hispanic Corridor (proto-Atlantic) concerns its effect as barrier between eastern Pacific and western Tethys which probably caused the Late Bajocian-Early Callovian East-Pacific Realm. Tethys expanded again into the eastern Pacific coincident with the mid-Callovian birth of the true central Atlantic ocean (southern North Atlantic). The transgression-regression curve for the Neuquen Basin and other parts of the Andes resembles the world sea level curve. The major radiation (*) and migration (→) events occurred at around peak sea level.

Radiation, Migration and Sea Level Changes

As can be seen from Figs. 4, 8 and 9, the principal evolutionary radiations in the Andes occured during the late Aalenian - early Bajocian and Late Bathonian. These events coincided with high sealevels throughout most of the Andean Jurassic area and, significantly, also with eustatic peaks of the world oceans. Times of moderate endemism during the early Bajocian and around the Bajocian-Bathonian and Callovian-Oxfordian boundaries, coincide with regressive phases or their onset (Fig. 9).

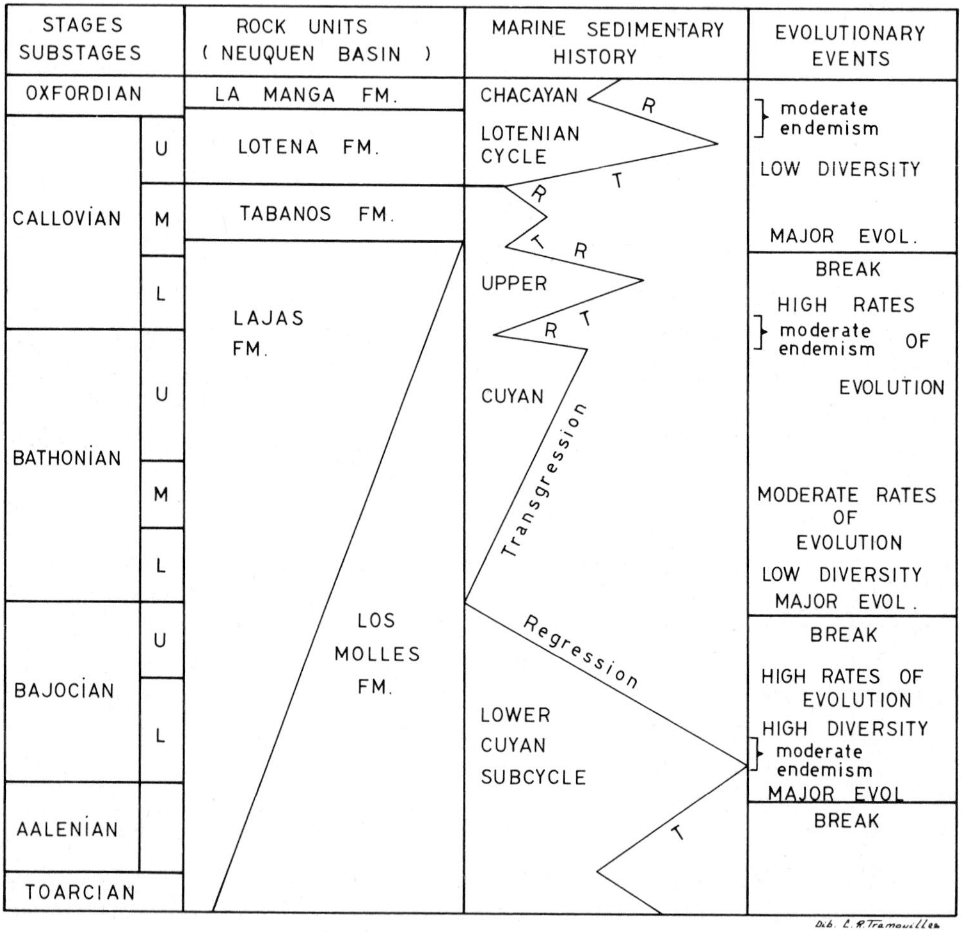

STAGES SUBSTAGES		ROCK UNITS (NEUQUEN BASIN)	MARINE SEDIMENTARY HISTORY	EVOLUTIONARY EVENTS
OXFORDIAN		LA MANGA FM.	CHACAYAN R	moderate endemism
CALLOVIAN	U	LOTENA FM.	LOTENIAN CYCLE T	LOW DIVERSITY
	M	TABANOS FM.	R T R	MAJOR EVOL.
	L	LAJAS FM.	UPPER T R	BREAK HIGH RATES moderate endemism OF EVOLUTION
BATHONIAN	U		CUYAN	
	M		Transgression	MODERATE RATES OF EVOLUTION
	L			LOW DIVERSITY MAJOR EVOL.
BAJOCIAN	U	LOS MOLLES FM.	Regression	BREAK HIGH RATES OF EVOLUTION
	L		LOWER CUYAN SUBCYCLE	HIGH DIVERSITY moderate endemism MAJOR EVOL.
AALENIAN			T	BREAK
TOARCIAN				

Dib. L. R. Tramouilles

Fig. 9: Sedimentary and evolutionary history of the Neuquen Basin during the Middle Jurassic.

The major pre-Callovian migration events through the Hispanic Corridor are obviously also dependent on sea level changes which must have effected this shallow, cratonic seaway. The curve for the "opening" and "closing" of the Hispanic Corridor (Fig. 8, centre) is based on biogeographic evidence only and illustrates its development as a barrier to migration in both directions. There is reasonably good correspondance with the eustatic and transgression-regression curves, considering the many structural and isostatic events expected to also effect such an extensive seaway. Middle Jurassic migrations through this seaway occurred repeatedly and in both directions; eastward mainly in the late Aalenian - earliest Bajocian and perhaps the early Callovian (Fig. 8, right). These eastward migrations coincided approximately with the sea level peaks.

These eastward migrations and the paleogeographic features of the long, shallow Hispanic Corridor in pre-Callovian times probably preclude the possibility of any significant and consistent westward current, an extension of the equatorial current as assumed by ENAY (1980).

ZUSAMMENFASSUNG

Das Neuquen-Becken stellt die südöstliche Bucht des Jurassischen Randmeeres dar, das später größtenteils ein Teil der Anden von Peru, Chile und Argentinien bekam. Der ziemlich vollständige Ammoniten-tragende Dogger interfingert zum Beckenrand mit kontinentalen Ablagerungen. Der marine Dogger umfaßt hauptsächlich den Cuya Sedimentations-Zyklus, mit Plienbachium-Oberbajocium Unterem Subzyklus und Oberbathonium-Mittelcallovium Oberen Subzyklus; darauf folgt der Obercallovium-Malm Loteno-Chacay Sedimentations-Zyklus.

Dogger-Ammonitina des Neuquen-Beckens setzen sich zusammen aus Elementen der Anden-Province, der Ost-Pazifik- und Tethys-Reiche, und aus kosmopolitischen genera. Zum Norden werden endemische Gattungen selten, während kosmopolitische und Tethysche Gattungen dominieren; wahrscheinlich aufgrund von Paläobreite, und korrespondierend mit Lithofazies-Wechsel von terrigen nach Karbonat dominiert. Die gesamte Anden-Provinz (oder Region) gehörte dem Tethys-Reich bis zum frühen Bajocium an, dann dem Ost-Pazifik-Reich bis zum frühen Callovium, und dann wieder dem Tethys-Reich.

Phyletische Clades und Grade des Dogger sind dokumentiert, mit wichtigen Radiationen im späten Aalenium, frühesten Bajocium und späten Bathonium; gleichzeitig mit hohem Wasserspiegel, regional wie auch global. Mehrere wichtige Stämme (clades) entsprangen wahrscheinlich in den Anden, einschließlich Neuquen-Becken, und verbreiteten sich dann zur westlichen Tethys durch den Hispanischen Korridor ("proto-Atlantik"). Die wichtigsten sind: Sonniniidae, mit *Euhoploceras* von der Andischen *Puchenquia (Gerthiceras)* abstammend; Sphaeroceratidae, mit *Chondroceras* von der ost-pazifischen, hauptsächlich andischen *Emileia (Chondromileia)* abstammend; und ? Reineckeiinae(?), vielleicht mit *Reineckeia (Rehmannia)* von *Neuqueniceras (Frickites)* abstammend. Die ostwärtige Passage durch den Hispanischen Korridor entstand wahrscheinlich durch das oekotonische Übergreifen von Faunen des -anden/Ost-Pazifik und der West-Tethys im Mixteca Terrain, Süd-Mexico. Die Macrocephalitinae entstanden aus Eurycephalitinae im südlichen Pazifik und breiteten sich westwärts nach Tethys aus. Die Haupt-Auswanderungen durch den Hispanischen Korridor stimmen zeitlich mit hohen eustatischen Wasserspiegeln überein, die augenscheinlich Barrieren in der flachen Seestraße überfluteten.

REFERENCES

Arkell, W.J., B. Kummel & C.W. Wright 1957: Mesozoic Ammonoidea. In R.C. Moore, ed., Treatise on Invertebrate Paleontology, Part L, Mollusca 4: L80-L437. Kansas Univ. Press.

Bartok, P.E., O. Renz & G.E.G. Westermann 1984, in press: The Siquisique Ophiolites, northern Lara State, Venezula: a discussion on their (?) Middle Jurassic Ammonites.- Geol. Soc. Amer. Bull.

Callomon, J.H. 1984: A review of the biostratigraphy of the post-Bajocian Middle and Upper Jurassic ammonites of western North America. In Westermann, ed., Jurassic-Cretaceous Biochronology and Paleogeography of North America.- Geol. Ass. Canada Spec. Pap. 27.

Cariou, E. 1980: L'Etage Callovian dans le centre-ouest de la France.- These Université Poitiers, 790 p.

Damborenea, S.E. & Mancenido, M.O. 1979: On the palaeogeographical distribution of the pectinid genus Weyla (Bivalvia, Lower Jurassic).- Paleogeogr., Paleoclimatol., Paleoecol. 27: 85-102.

Dellape, D.A., C. Mombru, G.A. Pando, A.C. Riccardi, M.A. Uliana & G.E.G. Westermann 1979: Edad y correlation de la Formation Tabanos en Chacay Melehue y otras localidades de Neuquén y Mendoza.- Obra Centen., Museo La Plata, 5: 81-105.

Dietl, G. 1978: Die heteromorphen Ammoniten des Dogger.- Stuttgarter Beitr. Naturkd., B, 33: 1-76.

--- 1981: On Macrocephalites (Ammonoidea) of the Aspidoides - Oolith and the Bathonian/Callovian Boundary of the Zollernalb (SW-Germany).- Stuttgarter Beitr. Naturkd., B, 68: 1-15.

Donovan, D.T., Callomon, J.H. & Howarth, M.K. 1981: Classification of Jurassic Ammonitina. In House, M.R. and Senior, J.R., eds., The Ammonoidea.- System. Ass. Spec. Vol. 18: 101-156.

Enay, R. 1980: Paléobiogéographie et ammonites Jurassique: "Rhythmes fauniques" et variations du niveau marin; voies d'echanges, migrations et domaines biogéographiques.- Livre Jubil. Soc. Géol. France, Mém. h.-s. 10: 161-181.

Hallam, A. 1977: Biogeographic evidence bearing on the creation of the Atlantic seaways in the Jurassic. In R.C. West, eds., Paleontology and plate tectonics.- Milwaukee Publ. Museum Spec. Publ., Biol. Geol. 2: 23-39.

--- 1978: Eustatic cycles in the Jurassic.- Paleogeogr., Paleoclimatol., Paleoecol. 23: 1-32.

--- 1983: Early and Mid-Jurassic molluscan biogeography and the establishment of the central Atlantic Seaway.- Paleogeogr., Paleoclimatol., Paleoecol. 43: 181-193.

Hillebrandt, A.v. 1970: Zur Biostratigraphie und Ammonitenfauna des Südamerikanischen Jura (insbes. Chile).- Neues Jahrb. Geol. Palaeontol. Abh. 136: 166-211.

--- 1973: Neue Ergebnisse über den Jura von Chile und Argentinien.- Münster. Forsch. Geol. Palaeontol. 31/32: 167-199.

--- 1981: Kontinentalverschiebung und die paläozographischen Beziehungen des Südamerikanischen Lias.- Geol. Rundschau 70: 570-582.

Hillebrandt, A.v. & G.E.G. Westermann 1984: Aalenian (Jurassic) Ammonite Faunas and Zones of the Southern Andes.- Zitteliana (in press).

Morton, N. 1972: The Bajocian ammonite Dorsetensia in Skye, Scotland.- Palaeontol. 15: 504-518.

Pavia, G. 1971: Ammoniti del Baiociano superiore di Digne (Francia SE, Dip. Basses-Alpes).- Boll. Soc. Palaeontol. Ital. 10: 75-142.

--- 1983: Ammoniti e biostratigrafia del Bajociano inferiore di Digne (Francia SE, Dip. Alpes-Haute-Province).- Museo Reg. Science. Nat. Torino, Monogr. II, 254 p.

Riccardi, A.C. 1983: The Jurassic of Argentina and Chile. In Moullade, M. and Nairn, A.E.M., eds., The Phanerozoic geology of the World II. The Mesozoic. B. Elsevier, Amsterdam, Oxford-New York-Tokyo, p. 201-263.

Senior, J.R. 1977: The Jurassic Ammonite Bredyia Buckman.- Palaeontology 20: 675-694.

Sandoval, J. 1983: Biostratigrafia y Paleontologia del Bajocense y Bathonense en las Cordilleras Beticas.- Tesis doctoral Univ. Granada, 613 p.

Schindewolf, O.H. 1965: Studien zur Stammesgeschichte der Ammoniten. Lfg. IV.- Akad. Wiss., Abh. math.-natur. Kl., Jahrg. 1965, 3: 137-238.

Stehn, G. 1923: Beiträge zur Kenntnis des Bathonian und Callovian in Südamerika. Beiträge zur Geologie und Paläontologie von Südamerika.- Neues Jahrb. Min., Geol. Paläont. Beil. Bd. 49: 52-158.

Stipanicic, P.N., G.E.G. Westermann & A.C. Riccardi 1975: The Indo-Pacific Ammonite Mayaites in the Oxfordian of the Southern Andes.- Ameghiniana 12: 281-305.

Thierry, J. 1976: Le Genre Macrocephalites au Callovien Inferieur (Ammonites, Jurassique moyen).- Mém. Géol. L'Univ. Dijon, 491 p.

Thierry, J. 1982: Tethys, Mesogée et Atlantique au Jurassique: quelques reflexions basees sur les faunes d'Ammonites.- Bull. Soc. Géol. France XXIV: 1053-67.

Westermann, G.E.M. 1956: Phylogenie der Stephanocerataceae und Perisphinctaceae des Dogger.- Neues Jb. Geol. Paläont., Abh., 103: 233-279.

--- 1964: El hammatoceratido Podagrosiceras athleticum Maubeuge y Lambert, del Bajociano inferior (Aaleniano) del Neuquen central, Argentina (Ammonitina, Jurasico).- Ameghiniana, 3: 173-180.

--- 1969: The ammonoid fauna of the Kialagvik Formation at Wide Bay, Alaska Peninsula. Part II, Sonninia sowerbyi Zone (Bajocian).- Bull. Amer. Paleontol., 57, 255: 225 p, 47 pls.

--- 1981: Ammonoid biochronology and biogeography of the Circum-Pacific Middle Jurassic. In M.R. House and J.R. Senior (eds.), The Ammonoidea.- Syst. Ass. Spec. Vol. 18: 459-498.

--- 1983: The Upper Bajocian and Lower Bathonian (Jurassic) ammonite faunas of Oaxaca, Mexico and West-Tethyan affinities.- Paleont. Mex. 46: 1-63.

--- 1984: The Late Bajocian Duashnoceras Assemblage (Jurassic Ammonitina) of Mixtepec, Oaxaca.- 3rd Latin Amer. Paleont. Conv., Mexico.

Westermann, G.E.G. & A.C. Riccardi 1972: Middle Jurassic ammonid fauna and biochronology of the Argentine-Chilean Andes. Part I: Hildocerataceae.- Palaeontographica, Abt. A, 140. 1-116, pls. 1-31.

--- 1976: Middle Jurassic ammonite distribution and the affinities of the andean faunas.- Primer. Congr. Geol. Chileno, Santiago de Chile, Vol. 1: C 23-39.

--- 1975: Edad y taxonomia del genero Podagrosiceras Lambert y Maubeuge (Ammonitina Jurásico medio).- Ameghiniana 12: 242-252.

--- 1979: Middle Jurassic ammonid fauna and biochronology of the Argentine-Chilean Andes. Part II: Bajocian Stephanocerataceae.- Paleontographica (A), 164: 85-188.

--- 1980: The Upper Bajocian ammonite Strenoceras in Chile: first circum-Pacific record of the Subfurcatum Zone.- Newsl. Stratigr., 9: 19-29.

--- 1982: Ammonoid fauna from the early Middle Jurassic of Mendoza province, Argentina.- J. Paleont. 56: 11-41.

Westermann, G.E.G., A.C. Riccardi, O. Palacios & C. Rangel 1980: Jurassico Medio en el Peru.- Inst. Geol. Min. Metal., Bol. 9: 1-47.

Westermann, G.E.G., R. Corona & R. Carrasco 1984: The Andean Neuqueniceras Fauna of Cualac, Mexico. In Westermann, ed., Jurassic-Cretaceous Biochronology and Paleogeography of North America.- Geol. Assoc. Can., Spec. Pap. 27:

DRASTIC CHANGES IN CARBONIFEROUS AMMONOID RATES OF EVOLUTION

Jürgen Kullmann

Tübingen

Abstract: Evolutionary rates permit four phases of ammonoid diversity development in ammonoids to be recognized during the Carboniferous: (A) Tournaisian, (B) Visean (excl. up.V_{3c}), (C) Namurian A and B (incl. up.V_{3c}), (D) Namurian C - Stephanian. All periods except C have a horotelic survivorship distribution, but different absolute rates; A and C are characterized by the predominance of short-lived, partly (C) tachytelic genera (longevity ~ 4 Myr, B of long-lived (~ 12 Myr), and D by the prevalence of both intermediate and long-lived genera (~ 4 Myr). Causal explanations for this evolutionary pattern are discussed.

INTRODUCTION

Because of their relatively rapid rate of evolution the ammonoids have always served as excellent guide fossils, and in the subdivision of larger stratigraphical units. However, they can also provide numerous data on ancient life conditions and the geological events which influenced their habitats.

The evolution of the ammonoids during Upper Devonian and Carboniferous times is of special interest. During these epochs several plate-tectonic movements, in connection with the Variscan orogeny, resulted in major changes in the area of ocean basins and epicontinental seaways. If it is true that there is a close relationship between the evolution of organisms and the geotectonically caused changes and shiftings of marine habitats an investigation of evolutionary changes in the ammonoids may provide a clear indication of the influences of Carboniferous geotectonic events on ammonoid habitats.

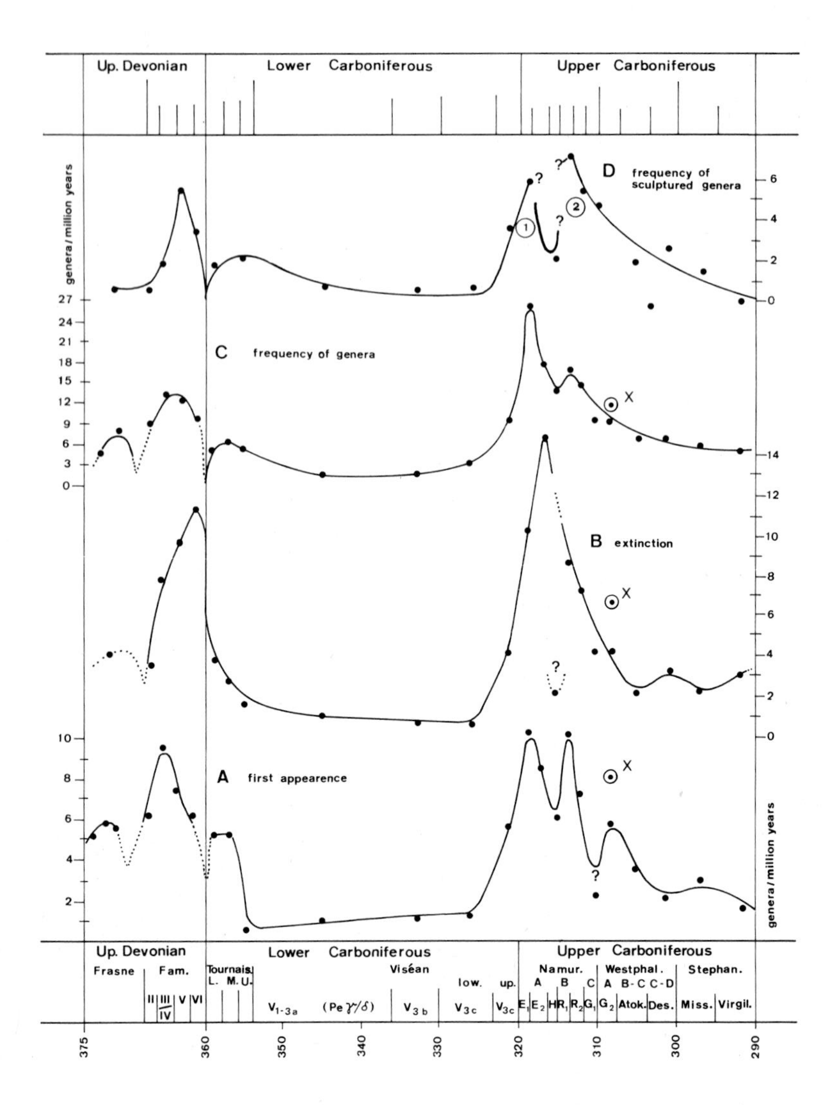

<u>Fig. 1:</u> Frequency distribution of the ammonoid genera during the Upper Devonian and Carboniferous; number of genera per 1 million years (Myr). Abscissa: absolute time scale, data of epochs according to Odin, 1982, stage duration estimated. Abbreviations of the stratigraphic units: Tournais.: Tournaisian (= Kinderhookian + low. Osagian); V_{1-3a} : low. Visean (= up. Meramecian-mid.Chesterian); E: *Eumorphoceras* stage, (E_1) Pendleian, (E_2) Arnsbergian (= up. Chesterian); H: *Homoceras* stage, Chokierian, Alportian; R: *Reticuloceras* stage, (R_1) Kinderscoutian, (R_2) Marsdenian (low. + mid. Hale, low. Morrowan); G: *Gastrioceras* stage, (G_1) Yeadonian (= up. Hale, low. Morrowan), (G_2) Westphalian A (= up. Morrowan); Atok.: Atokan; Des.: Desmoinesian; Miss.: Mississippian; Virgil.: Virgilian.

A: First appearance of new genera; -- B: Last occurrence of existing genera; -- C: Frequency of genera; -- D: Frequency of genera containing sculptured species (1: mainly forms with vertical or longitudinal (spiral) ornamentation; 2: mainly forms with coarse ornamentation or sculpture). — x: Orulganitidae, a peculiar family of smooth goniatites of lower Westphalian age restricted to NE Siberia.

For several years the ammonoid volume (part L) of the "Treatise on Invertebrate Paleontology" has been in revision. It provides (M.R. House, Devonian goniatites, see HOUSE, 1981, O.H. Schindewolf, D. Korn & J. Kullmann, Clymeniids, J. Kullmann, Carboniferous goniatites, see KULLMANN, 1981) a detailed survey on the development of the diversity of the Carboniferous ammonoids and their Upper Devonian predecessors. Data from these works are the basis of this investigation on the taxonomic rates of evolution of Carboniferous ammonoids.

1. THE DISTRIBUTION OF AMMONOIDS DURING THE CARBONIFEROUS

During the Famennian a little more than 50 ammonoid genera can be recognized. At the end of the Devonian all the clymeniids and all goniatites except the genus *Imitoceras* became extinct. After this extinction, immediately before the beginning of the Carboniferous, *Imitoceras* first gave rise to *Acutimitoceras*, and after that in rapid succession to further genera of the family Prionoceratidae.

The stratigraphic balance of all Carboniferous genera is as follows:

"leftover" from the Devonian (out of > 50 in the Famennian)	2
first appearance in the Carboniferous	212
extinction in the Carboniferous	200
"holdover" from the Carboniferous	14

The relationship of the Lower and Upper Carboniferous (in the sense of West European usage) is quite distinctive. According to the new absolute dating given by ODIN (1982) the Lower Carboniferous comprises 40 million years (Myr), the Upper Carboniferous only 30 Myr. The diversity distribution of ammonoids is in inverse proportion to these timespans.

During the Tournaisian (which is considered a short interval) a greater number of genera appeared, which had already acquired the key characters typical for the whole of the Lower Carboniferous. However, the Visean interval, the duration of which has been estimated by Odin at 35 Myr, appears to be a time of stagnation; the evolution of the ammonoids during this long period can be almost termed "arrested evolution".

However, this situation changed suddenly within the higher portion of the Upper Visean. At the end of the middle portion of the Upper Visean new types appeared which were first distinguished by their conch shape and sculpture (KULLMANN, KORN & PITZ, 1983), and in the upper part of the Visean also in the configuration of their suture-lines (Fig. 1A). Diversity (Fig. 1C) increased quickly at the end of the Visean, and at the beginning of the Namurian a first diversity maximum arose, which was recruited above all from newly changed conch shapes and suture-lines, more than from sculpture (Fig. 1C and D; Fig. 4).

Boundary Lower/Upper Carboniferous:	up.V_{3c} /E_1	low.V_{3c} /up.V_{3c}
"leftover" from the Devonian	2	2
first appearance in the Lower Carboniferous	75	57
extinction in the Lower Carboniferous	54	41
"leftover" from the Lower Carboniferous	23	18
first appearance in the Upper Carboniferous	137	155
extinction in the Upper Carboniferous	146	159
"holdover" from the Upper Carboniferous	14	14

Table 1: Number of genera (and subgenera) during the Lower and Upper Carboniferous. Left column: subdivision in the conventional extent - the uppermost Visean (upper Bollandian, upper V_{3c}, = P_2, = Go γ) as well as the lowermost Namurian (Pendleian, E_1) correspond to beds belonging in the middle portion of the Chesterian in North America. Right column: stratigraphic subdivision according to RUZHENTSEV (1958) and RUZHENTSEV & BOGOSLOVSKAYA (1971), which corresponds to the boundary lower/upper Bollandian (lower V_{3c}/upper V_{3c}, P_1/P_2, Go $\beta|\gamma$) within the uppermost Visean, equivalent about to the boundary between lower and middle Chesterian in North America.

At latest in the Upper Namurian this diversity maximum ceased to exist. A small maximum of new genera (Fig. 1A) followed at the beginning of the Westphalian. Another, hardly noticeable but probably more important, maximum arose at the beginning of the Stephanian; at a time, however, in which the ammonoids were clearly on the decline.

2. THE AVERAGE LONGEVITY OF THE CARBONIFEROUS AMMONOIDS

If the "holdover" genera of the Denovian and the "leftover" genera of the Permian are not counted, the average longevity of the Carboniferous ammonoid genera amounts to 6.8 Myr (Fig. 2). More than a quarter (28.5 %) of the genera have a longevity less than 2 Myr, another two quarters each (23.5 %, 21.5 %, resp.) 2 - 4 Myr and 4 - 8 Myr. The last quarter (26.5 %) contains the "living fossils" among the goniatites, the genera with a longevity of 8 - 40 Myr.

Considering Lower and Upper Carboniferous separately (see Tab. 1), totally different longevities of the ammonoid genera for the Lower and Upper Carboniferous can be observed. Using the boundary proposed by RUZHENTSEV (1958; with the assignment of the uppermost Visean, upper V_{3c} to the Namurian see explanation of table 1) the average longevity of the Lower Carboniferous genera amounts to 12 Myr, of the Upper Carboniferous to only 4.75 Myr.

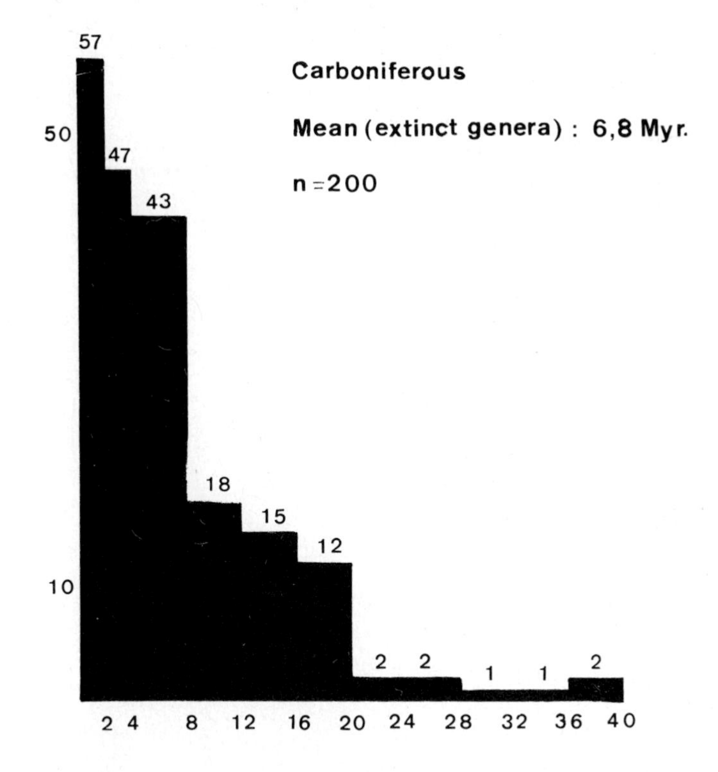

Fig. 2: Frequency distribution of all Carboniferous ammonoid genera except holdover genera. Histograms based on longevity of the Carboniferous genera; mean of the extinct genera: 6.8 Myr. Abscissa: Myr.

It is obvious that at the end of the Lower Carboniferous (beginning with the uppermost Upper Visean) there was not only an unusual increase in diversity but also a shifting of the predominant type of ornamentation (Fig. 3). Among Paleozoic ammonoids, three types of shell surfaces can be distinguished:

(1) The surface may be smooth, covered with fine, inconspicuous growth lines or longitudinal lines (Fig. 3, 1a-c);

(2) the surface may be covered by prominent growth striae, or may bear prominent spiral, lirate or reticulate ornamentation (Fig. 3, 2a-c);

(3) or the shell surface may be sculptured by ribs, nodes, furrows, at least during early or later ontogenetic stages (Fig. 3, 3a-d).

The percentage of the conspicuously ornamented (2) resp. coarsely sculptured (3) genera increased in the uppermost Visean slowly from about 30 % to more than 50 %; beginning with the upper Namurian the ammonoids developed increasingly coarse sculptures (Fig. 4).

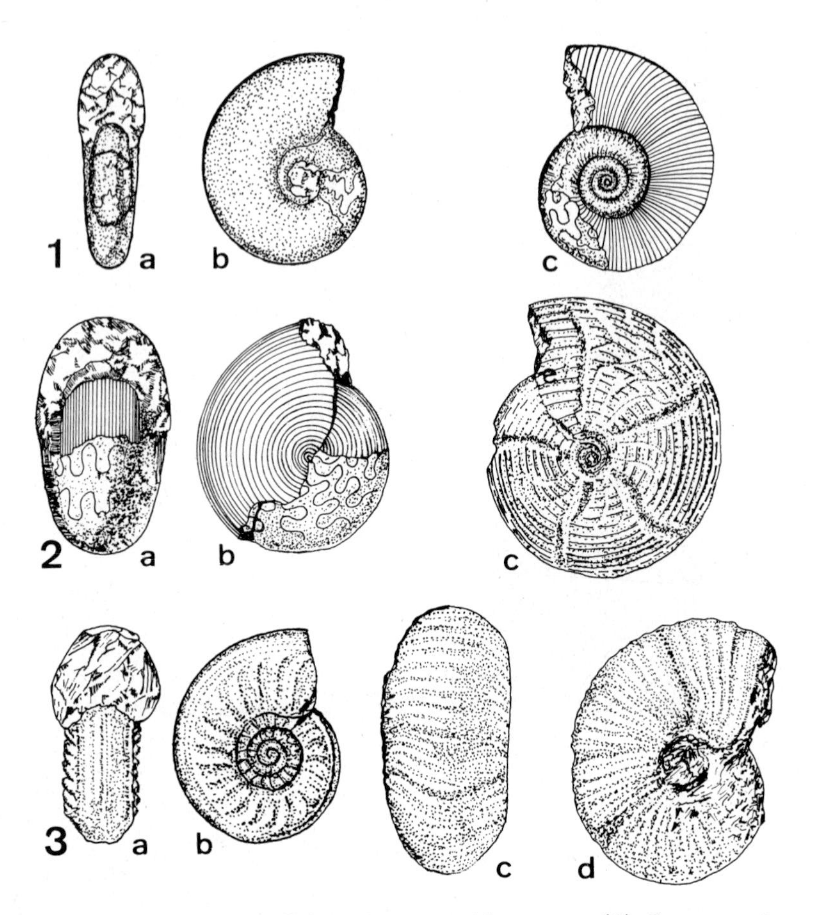

Fig. 3: Shell ornamentation in Paleozoic ammonoid genera. (1) Test smooth, covered with fine, inconspicuous growth lines or longitudinal lines; a, b: *Boesites* , c: *Epicanites* -- (2) Shell surface covered by prominent growth striae, or with prominent spiral, lirate, or reticulate ornamentation; a,b: *Agathiceras*, c: *Lusitanites* -- (3) Shell surface sculptured by ribs, nodes or furrows; a,b: *Eumorphoceras* , c,d: *Pericyclus*. (After MILLER & FURNISH, 1957, redrawn; all figures slightly enlarged, except Fig. 3c,d).

On the basis of these different diversity developments (see Fig. 1 C) the Carboniferous can be subdivided in four portions: a phase of radiation, characterized by a considerable increase of diversity, each in the Tournaisian and in the Namurian, was followed by a phase of lower diversity or even evolutionary stagnation.

Thus, four periods can be distinguished:

(A) Tournaisian,
(B) Visean (without upper V_{3c}),
(C) Namurian A and B (including upper V_{3c}) and
(D) Namurian C through Stephanian.

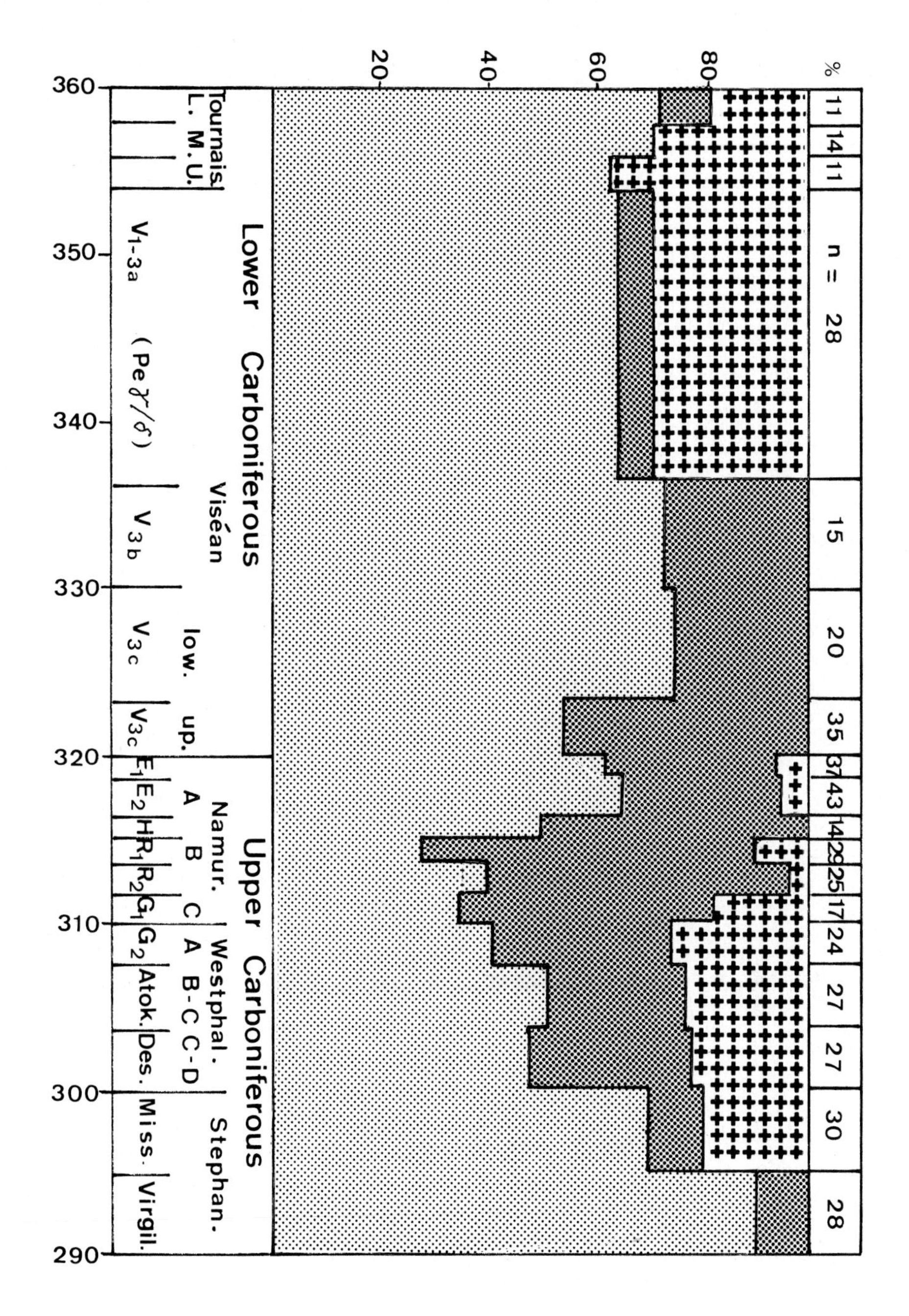

Fig. 4: Percentage of ornamented ammonoid genera of the Carboniferous. (1) Lower portion: Test smooth, covered with fine, inconspicuous growth lines or longitudinal lines. -- (2) Middle portion: Shell surface covered by prominent growth striae, or with prominent spiral, lirate, or reticulate ornamentation. -- (3) Upper portion: Shell surface sculptured by ribs, nodes or furrows.

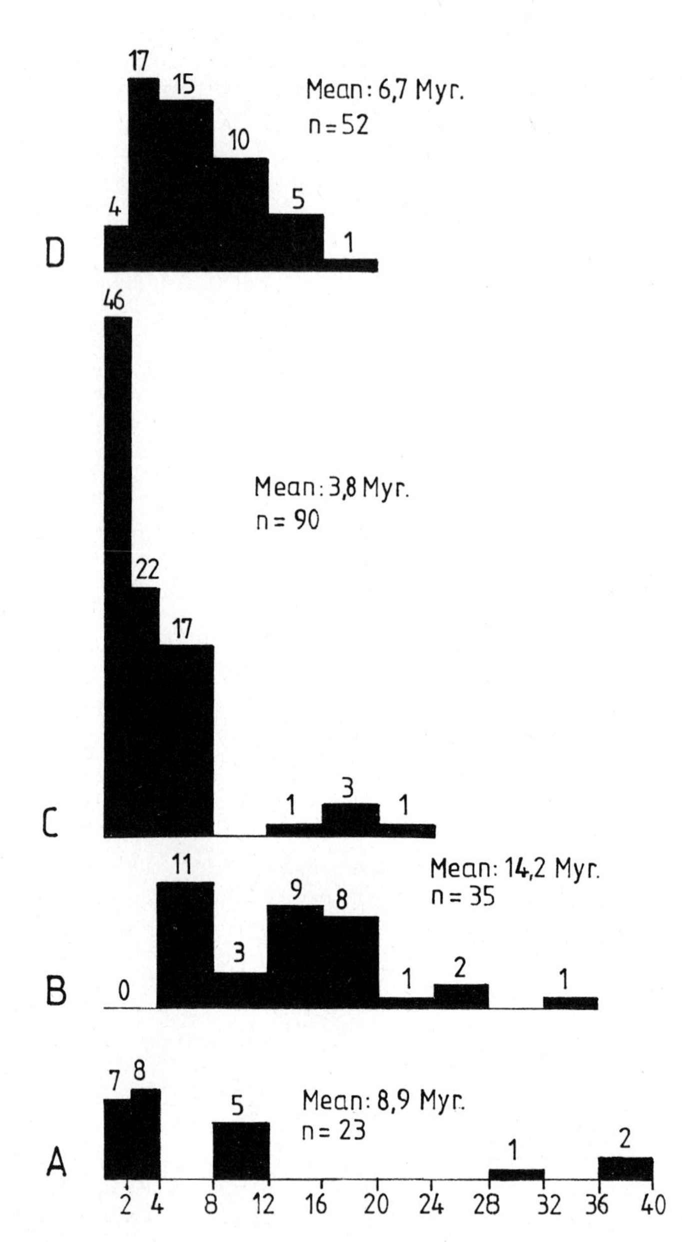

Fig. 5: Frequency distribution of the Carboniferous ammonoid genera. Histograms based on the longevity of the genera, in Myr (abscissa). A: Tournaisian, duration: ~6 Myr, average longevity: 8.9 Myr; B: Visean (excl. up V_{3c}), duration ~30 Myr, average longevity: 14.2 Myr; C: Namurian A and B (incl. up. V_{3c}), duration: 12 Myr, average longevity: 3.8 Myr; D: Namurian C - up. Stephanian, duration: 22 Myr, average longevity: 6.7 Myr.

These periods exhibit quite different figures for average longevity (Fig. 5); in the phases of radiation the goniatite genera have a considerably shorter lifetime, namely

8.9 Myr in the Tournaisian (Fig. 5A), 3.8 Myr in the Namurian (Fig. 5C). The following phases show an average longevity of 14.2 Myr in the Visean (Fig. 5B) and 6.7 Myr in the Westphalian-Stephanian (Fig. 5D).

Subdividing the goniatite genera according to their average longevity into 3 categories

(1) short-lived (O-4 Myr) genera,
(2) average ranging genera (4-12 Myr) and
(3) long-lived (12-4O Myr) genera (text-fig. 6),

shows that the first and third portions are characterized by the short-lived genera, whereas the second and fourth portions contain the longer lived genera. Periods A and C, the phases of radiation or, in Schindewolf's terminology (SCHINDEWOLF; 195O, p. 229) the "typogenetic phases", are characterized by 65 - 75 % short-lived genera and 5 - 15 % long-lived genera. In the phases of stagnation B and D (Schindewolf's phase "typostasis") the relation of short-lived genera to long-lived genera changes; in the case of B there are almost no short-lived genera, and in D a decline of the short-lived genera from 76 % to 4O % is compensated for by an increase of the long-lived genera from 5 to 12 %.

In terms of G.G. Simpson's categories of taxonomic rates of evolution (SIMPSON, 1953, p. 313), all these periods except C exhibit a horotelic distribution. Only in C is the prevalence of tachytelic genera obvious: about 5O % of all genera have a lifetime of less than 2 Myr, which means, that in this case excessively fast evolution took place which moved with tachytelic rates.

In the Upper Carboniferous (C and D) the rate of origination of genera declined only slowly, and the percentage of the long-lived genera increased reciprocally (a stagnation was reached no earlier than the Lower Permian). Fig. 4 shows that the percentage of sculptured forms remains almost constant till the Stephanian.

3. CAUSAL EXPLANATIONS OF THE EVOLUTIONARY PATTERNS OF THE CARBONIFEROUS AMMONOIDS

A major source of difficulty in interpreting ammonoid evolution during the Carboniferous is the pervasive lack of absolute dates for the individual stages. Fairly certain at present are the absolute dates of the Devonian/Carboniferous boundary (35O-365 Myr, used herein: 36O Myr; cp. ODIN, 1982 and GUTSCHICK & SANDBERG, 1983), the Visean/Namurian boundary (315-33O Myr, herein: 32O Myr) and the Stephanian/Asselian boundary (285-3OO, herein: 29O Myr). The Tournaisian/Visean boundary is poorly dated, but according to ODIN (1982) is seems to be only little younger than the Devonian /Carboniferous boundary; Visean time seems to be seven times as long as the Tournaisian.

SANDBERG et al. (1983) base their absolute ages on stratigraphical events occurring during the timespan from Middle Devonian till Late Mississippian of the Western United States, with the assumption of an equal temporal duration of conodont zones. They argued "for the approximately equal length of Late Devonian conodont timespans on the basis of steady, rapid evolution of a single phylogenetic lineage in favorable tropical environments". They calculated the average timespan of Upper Devonian conodont zones to be O.5 Myr, a figure "obtained by dividing the radiometric timespan of 13-15 m. y. for the Late Devonian by 27 conodont zones" (SANDBERG et al., 1983, p. 694). The timespan for Early Carboniferous conodont zones is assumed to be 1.5 Myr. In the light of Odin's dating, who supposes the Tournaisian to be 5 Myr long, the 1O conodont zones for the Lower Mississippian (Kinderhookian and Osagean = Tournaisian) would have had the same length as those in the Devonian, O.5 Myr. In any case, for the rest of the Lower Carboniferous remains 25 Myr, which follows from Sandberg's et al. calculation, or as Odin assumes, 35 Myr, a period which comprises only 4 conodont zones.

The evolution of the goniatites coincides largely with that of the conodonts (cf. CLARK, 1983); so as in the Famennian (considering clymeniid evolution), in the Tournaisian (evolution of the Prionoceratidae, Muensteroceratidae and Pericyclidae), and in the Visean: extremely slow evolution, almost stagnation till the uppermost Visean.

There is no doubt, however, that the length of the individual zones are partially unequal; the view that continuing evolutionary conditions lead to a continuous rate of evolution (which could be used as a chronometer) is not supported by most evolutionary biologists. Thus absolute dates simply based on zone sequences seems to be useful only for rather inaccurate estimates.

The strong changes in diversity at the Devonian/Carboniferous boundary, as well as at the end of the Lower Carboniferous and during the Namurian and Westphalian, represent almost certainly reactions of the organisms to paleoecological and possibly major biogeographical changes. The beginning of the Carboniferous is characterized by a widespread change in the style of sedimentation, which can be seen not only in the European Variscides but also to a certain extent in North Africa and North America. In Western Europe, the Devonian Period with a great many variations in facies and sedimentation is interrupted at the end of the Devonian by a transgressive phase; it is followed by a long-lasting period of sedimentary uniformity (A and B, Fig. 5,6), in which black shales, lydites and nodular limestones prevail, e. g. by the "stage of nivellation" observed in the Cantabrian Mountains of Northern Spain (KULLMANN & SCHÖNENBERG, 1975) or the "stagnation phase" in the "Rheinische Schiefergebirge" (FRANKE et al., 1978).

At the end of the long period of the Visean, a widespread new change in sedimentary conditions can be observed. In the uppermost Visean and lower/middle Namurian (C, Fig. 5,6), the transformation of the uniform basinal facies into flysch sedimentation

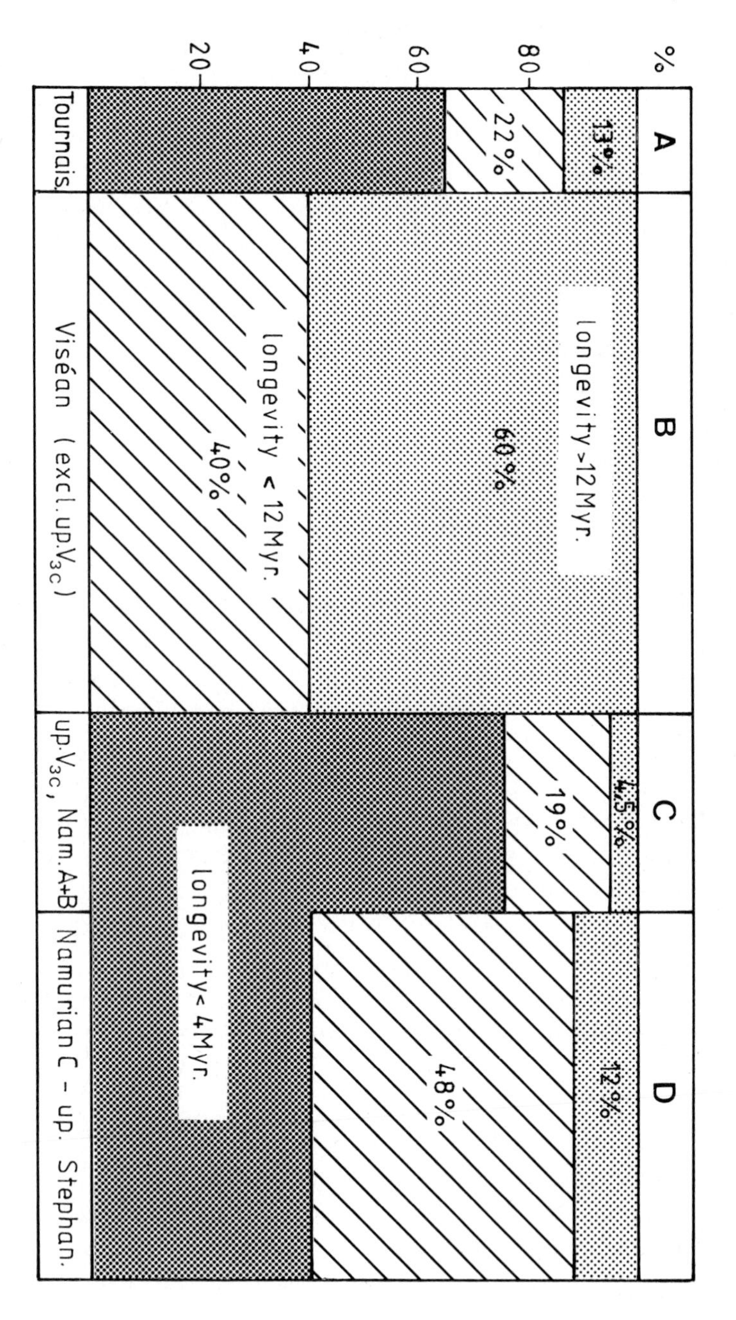

Fig. 6: Percentage of short-lived (0-4 Myr) and long-lived (> 12 Myr) Carboniferous ammonoid genera.

took place; as a consequence of vertical movements resulting in marine regressions the facies variations started again in many Variscan orogens. This sedimentary change marks the beginning of orogenic dynamics of the "differentiation stage" (KULLMANN & SCHÖNENBERG, 1975), preluding the Variscan orogeny (D, Fig. 5, 6).

The temporal coincidence of these geodynamic events and the changes in ammonoid diversity suggests geodynamic factors as a possible cause for the evolutionary pattern of the Carboniferous ammonoids. In connection with plate-tectonic events, which led to eustatic rises of sea level, the major turning-points of ammonoid evolution may be interpreted as due do drastic changes in the size of the ammonoid realm. Since speciation seems to be correlated with habitat area and geographic distances, any change in the configuration of the area of marine realm by geodynamic events may result in drastic changes in the diversity of pelagic organisms.

4. CONCLUSIONS

1. After a precipitous decline in the number of genera at the end of the Devonian Period, ammonoids proliferated again during the Tournaisian; the average duration was 8.9 Myr, with more than 60 % genera being short-lived (longevity: 0 - 4 Myr).

2. The long period of the Visean (35 Myr) was a timespan of evolutionary stagnation, the average duration being 14.2 Myr, being dominated (60 %) by long-lived genera (longevity: 12-40 Myr).

3. The most important phase of radiation of all Late Paleozoic ammonoids was the timespan from uppermost Visean till middle Namurian; the average duration here was only 3.8 Myr. This interval is further characterized by a non-horotelic survivorship distribution, with 75 % short-lived genera, including a high percentage of tachytelic forms.

4. The period upper Namurian till the end of the Carboniferous is characterized by the predominance of intermediate and long-lived ammonoid genera, with an average duration of 6.7 Myr.

5. The drastic changes in ammonoid diversity at the beginning of the Carboniferous and in the uppermost Visean/Lower Namurian can be interpreted as resulting from plate-tectonic movements which led to major size changes in the areal extent of marine habitats due to eustatic sea level rises.

Acknowledgments. For their helpful comments on the manuscript and many valuable suggestions I would like to thank Arthur J. Boucot (Oregon State University, Corvallis), George R. McGhee, Jr. (Rutgers, State University of New Jersey) and Peter Ward (University of California, Davis).

REFERENCES

Clark, D.L. 1983: Extinction of conodonts.- Jour. Paleont. 57, 652-661, 10 figs., Lawrence, Kans.

Franke, W., Eder, W., Engel, W. & Langenstrassen, F. 1978: Main Aspects of Geosynclinal Sedimentation in the Rhenohercynien Zone.- Z. deutsch. geol. Ges. 129, 201-216, 7 figs., Hannover.

Gutschick, R.C. & Sandberg, C.A. 1983: Mississippian continental Margins of the conterminous United States.- Soc. Econ. Paleont. Miner. Spec. Publ. 33, 79-96, 7 figs.

House, M.R. 1981: On the origin, classification and evolution of the early Ammonoidea.- In: House, M.R. & Senior, J.R. (eds.): The Ammonoidea.- Syst. Assoc. spec. vol. 18, 1-36, 7 figs., London.

Kullmann, J. 1981: Carboniferous Goniatites.- In: House, M.R. & Senior, J.R. (eds.): The Ammonoidea.- Syst. Assoc. spec. vol. 18, 37-48, 1 fig., London.

Kullmann, J., Korn, D. & Pitz, T. 1983: Sulcogirtyoceras Ruzhentsev - eine weitverbreitete skulptierte Goniatiten-Gattung des hohen Unterkarbons.- Neues Jb. Geol. Paläont. Mh. 1983, 544-556, 5 figs., 1 tab., Stuttgart.

Kullmann, J. & Schönenberg, R. 1975: Geodynamische und paläökologische Entwicklung im Kantabrischen Variszikum (Nordspanien). Ein interdisziplinäres Arbeitskonzept.- Neues Jb. Geol. Paläont. Mh. 1975, 151-166, 2 figs., Stuttgart.

Miller, A.K. & Furnish, W.M. 1957: Paleozoic Ammonoidea (excl. Clymeniina). In: Moore, R.C. (ed.): Treatise on Invertebrate Paleontology. pt. L, Ammonoidea, 1-36, 47-79, figs. 1-37, 46-123, New York.

Odin, G.S. 1982: The Phanerozoic Time Scale revisited.- Episodes 1982 (3), 3-9, 5 figs.

Ruzhentsev, V.E. 1958: Namyurskiy yarus v mirovoy stratigraficheskoy shkale.- Byull. mosk. obshchv. ispyt. prirod., otd. geol. 33 (5), 1 fig., 1 tab., Moskva.

Ruzhentsev, V.E. & Bogoslovskaya, M.F. 1971: Namyurskiy etap v evolyutsii ammonoidey. Rannenamyurskiye ammonoidei.- Trudy Paleont. Inst. Akad. Nauk SSSR 133, 382 pp.,89 figs., 40 pl., Moskva.

Sandberg, C.A., Gutschick, R.C., Johnson, J.G., Poole, F.G. & Sando, W.J. 1983: Middle Devonian to Late Mississippian geologic history of the Overthrust belt region, western United States.- Rocky Mountain Assoc. Geol., In: Powers, R.B. (ed.): Geologic Studies of the Cordilleran Thrust Belt, 2, 691-719, 17 figs., Denver.

Schindewolf, O.H. 1950: Grundfragen der Paläontologie. Geologische Zeitmessung. Organische Stammesentwicklung. Biologische Systematik.- 506 pp.,332 figs., Stuttgart.

Simpson, G.G. 1953: The Major Features of Evolution.- Columbia Biol. Ser. 17, 434 pp.,51 figs., New York and London.

AMMONITE SHELL FORM AND TRANS-GRESSION IN THE BRITISH LOWER JURASSIC

D. T. Donovan

University College London

Abstract: The early Jurassic transgression has been traced in detail in southern England. Shell forms in Ammonitina have been classed in eight groups, and the occurrence of these groups is related to the progress of the transgression. Unkeeled shells are replaced by forms with carinate, bisulcate venters as the transgression proceeds, and diversity increases; this is interrupted by regression during the *Oxynotum* Zone, during which oxycones are dominant. Renewed transgression again shows unkeeled followed by carinate, bisulcate shells, which are then replaced by the Eoderocerataceae at the beginning of the Pliensbachian. Regression in the late Pliensbachian again brings oxycones.

Sutural patterns are classified into two contrasting types, sparse and dense, the latter mainly occurring in transgressive episodes.

INTRODUCTION

An offlap - onlap curve has been drawn for the Hettangian to Pliensbachian stages inclusive. This refers specifically to the London Platform in southern Britain and is not intended to be a 'sea level curve' of general applicability. The question of eustatic control is not considered here. The curve for the Hettangian to Lower Pliensbachian is based on the detailed subsurface information presented by DONOVAN, HORTON & IVIMEY-COOK (1979). The curve for the Upper Pliensbachian (Domerian Substage) is less well controlled because borehole information is not so good, and because of the erosion which affected the top of the Lower Lias (DONOVAN et al, op. cit. fig. 2).

The overall picture is one of transgression interrupted by two main regressive epi-

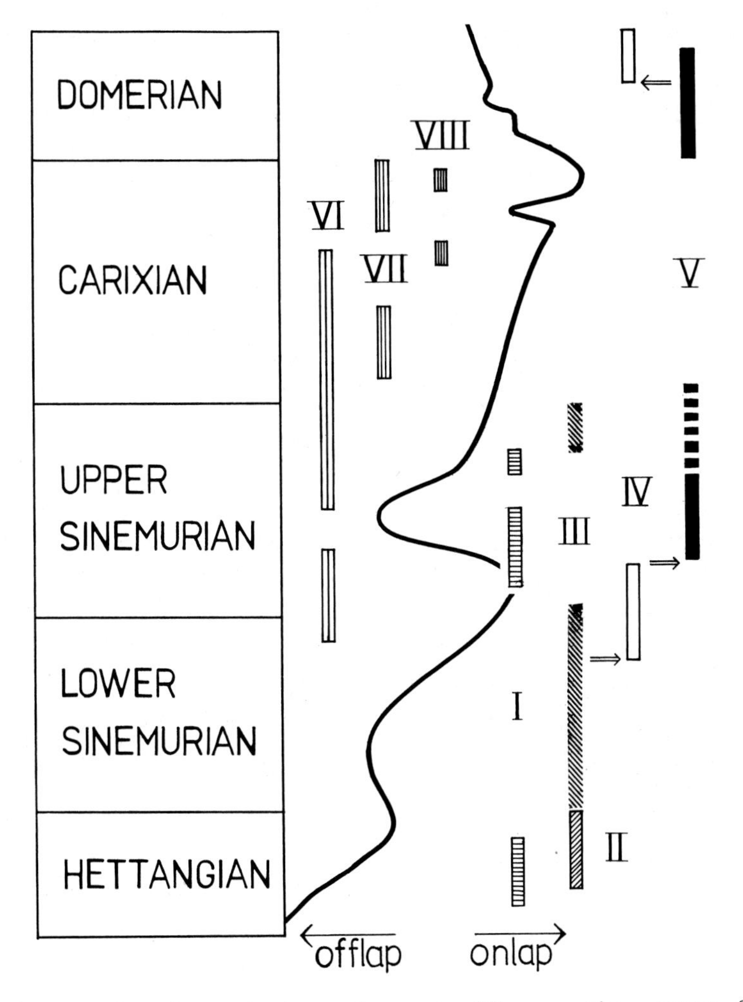

Fig. 1. Curve indicating relative onlap and offlap on the western flank of the London Platform, southern England, and the stratigraphical ranges of the ammonite shell form groups I - VIII in the same area. For explanation of the shell form groups, see text.

sodes. The first regression, in *oxynotum* Zone time, gave rise to a widespread non--sequence throughout much of southern England. The second resulted in the condensed deposits of the Spinatum Zone (the Marlstone Rock-Bed of many English localities). The two chief periods of transgression are referred to here as the first and second transgressions respectively.

The occurrence of different shell forms has been plotted alongside the offlap--onlap curve (Figs. 1, 2). It is emphasised that the ranges shown apply to the area studied and are not necessarily applicable elsewhere. Very rare occurrences (one or two specimens at isolated horizons) are in general not included in the ranges shown.

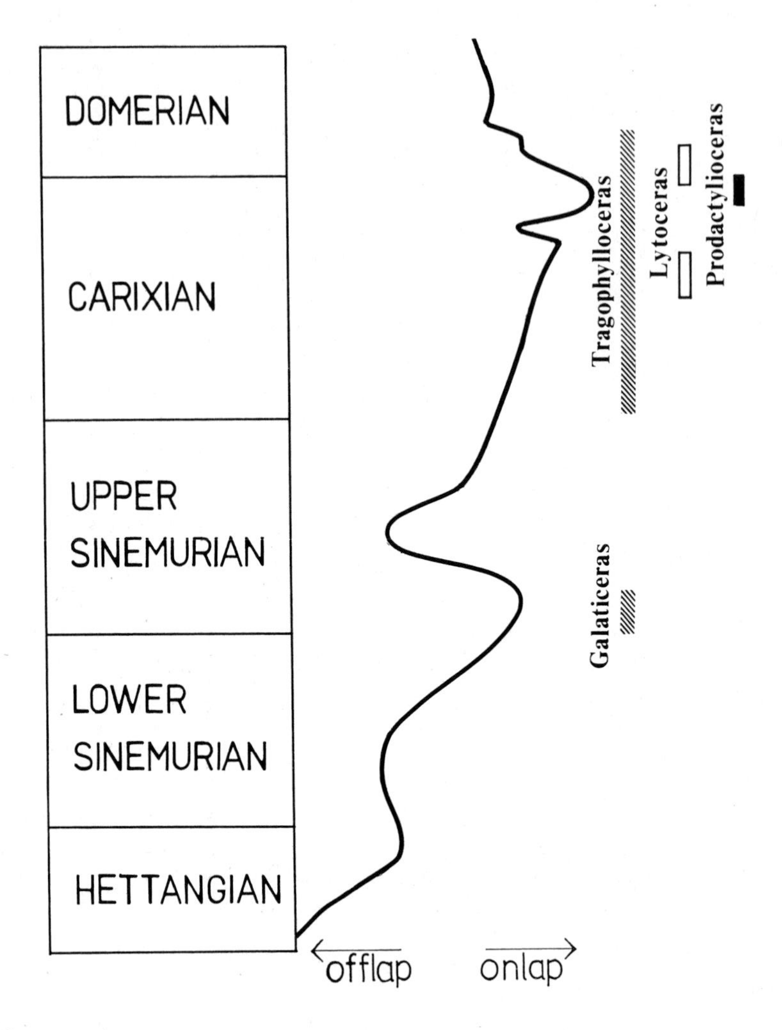

Fig. 2. Curve indicating onlap and offlap, as in Fig. 1, and the stratigraphical ranges of Tethyan ammonite genera in southern England.

GROUPING OF AMMONITES BY SHELL FORM

An attempt was made to group early Jurassic ammonites by their shell forms using the W : D plot of RAUP (1967). This was not wholly successful, for two reasons. First, RAUP's measure W, the whorl expansion rate, based on the ratio of successive radii, is not easy to measure, and is thus subject to error. Second, the early Jurassic ammonites of shelf areas, as a whole, occupy a restricted part of the RAUP W : D plot, lying in a band close to the line whose equation is W = 1/D (on the other side of which the whorls are no longer in contact). For forms with W 2 the use of this plot approximates to classifying them by their umbilical percentages.

There are also, of course, important features which cannot be shown on such a plot. The features used here are the presence or absence of ribs, and of spines, on the whorl flanks; and the nature of the venter: smooth, with a keel, or with keel and flanking grooves (carinate-bisulcate). A few genera have ribs crossing the venter, which may form forwardly-pointing chevrons. Ribs are almost always approximately straight and radial. Bifurcating or branching ribs are almost unknown; they are present in the later schlotheimiids and, weakly, in some oxycones.

Sutural characters also need to be considered in characterising shell form. We are not concerned with ontogenetic development or with the identification of elements, but with features which may have been relevant for the palaeoecology of the ammonite. Sutures are therefore characterised in terms of two extreme conditions. On the one hand are sutures with comparatively simple elements arranged so that relatively wide bands of plain shell separate successive suture lines. At the other extreme are sutures with complicated elements arranged so that an 'all-over' pattern results, with no well-defined areas between successive sutures. These patterns are here termed sparse and dense respectively. These patterns are thought to have had functional significance, the dense type affording more support to the shell-wall against external pressure than the sparse. Possibly they also had significance in terms of the area of septal surface available for 'wetting' (cf. DENTON & GILPIN-BROWN 1966, p. 742). The spacing of septa does not vary much in the genera investigated, being usually within a few integers either side of 2O per whorl.

Summary of the Groups Recognised

I Serpenticones, ribbed, venter smooth or with weak, poorly-demarcated keel. Sutural pattern sparse. Typical genera *Caloceras*, *Alsatites*, *Echioceras*.

II Fairly evolute shells (' planulates') with close, sharp ribs which form chevrons on the venter. Large shells become smooth. Sutures may be simple on early whorls or small individuals, but usually become complex and densely-patterned at larger sizes. This shell form occurs only in Schlotheimiidae.

III Serpenticones, umbilicus greater than 45%. Ribbed, with carinate, bisulcate venter. Sutural pattern usually sparse. Typical genera *Coroniceras*, *Arnioceras*, *Paltechioceras*.

IV Shells with umbilicus between 35% and 45% (D = 0.35 - 0.45). Ribbed with carinate bisulcate venter. Sutural pattern sparse. Typical genera *Asteroceras*, *Pleuroceras*.

V Oxycones and near-oxycones. Ribs usually weak, or absent. Sutures generally forming fairly dense patterns, though with some variability. Typical genera *Eparietites*,

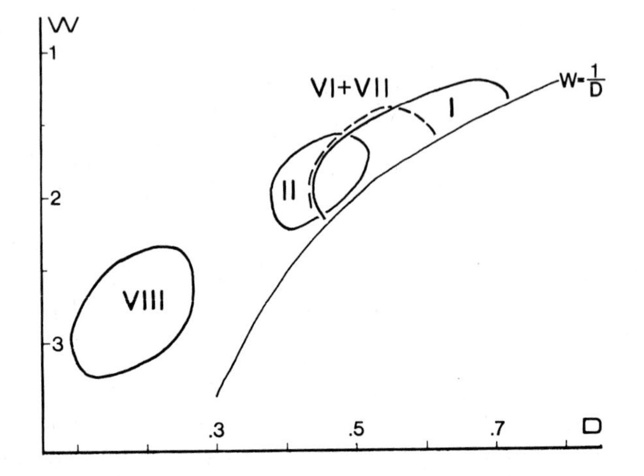

Fig. 3. Areas occupied by certain shell form groups on a plot of W : D. For definition of W and D see Raup (1967). For explanation of the shell form groups denoted by roman numerals, see text.

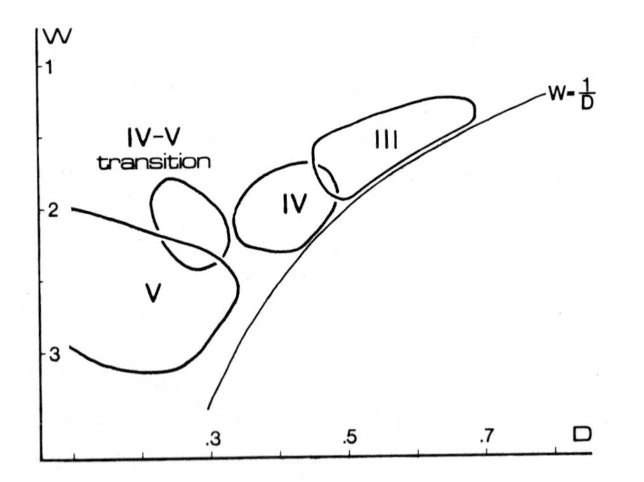

Fig. 4. Areas occupied by shell form groups with strongly carinate venters on a plot of W : D. For definition of W and D see Raup (1967). For explanation of the shell form groups, see text. The area shown as 'IV-V transition' corresponds to the genus *Eparietites*.

Oxynoticeras, *Amaltheus*.

VI Serpenticones, ribs which usually bear one or two spines, venter smooth or with ribs weakly passing over it. Sutural pattern usually dense. Typical genera *Microderoceras*, *Bifericeras*, *Crucilobiceras*, *Acanthopleuroceras*.

VII Serpenticones, ribbed, spines weak or absent. Ribs usually cross venter fairly strongly. Sutural pattern dense. Typical genera *Uptonia*, *Aegoceras*.

VIII Sphaerocones. *Liparoceras* and its subgenera.

The areas occupied by the groups on W : D plots are shown in Figs. 3, 4. It will be noted that the few smooth ammonites of the early Jurassic have not been classified.

RELATION OF SHELL FORM, TRANSGRESSION AND REGRESSION

The first transgression shows a clear sequence of forms. The early stages are characterised by evolute forms with smooth venters or, at the most, weak ventral ornament of group I. These are joined by the schlotheimiids of group II. The start of a period of stasis, or slight regression, is accompanied by the appearance of arietitids of group III. Their appearance was due to an extension of their range northward from the Tethys which seems to have been a rapid event throughout north-west Europe. These carinate, bisulcate shells dominate the scene for some time. As the transgression resumed, more rapidly expanding shells of group IV make their appearance, the earliest being *Euagassiceras*. Shells of groups III and IV co-exist for some time.

The sequence of groups III, IV and V towards more rapidly expanding and involute shells is, of course, an evolutionary sequence within the Superfamily Psilocerataceae. This evolution probably took place within north-west Europe. Shells of group IV characterise the maximum of the first transgression along with the earliest local representatives of group VI. However, the regression of the *oxynotum* Zone coincided with the replacement of group IV by group V. The data are not sufficiently precise to say whether the oxycones of group V evolved in response to the regressive situation.

Shells of group I (*Epophioceras*, *Echioceras*) re-appear during the regressive phase centred upon the *oxynotum* Zone. With resumed onlap they are succeeded by shells of group III (*Paltechioceras*). These become extinct at the end of the Sinemurian as the transgression proceeds.

Oxycones (group V) (*Amaltheus* and allied genera) characterise the regressive phase in the early Domerian, just as they had characterised the late Sinemurian regression.

There is, however, an evolutionary change from these into more open shells (*Pleuroceras*) which fall in group IV (HOWARTH 1958).

The remaining shell groups are restricted to members of the Superfamily Eoderocerataceae. In general these forms are restricted to the periods of greatest onlap. They do not appear in Britain at all until the *turneri* Zone (*Microderoceras*, *Promicroceras*) when the first transgression was well advanced. They disappear during the maximum of the *oxynotum* Zone regression. They then continue throughout the second transgression until the commencement of regression at the end of the Carixian. Most of these forms belong to our group VI, but group VII is common in the *jamesoni* Zone (*Uptonia*, some *Polymorphites*) and the *davoei* Zone (*Aegoceras*).

Sphaerocones (group VIII) become common only when the second transgression is near its maximum, disappear during the *maculatum* - early *capricornus* Subzone regression, and reappear briefly until the Domerian regression sets in.

It is only to a very limited extent that one can see any similarity of sequence of shell forms during the first and second transgression respectively. There is a change from shells with plain venters to carinate, bisulcate forms early in both transgressions, but while keeled forms dominate the rest of the first transgression, in the second they disappear at the end of the Sinemurian. Both main regressive phases are characterised by oxycones.

The distribution of genera belonging to 'tethyan' families is shown in Figure 2, although these are not classified in the shell form groups used in the earlier part of this paper. In the first transgression there is only *Galaticeras* (HOWARTH & DONOVAN 1964) which appears for a brief interval, in southern England only, at the maximum of the transgression. The second transgression, which extended further than the first, has three genera of which *Tragophylloceras* has the longest range. *Lytoceras* comes in when the transgression is well advanced, and disappears again during the *maculatum* - *capricornus* regression. After the latter it re-appears briefly, along with *Prodactylioceras* (Dactylioceratidae are common members of British faunas only in the Toarcian - for the present purpose they are classified as Tethyan forms), until the Domerian regression sets in.

EVOLUTION AND TRANSGRESSION

The phylogenetic pattern of early Jurassic ammonites was characterised by well-marked periods of diversification (DONOVAN, CALLOMON & HOWARTH, 1981, Fig. 2).

The earliest, during the Hettangian, took place in the Tethys and the present data are not relevant. The diversification within the Arietitidae, giving rise to the subfamilies Agassiceratinae and Asteroceratinae, occurred in northern Europe during the *semicostatum* Zone and corresponds with renewed transgression following stillstand in the earliest Sinemurian.

The diversification of the Eoderocerataceae, no less than four families, arising from the root stock of the Eoderoceratidae, coincided with the early stages of the second transgression; but there is little to suggest that this diversification took place in the area under consideration. Of course, the transgression may have been an event of more general importance which controlled evolution elsewhere.

The correspondence of oxycones with episodes of regression has already been noted. The data are not good enough to show whether the oxycone shell evolved as a response to regressive conditions. Any such proposal for the Domerian would have to explain the development of non-oxycone species (*Amaltheus subnodosus*, *A. gibbosus*), leading to *Pleuroceras*, contemporaneously with oxycene shells.

These data are in agreement with the relationship proposed by NEWELL (1963) that regression should lead to extinction and thus reduction in diversity, whereas transgression results in radiation and consequent increase in diversity. A general relationship between ammonite diversity and the extent of continental flooding has been shown by KENNEDY (1977) for the Mesozoic, and it is substantiated in more detail by the present evidence. The matter has recently been discussed and summarised by HALLAM (1981, pp. 225-230).

SUTURAL CHARACTERS

Sparse sutural patterns (see above) characterise the Psilocerataceae, while dense patterns are the rule in Eoderocerataceae. While we may observe that dense patterns replace sparse ones as the transgression proceeds, this is equivalent to the statement that Eoderocerataceae replace Psilocerataceae, and is no help in deciding whether it was the sutural pattern or some other adaptation which rendered Eoderocerataceae more suited to the latter transgressive stages. However, the existence of shell forms of group VI with dense patterns (*Microderoceras*, *Xipheroceras*) during the maximum of the first transgression lends support to the idea that dense patterns were particularly suited to transgressive conditions, while conversely Eoderocerataceae of group IV (*Pleuroceras*), which dominate the later Domerian regression, had mainly sparse patterns, an exception to the rule for the superfamily.

CONCLUSION

In view of our lack of knowledge of ammonite mode of life, we may hardly expect to find satisfactory explanations of any correlations that may be observed. Certainly there is no correlation of different shell forms or different genera with sedimentary facies. This is shown strikingly by the well-known fossiliferous bed, the *jamesoni* Limestone of the area of Radstock, Avon (DONOVAN; 1984), which is a condensed limestone deposited in shallow water. It contains in abundance all the usual genera found in the normal mudstone facies of the *jamesoni* and *ibex* Zones, including *Tragophylloceras* and *Lytoceras*. The same observation is made with regard to other parts of the Jurassic by LEHMANN (1981, p. 173).

The relationship between area and diversity is easy to understand for benthonic organisms; extension of the area of the sea may lead to increase in the number of habitats or ecological niches. The relationship is less easily comprehensible for ammonites, for which it appears to hold in the present case, if they are regarded as pelagic in common with the majority of authors. It is true that some authors have regarded them as benthonic (most recently LEHMANN 1981, p. 179), but a strictly benthonic habit is not accepted by the present writer and has also been rejected by TINTANT et al. (1982, p. 953). KENNEDY & COBBAN (1976, p. 14) warn against assuming a single life style for all ammonites, pointing out that their diversity is likely to indicate a variety of feeding strategies. KENNEDY & COBBAN also suggest that ammonites exploited the lower levels in the marine food web, and this could have made them sensitive to changes in the benthonic fauna consequent on transgression or regression.

REFERENCES

Denton, E.J. & Gipin-Brown, J.B. 1966: On the bouyancy of the Pearly Nautilus.- J. mar. biol. Ass. U.K., 46, 723-759.

Donovan, D.T. & Kellaway, G.A. 1984: Geology of the Bristol District: The Lower Jurassic rocks.- Mem. Geol. Surv. G.B.

Donovan, D.T., Callomon, J.H. & Howarth, M.K. 1981: Classification of the Jurassic Ammonitina. In: The Ammonoidea (ed. M.R. House & J.R. Senior), Systematics Ass. Special Vol. 18.- London & New York: Academic Press, 101-155.

Donovan, D.T., Horton, A. & Ivimey-Cook, H.C. 1979: The transgression of the Lower Lias over the northern flank of the London Platform.- J. Geol. Soc. London, 136, 165-173.

Hallam, A. 1981: Facies interpretation and the stratigraphic record.- Oxford & San Francisco: W.H. Freeman & Company. xii + 291 p.

Howarth, M.K. 1958: A monograph of the ammonites of the Liassic Family Amaltheidae in Britain. Part 1.- London: Palaeontographical Society.

Howarth, M.K. & Donovan, D.T. 1964: Ammonites of the Liassic family Juraphyllitidae in Britain. Palaeontology, 7, 286-305.

Kennedy, W.J. 1977: Ammonite evolution. In: Patterns of Evolution (ed. A. Hallam).- Amsterdam: Elsevier, 251-304.

Kennedy, W.J. & Cobban, W.A. 1976: Aspects of ammonite biology, biogeography, and biostratigraphy.- Spec. Pap. Palaeontol. 17.

Lehmann, U. 1981: The ammonites. Their life and their world.- Cambridge: University Press.

Newell, N.D. 1963: Crises in the history of life. Sci. American, (February, 1963), 76-92.

Raup, D.M. 1967: Geometric analysis of shell coiling: coiling in ammonoids.- J. Paleont. 41: 43-65.

Tintant, H., Marchand, D. & Nouterde, R. 1982: Relations entre les milieux marins et l'évolution des Ammonoidés: les radiations adaptives du Lias.- Bull. Soc. géol. France, (7), 24: 951-961.

THE ROLE OF THE ENVIRONMENT IN THE NAUTILACEA

Henri Tintant and Mpumbu Kabamba

Dijon

Abstract: The classification of the superfamily Nautilaceae, which includes all post-Triassic nautiloids, rests at present on purely typological criteria. The families and genera are based on the presence of well-defined morphological characteristics. Recent studies, however, have shown that many of these characteristics are related of the mode of life of the animal, and are consequently susceptible to repeated occurrences in different lineages.

The development of the ornamentation, the form of the suture line and the position of the siphuncle all appear to be closely linked to water depth and its degree of agitation. Consequently only the knowledge of the evolutionary lineages can lead to a satisfactory classification of this group.

Résumé: La classification de la super-famille des Nautilacés, qui regroupe les nautiles post-triasiques, repose actuellement sur des critères purement typologiques: les familles et les genres y sont fondés sur la possession de caractères morphologiques bien définis. Cependant les études récentes montrent que beaucoup de ces caractères sont liés au mode de vie de l'animal, et sont par suite susceptibles d'apparaître de facon iterative dans des lignées différentes.

Le développement de l'ornamentation, la forme de la ligne de suture et la position du siphon semblent en rapport étroit avec la profondeur de l'eau et son agitation. Seule la connaissance de l'évolution des lignées peut conduire à une classification satisfaisante de ce groupe.

INTRODUCTION

The superfamily Nautilaceae comprises a homogeneous, probably monophyletic group, which includes all the post-Triassic nautiloids. The nautiloid type of body plan, which first appeared in the Upper Devonian, continued through the later Palaeozoic and the Triassic, diversifying to form numerous species grouped into more than 130 genera, 16 families and 4 superfamilies (KUMMEL; 1964). At the end of the Triassic, the Nautilida, like the ammonites and probably for the same reasons, experienced a profound crisis resulting in the almost total disappearance of its taxa. Only a single genus, *Cenoceras*, continued into the Lower Jurassic (KUMMEL, 1953), and it became the origin of an important adaptative radiation lasting until present-day *Nautilus*. More than 1000 species have been described, which probably only represent a small part of the group, and these have been placed in about 30 genera distributed among 1 to 6 families, depending upon the author.

Although some of the genera date back to the 19th century, the majority of them, and their classification into well-defined families, are due to L.F. SPATH (1927). This classification has been refined and completed, but retained in broad outline, in more recent studies by B. KUMMEL (1956, 1964). The classification advocated in the U.S.S.R. by V.B. Shimansky agrees fairly well at the generic level, but differs in the family grouping (SHIMANSKY 1957, 1975).

These classifications are essentially typological, in the sense that they rest solely upon the consideration of a certain number of morphological characters. These are utilised *a priori* for the definition of genera and families, independent of all phylogenetic studies of lineage evolution, which generally remains very poorly known.

Among these characters, those most frequently used (in decreasing order of the importance usually attributed to them) are:

- shell ornamentation
- the suture line
- shell coiling and whorl cross-section
- position of the siphuncle.

However, the importance of these characters has been critically examined in studies of post-Triassic nautiloid populations undertaken by one of us (H.T.) during the last fifteen years, and in more limited but very detailed analysis of Upper Cretaceous Nautilacea from Madagascar by the other author (M.K.). We have shown that many of these characters demonstrate clear adaptative significance, and that they may appear independently and in a repetitive fashion in very different lineages -- which makes their taxonomic use, at the very least, suspect.

1. THE FLUTING OF THE SEPTA

The shape of the septa is very variable in the Nautilacea, including both "watch glass" forms, with almost straight suture lines, and forms which are very plicate. In the latter case, the suture line may comprise as many as 5 lobes which are sometimes very deep, generally rounded at the top, occasionally angular (TOBIEN, 1964), but are never incised. Contrary to what might be expected, the simplest form is not the most primitive.

The first species of the Lower Lias (genus *Cenoceras* s.s.) have septa with 3 lobes, a deep lateral lobe (L) and less marked external (E) and internal (I) lobes. Starting from the ancestral form, septal evolution was able to progress in two opposing directions; either by simplification, leading to almost planar septa *(Eutrephoceras)* or by becoming more resulting in deeper and more numoerous lobes.

From the middle of the 19th century, those nautiloids with sinuous septa have been distinguished from *Nautilus* and placed in the genus *Hercoglossa* (CONRAD, 1886). For a long time this name was used for all the post-Triassic Nautilacea with very folded septa. However, it was soon apparent that such a classification united two groups with markedly different origins and ages. The first, from the Middle and Upper Jurassic, was clearly derived from *Cenoceras*, probably in several waves, and now constitutes the genus *Pseudagonides* SPATH (1927). This group ended at the Jurassic - Cretaceous boundary with forms with 5 deep and angular lobes *(Pseudonautilus* MEEK, 1876). The second group, for which the name *Hercoglossa* must be retained, is restricted to the latest Cretaceous (from the Campanian) and the Eocene. It probably originated from *Eutrephoceras* via the intermediate form *Cimomia* CONRAD (1866), and continued until the Miocene, represented by the genus *Aturia*, with an angular lateral lobe and a dorsal siphuncle (fig. 1).

Pseudagonites is distinguished from *Hercoglossa* by the frequent presence of a fairly pronounced external lobe, however, when this character is absent, the two genera can be morphologically indistinguishable.

This homeomorphy appears to be the result of parallel adaptation to the same mode of life. If the distribution of *Pseudaganides* in Jurassic sediments is studied, it can be seen that they are usually associated with basinal facies rich in pelagic faunas. The same can be said of the Upper Cretaceous *Hercoglossa*, and it is probably this localization which allowed them to escape the extinction which hit many other groups at the Cretaceous - Tertiary boundary (TINTANT, 1982). In the Tertiary, *Aturia* is similarly confined to the deeper marine environments.

In contrast, the groups with simplified septa, like *Eutrephoceras*, have a much greater range and appear to show a preference for shallower internal platform zones.

It seems, therefore, that the shape of the suture line is an adaptive response to variations in oceanic water depth. A very plicate septum improved mechanical resistance to implosion and probably facilitated the exchange of cameral liquid with the exterior. Such a plication is therefore susceptible to repetition in an iterative fashion in all the lineages adapting to a relatively deep water mode of life. A similar evolutionary trend occurred at the top of the Lower Cretaceous, in the later *Heminautilus*, a genus which

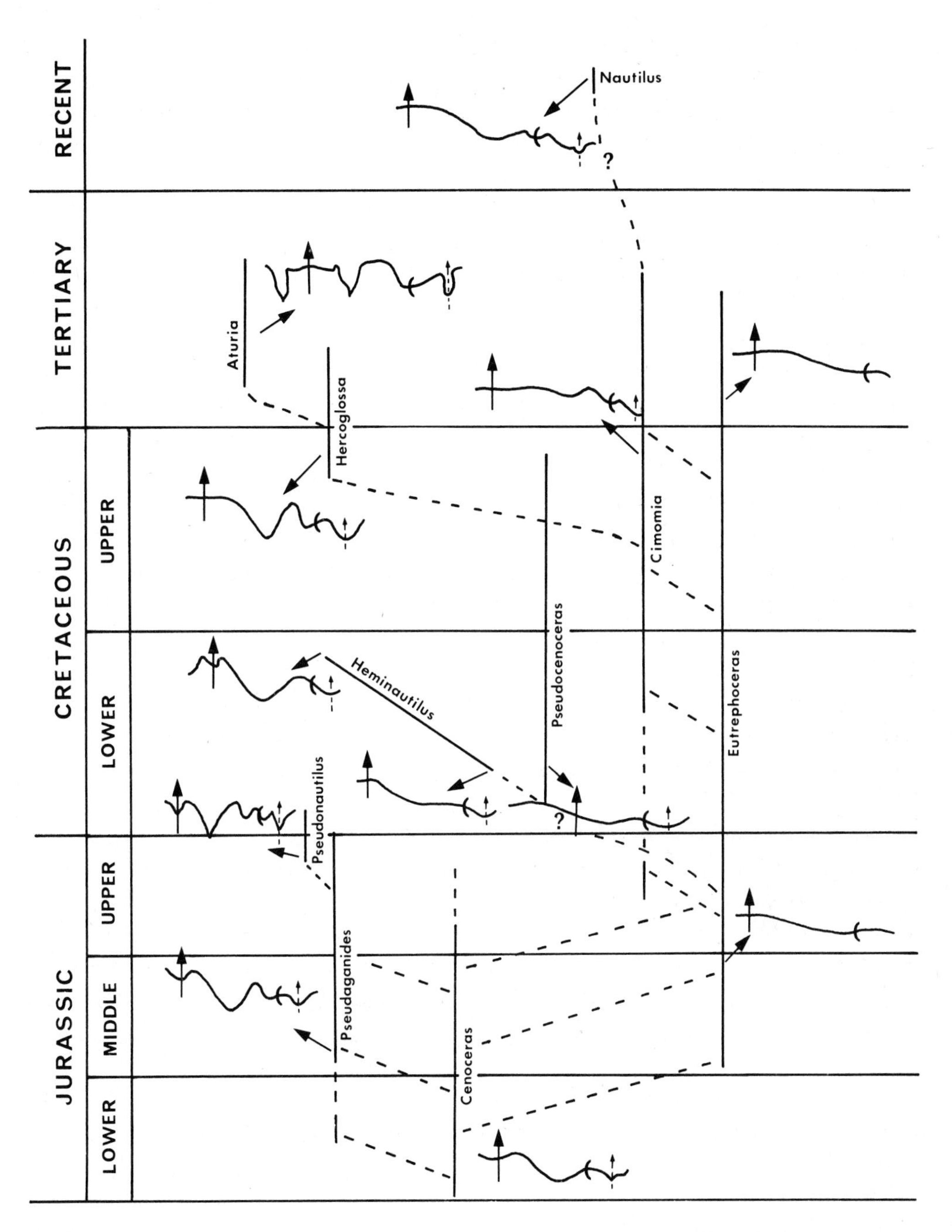

Fig. 1: Diagram showing the evolution of the suture line in the principal lineages of Nautilaceae.
The intensity of plication, which is probably related to adaptation to deeper water environments, increases from right to left. The broken lines signify probable phylogenetic relationships between genera.

had an origin completely independent of the two mentioned above (CONTE, 1980).

Finally, in present-day *Nautilus*, which is able to live at depths exceeding 600 m, the septa, and notably the lateral lobe, are much more plicate than in its immediate ancestors (*Eutrephoceras* or, more likely, *Cimomia).*

2. ORNAMENTATION

The first Nautilacea of the Lower Lias have a reticulate ornament, with radial growth lines cut by fine longitudinal ribs. Very quickly, from the Pliensbachian onwards, this ornamentation began to be restricted, in a palingenetic manner, to the first whorl of the shell, where it can be seen in all the subsequent genera, including present-day *Nautilus*. This recapitulatory feature confirms the homogeneity of the superfamily.

Most of the post-Liassic nautiloids have a smooth shell or an ornament of fine longitudinal ribs. However, in some species ornamentation is frequently observed, which comprises fairly strong radial ribbing, restricted to the body chamber or the last whorl of the phragmocone. Hyatt, in 1886, created the genus *Cymatoceras* for these ornate forms. However, it was very soon established that these species present important differences in their suture lines, their cross-sectional shape and the position of the siphuncle. Consequently *"Cymatoceras"* was divided into about 10 more homogeneous genera, which were united by SPATH (1927) in the family Cymatoceratidae. However, these ornamented genera exhibit clear convergence with the contemporaneous smooth forms. The studies undertaken by WIEDMANN (1960), SHIMANSKY (1975), TINTANT (1970) & KABAMBA (1983) have shown that in reality, each smooth genus is able to produce forms with ribs, and even transverse folds ("genus" *Anglonautilus* SPATH, 1927), by the strengthening of the growth lines. Thus for a true phylogenetic classification the grouping of all the ribbed forms in a completely artificial family, Cymatoceratidae, must be abandoned. The ornamented forms must be grouped with their contemporaneous smooth taxa, in which they may represent either specialised species of derived subgenera and genera.

Here again the appearance of ornamentation appears closely dependent upon environmental conditions. The study of the distribution of Cretaceous ornamented species with simple suture lines (genus *Cymatoceras*) shows that they occupied a more internal position on the U. Cretaceous transgressive platforms than the smooth *Eutrephoceras* from which they are derived (KABAMBA, 1983). In the same way, the distribution of species of *Cymatonautilus* in the Upper Jurassic is controlled by the phases of internal platform development (TINTANT, 1983).

From this the idea follows that the development of ornamentation in the nautiloids, as in the ammonoids (e.g. TINTANT, MARCHAND and MOUTERDE, 1982), was an adaptive response to shallower and more agitated environments, and perhaps also for defence against certain predators. The smooth forms, on the other hand, correspond to more tranquil, but not necessarily deeper environments. This is the case in the Arabian Middle Jurassic, in which the accompanying faunas indicate a very shallow but tranquil environment, and where the numerous nautiloida currently being studies have very simple septa, but have no sign of radial ornament.

A possible objection to this hypothesis is the existence of Nautilacea (genus *Paracymatoceras* Spath) which exhibit at the same time very sinuous suture lines and strong radial ribbing. However, this only shows that the adaptation is not an obligatory and necessary morphological response to the environment, as would be required from a Lamarckian concept. It is equally constrained by the genetic structure of the organism and affected by various selection pressures, which sometimes invoice very complex strategies.

In this case we are probably dealing with a group derived from smooth deep water forms with very plicate Pseudaganides-like sutures, which, in moving into shallow agitated water saw the development of ornamentation without losing its sutural characteristics. The latter were becoming obsolescent, but not detrimental.

3. THE POSITION OF THE SIPHUNCLE

The recent observations of M. KABAMBA (1983) on the position of the siphuncle in the closely related genera *Eutrephoceras* and *Cymatoceras* in the Cretaceous has clearly established the adaptative significance of this character. In the Lower Cretaceous these two genera comprise exclusively species with a dorsal siphuncle. Then, from the Aptian onwards, species appeared with central siphuncles. Subsequently in the Upper Cretaceous, species with sub-ventral siphuncles became predominant, and even almost exclusive in *Cymatoceras*. Nevertheless throughout the period a certain number of species of *Eutrephoceras* retained almost dorsal siphuncles.

This migration of the siphuncle appears to be closely related to the expansion of these two genera into the very shallow epicontinental seas produced by the Upper Cretaceous transgression, and probably corresponds to an adaptation to this environment. The dorsal position of the siphuncle probably favours a relatively deep water mode of life by avoiding the decoupling of the siphuncle and by facilitating the pumping of intercameral liquid necessary for vertical movement. In shallow water this mechanism would become redundant and selection pressures would no longer be operative, allowing the siphuncle to move without difficulty into a ventral position.

At the end of the Upper Cretaceous, the major regression which characterizes the epoch resulted in a rapid and considerable reduction in shallow epicontinental seas. This caused, not only the disappearance of *Cymatoceras*, but also of all the *Eutrephoceras* with ventral siphuncles. Only rare species retaining siphuncles in dorsal position persisted onwards until the Miocene. These were naturally accompanied by groups with plicate septa *(Cimonia, Hercoglossa, Aturoidea)*, which passed the Cretaceous - Tertiary boundary without adverse effects.

4. OTHER CHARACTERS

The other morphological characters of the Nautilacea, which have not yet been the subject of as detailed analysis, appear to be similarly influenced by the environment. This is the case for the whorl section, which is generally sphaerocone in the deep water groups, but may become more or less compressed and tend towards an oxycone morphology. Examples of the latter are seen in the Liassic species *Cenoceras araris* (Dum.) and the Upper Cretaceous to Eocene genus *Deltoidonautilus*; forms which all appear to be restricted to very shallow water environments.

Similarly, forms with relatively large umbilici, for which SPATH (1927) proposed the genus *Ophionautilus*, equally seem to be species characteristic of shallow water.

Finally, shell size may also be closely dependent upon life conditions. The giant forms of the genus *Paracenoceras* are found in platform limestones on the inner side of coral barriers. Conversely, dwarf forms of the same genus have been collected in deep water facies. The genus *Pseudaganides*, characteristic of the same deep water environments, is also unusually very small.

CONCLUSION

This brief analysis demonstrate that most of the morphological characters previously used in a typological manner, for the definition of genera and species of the Nautilacea, also correspond to well defined adaptative responses to environmental conditions. Consequently, they are susceptible to appear independently, in an iterative fashion, in very different and stratigraphically widely-separated lineages.

A biologically satisfactory classification of the Nautilaceae, therefore, can not be based exclusively on the consideration of these characters, as it would be dangerous to give them an a priori precise taxonomic value. Instead, classification must be based on stratigraphical data deduced from the evolution of lineages. This work, on which we have been engaged for several years, is far from being finished, but it has already led us to propose a phylogenetic classification of the superfamily Nautilaceae (TINTANT & KABAMBA, 1984). This classification in still provisional, but is capable of providing a model to guide future research.

The stability of the nautiloid type of organisation, which presents an effective and orginal solution to the problem of flotation, opposes any great plasticity of the various characters of the shell, the siphuncle and the suture line in particular. The plasticity permitted can be seen in the iterative adaptation of these cephalopods to various niches, resulting in frequent speciation. As a result there are numerous species, with generally fairly brief ranges (usually less than a stage and sometimes shorter) allowing the utilisation of the group in biostratigraphy.

REFERENCES

Conrad, T.A. 1866: Observations on Recent and fossil shells, with proposed new genera and species.- Am. Tourn. Conch. vol 2, p. 101-103.

Conte, G. 1980: Heminautilus sanctaecrucis, nouvelle espèce de Nautiloide. Géobios, vol. 13, part 1, pp. 137-141, 1 pl.

Kabamba, M. 1983: Les nautiles du Crétacé supérieur de Madagascar. Contribution à l'étude des Nautiloidés du Crétacé. Thèse de 3ème cycle Dijon, 136 p., 7 pl.

Kummel, B. 1953: The ancestry of the family Nautilidae. Breviora, n° 21, pp. 1-7.

--- 1956: Post-Triassic Nautiloid Genera. Bull. Mus. Comp. Zool. Harvard. vol. 114, n° 7, pp. 224-494, pl. 1-28, fig. 1-35.

--- 1964: Nautilida, in Moore, R.C., Treat. Invert. Paleont., part. K, Mollusca 3, pp. 383-466, fig. 286-337.

Meek, F.B. 1876: Report on the invertebrate Cretaceous fossils from Nebraska Territory - U.S. Geol. Surv. Terr., vol. IX, 629 p., 45 pl.

Shimansky, V.N. 1957: Sistematika i Filogenia otyada Nautilida. Mosk. Obshch. Ispyt. Prir. Bull., vol. 32, pp. 105-120.

--- 1975: Melovye Nautiloidei. Acad. Sci. URSS, Trans. Pal. Inst. vol. 150, pp. 1-208, pl. XXXIV.

Spath, L.F. 1927: Revision of the Jurassic cephalopod fauna of Kachh (Cutch). Mem. Geol. Surv. India, Pal. Indica, N.S. vol. IX, part. 1, pp. 1-71.

Tintant, H. 1970: Les nautiles à côtes du Jurassique. Ann. Paléont. Invert. vol. LV, part 1, pp. 53-96, pl. A-E.

--- 1982: Pourquoi les nautiles ont-ils survécu aux ammonites? 9th R.A.S.T., Paris, p. 602.

--- 1983: Un cas de parallélisme évolutif synchrone chez les nautiles à côtes du Jurassique. Bol. Soc. Géol. Portugal (vol. de Homen. ao Prof. Dr. C. Teixeira), vol., XXII, pp. 63-69, pl. I-II.

--- et Kabamba, M. 1984: Le nautile, fossile vivant ou forme cryptogène? Essai sur l'évolution et la classification des Nautilacés. Bull. Soc. Zoll. France, (in press).

---, Marchand, D. et Mouterde, R. 1982: Relations entre les milieux marins et l'évolution des ammonoidés au Jurassique: les radiations adaptatives du Lias. Bull. Soc. Geol. Fr., 7th series, n° 5-6, pp. 951-961.

Tobien, H. 1964: Über Suturen nautiliconer Nautiloidea (Cephalopoda). Notizblatt Hess. Landesamt Bodenforsch., vol. 93, pp. 47-60, 23 fig.

Wiedmann, J. 1960: Zur Systematik jungmesozoischer Nautiloidea. Palaeontographica Bd. 115, Abt. A, pp. 144-206, 11 pl., 26 fig.

PART 2

SEDIMENTARY TRENDS IN MARGINAL EPICONTINENTAL BASINS

Biological responses to sea-level changes are higher order phenomena. It is the physical environment which is affected at first, and these changes are registered by the response of sediments. Clearly, the local sedimentological changes depend on the relation between mean water depth of the basin and the magnitude of sea-level changes. Therefore, one expects that the impact of sea-level fluctuations should be strong in shallow marginal seas which lack major submarine currents. All four papers presented here refer to this situation, and they all agree that the environment is continuously altered during sea-level changes in terms of extent and 'quality' of local bottom conditions. Technically there are three ways to analyze and model such basin configurations:

- ** *one can choose a fixed space coordinate, a single vertical section, and study how the sediments evolve from specified initial conditions. As G. Einsele illustrates, this strategy provides models for the depth zonation of a basin and allows to analyze idealized basin situations under various initial and boundary conditions;*
- ** *another strategy is to fix the relative position of a vertical section in terms of water depth or another hydrodynamic parameter. This allows to analyze how sea-level changes alter, e.g., accumulation rates at a fixed depth zone, a method applied by McGhee & Bayer;*
- ** *or one traces 'marker beds' of known position within a sea level cycle throughout the basin and records their sedimentological composition in terms of proximity gradients. An attempt in this direction is given by K. Brandt.*

W. Ricken reminds us of the fact, that our record may be heavily distorted. Besides synsedimentary processes like bioturbation, which may destroy all primary sedimentary structures, it is diagenesis and even the effect of weathering which overprint and bias periodic and cyclic patterns.

The probably most important aspect is that pre-Pleistocene eustatic cycles are rather slow (1 Ma) and of minor magnitude (50 m) allowing for a quasi equilibrium between hydrodynamic regime and sediments (G. Einsele), or for a gradual shift of substrate conditions through time. On the other hand, the different phases of a sea-level cycle are not equally represented: The speed of change reaches a minimum near the extrema of the cycle providing time for extensive formation of 'marker beds' such as black shales and oolitic horizons (McGhee & Bayer). 'Marker beds', traceable throughout the basin (K. Brandt, W. Ricken), provide a generic division of sedimentary sequences and may be related to chronostratigraphic boundaries.

RESPONSE OF SEDIMENTS TO SEA-LEVEL CHANGES IN DIFFERING SUBSIDING STORM-DOMINATED MARGINAL AND EPEIRIC BASINS

Gerhard Einsele

Tübingen

Abstract: Even "short term" or "third order" pre-Pleistocene sea-level changes had time periods roughly 100 times longer than the Pleistocene oscillations, although amplitudes were generally considerably smaller. Therefore, during nonglaciated periods marginal sea floor topography (that is, the shelf profile) as well as sediment distribution in persistent, storm-dominated basins tend to approximate equilibrium conditions in relation to certain (changing) sea-level stands. However, if the mean rate of sea-level fall is faster than subsidence, submarine erosion below the lowermost stand of sea-level strongly affects sedimentary processes and final aggradation. This is particularly the case in slowly subsiding epeiric or epicontinental seas with a small influx of allochthonous material from land sources. Whereas the sediments of emerging parts of these basins are more or less eroded by fluvial processes and subaerial chemical dissolution (requiring long exposure!), the extensive reworked submarine horizons (containing lithified concretions and associated with proximal tempestites) are preserved. Basinward, these horizons truncate regressive sequences and finally pinch out. Toward the shoreface they rest on transgressive sequences and represent hiatuses or condensed horizons of increasing time span. "Sea level curves" derived from sediment thicknesses thus often become strongly distorted in comparison with the time-related curves. In deeper parts of the basin, where eroded sediments are redeposited, the regressive phase of the "sea level curves" appears to be "longer" than the transgressive one. This effect is magnified, if sediment influx during regressive phases increases.

These rather complicated relationships can be quantitatively demonstrated with a number of simplistic models using assumptions which are considered characteristic of ancient sea level changes and the subsidence of marginal shelf seas or epicontinental basins. Finally some typical sections from the epicontinental Jurassic sea of Southern Germany are discussed and compared with the results of the models.

1. INTRODUCTION

For the purpose of this publication, it is not crucial to determine whether the discussed sea level phenomena are global in extent or only regional relative rises and falls of sea level, which may be caused by local tectonic or depositional processes (e.g. PITMAN, 1978). Therefore, the different opinions brought forward by several authors (e.g.

PITMAN, 1978; DONOVAN & JONES; 1979; WATTS & STRECKLER, 1979; SOUTHAM & HAY, 1981; MÖRNER, 1982; WATTS, 1982; PITTMAN & GOLOVCHENKO, 1983; HALLAM, 1984) to explain short term and long term changes of sea level in pre-Pleistocene times are not discussed here. However, it is assumed that, regardless of tectonic setting, sea level changes can have similar amplitudes and time periods.

Hence, within the guide lines of these assumptions, the main purpose of this paper is to check and compare the response of sediments in two important types of sedimentary basins:

(1) Shelf seas at continental margins and

(2) epeiric or epicontinental seas.

For shelf seas, sea level changes and their effects have recently been studied rather intensively by means of "seismic stratigraphy" (e.g. PAYTON, 1977), but their effects in ancient epeiric seas (which are now exposed) are less well known. In these cases, seismic stratigraphy often cannot be applied with much success. Substantial synsedimentary tectonic movements are rare or missing, and subsidence is slow. Therefore, angular unconformities are absent or difficult to find. For this reason, some simplistic transgression-regression models are used instead in order to simulate some important processes and enable a better understanding of the effects of sea level changes.

It is not possible to include in a model all of the factors which control sedimentation in shallow basins. The models presented below therefore are restricted to physical processes between the coast and the zone somewhat deeper than storm wave base. In addition, they are valid only for storm-dominated seas in predominantly humid, temperate climates. Sediments are chiefly clastic or bioclastic, and on the average sediment supply is sufficient to compensate for subsidence, i.e. over a long period of time, water depth of the (shallow) basins remains approximately constant.

Finally, the results from the models are applied in a general way to the Jurassic sequence of southern Germany. Additional articles in this volume present more detailed descriptions of parts of this sequence and fauna.

2. DURATION OF SEA LEVEL OSCILLATIONS

The reaction of sediments to sea level changes in the coastal and wave-influenced zone depends highly on the duration of sea level oscillations. Therefore, it is necessary to discuss briefly our knowledge of this rather controversial subject.

The period of the large Pleistocene sea level oscillations is of the order of 0.02 to 0.1 Ma (e.g. HAYS, IMBRIE & SHACKLETON, 1976) and their amplitude may be greater than 100 m. In contrast, the duration of the pre-Pleistocene oscillations is 100 to 10.000 times longer; their amplitude often appears to be much smaller if long term cumulative effects of repeated rises and falls are excluded. According to VAIL, MITCHUM & THOMSON (1977) or HARDENBOL, VAIL & FERRER (1981), the so-called "short term" or "third order" sea level variations lasted 1 to 10 Ma, HALLAM's (1978, 1981 a and b) average cycle for the Jurassic sequence is about 3 Ma, the Cretaceous cycles cover 4-5 Ma (COOPER, 1977; KAUFFMAN, 1977), and VAIL & HARDENBOL (1979) describe an average period of 4 Ma for Tertiary transgression-regression cycles. The magnitude of these sea level changes is difficult to determine. It is assumed often to reach more than 20 m but usually less than 100 m.

"Long term" cycles of the first and second order (VAIL, MITCHUM & THOMSON, 1977) which represent trends of at least 100 Ma or oscillations of the order of tens of Ma, respectively, are not considered here. In the following models only "short term" ("third order") cycles are discussed. For the sake of simplicity it is assumed that they usually have a constant period of 4 Ma and an amplitude of 50 m.

3. DIFFERENCES BETWEEN MARGINAL SHELF SEAS AND EPEIRIC SEAS

Recent work on sea level variations and seismic stratigraphy (e.g. Vail and coworkers in PAYTON, 1977; PITMAN, 1978; HARDENBOL et al., 1981, WATTS, 1982, PITMAN & GOLOVCHENKO, 1983) is concerned chiefly with passive continental margins. If we want to study the results of sea level changes in epeiric seas, we have to take into account some important differences between these two environments (Table 1). At least during the first 50 to 100 Ma of the development of an ocean with passive margins, the rate of subsidence in large parts of the marginal basins (i.e. on the shelves on top of thinning and differentially subsiding continental crust) is greater than the rate of the pre-Pleistocene sea level fall of a regressive half-cycle. In contrast, slowly subsiding epeiric or epicontinental basins tend to have a lower rate of subsidence than that of the sea level fall. As HARDENBOL et al. (1981) have also pointed out, in the latter case large areas covered with shallow marine sediments can be exposed to subaerial and submarine erosion. At such times, cutting of channels is common, which cause funnelling of eroded materials to lowstand deltas and fans. However, if the sea level falls at a rate slower than subsidence and, at the same time, the rate of sedimentation is low (lower than the rate of subsidence), these phenomena can hardly be of great importance.

Table 1. Some characteristics of marginal and epeiric shallow seas

	Marginal Shelf Sea (Atlantic-type) e.g. Jurass., Cretac.	Epeiric Shallow Sea e.g. Jurassic of s. Germany
Tectonic Setting	thinning marginal continental crust	continental crust
Morphology	high relief with shelf break and adjoining slopes, submarine canyons, deep sea fans	low relief, no distinct shelf
Hydrographic conditions	predominantly high energy (deep storm wave base, strong ocean currents, often (high tides)	predominantly low energy (shallower storm wave base, weaker currents, often only low tides)
Rate of subsidence (SUB) and sea level fall (SLF)	Pleistocene cycles: SLF > SUB "short term" pre - Pleistocene cycles: SLF < SUB[1]	not applicable SLF > SUB
Consequences of sea level fall	For sedimentation rate < SUB: little subaerial and submarine erosion	strong subaerial and submarine erosion, cutting of channels, low-stand deltas and fans

4. EQUILIBRIUM CONDITIONS AND PRESERVATION OF SHALLOW-MARINE SEDIMENTS OF PLEISTOCENE AND OLDER SEA LEVEL OSCILLATIONS

As we have seen, "short term" sea level oscillations of pre-Pleistocene times have lasted 100 times longer than Pleistocene oscillations. This great difference has some important consequences. However, in one important respect, the Pleistocene oscillations on the continental shelves can be compared with the pre-Pleistocene sea level changes in epicontinental seas. That is, sea level fall was definitely much faster than the rate of subsidence.

---

1) During the first 50 to 100 Ma after rifting and at some distance seaward from the hinge line

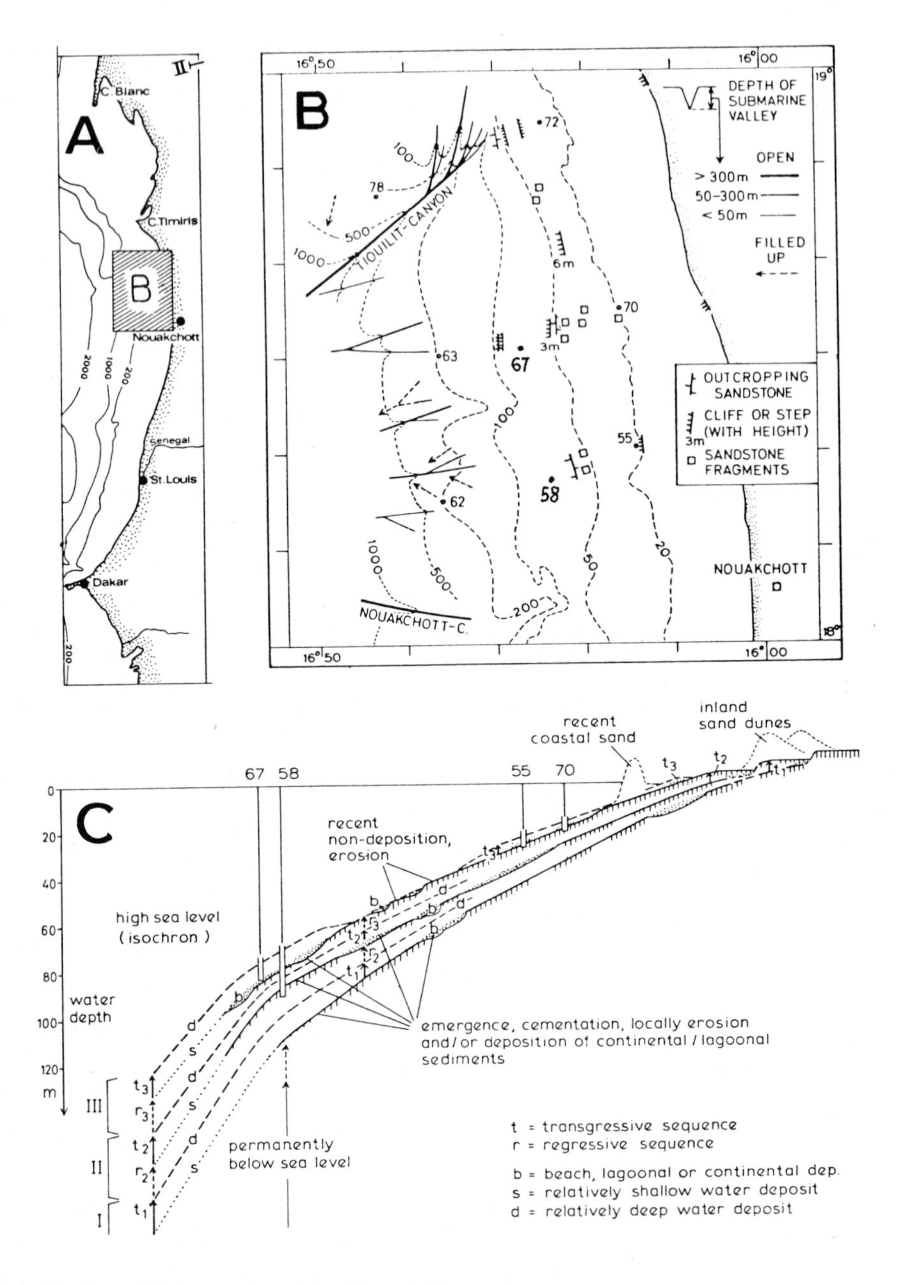

Fig. 1: A. Area of investigation, West Africa.
B. Selected core locations as well as submarine valleys and outcropping sandstone at the shelf edge and upper slope of Mauritania. Note the different sizes and partial filling of valleys as detected by reflection seismic work.
C. Generalized hypothetical cross section of shelf and coastal sediments of three Quaternary sea-level oscillations (I, II, III) as well as interpretation of the stratigraphic sequence of the 4 vibrocores (r=regressive, t=transgressive sequence). Comparatively good preservation of emerged carbonate-bearing shelf sand layers by cementation under arid climate.

It has often been pointed out (e.g. CURRAY, 1964; SWIFT et al., 1971; SWIFT, 1974) that the shelf sediments deposited since the last transgression on the continental margins are generally not yet in equilibrium with the presently existing hydraulic regime. Over wide areas they are still influenced by the pattern of sediment distribution of the last low sea level stand about 18 000 years ago. In order to describe this discrepancy, terms such as "relict sediments" or, if reworked, "palimpsest sediments" are used. In addition, the last few regressive phases of the Pleistocene did not last long enough to remove the emerged sediments deposited by the former transgressions. Therefore, many of our present shelf sediments are difficult to understand and do not provide a good key to the past. There are only a few cases in which Pleistocene transgressions and regressions are well documented in shallow marine sediment sequences.

Figs. 1 and 2 show an example from the shelf of Mauritania. Because of the substantial contribution of wind blown sand, the Quaternary sedimentation rate on this shelf was relatively fast (18 cm/1000 a). In addition, during the regressive phases, part of the emerged shallow marine sediment, which was from an older transgressive or regressive sand sheet, was cemented to beach rock under the influence of a warm, arid climate. Thus part of the sediments from an earlier transgression-regression (T-R) cycle were preserved from dissolution and reworking or mixing with sediments of the Holocene transgression. Therefore several cores obtained by vibrocoring show a more or less undisturbed Holocene transgressive sequence (including some storm layers) on top of one or several beach rocks representing former coastal sediments (Fig. 1c, r_2 and r_3, and Fig. 2; see also EINSELE et al., 1977). Even on land, "relict" marine sediments are widely distributed in this area (Fig. 1c). They represent peaks of former transgressions, when the sea stood 1 to 3 m above the present datum line for some hundred or thousand years.

For a T-R cycle of 4 Ma duration, the situation is quite different. There is much more time to reach a state of equilibrium between the hydraulic regime and sedimentation in different water depths. At some depth below the sea floor, submarine diagenesis starts to indurate sediments during one half-cycle. In addition, all sediments above the lowest stand of sea level are exposed to chemical dissolution and mechanical erosion for a period between 0 and 4 Ma. If we assume a chemical dissolution rate of 40 m/Ma for carbonate and 10 m/Ma for siliciclastic sediments (HOHBERGER & EINSELE, 1979) in a humid, temperate climate, then there is sufficient time to remove a considerable part or even most of the sediments deposited during the higher stands of sea level just by dissolution. This effect is demonstrated in a semiquantitative to quantitative manner in Fig. 5, which will be discussed later along with other phenomena.

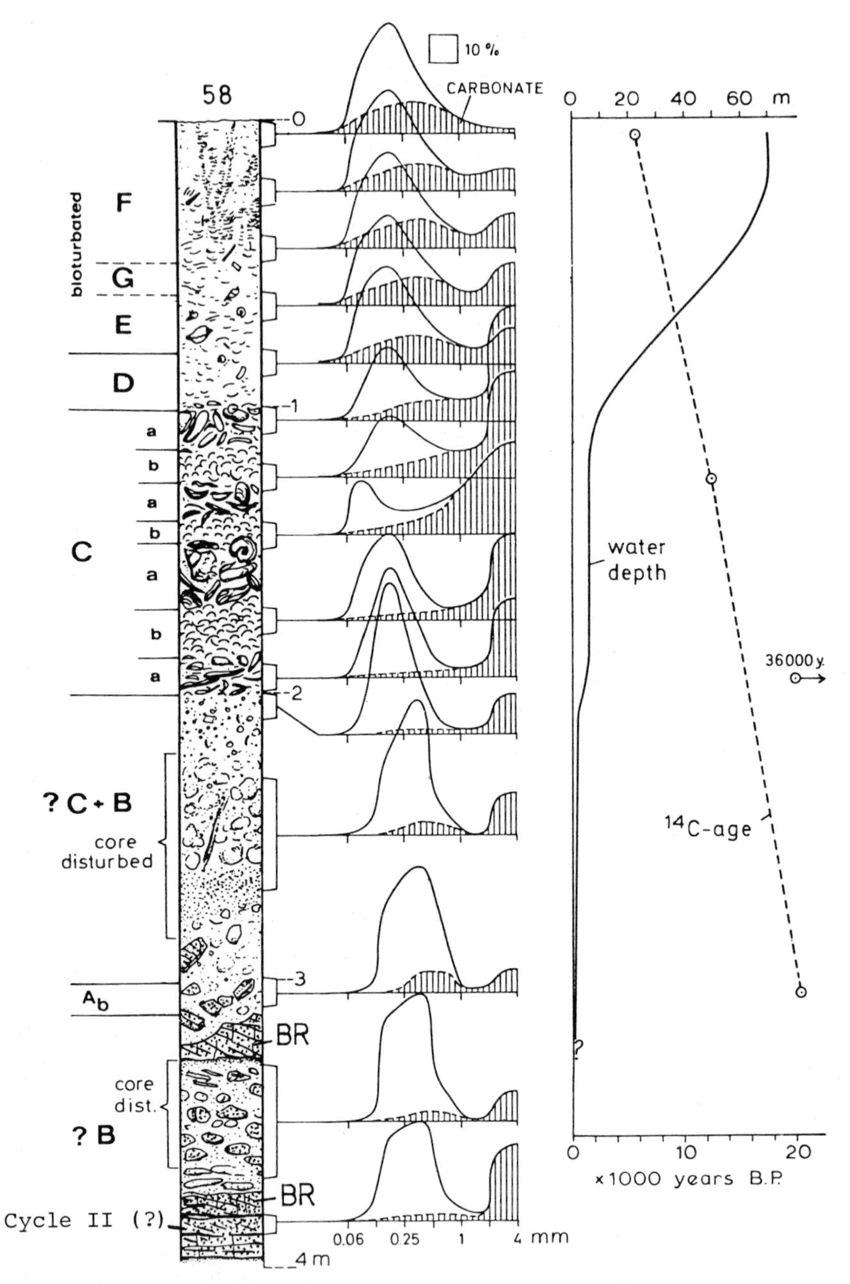

Fig.2: Vibro-core 58 from a depth of 70 m (compare Fig. 1B) showing beach rock (BR) formed during the last regression and the following transgressive sequence. A through F = faunal associations indicating increasing water depth during deposition of sediment. Section C between 1 and 2 m depth in core contains reworked shell material from protected embayments or lagoons (a= Ostrea shell beds, b= Cardium shell beds). These layers are interpreted as storm deposits. Frequency curves show grain size distribution and carbonate content in different size fractions. Water depth during sediment accumulation determined from faunal associations; ^{14}C analyses were performed on shell material. (From EINSELE et al., 1977).

5. SUBMARINE SEDIMENT ACCUMULATION AND EROSION DURING A TRANSGRESSION-REGRESSION CYCLE

5.1. Without subsidence

The interrelationship between sediment supply, subsidence and sediment accumulation at different locations within a basin as modified by rising and falling sea level is rather complicated (see also SWIFT, 1974; PITMAN, 1978; PITMAN & GOLOVCHENKO, 1983). Therefore, we begin here with a simple model (Fig. 3A) with the following assumptions:

1) There is no subsidence,
2) The water body of the basin is storm-dominated (tides and tidal currents are neglected), and the depth range of storm waves remains constant,
3) Sediment supply is predominantly allochthonous from neighboring land areas, and sediments are evenly distributed,
4) Sediment buildup stops at an equillibrium profile (Fig. 3A) which is controlled by the storm wave base (STWB) and, near the coastline, the transformation of wave energy into surface and bottom currents (SWIFT, 197O, 1978). The same equilibrium profile is shaped by storm wave erosion during regressions.

If the T-R cycle starts at the lowermost stand of sea level (LSL), then with rising sea and therefore a rising STWB, a wedge-like transgressive sediment body can be built up until the highest stand of sea-level (HSL) is reached, when a new, landward shifted and elevated equilibrium profile is formed. During this process, shoreface erosion may truncate older deposits, especially if the basin is sediment starved (SWIFT, 1978). During the subsequent regression, this sediment wedge can be partly removed via channels or entirely removed by the falling STWB. At the end of the cycle (and without subsidence) when the LSL is reached again, all or most of the transgressive sediment is cannibalized and swept into deeper water below the LSL-STWB. There, sediments can be moved further only by means of bottom currents.

5.2. With subsidence

Shallow sedimentary basins which receive and store sediments cannot persist for longer geological periods without subsidence or corresponding sea level rise. Therefore, we assume for the next steps of our discussion that

1) There is a constant rate of subsidence (SUB) existing equally across the basin.
2) The mean rate of sea level rise (SLR and sea level fall (SLF) is equal to the rate of subsidence: SLR = SLF = SUB.

3) Or: SLR as well as SLF are greater than SUB.

If the model of Fig. 3A includes subsidence for assumption (2) we get Figs. 3B and C. Starting with LSL at the datum line, Fig. 3B shows the buildup of a transgressive sequence (TS) to the storm controlled equilibrium line, as seen in Fig. 3A. Due to ongoing subsidence, this sequence now becomes considerably thicker than before, provided there is sufficient sediment supply (SS) for aggradation. AT the same time the shoreline (SH) migrates landward, as shown in Fig. 3A. Since subsidence is still proceeding, the subsequent SLF and its falling STWB cannot erode the transgressive sediment wedge (Fig. 3C). Instead, a locally reduced regressive sequence (RS) is deposited and the shoreline profile may again translate seaward. The model in Fig. 3C may be useful for marginal shelf seas with SUB > 20 m/Ma, because then SUB tends to be as great or greater than SLF (compare Table 1). If the amount of SUB increases from a coastal hinge line toward the sea, the complications described by PITMAN & GOLOVCHENKO (1983) have to be taken into account.

Fig. 3D demonstrates the situation listed as assumption 3 above. SLR and SLF are the same as in Figs. 3B and C, but now subsidence is much lower (SUB = 1/4 SLF or SLR). As a result, the transgressive sediment wedge is no longer protected from erosion by sufficient SUB during SLF. The upper portion of the wedge which is slightly below LSL will be eroded and redeposited in deeper water. In water shallower than STWB at LSL, hardly any sediment of the regressive sequence (RS) will be deposited. Furthermore, in contrast to the model in Fig. 3C, the more elevated parts of TS between the lowest and highest shoreline (LSH and HSH) are exposed to subaerial erosion, if they are not covered by terrestrial deposits during the regressive phase. This model applies to epicontinental seas with SUB < SLF, where SUB may often be of the order of 5-15 m/Ma.

6. STORM WAVE ACTION AND DEPTH OF SUBMARINE EROSION

The models in Fig. 3A and D show the importance of submarine erosion on sediment sequences affected by T-R cycles. KLÜPPEL (1917) has described cyclic sequences in the Jurassic of Eastern France (Lorraine) which contain an eroded top layer that he explained by marine regressions and transgressions caused by repeated tectonic uplift and subsidence. In the meantime many workers have studied the phenomena of submarine erosion, omission, and condensing of layers (see e.g, BRANDT, BAYER et al., BAYER & McGHEE; all in this volume). Therefore, only a few important points have to be mentioned.

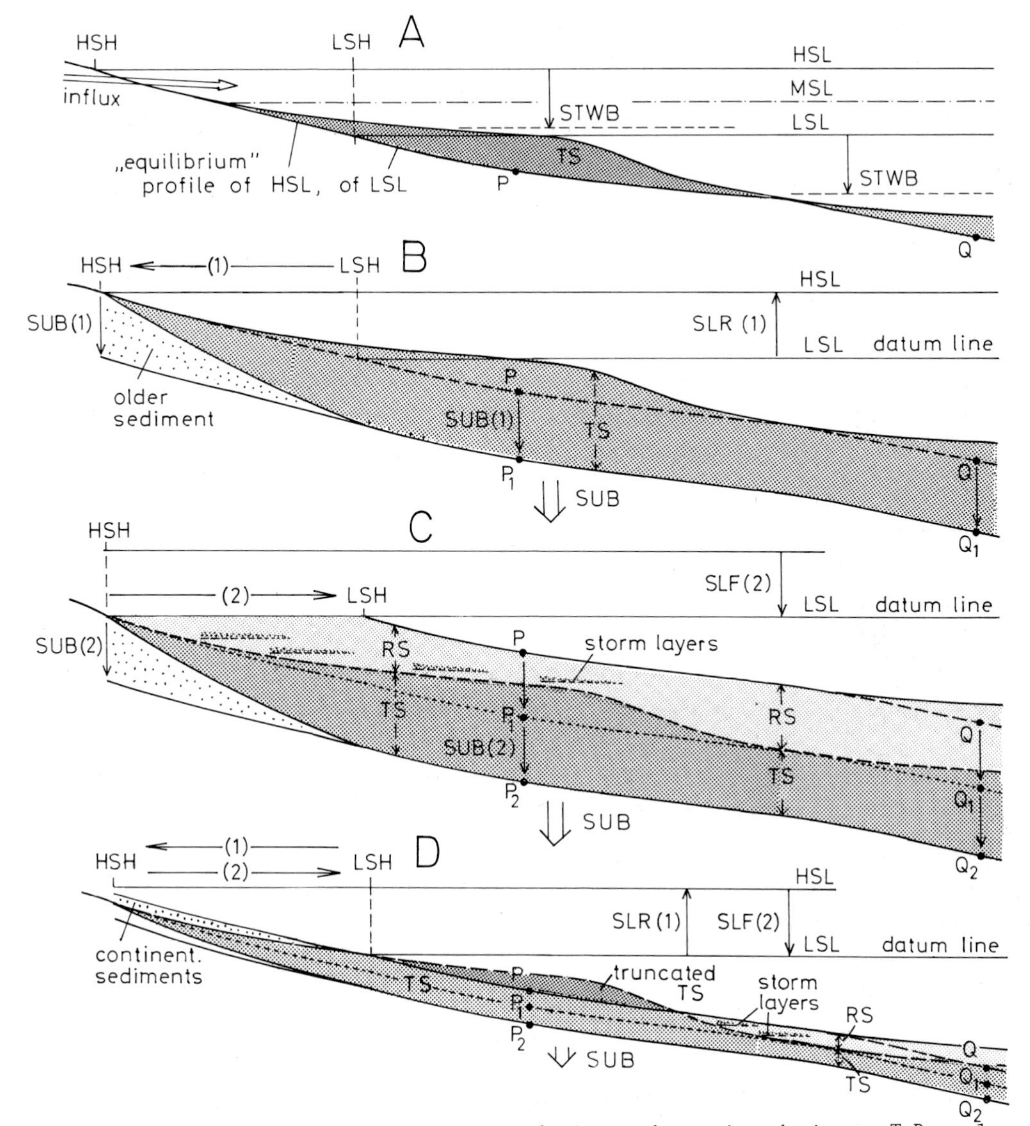

Fig. 3. Simple models for sediment accumulation and erosion during a T-R cycle between the highest shoreline (HSH) at high sea level (HSL) and the storm wave base (STWB) of low sea level (LSL) with lowest shoreline (LSH). TS= transgressive sequence sequence, RS= regressive sequence. P, P_1, and P_2 or Q, Q_1, and Q_2 = trace of subsiding points (in respect to datum line) after subsidence during transgression (1) and regression (2).

A. Basin without subsidence, after the end of one T-R cycle. All sediment is removed from the near-shore zone (except some "transgression conglomerate" not shown in Fig.).

B. Basin, in which the rate of subsidence (SUB) is equal to the mean rate of sea level fall (SLF) and sea level rise (SLR). Transgressive phase.

C. Same as B, but during subsequent regressive phase. No erosion of TS.

D. Same as D, but with lower rate of subsidence (SUB= 1/4 SLF). Submarine erosion of large portions of TS.

For further explanation, see text.

In cohesionless sediments, the depth of erosion and range of sediment transport will be determined by the wave and current energy required to move particles of a certain hydraulic behavior. Then some kind of an equilibrium profile of the sea bottom is comparatively easily achieved. For example, storms pick up loose material and form tempestites which can be different in nature, for example proximal types with repeated amalgamation (Fig. 2) or distal ones with similar characteristics as turbidites (see e.g. AIGNER or SEILACHER, in EINSELE & SEILACHER, 1982). Bottom currents can cut shallow channels which allow transportation of material into deeper water (compare Fig. 1B; or see RICKEN, this vol.).

In cohesive sediments, which gain in strength with increasing depth of burial, as well as in diagenetically indurated materials such as carbonate layers or concretions, the depth of erosion will usually stop at a particularly firm or hard layer. Unfortunately, we do not know very much about the relationship between a given wave or current energy and the resulting depth of submarine erosion in this type of sediment. We observe, however, that many reworked horizons rest on carbonate concretions or contain reworked, bored concretions of different nature (see e.g. BRANDT, or BAYER & McGHEE, this volume). So if we know the depth below the sediment-water interface at which these concretions were formed, we can estimate the depth of erosion. However, there are some significant uncertainties involved. The growth of concretions can be influenced by several

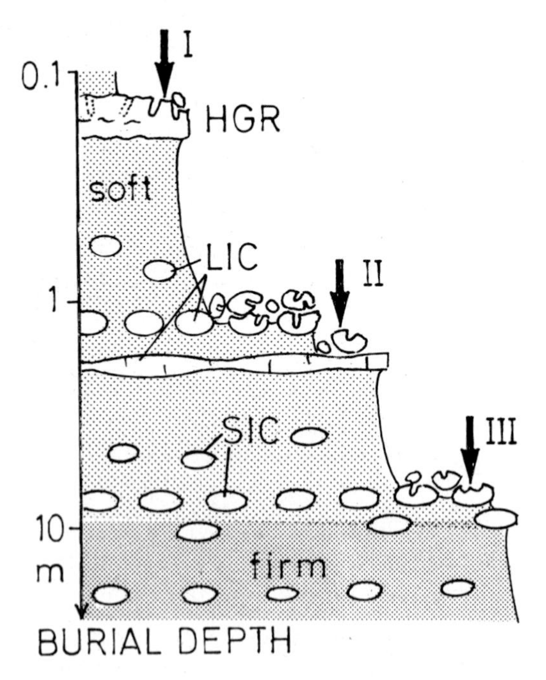

Fig. 4: Steplike submarine erosion and estimation of erosional depth according to the occurrence of hardgrounds (HGR), limestone concretions (LIC), or ankeritic and sideritic concretions (SIC).

factors, e.g. by the changing chemistry of sea and pore water, or also by the rate of sedimentation (CAMPOS & HALLAM, 1979; JEANS, 1980; GAUTIER, 1982).

Nevertheless, in Fig. 4 an attempt has been made to show some general guidelines for the formation of concretions. Beachrock or hardgrounds can be generated directly on the sea bottom or within a few centimeters or decimeters of the interface (PURSER, 1969), but they require special environmental conditions such as emergence or warm sea water and high salinity. Limestone concretions as well as pyrite (and phosophorite) grow in the zone of sulphate reduction, up to several tens of meters below the sea floor (RAISWELL, 1971; SRAMEK, 1978; JEANS, 1980). Concretions of ankerite or siderite already need an environment in which the sulphate content of the pore water is used up (e.g. IRWIN et al., 1977). If their shape is rather flat and their carbonate content comparatively low, several authors (e.g. CAMPOS & HALLAM, 1979, or GAUTIER, 1982) assume that they grow about 10 m or more below the sediment-water interface. Consequently, if such concretions are found in reworked horizons (particularly at the base), one can at least say that several meters of sediment have been removed. In a sequence with alternating soft and hard layers, erosion may proceed stepwise as indicated in Fig. 4. In this situation, an equilibrium profile of the sea floor, as discussed above can hardly be expected. In the present oceans, according to the results of the Deep Sea Drilling Project, firm clayey or muddy sediments and lithified pelagic limestones are usually encountered more than 100 m below the sea floor. However, it is assumed that lithification of fine grained limestones may also occur at shallow burial depth, e.g. in sediments rich in organic matter (EINSELE & MOSEBACH, 1955; KÜSPERT, 1983) or in sediments containing substantial amounts of unstable carbonate minerals, which frequently occur in shallow seas and favour early diagenesis.

7. CHRONOSTRATIGRAPHIC TRANSGRESSIONS-REGRESSION SEQUENCES NEAR THE MARGINS OF EPEIRIC BASINS

The results of the preceding chapters are now summarized for a typical epicontinental basin with the following characteristics (Fig. 5):

1) Slow, continuous subsidence of the order of 5 to 20 m/Ma,
2) Sea level fall (SLF) is faster than subsidence (SLF > SUB),
3) The duration of a T-R cycle remains 4 Ma throughout the sequence or can fluctuate within the sequence between 4 and 2 Ma.
4) Long term, basin-wide sediment accumulation (below the storm wave base of LSL) approximately equals subsidence.

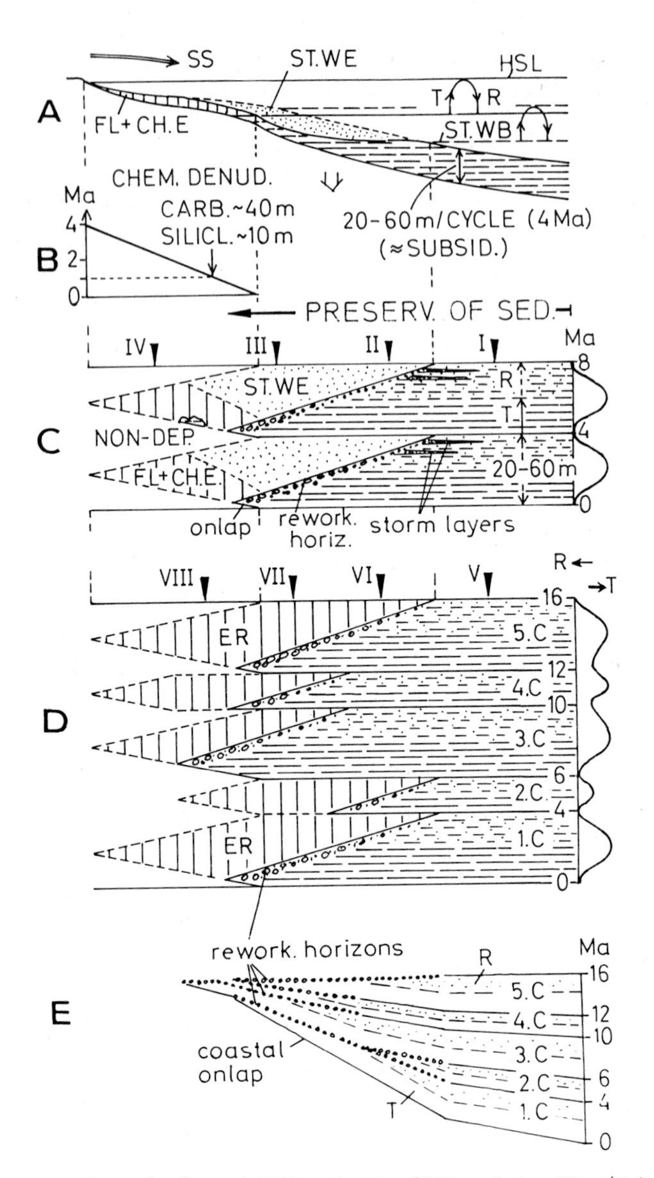

Fig. 5:
A: Marginal part of a slowly subsiding basin (SUB = 5 to 20 m/Ma) subjected to T-R cycles with periods of 2 to 4 Ma, SLF SUB, and basin-wide sediment accumulation SA= SUB. FL + CH.E = fluvial + chemical erosion (dissolution), STWE= (submarine) storm wave erosion, ER = both fluvial and chemical erosion.

B: Increasing time span for chemical denudation of carbonate and siliclastic deposits toward the highest shoreline (model C, rates of denudation in m/Ma).

C and D: Chronostratigraphic sections (linear time scale) across A; C with constant and D with changing time period of T-R cycles.

E: Normal stratigraphic cross section of the T-R cycles shown in D. The upper boundary of the sequence is assumed to lie horizontally, and thickness increases toward the center of the basin. Note the landward merging of reworked horizons of the different cycles (1st to 5th C). The coastal onlap at the base of the sequence is composed of the transgressive phases of cycles 1 and 3.

As we have already seen (Fig. 3), the sediment cover of the T-R cycles thins landward because, among other factors, the time available for sediment accumulation approaches zero landward as sea level rises. However, the time available for erosional processes increases toward the highest shoreline and can become equal to the period of one T-R cycle (4 Ma, Fig. 5B). Therefore it is likely that all sediment will be removed by chemical and fluvial erosion, especially near the highest shoreline (Fig. 5C). The rest of the sediment will be eroded by wave action during the regressive phase and be left behind as storm layers and reworked horizons as described above. A simplified chronostratigraphic section drawn through the marginal part of such a basin for a time span of 8 Ma may have the resulting shape shown in Fig. 5C. Only small relicts from a formerly extensive transgressive onlap remain, while submarine erosion surfaces cover large areas. Sediment columns studied at location I show a complete succession of transgressive and regressive sequences. At location IV, all marine sediments have been mechanically eroded or dissolved by meteoric water. At locations II and III, the transgressive sequence is more or less preserved, whereas the regressive sequence is partly or totally removed.

The time spans of the hiatuses as well as those of the corresponding condensed horizons depend upon the position of the investigated site within the former depositional basin. Usually the reworked horizons will be found below the ancient LSL, because the emerged areas of the transgressive sequence are more or less gone. Therefore we have to expect both truncated regressive (more seaward) and truncated transgressive (more landward) sequences of purely marine sediments.

Tempestites are formed when the storm wave base touches soft ground, hence chiefly in combination with shallow erosion. Therefore they are best preserved near the top of a regressive sequence (between locations I and II in Fig. 5C). One should be careful not to mistake them for more distinct reworked horizons, since amalgamated storm layers may look similar to condensed horizons. Tempestites can also be deposited further landward, but they are removed by continuing erosion. In the reworked horizons, evidence of deep erosion (such as exhumed lithified concretions) increases toward the coastline (Fig. 5C, location III). The model in Fig. 5C applies to T-R cycles of equivalent duration and amplitude. If these two factors are variable, the resulting chronostratigraphic section across the marginal part of a sedimentary basin may become like that shown in Fig. 5D. Here, the second, shorter cycle did not reach the same transgressive level as the first one, and the regressive sea level stand of the fourth cycle was higher than that of the first two cycles. If sediment columns at different sites are investigated, as was done in Fig. 5C, the number of hiatuses and reworked horizons can be seen to differ from one location to another. However, it is interesting to note that, for example, at site VI cycles 3 and 4 may be overlooked, or at site VII the second cycle is missing, and at site VIII only the third cycle is present. In Fig. 5E the same situation is shown in a normal stratigraphic cross section. Here, the reworked horizons of the

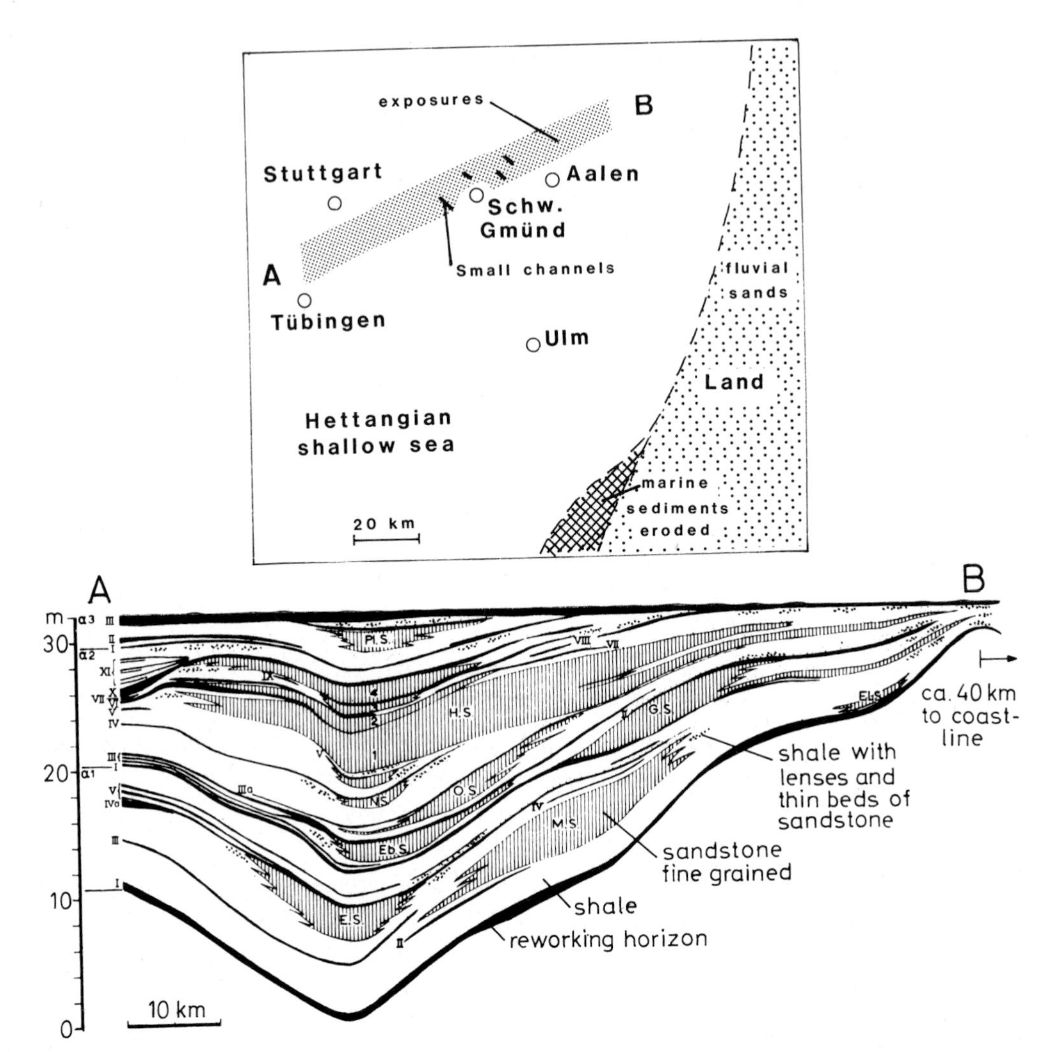

Fig. 6: Generalized cross section through marine sediments of a shallow Liassic (Pliensbachian, Hettangian and Sinemurian) epicontinental basin of southern Germany (after BLOOS 1976). Only two or three out of a large number of reworked horizons can be traced along the entire section. Note that some of the sand bodies can be interpreted as fills of wide and shallow channels cut into shales.

different T-R cycles shown in Fig. 5D all merge toward the coast in a similar manner as has already been pointed out by VAIL & TODD (1981). These local differences are one of the major reasons why the correlation of T-R cycles and the determination of their number for a certain time span are so difficult.

In a very careful study on Hettangian and Lower Sinemurian shallow marine sediments of southern Germany, BLOOS (1976) has shown that the number of reworked horizons decreases from the more central part of the Liassic basin towards the coastline (Fig. 6). Only two or three of these horizons (e.g. 2I and 3III), within a sequence representing approximately 2 to 3 Ma are thought to have been generated by marine regressions (HALLAM, 1981), but possibly more of these horizons represent T-R cycles. Chronostratigraphic sections published in PAYTON (1977, e.g. TODD & MITCHUM, VAIL, MITCHUM & THOMPSON) represent much larger areas and time spans 10 to 50 times longer than the example from the Lower Liassic mentioned above.

8. THICKNESS OF TRANSGRESSION-REGRESSION SEQUENCES AT CERTAIN LOCATIONS WITHIN A SUBSIDING BASIN

In addition to the previous section presenting a general survey of the response of sediments to T-R cycles, we now try to study the situation at certain locations within a subsiding, shallow-marine basin. To do this we again use the assumptions (2) and (4) of chapter 5.1 (storm-dominated hydraulic regime; predominantly storm controlled morphology of sea bottom tending towards equilibrium conditions) as well as assumptions (3) and (4) of chapter 7 (period of T-R cycle 4 Ma; long term and basin-wide sediment accumulation SA=SUB, where SUB remains constant).

In order to compare shelf seas of passive continental margins and epeiric seas (Table 1) the following additional assumptions are made:

I. Shelf seas on continental margins

1) Sediment supply SS = SUB = 50 m/Ma
2) Sea level fall SLF 25 m/Ma, (SLF < SUB)
3) Storm wave base STWB = 50 m

II. Epeiric seas

1) SS = SUB = 10 m/Ma
2) SLF = 25 m/Ma, (SLF > SUB)
3) STWB = 20 m

The magnitude of the sea level changes is assumed to be the same in both settings, namely 50 m (i.e. 25 m/Ma for one half cycle) and the curve is assumed to follow a regular sinusoidal oscillation. There may be more or less constant or strongly changing sediment influx of allochthonous or chiefly autochthonous (biogenic) materials (low input during transgressions, high input during falling sea level). As compared to the present oceans, the depth of the storm wave base in the models I and II may not be deep enough, to represent high-energy coastal basins. However, it may be a reasonable estimate for basins of a lower to medium energy level, especially for epeiric seas (see also HALLAM, 1981b). From a great number of possible locations between the highest shoreline and the deepest part of the basin we choose only three points P and Q_a and

Q_b (Figs. 3 and 7 through 10), point P being slightly below the storm wave base (STWB) of mean sea level (MSL), and the other two (Q_a and Q_b)well below STWB of the lowest sea level (LSL). Point P lies in the area in which truncated sequences with reworked horizons can be expected, points Q_a and Q_b will be devoid of submarine erosion. Points some distance landward of P would not be worth investigation, because here hardly any marine sediments are preserved (compare Fig. 5).

The graphical treatment of sea level variations under the four alternatives listed above is shown in Figs. 7 through 10. In all four Figures, the T-R cycles start at mean sea level (MSL), while time is plotted on the abscissa. Due to continuing subsidence, each point in the graph (or at a certain position in the basin), for instance P, migrates downward with time to P_1. However, if sediments accumulate at this point, the sea floor (SF) may only fall to a point shallower than the "base line" between P and P_1, or even rise as compared to its elevation at time zero and reach Point P_1'. Sediment accumulation (SA) is prevented or erosion into older sediments is permitted down to the oscillating storm wave base (STWB). The resulting sediment columns for points P and Q within the basin are shown on the righthand side of the T-R graphs including the corresponding curves for water depth (WD) and sediment accumulation (SA) or sediment supply (SS). The latter one may differ from SA, when erosion takes place or material elsewhere eroded is redeposited (see also Fig. 3).

MODEL I,1 (MARGINAL SHELF SEA, CONSTANT SEDIMENT INFLUX)

The graph in Fig. 7B is valid for point P situated somewhat below the STWB of mean sea level (MSL). Since SS = SUB, the sea floor at point P tends to maintain a steady elevation. However, during the second half of the regression (time 0 to 1 Ma), part of SS cannot accumulate because of the lowered STWB. Rather, the sediment will bypass point P and be deposited in deeper water (increased sediment accumulation SA at sites Q_a and Q_b). With the onset of transgression, sediment accumulation (SA) will be equal to SS. Consequently, the sea floor will stay in the same position and reach point P_1' after 4 Ma (end of the first cycle). Therefore, the second T-R cycle will start at a point (P_2) somewhat lower than the first one (P), and the subsequent sedimentation will not longer be affected by storm wave erosion.

If the sediment supply (SS) is generally the same at locality Q_a as at P, the only difference between the sediment columns at P and Q_a is an increased sediment accumulation (SA) in the time span between 0 and 1 Ma. If SS decreases at site Q_b to a half of that at site P (in terms of sedimentation rate or accumulated material per unit area and time, the thickness of the sediment column will become correspondingly smaller. This is reasonable because the allochthonous material must be transported over a greater distance from the land source. At both locations, Q_a and Q_b, the "sea level curve", whose shape is based on the thicknesses of the T-R sequences, still resembles the assumed sinusoidal curve of the true sea level.

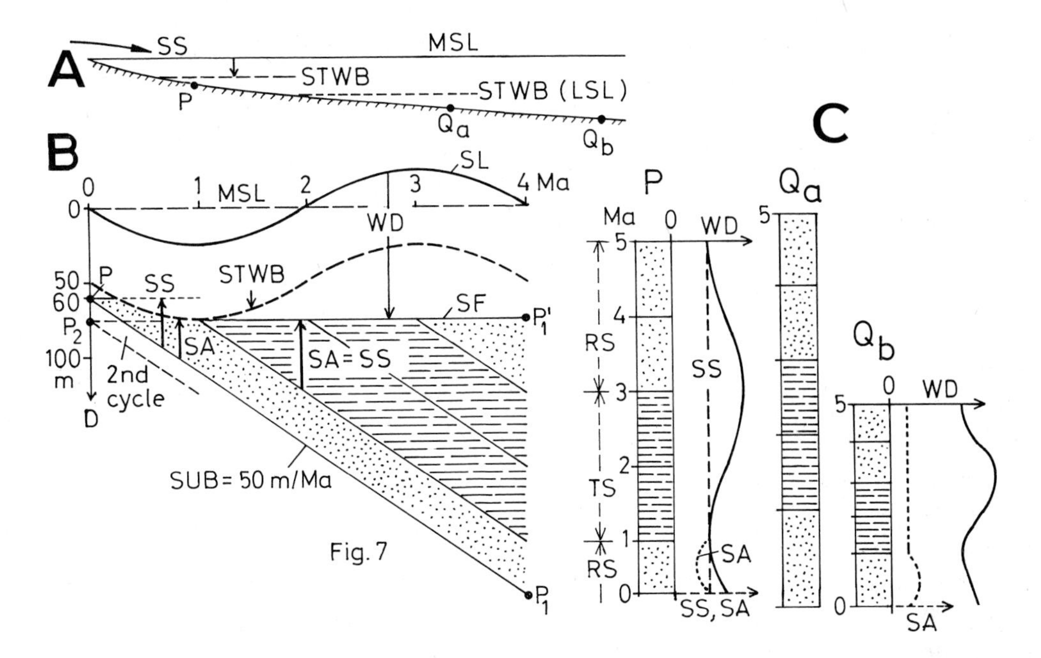

Fig. 7: Model I,1: Continental shelf, constant sediment supply SS = subsidence SUB = 50 m/Ma. Sea-level fall = 25 m/Ma < SUB

Fig. 7 through 10: General explanations. Accumulation and erosion of sediment by falling storm wave base at different locations in a marginal basin during a T-R cycle. A= cross section of basin with mean sea level (MSL), storm wave base (STWB) and sites P, Q_a and Q_b . B = idealized T-R cycle of 4 Ma, simultaneous subsidence (SUB) of point P to position P_1, and sediment accumulation (SA); (SS= sediment supply, SF= sea floor, WD= water depth). C= sediment column at sites P and Q with (non-linear) time scale showing transgressive (TS) and regressive (RS) sequences. D (only in Figs. 9 and 10)= changing sediment supply during the T-R cycle. For further explanation see text.

MODEL II,1 (EPEIRIC SEA, CONSTANT SEDIMENT INFLUX)

At location P the falling sea level and its STWB cut deeply into the pre-existing sediments (Fig. 8). From about 1 Ma onward new sediment is deposited which keeps the sea floor at a constant elevation. Hence, the next cycle starts at P_2 at the same level as P_1 and just escapes further submarine erosion (in the model the STWB just touches the seafloor, so at this point tempestites may occur). The sedimentary column at point P is strongly reduced by submarine erosion (the hiatus comprises the time span from -1 to +1 Ma), whereas at site Q_a, due to redeposition of the eroded material from site P, this period is represented by a thick regressive sequence. Therefore, the shapes of the inferred "sea level curves" differ markedly between locations P and Q. At P we see the typical "rapid regression" as postulated by VAIL and his coworkers (in PAYTON, 1977), but at Q, the regression proceeds "more slowly" than the transgression. In reality, both curves stem from a symmetrical sinusoidal curve of the true sea level change.

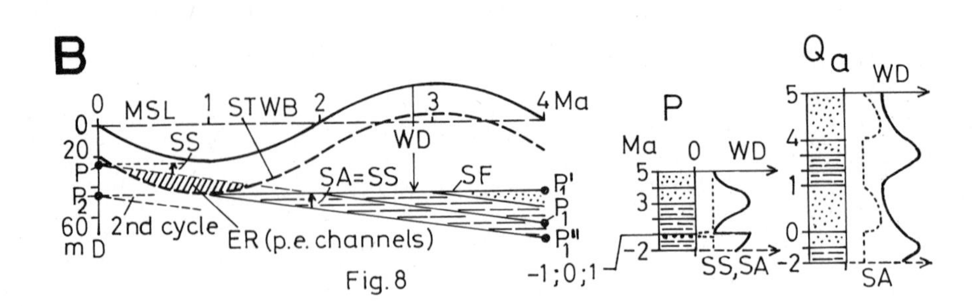

Fig. 8: Model II,1: Epeiric sea, constant sediment supply SS = subsidence SUB = 10 m/Ma. Sea-level fall = 25 m/Ma > SUB

MODEL I,2 (MARGINAL SHELF SEA; CHANGING SEDIMENT INFLUX)

In this case it is assumed that sediment influx from land sources and emerged shallow marine sediments is four times greater during regressions than during transgressions (Fig. 9). In the long run, however, the mean sediment supply (MSS) will compensate for subsidence. Consequently, the amount of sediment which cannot accumulate at site P during the first million years and settle in deeper water at site Q will increase in comparison to model I,1. Later, during the transgressive phase, sediment accumulation at P will decrease and hence the sea bottom will be lowered. Again the second cycle will start at a lower level than the first one and can no longer be affected by submarine erosion. In contrast to model I,1, the regressive sequences at both P and Q_b (when the sedimentation rate is reduced to a half of that at P) have become considerably thicker than the transgressive ones, especially at the deeper location, Q_b. Only the regressive sediments deposited during the first million years are somewhat reduced, and distinct omissions and reworked horizons cannot be expected.

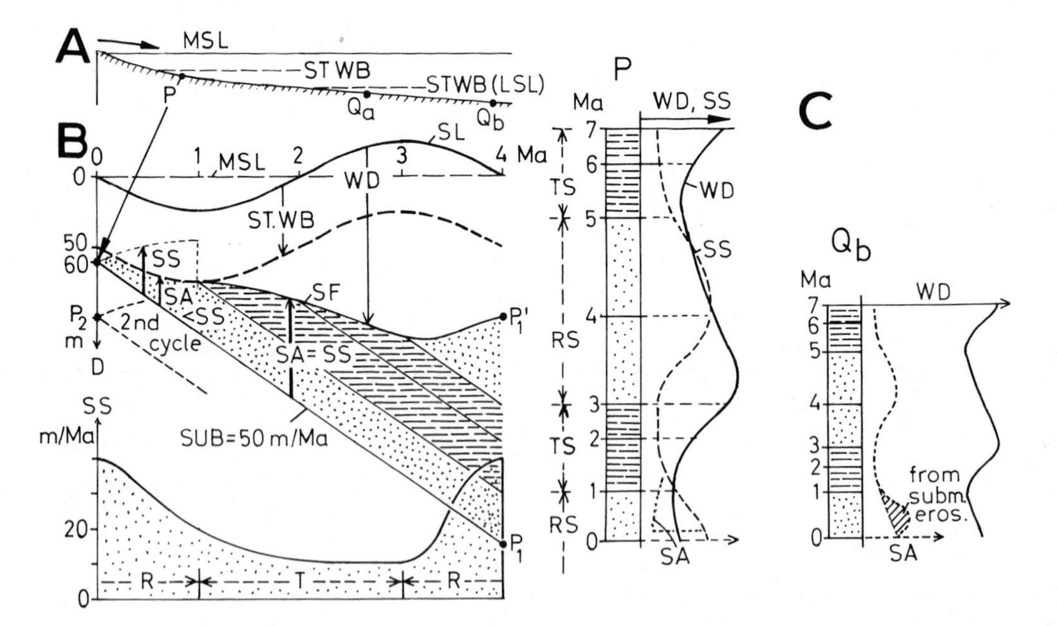

Fig. 9: Model I,2: Continental shelf, changing sediment supply during T-R cycle; otherwise as model I,1 (Fig. 7)

MODEL II,2 Epeiric sea, CHANGING SEDIMENT INFLUX)

For sediment influx into an epeiric sea, the same assumptions are made as in model I,2. At location P, during the first million years we have both nondeposition and submarine erosion (Fig. 10). A regressive sequence of roughly half a million years from the older sediments will be truncated. Therefore, the hiatus will comprise a total of about 1.5 Ma. Because of a greater SS from land sources, the deeper water location Q_a will receive more sediments than in model II,1. Hence, there we find much thicker regressive sequences than transgressive counterparts. Correspondingly, the shape of the inferred "sea level curve" again varies considerably from location to another. For location P it is important to note that, in contrast to all previous models, the second and the subsequent cycles will also be affected by submarine erosion during the second phase of regression. The depth of this erosion is, however, less pronounced than during the first cycle.

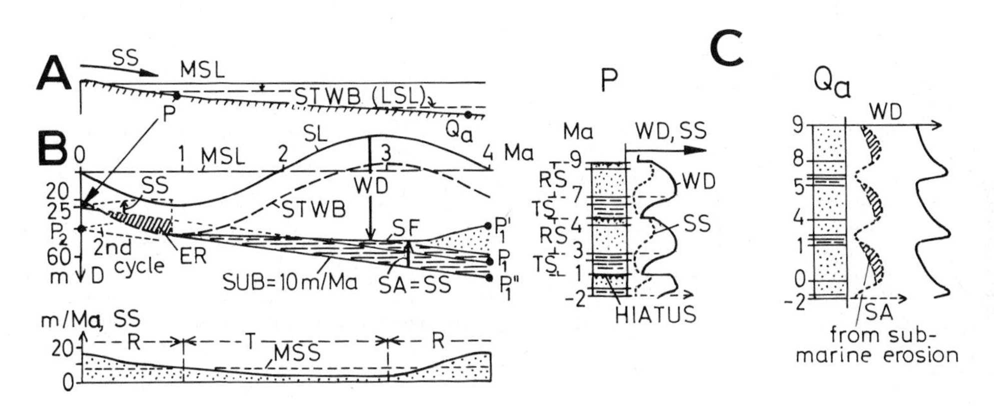

Fig. 10: Model II,2: Epeiric sea; changing sediment supply during T-R cycle, otherwise as model II,1 (Fig. 8).

9. CONSEQUENCES OF THE MODELS

Of course the few models discussed here cannot cover the variety of cases actually found in nature. One of the main shortcomings is that they neglect tide-dominated shallow-marine environments, but these are not common in epeiric seas (HALLAM, 1981b). In the case of wave-dominated shallow seas, the models may be useful in understanding some of the most important reactions of sediments to sea level changes. In summarizing the previous deductions, we can list the following points:

1) Even "short term" or "third order" pre-Pleistocene sea level oscillations, as described thus far, have a period (a few million years) roughly 100 times longer than the Pleistocene oscillations. Therefore, shallow marine sediments of older shelf and epicon-

tinental seas responded, in some respects, differently from Pleistocene sediments on the present continental shelves. They had more time to adapt to a kind of equilibrium between the hydraulic regime and sea-bottom morphology. Emerging parts could be more or less eroded during regressions; submarine erosion, reworking and condensing lasted for a long time (on the order of 1 Ma) and left distinct reworked and condensed horizons.

2) However, extensive submarine erosion can occur only if sea level fall is faster than subsidence. Therefore, slowly subsiding shallow basins (e.g. 10 m/Ma) such as epeiric seas are more sensitive to a certain type of sea level oscillation (e.g. with a period of 4 Ma and an amplitude of 50 m) than more rapidly subsiding basins (of the order of = 50 m/Ma). Shelves of passive continental margins belong in this latter category during their first 50 to 100 Ma of development.

3) In field exposures reworked horizons, if they can be traced over long distances, are probably the best indicator for regressions (or tectonic uplift, or a changing hydraulic regime). They occur below the lowest stand, which is also where they are chiefly preserved. Truncation starts to affect regressive sequences and, particularly in a landward direction, can cut down into older transgressive sediments and generate hiatuses and condensing phenomena of increasing time span.

4) Reworked horizons, especially in epeiric seas, often rest on diagenetically indurated layers of carbonate concretions. In many cases, they indicate submarine erosion to a depth of at least a few meter, sometimes even 10 m and more.

5) If there is little submarine erosion, reworked horizons may be accompanied (or chiefly underlain) by proximal storm layers (tempestites). There is no sharp distinction between amalgamated tempestites and reworked horizons.

6) In a section from deeper water to the shoreline of an ancient basin, the number of reworked horizons which occur within a certain time span may change, especially if the periods and amplitudes of sea level oscillations vary.

7) The response of sediments to T-R cycles differs from site to site, even when the sediment supply compensates for subsidence over long time periods. However, at a site receiving constant sediment influx which is affected by storm wave erosion (location P in Figs. 8 and 10), some of the regressive sequences can become truncated and thinner than the transgressive sequences. In such a case, the "sea level curves" inferred from sediment thicknesses are distorted and show "rapid regression" (Fig. 8). At deeper sites well below the lowest storm wave base (Point Q in Fig. 8), the regressive sequences become thicker than the transgressive ones. Consequently the inferred "sea level curves" now demonstrate much "longer" regressive phases.

If the sediment influx during regressions is substantially greater than during transgressions, then at location P each T-R cycle will leave behind a reworked horizon (Fig. 10). The distortions of the true sinusoidal sea level curve, as mentioned above, will be even more pronounced. At location P, we can observe repeatedly "abrupt regressions", and at site Q , the regressive sequences become much "longer" than the transgressive counterparts.

8) In basins with greater subsidence (SUB < SLF), where thicker sediment layers can accumulate over the same time span, most of the phenomena described above cannot be observed. There is no big difference between the thicknesses of transgressive and regressive sequences, and the "sea level curves" inferred from the thicknesses of the sediment columns are more or less symmetrical (Figs. 7 and 9). Only in the case of increased sediment influx during regressions (at the deeper site, Q_b, in Fig. 9) can the regressive sequences grow in thickness and hence generate some asymmetry in the "sea level curve". In such basins, however, other facies changes such as differing composition and fauna of sediments may enable the recognition of T-R cycles (see e.g. KAUFFMAN, 1977).

10. SOME EXAMPLES FOR THE RESPONSE OF DIFFERENT LITHOLOGIES TO T-R CYCLES IN THE JURASSIC OF SOUTHERN GERMANY

As for example DUFF et al. (1967) and HALLAM (1981b) have already pointed out and as was shown more quantitatively in the previous sections, ancient sea level changes and their effects on sediments can be studied best in marginal parts of slowly subsiding shallow-marine basins.

The Jurassic sea of southern Germany presents a good example as does the Jurassic of the British Isles (HALLAM 1978). Subsidence was of the order of 5-25 m/Ma, and the total Jurassic sequence is exposed some tens of kilometers up to about 100 km away from the ancient shoreline (comp. Fig. 6 and paleogeographic maps in BRANDT and BAYER & McGHEE, this volume). Although it has been known for a long time that shallow marine conditions prevailed in this basin, the well-bedded sediments and the occurrence of certain "marker horizons" were never satisfactorily explained, except for a few interesting studies (e.g. KLÜPFEL, 1917; ALDINGER, 1965). Therefore, in this and especially in the following articles some parts of the Jurassic sequence are described. In the context of this paper, only a brief survey of different, somewhat idealized sediment sequences affected by T-R cycles is possible. My interpretations of given sediment sequences are tentative and should be checked further by field and laboratory work.

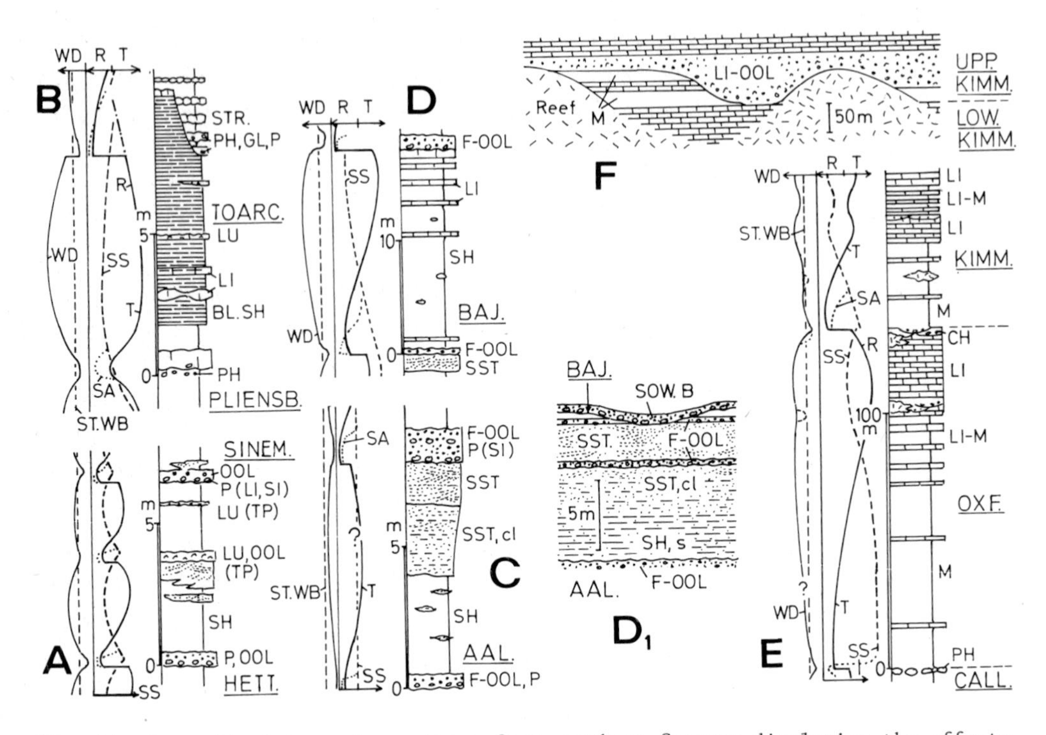

Fig. 11: Generalized Jurassic sections from southern Germany displaying the effects of T-R cycles (or repeatedly changing hydraulic regime); for further explanations see text.
WD = water depth, R = regression, T = transgression, SS = sediment supply, SA = sediment accumulation, STWB = storm wave base. SH = shale, BL.SH = black shale marl), Li = Limestone,, LU = Lumachelle, SST = sandstone, cl = clayey, M = marl, TP = Tempestite (storm layer), P = pebbles of carbonate, OOL = oolithic (calcareous), F-OOL = iron oolith, SI = siderite, PH = Phosphorite, GL = glauconite, STR = stromatolites, CH = channels (shallow).

The examples presented in Fig. 11 show the following typical sections described in order of increasing sedimentation rate):

A. Shale - sandstone - detrital limestone - shell bed seauence of Hettangian-Sinemurian age (for further details see BLOOS, 1976 and 1982). The limestones containing carbonate concretions and ooids are explained as reworked horizons with substantial preceding submarine erosion; other shell-bearing beds represent more or less amalgamated storm layers. Sands and sandy deposits were probably deposited during regressive phases.

B. Truncated shale - black marly shale - fine grained limestone sequence of Pliensbachian - Toarcien age (for further details see e.g. EINSELE & MOSEBACH, 1955; JOACHIM, 1970; KÜSPERT, 1983). In the more landward parts of the basin, the upper Pliensbachian sediments contain a fairly distinct reworked horizon with phosphorite nodules (BRANDT, this volume). The overlying black marly shales with some intercalated limestones can be traced over large areas of Europe (see e.g. RIEGRAF, 1982) Most likely they were deposited during the peak of the lengthy Liassic transgression (HALLAM, 1981). In the upper part of these Toarcian black shales, bedding becomes less regular and shell beds form an important part of the sequence. Finally come light gray Upper Toarcian marls and shellbearing limestones, some of them containing intraformational pebbles, stromatolites, phosphorite nodules and glauconite (BROCKERT 1959). The contact at the top of the black shales is sharp. Submarine erosion either stopped at fine-grained limestones of the black shale sequence, or cut wide shallow channels up to 10 m deep into the black shales (ETZOLD, 1980).

C. The generalized example from the Aalenian sequence is similar to that described in example A. The main differences are that now the proportions of quartz sands and iron ooids as well as sideritic pebbles are greater. It appears that the regressive sequence is thicker than the transgressive one, as was postulated above for cases in which sediment supply during sea level fall increases. FREYBERG (1962) assumed that the iron ooids were formed in shallower water and were redeposited in somewhat deeper parts of the basin. In addition, sedimentary structures in the sands were interpreted by tidal action (WERNER, 1959, for further details see BAYER & McGHEE, this vol.).

D. The example from the Bajocian starts with an oolitic ironstone bed on top of a sandstone layer, hence with a similar, but less pronounced succession than in the previous sediment column. The overlying shales and thin, parallel bedded limestones probably represent a transgressive sequence which is truncated and overlain by a second oolitic ironstone layer with pebbles. In this case, the regressive sequence is nearly missing, and the shape of the inferred "sea level curve", therefore, has become strongly asymmetrical with a "rapid regression" (compare BAYER et al., this vol.). At the same level FRANK (1945) has described a strange feature which here is reinterpreted as a shallow channel cutting into an oolitici ironstone bed (D_1). The channel fill is a typical reworked horizon with intraformational pebbles, ferriferous ooids, etc..

E. The Oxfordian and Kimmeridgian marl-limestone sequence begins with a comparatively inconspicuous reworked horizon containing phosphorite. Probably due to a growing biogenic carbonate production, sedimentation rate increased considerably and may have reached an order of 20 m/Ma. However, at the same time, an apparently global sea level rise prevented the basin from being filled up with sediment. On the contrary, during Oxfordian times the sea covered larger areas than before (ZIEGLER, 1978). As a result, sediment input from land sources decreased so that the higher portion of the Oxfordian sequence became carbonate-dominated. However, at the lower part and particularly at the top of this rhythmic limestone-marl succession (EINSELE,

1982; RICKEN, this vol.), the regular bedding was disturbed by the accumulation of bioclastic detritus and by shallow channels. Whereas the first shallowing might have been caused by too rapid sediment buildup, the second is more likely to have been generated by sea level fall. This interpretation is supported by the occurrence of widespread fossil beds and the renewal of marl deposition in the lowermost Kimmeridgian.. In the Kimmeridgian, a similar change from marls to limestone-marl sequences can be observed. It is yet to be determined whether or not some smaller variations in the contribution of carbonate to the total sediment volume is caused by minor sea level oscillations or some other factors. The increasing abundance of reef limestones at many localities hampers a more detailed study of this section.

F. At the transition from the Lower to the Upper Kimmeridgian, however, channels were cut into reefal structures as well as into bedded lime muds and filled with oolitic detrital calcareous sands (SCHNEIDER, 1957; REIF, 1958). This phenomenon was and may be still interpreted as the consequence of tidal action, but again sea level fall cannot be excluded as an important factor.

11. DISCUSSION

The results gained by idealized models were summarized in section 9. With the examples from the Jurassic of southern Germany we may draw the following conclusions:

1) All sections in Fig. 11 A through E show truncated regressive sequences with reworked horizons. (In the upper part of section D, only a small portion of this sequence has been saved from erosion). Submarine erosion often generated small channels or wide shallow depressions. Consequently, all exposures studied lie within a zone where the storm wave base touched the sea floor, but where the sea floor did not emerge during low stands of sea level (location P in Figs. 7 through 10). To permit storm wave erosion, sea level fall during regressions must have been faster than subsidence. Since many exposures in a wide band along the ancient coastline display these characteristics, the basin floor appears to have been rather smooth and only slightly inclined. Jurassic sediments which were penecontemporaneously exposed to subaerial erosion are little known, as is in agreement with the models of Fig. 5.

2) In some cases of Fig. 11 the transgressive sequences are thicker than the regressive ones (D and E) or viceversa (B and C). In all cases, "sea level curves" derived directly from the geologic column and the thickness of the transgressive and regressive sequences are more or less distorted and suggest a "rapid" sea level fall similar to the models of Fig. 8 and particularly of Fig. 10. This is again explained by sedimen-

tary processes under the conditions of location P as mentioned above. In addition, the repeated occurrence of "rapid regressions" indicates that sediment influx during regressions was probably greater than during transgressions.

3) The lowest sea level stands are documented by the reworked horizons. Whether they coincide with the base of these layers or with features within these horizons is not known. In any case, the base of reworked and redeposited material represents the deepest erosional face which often is marked by exhumed concretions.

4) In the Jurassic basin of southern Germany, water depth generally fluctuated around the storm wave base. Only during the peaks of transgressions (Fig. 11 B, D and part of E) did storm wave action not affect sediments in large areas and for longer periods.

5) Sediment supply from land sources was generally very low, but during certain intervals it increased considerably, for example in the Lower Aalenian (not represented in Fig. 11). Particularly in the Lower Oxfordian (Fig. 11 E), due to the onset of high biogenic carbonate production, the sedimentation rates became rather high and could compensate for an accelerated (?) transgressional trend during that time period.

6) From this work and other articles in this volume it is not yet possible to determine the number and magnitude of sea level oscillations which can be found in the Jurassic sequence of southern Germany. The examples shown in Fig. 11 represent T-R cycles of 1 to 4 Ma duration with possible amplitudes of the order of 10-20 m, possibly up to 50 m.

It cannot be excluded that smaller scale T-R cycles occurred, because the number of reworked horizons in certain sections is greater than one for every 1 to 4 Ma (compare Fig. 6). As pointed out earlier (Chapter 7, Fig. 5 D) the reworked horizons of minor sea level oscillations cover only limited areas. Therefore, in these cases it is difficult to distinguish between "normal" submarine erosional processes and wave base erosion below a falling sea level.

ACKNOWLEDGEMENTS

Dr. A. .Hallam (Birmingham), Linda Hobert and W. Ricken (Tübingen) critically read the manuscript and made many valuable suggestions.

REFERENCES

Aldinger, H. 1965: Über den Einfluß von Meeresspiegelschwankungen auf Flachwassersedimente im Schwäbischen Jura.- Tschermacks mineral. u. petrogr. Mitt., 10, 61-68.

Bloos, G. 1976: Untersuchungen über Bau und Entstehung der feinkörnigen Sandsteine des Schwarzen Jura alpha (Hettangium und tiefstes Sinemurium) im schwäbischen Sedimentationsbereich.- Arb. Inst. Geol. Paläont. Univ. Stuttgart, N.F. 71, 1-269, Stuttgart.

Bloos, G. 1982: Shell beds in the lower Lias of South Germany - facies and origin. In: Einsele, G. & Seilacher, A. (eds.): Cyclic and event stratification, p. 223-239, Springer, Berlin - Heidelberg - New York.

Brockert, M. 1959: Zur Ammonitenfauna und Stratigraphie des Lias żeta in Baden-Württemberg. Diss. Univ. Tübingen, 146 S. (unpublished).

Campos, H.S. & Hallam, A. 1979: Diagenesis of English Lower Jurassic limestones as inferred from oxygen and carbon isotope analysis. Earth and Planetary Sci. Letters, 45, 23-31, Amsterdam.

Cooper, M.R. 1977: Eustacy during the Cretaceous: its implications and importance. Palaeogeography, Palaeoclimatology, Palaeoecology 22, 1-60, Amsterdam.

Curray, J.R. 1964: Transgressions and regressions. In:Miller, R.L. (ed.):Papers in marine geology: Shepard Commemorative Volume, p. 175-203, Macmillan.

Donovan, D.T. & Jones, E.J.W. 1979: Causes of world-wide changes in sea-level. J. Geol. Soc. London, 136, 187-192.

Duff, P. Mcl. D., Hallam, A. & Walton, E.K. 1967: Cyclic sedimentation. Developments in Sedimentology 10, 280 p., Elsevier, Amsterdam.

Einsele, G. 1982: Limestone-marl cycles (periodites): Diagnosis, significance, causes - a review. In: Einsele, G. & Seilacher, A. (eds.):Cyclic and event stratification, p. 8-53, Springer, Berlin - Heidelberg - New York.

Einsele, G., Elouard, P., Herm, D., Kögler, F.C., Schwarz, H.U. 1977: Source and biofacies of late Quaternary sediments in relation to sea level on the shelf off Mauritania, West Africa.- "Meteor" Forschungs-Ergebnisse, Reihe C, No. 26, 1-43, Berlin - Stuttgart.

Einsele, G. & Mosebach, R. 1955: Zur Petrographie, Fossilerhaltung und Entstehung der Gesteine des Posidonienschiefers im Schwäbischen Jura.- Neues Jb. Geol. Paläont., Abh. 101, 319-430, Stuttgart.

Einsele, G. & Seilacher, A. 1982: Cyclic and event stratification. 536 p. Springer, Berlin - Heidelberg - New York.

Etzold, A. 1980: Erläuterungen zu Blatt 7126 Aalen. Geolog. Karte 1 : 25 000 von Baden-Württemberg. 234 S., Geol. Landesamt, Stuttgart.

Frank, M. 1945: Die Schichtenfolge des mittleren Braunen Jura (gamma/ delta, Bajocien) in Württemberg.- Jahresber. und Mitt. Oberrhein. geol. Verein, N.F. 31, 1-32, Stuttgart.

Freyberg, B. von 1962: Eisenerzlagerstätten im Dogger Frankens.- Geol. Jb. 79, 207-254, Hannover.

Gautier, D.L. 1982: Siderite concretions: indicators of early diagenesis in the Gammon Shale (Cretaceous).- J. Sed.-Petrol. 52, 3, 859-871, Sept. 1982.

Hallam, A. 1978: Eustatic cycles in the Jurassic. Palaeogeogr. Palaeoclimatol. Palaeoecol. 23, 1-32.

Hallam, A. 1981a: A revised sea-level curve for the early Jurassic.- J. geol. Soc. London, 138, 735-743.

Hallam, A. 1981b: Facies interpretation and the stratigraphic record. 291 p., Freeman, Oxford and San Francisco.

Hallam, A. 1984: Pre-Quaternary sea-level changes. Ann. Rev. Earth Planet. Sci. 12, 205-243.

Hardenbol, J., Vail, P.R., Ferrer, J. 1981: Interpreting paleoenvironments, subsidence history and sea level changes of passive margins from seismic and biostratigraphy. Oceanologica Acta, 1981. Proceed. 26th Internat. Geol. Congr., Geology of continental margins symposium, Paris 1980, 33-44.

Hays, Y.D., Jmbrie, J. & Shackleton, N.J. 1976: Variations of the earth's orbit: pacemaker of the ice ages.- Science 194, 1121-1132.

Hohberger, K. & Einsele, G. 1979: Die Bedeutung des Lösungsabtrags verschiedener Gesteine für die Landschaftsentwicklung in Mitteleuropa.- Z. Geomorph. N.F., 23, 361-382, Berlin - Stuttgart.

Irwin, H., Coleman, M., Curtis, C.D. 1977: Isotope evidence for several sources of carbonate and distinctive diagenetic processes in organic-rich Kimmeridgian sediments.- Nature 269, 209-213.

Jeans, C.V. 1980: Early submarine lithification in the Red Chalk and lower Chalk of eastern England: a bacterial control model and its implications.-Proc. Yorkshire Geol. Soc. 43, part 2, No. 6, 81-157.

Joachim, H. 1970: Geochemische, sedimentologische und ökologische Untersuchungen im Grenzbereich Lias delta/epsilon (Domerium/Toarcium) des Schwäbischen Jura.- Arb. Geol.-Paläont. Inst. Univ. Stuttgart, N.S. 61, 243 S.

Kauffman, E.G. 1977: Cretaceous facies, faunas, and paleoenvironments across the Western Interior Basin. The Mountain Geologist 14, 75-274.

Klüpfel, W. 1917: Über die Sedimente der Flachsee im Lothringer Jura.- Geolog. Rundschau 7, 97-109, Leipzig.

Küspert, W. 1983: Faziestypen des Posidonienschiefers (Toarcium, Süddeutschland), eine isotopengeologische, organisch-chemische und petrographische Studie. Diss. geowissensch. Fakultät Univ. Tübingen, 232 S.

Mörner, N.A. 1982: Sea level changes as an illusive "Geologic index".- Bull. INQUA Neotectonic Commission, No. 5, 55-64, Stockholm.

Payton, C.E. 1977: Seismic stratigraphy-applications to hydrocarbon exploration.-Amer. Assoc. Petrol. Geologists, Memoir 26, 516 p. Tulsa.

Pitman, W.C. III 1978: Relationship between eustacy and stratigraphic sequences of passive margins.- Geol. Soc. Amer. Bull. 89, 1389-1403.

Pitman, W.C., III, & Golovchenko, X. 1983: The effect of sealevel change on the shelfedge and slope of passive margins.- Soc. Econ. Paleont. Mineralog., Spec. Publ. 33, 41-58.

Purser, B.H. 1969: Syn-sedimentary marine lithification of Middle Jurassic limestones in the Paris Basin.- Sedimentology 12, 205-230.

Raiswell, R. 1971: The growth of Cambrian and Liassic concretions.- Sedimentology 17, 147-171.

Reiff, W. 1958: Beiträge zur Geologie des Albuchs und der Heidenheimer Alb (Württemberg).-Arb. Geol.-Paläont. Inst. Techn. Hochschule Stuttgart, N.F., Nr. 17, 142 S.

Riegraf, W. 1982: The bituminous lower Toarcian at the Truc de Balduc near Mende (Départment de la Lozère, S-France). In:Einsele, G. & Seilacher, A. (eds.):Cyclic and event stratification, 506-511, Springer, Berlin - Heidelberg - New York.

Schneider, J. 1957: Stratigraphie und Entstehung der Zementmergel des Weißen Jura in Schwaben.- Arb. Geolog.-Paläont. Inst. Techn. Hochschule Stuttgart, N.F., Nr. 11, 1-94.

Southam, J.R. & Hay, W.W. 1981: Global sedimentary mass balance and sea level changes. In: Emiliani, C. (ed.):The oceanic lithosphere. The Sea, v. 7, 1617-1684, Wiley, New York.

Šrámek, J. 1978: Relative age of diagenetic carbonate concretions in relation to the sediment porosity.- Acta Universitatis Carolinae.- Geologica Kratochvil, vol. No. 3-4, 307-321, Praha.

Swift, D.J.P. 1970: Quaternary shelves and the return to grade.- Marine Geology 8, 5-30.

Swift, D.J.P. 1974: Continental shelf sedimentation. In:Burk, C.A. & Drake, C.L. (eds.): The geology of continental margins. 117-133, Springer, Berlin.

Swift, D.J.P. 1978: Continental-shelf sedimentation. In:Fairbridge, R.W. & Bourgeois, J.(1978): The Encyclopedia of Sedimentology, 190-196, Dowden, Hutchinson & Ross, Stroudsburg.

Swift, D.J.P., Stanley, D.J., Curray, J.R. 1971: Relict sediments on continental shelves: a reconsideration.- J. Geology 79, 322-346, Chicago.

Todd, R.G.& Mitchum, R.M. Jr. 1977: Seismic stratigraphy and global changes of sea level, part 8: Identification of upper Triassic, Jurassic, and lower Cretaceous seismic sequences in Gulf of Mexico and offshore West Africa.- In:Payton, Ch.E. (ed.):Seismic stratigraphy - applications to hydrocarbon exploration.- Am. Assoc. Petrol. Geolog., Mem. 26, 145-163.

Vail, P.R. & Hardenbol, J. 1979: Sea-level changes during the Tertiary.- Oceanus 22, 71-79.

Vail, P.R., Mitchum, R.M. & Thomson, S., III 1977: Seismic stratigraphy and global changes of sea level, part 3: relative changes of sea level from coastal onlap.- In:Payton, C.E. (ed.): Seismic stratigraphy - applications to hydrocarbon exploration. Americ. Assoc. Petrol. Geolog., Memoir 26, 63-82.

Vail, P.R. & Todd, R.G. 1981: Northern North Sea Jurassic unconformities, chronostratigraphy and sea-level changes from seismic stratigraphy. Petroleum geology of the Continental Shelf of North-West Europe, Chapt. 19, 216-235, Inst. of Petroleum, London.

Watts, A.B. 1982: Tectonic subsidence, flexure and global changes of sea level. Nature 297, 469-474.

Watts, A.B. & Streckler, M.S. 1979: Subsidence and eustasy at the continental margin of eastern North America. In:Talmani, M., Hay, W.F., Ryan, W.B.F. (eds.): Deep drilling results in the Atlantic Ocean: continental margins and paleoenvironment, Maurice Ewing Ser. 3, 218-234.

Werner, F. 1959: Zur Kenntnis der Eisenoolithfazies des Braunjura beta von Ostwürttemberg. Arb. Geolog.-Paläont. Inst. T.H. Stuttgart, N.F., Nr. 23, 169 S.

Ziegler, P.A. 1978: North Sea rift and basin development. In: Ramberg, I.B. & Neumann, E.-R. (eds.): Tectonics and geophysics of continental rifts, 249-277. Reidel, Dordrecht.

THE LOCAL SIGNATURE OF SEA-LEVEL CHANGES

G.R. McGhee, Jr. & U. Bayer

Tübingen

Abstract: Black shales and iron-stones are often considered as the signatures of regressions and transgressions. Their occurrence and the manner in which they relate to transgressive-regressive cycles is elucidated here by examples from the Devonian and the Jurassic.

Sea level changes cause local changes in sedimentation rates -- an example is here given from the South German Jurassic which indicates an average periodicity of 4 Myr; a periodicity which also appears likely for our Devonian example.

There is no one-to-one correspondence between global sea-level changes and local transgressive-regressive cycles. These relationships are discussed using a simple mathematical model which allows, in addition, the prediction of the position of rhythmic sequences within larger regressive-transgressive trends.

INTRODUCTION

Transgressive-regressive sequences and associated cyclic sedimentation patterns are of interest to both sedimentologists and paleontologists -- especially if the cycles are driven by some global mechanism like eustacy or plate tectonics (HARLAND & HEROD, 1975; MACKENZIE & PIGOTT, 1981; SHANMUYAM & MAIOLA, 1982). On one hand such cycles (should) cause environmental changes by either altering the extent of (e.g.) shelf areas or by changing sediment supply and lithology, i.e. the facies. Therefore, they are capable of driving evolution and of causing ecological substitutions. On the other hand such physical 'events' provide the possibility of correlation independent of biostratigraphy -- an aspect first emphasized in 'seismic stratigraphy' (VAIL et al., 1977) and on a much smaller scale in 'event stratigraphy' (cf. EINSELE & SEILACHER, eds. 1982).

However, the relationship between global sea-level changes, local transgressive-regressive sequences and 'marker lithologies' is not a simple one-to-one correspondance (VAIL, HARDENBOL & TODD, 1984). It is the intention of this paper to elucidate some aspects of these relationships.

1) DEVONIAN BLACK SHALE CYCLES

Strata of Devonian age are excellently exposed in extensive outcrops in New York State, and comprise the standard reference section of the Devonian for the eastern United States (RICKARD, 1975). The strata of Middle and Late Devonian age were deposited in the classic progradational "Catskill Delta" complex. These consist of a thick wedge of clastic sediments which slowly prograded from eastern to western New York during this span of time, leaving a classic facies pattern of overall "regression" due to basin filling from erosion of the Acadian Highlands in the east (Fig. 1).

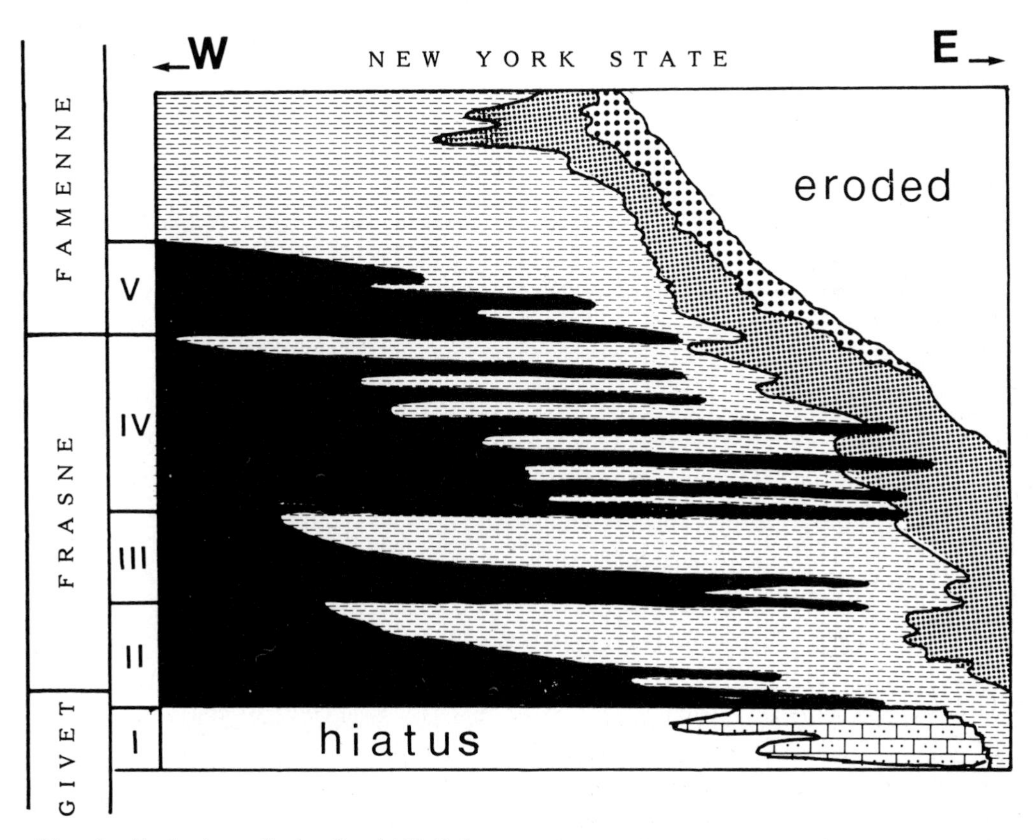

Fig. 1: Evolution of the Catskill Delta complex in New York State (Upper Devonian). Generalized lithology: black shales (black), claystones (dashed), sand and pebbles (points). Modified from RICKARD (1975).

In Fig. 1 the facies relationships in the late Middle and Late Devonian Catskill Delta sediments can be seen to be relatively simple -- terestrial coarse sands and red-beds grade into shallow water sands, which grade into deeper water silts, and finally into black shales (muds), which were deposited in the anoxic deeper water region of the basin axis.

This relatively simple facies pattern is the result of the relatively simple tectonic development of the Catskill Delta basin in New York, and reflects the interplay between gradual basin subsidence due to extensive sedimentary deposition, and the gradual filling of the basin due to the fact that the rate of subsidence was less than the overall rate of deposition.

The overall pattern of progradation was punctuated, however, by numerous reversals in the pattern of sediment deposition. During these reversals, or "cycles", the black basinal muds -- usually confined to the deep water basin axis region in western New York -- were expanded greatly in areal extent throughout the entire basin. In many cases these black shale extensions can be traced far to the east (Fig. 1), overlying shallow water sands and silt deposits, and in some cases even interfingering with terrestrial red bed deposits (SUTTON et al., 1970). During each cycle of black shale deposition stagnant, oxygen depleted waters extended into previous shallow water, well oxygenated regions which normally supported a diverse benthic fauna and flora (McGHEE, 1982; McGHEE & SUTTON, 1981). With the onset of anoxic conditions the local benthic fauna was largely extinguished, and within the black shales themselves are found only organisms which are interpreted to have lived within the water column itself (nektic, planktic, and epiplanktic modes of life) and thus sank to the oxygen depleted waters below only after death.

The pulses of black shale deposition were produced by episodic periods of relative sea level rise within the stratified waters of the Appalachian basin. With each sea level rise the stagnant waters of the basin center transgressed and onlapped the shallow shelf regions of the Catskill Delta. The ultimate cause of each transgressive cycle is not clear, as each deepening event within the Appalachian Basin could have been produced by a true eustatic change in global sea level, or by an increase in the subsidence rate of the basin (BYERS, 1977).

Of further interest is, however, the fact that the transgressive cycles are hierarchical structures -- that is, that smaller scale cycles can be discerned within larger scale cycles. In the New York strata, five major transgressive events can be seen (Fig. 1). The first of these cycles (labelled "I" in Fig. 1) was a major transgressive event which occurred in the late Givetian. During this transgression the shoreline in New York shifted far to the east, and throughout most of the basin clastic sedimentation ceased. The result was a major non-depositional hiatus (a "starved basin") in the deeper basinal regions, and the deposition of the unusual Tully Limestone in shallower water regions. The other four transgressions produced major black shale tongues in New York, which interfinger with shallower water deposits in the east. These shale units are, in ascending stratigraphic order, the Geneseo Shale (II), the Middlesex Shale (III), and the Rhinestreet shale (IV) of Frasnian age, and the Dunkirk Shale (V) of Famennian age. From the large black shale units, however, further (and thinner) black shale tongues can be traced tens of kilometers to the east. The number of these smaller transgressive cycles increased during the span of the Frasnian, in that only two major tongues can be traced eastward from the Middlesex Black Shale (III) of mid-Frasnian age, whereas at least six tongues extend from the major black shale unit produced by the next transgressive cycle (the Rhinestreet Black Shale, IV).

The pattern of transgression and regression as reconstructed in the eastern United States is compared with that of Europe in Fig. 2. Two major transgressive-regressive cycles occurred in both Europe and eastern North America:

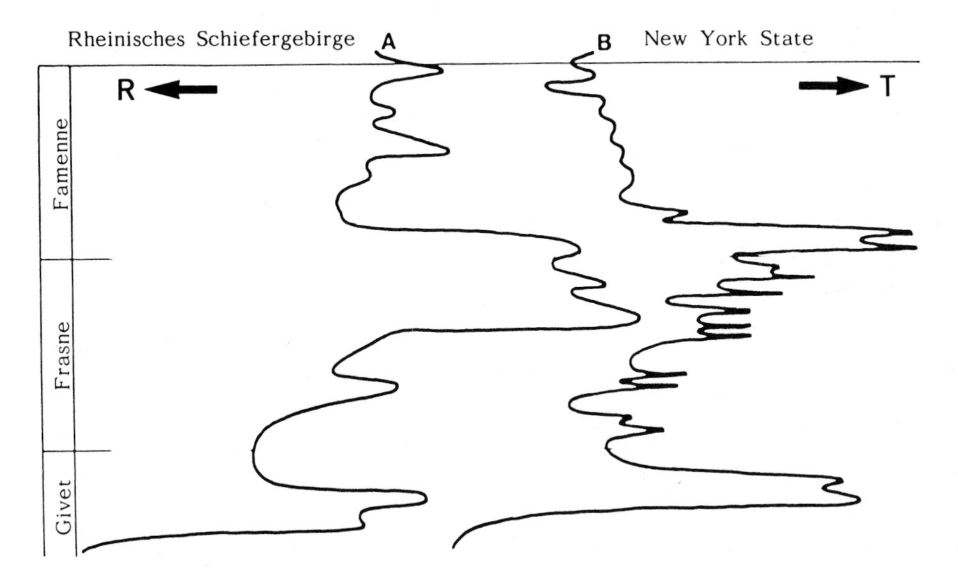

Fig. 2: Upper Devonian transgressive-regressive cycles: A in the Rheinisches Schiefergebirge (Germany, modified from KREBS, 1979) and B in the New York State.

1) a major, and sudden, relative sea level rise in the late Givetian, and

2) a more gradual overall rise in relative sea level which occurred throughout the Frasnian, but which was more pronounced in the late Frasnian and early Famennian.

Periods of relative sea level fall occur in the latest Givetian -- earliest Frasnian, and mid to late Famennian in both Europe and eastern North America. Similar cycles of transgression and regression in eastern North America and Europe could have been tectonically produced (perhaps by periodic uplift and quiescence of the Acadian-Caledonian Mountains), as these two regions of the world were parts of the same continent during the Devonian.

The smaller scale cycles are more problematical , however. While several smaller cycles can be seen in the German transgression-regression curve, it is impossible to correlate these with potentially equivalent subcycles in New York given our present data.

These smaller cycles may be only local in nature, produced by local tectonic events particular to each particular basin. The possibility remains open, however, that they may have been the result of larger scale tectonic events which affected the entire Laurussian Continent during the Devonian -- and thus may have occurred simultaneously in Germany and New York -- or, of course, that they reflect true small-scale eustatic events. Finally, the differences may be due to different levels of resolution. The New York black shales provide especially sensitive "markers" for transgressive events. However, they are not themselves results of the sea-level high-stand -- in deeper parts of the basin the anoxic situation continued throughout the interval. It is only the temporal spreading and shrinking of the anoxic area which provides the exceptional resolution. Similar changes in area of deposition of mud which is rich in organic carbon have been discussed in detail by ARTHUR & JENKYNS (1981) and SCHLANGER et al. (1981).

2) JURASSIC CYCLIC ACCUMULATION RATES

Sedimentary cycles are a well known phenomenon in the Jurassic of Central Europe (KLÜPFEL, 1917; HALLAM, 1961; EINSELE; BAYER et al., this volume). These lithological cycles, however, occur with widely differing magnitudes and are irregularily distributed within the stratigraphic column (Fig. 3). In the Liassic the sedimentary cycles correspond rather well with the transgressive-regressive cycles proposed by VAIL, HARDENBOL & TODD (1984) as defined by the temporal distribution of global unconformities and condensed stratigraphic sections.

The "typical" Lower and Middle Jurassic cycles begin with clays, which correspond roughly to the early regression phase of Vail's chart. The clays are followed by sandstones with an erosional upper surface (the "roof bed"). In the early transgressive phase rhythmic marl-clay sequencies occur, sometimes intercalated with black shales which most likely represent the maximum phase of the transgression (HALLAM, 1975; HALLAM & BRADSHAW, 1979). However, such "complete" sedimentary cycles are developed only in the Hettangian - lower Sinemurian, and to some extent in the Bajocian. Other cycles usually preserve only part of the overall cycle, either the transgressive phase (in the upper Liassic) or the regressive phase (in the lower Middle Jurassic).

If we examine the sedimentary a c c u m u l a t i o n r a t e s of the various stratigraphic sequences, however, a strikingly regular pattern appears. Here the accumulation rates given in Fig. 3 have been calculated using the time scale of VAN HINTE (1976) as the cycles based on global unconformities (VAIL, HARDENBOL & TODD, 1984) are also calculated from this time scale. However, the essential pattern does not change substantially if another time scale is used (e.g. HARLAND et al., 1982).

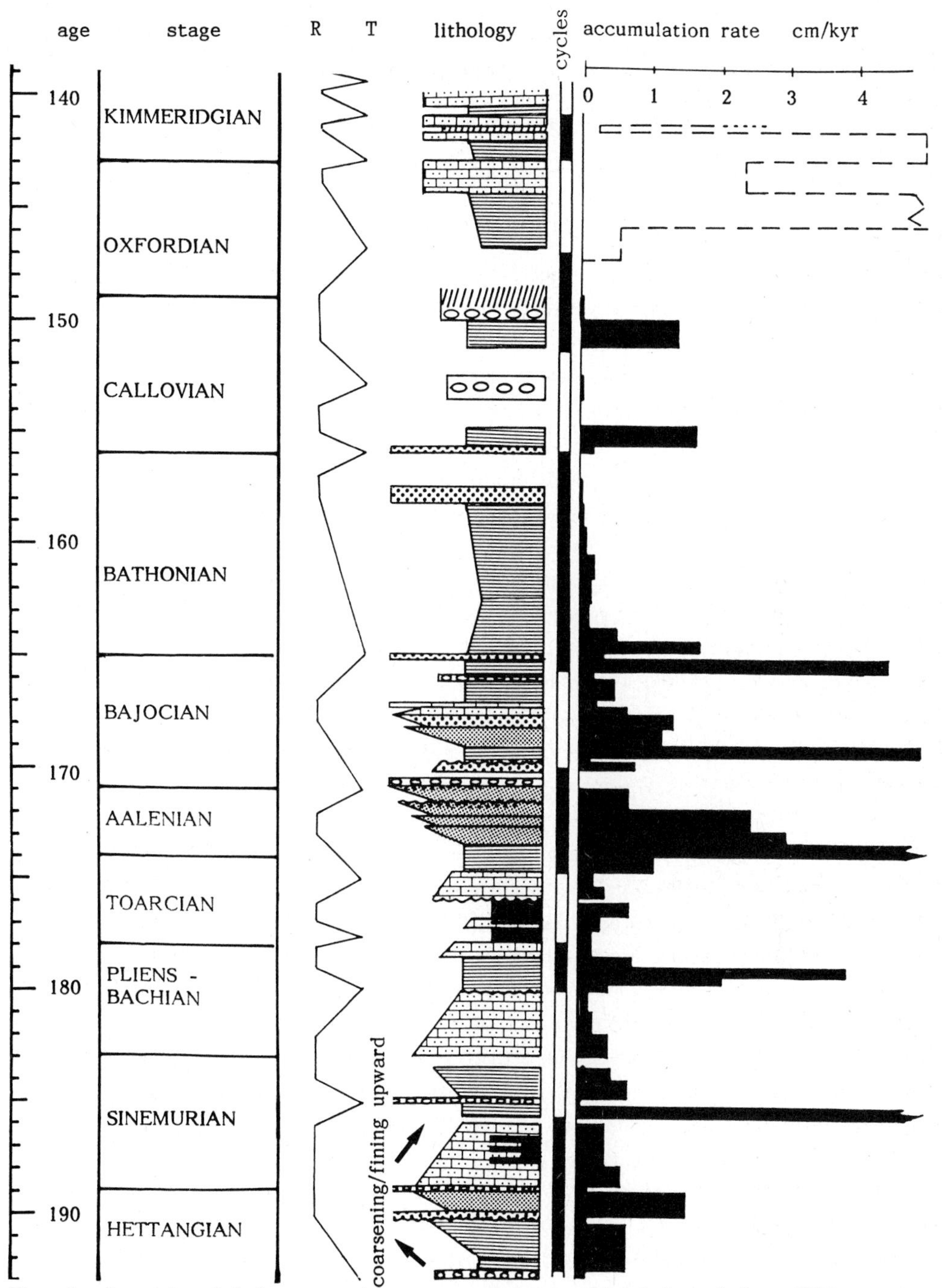

Fig. 3: Jurassic global regressive-transgressive trends (adapted from VAIL et al., 1984), generalized lithology of the South German Jurassic and accumulation rates. The accumulation rates are calculated for the thickest known sequences of each stage within the basin. Lithologies: black shales (black), iron-stones (heavy stippled), sands (stippled), clays (lines), reworked beds (circles), glauconite (vertical lines), missing intervals (without signature), marl-limestone rhythms ('brick' pattern).

Accumulation rates provide an additional signal to the pattern of lithological cycles. The maximum accumulation rate occurs during early regression and coincides with the accumulation of monotonous dark clays. Furthermore, these clays are geographically widespread -- throughout the entire south German basin. Accumulation rates then decline as the sediments become first sandy and later marly.

We may further note here that the temporal spacing of the maximal accumulation rates is remarkably regular, with a general periodicity of 4 to 6 Myr. This is especially interesting as VAIL & HARDENBOL (1979) have recognized a 4 Myr cycle in the Tertiary and COOPER (1977) and KAUFFMAN (1977) recognized the same 4 Myr periodicity in the Cretaceous (cf. EINSELE, this volume). A similar order of cycles (or multiples thereof) can even be infered from the Devonian transgression chart (Fig. 2).

Another remarkable point is the close temporal correspondence between the global seismic unconformities (transgressive--regressive cycles in Fig. 3) and the periodic pattern of maximum sediment accumulation. However, if one uses only the occurrence of i r o n - o o l i t i c beds and horizons for comparison with the global seismic unconformities (Fig. 4) an even better correspondence follows. Throughout the south German basin several typical "marker beds" are developed, which are either glauconitic (mainly in the upper Jurassic) or which contain scattered iron-oolites (chamosite or goethite). A characteristic feature of these beds is that they are thin (a few dcm) but also widely distributed geographically. These beds correlate very well with the condensed sections of the seismic stratigraphic chart -- phases of non-deposition during the transgression.

In addition to these thin but extensive iron-oolithic beds, horizons with thick iron-ore deposits are also developed. These further appear to correlate with the sequence of global unconformities of the seismic chart. Typical offlap patterns are here usually developed: cross-bedded iron-ores occur at the basin margin and then spread into the basin during the regression where they reach a maximal geographic distribution (BAYER & McGHEE, 1984, this volume; BAYER et al., this volume). This closely corresponds to the interpretation of VAIL, HARDENBOL & TODD (1984), who associated the global unconformities with the maximum rate of sea-level fall rather than with the maximum low stands. The eustatic chart of VAIL et al. (1984) is given in Fig. 4 and compared with HALLAM's (1978, 1981) Jurassic chart.

The stratigraphic position of other European iron ores is given in Fig. 4. They occur in similar stratigraphic positions as the south German ones, although they are also scattered more widely. To examine this apparent scatter it would be necessary to analyse each local basin configuration in more detail, as in the case of the south German basin iron oolites occur distributed throughout many sections of the Jurassic sequence as soon as one approaches the ancient coastal areas. An important aspect is that iron-oolitic beds

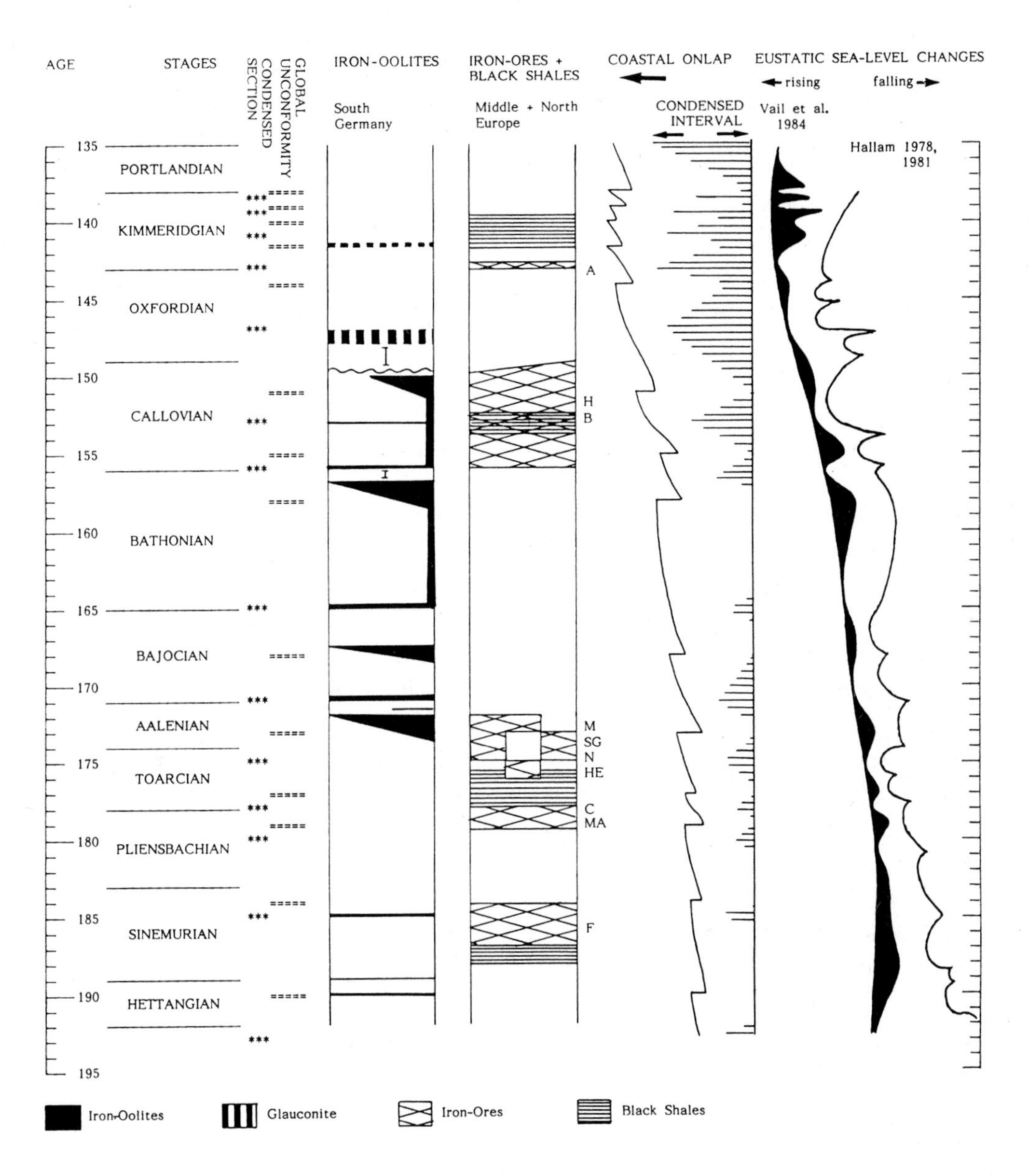

Fig. 4: The occurrence of iron-oolites in the South German Jurassic and of iron-ores and black shales in Middle and North Europe. For comparison the data from seismic stratigraphy (global unconformities, coastal onlap, condensed section) and the current global sea-level curves are given.

Iron-ores (letters): A: Abbotsbury iron-ore (England), B: Blumberg (South Germany), C: Cleveland ironstone (England), F: Frondingham Ironstone (England), H: Herznach (Switzerland), HE: Raasey ironstone (Scotland), M: Minette (Lorrain) and South German Aalenian iron-ores, Ma: Pecten ironstone (England), N: Northampton ironstone (England).

per se may not always be markers of regression. Only secondarily sorted (documented in cross-bedding) and accumulated iron-ores correspond to the regression (and here mainly to the most rapid fall of sea-level -- causing reworking and sorting; cf. EINSELE, BAYER & McGHEE, BAYER et al., this volume), i.e. it is the secondary "Lagerstätte" which is characteristic for the regressive phase. The differences between Vail's and Hallam's eustatic charts (Fig. 4) may in part be due to the different interpretation of sediments like ironstones. As in the case of Devonian black shales it is the variable area covered by the marker lithology which characterizes the transgressive-regressive cycle.

While the black shales in the Devonian example show a rather continuous cyclic pattern, the black shales of the Jurassic are only sporadic events which show no obvious relationship to the marker events of seismic stratigraphy. However, one relationship can be inferred from Fig. 4. The black shales appear in intervals where the seismic unconformities are clustered, i.e. where a higher frequency of sea-level changes is likely. In some cases the black-shales themselves show a fine rhythmic pattern and are further associated with rhythmic clay-marl occurrences -- either in stratigraphic sequence (South German Sinemurian black shales), or co-occuring regionally (black shales in the English Kimmeridgian / clay-marl rhythms in South Germany). On one hand this association would point to a primary control for clay-marl rhythms in the rock record (cf. RICKEN, this volume, for a discussion of diagenetic overprint) on the other hand the question arises as to whether the occurrence of small scale cycles is only characteristic for certain portions of larger eustatic cycles -- a point which will be discussed in the next section.

3) EUSTATIC AND TRANSGRESSIVE-REGRESSIVE CYCLES

Cyclic sedimentary records are a well established fact (Figs. 1,3), however it is often not possible to infere global sea-level changes even if data from different regions are available (Figs. 2,4). A now classical example is the interpretation of the global unconformities found in seismic stratigraphy (VAIL et al. 1977 etc.). The original sealevel interpretation of coastal onlap patterns has been criticised (for a summary of problems see SCHLANGER et al., 1981). However, in a recent paper VAIL, HARDENBOL & TODD (1984) have presented a new interpretation of their data (Fig. 4). The essential new aspect is that the rate of eustatic change is more important than the high and low sea--level stands, and further that the eustatic record is locally deformed by subsidence and variable accumulation rates. The effect of local disturbance, the dislocation of the extrema in the transgession and regression with respect to the eustatic high and low stands can be studied by rather simple mathematical models:

As a first approximation the eustatic cycle can be assumed to be a sinusoidal signal. If one considers in addition local subsidence, then a rising sea-level and a positive subsidence act together in increasing depth (D). A falling sea-level decreases depth; however, as depth is increased by subsidence, the simplest mathematical model for depth would be:

$$D = E + S; \quad D\text{: depth; } E\text{: eustatic change, } S\text{: subsidence}$$

By further assuming that S is a linear function of time ($S = at$), and that the eustatic signal is a sine function:

$$E = \sin(t) \quad \text{for a rising sea-level}$$
$$E = -\sin(t) \quad \text{for a falling sea level}$$

we can derive the simple model

$$D_{rise} = \sin(t) + at \tag{1a}$$

$$D_{fall} = -\sin(t) + at \tag{1b}$$

The two functions coincide in the extrema after a shift along the time axis, i.e. they describe a continuous periodic signal. What we would like to study now is the duration time of regressive and transgressive phases, and further, the rate of such change. To do this we have to take the derivatives of the equations (1):

$$\dot{D}_{rise} = \cos(t) + a \tag{2a}$$

$$\dot{D}_{fall} = -\cos(t) + a \tag{2b}$$

To determine the duration time of the two phases we have to find the temporal position of the extrema, i.e.

$$\begin{aligned} &\text{transgressive:} \quad \cos(t) = -a \quad \text{--->} \quad t_p > \pi/2 \\ &\text{regressive:} \quad \cos(t) = a \quad \text{--->} \quad t_p < \pi/2 \end{aligned} \tag{2c}$$

determines the values of t, and it becomes clear that the duration time of the transgression increases if the local basin is subsiding while the duration time of the regressive phase is shortend.

Another important aspect is the rate of change. To find the maximal rates we have to differentiate equations (2) once more:

$$\ddot{D}_{rise} = -\sin(t) \tag{3a}$$

$$\ddot{D}_{fall} = \sin(t) \tag{3b}$$

and in both cases the point of maximal rate of change follows from

$$\sin(t) = 0 \quad \text{--->} \quad t = 0. \tag{3c}$$

This point (the inflection point), therefore, is stable and will not be affected by linear subsidence. However, the rates of change are not stable, they are found by inserting $t = 0$ into equations (2)

$$\dot{D}_{rise} = 1 + a \quad = V_{transgression} \tag{4a}$$

$$\dot{D}_{fall} = a - 1 \quad = V_{regression} \tag{4b}$$

and it is clear that the rate of change (depth increase) is much faster during the transgression than shallowing during the regression if linear subsidence is added to the eustatic signal.

Finally we can analyse the absolute differences between the low and high stand for the two phases by inserting the results (2c) into equations (1) and the result is that the absolute distances vanish in the regressive phase with increasing subsidence, an aspect which is illustrated in Fig. 5.

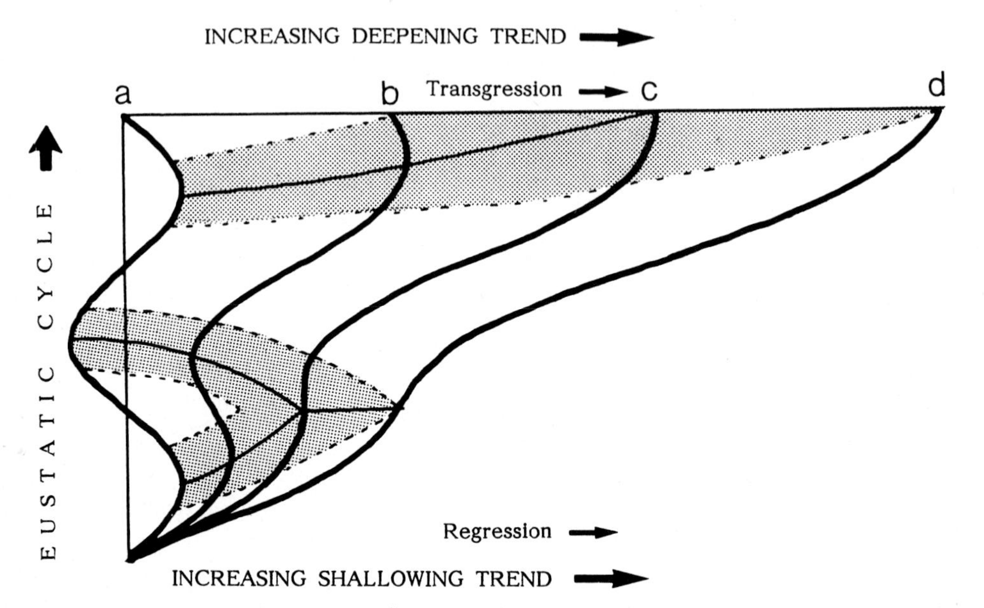

Fig. 5: Deformation of a symmetric "eustatic" signal by a local linear trend (e.g. subsidence). The curves 'a' to 'd' correspond to an increase in the linear trend. The curves can either be interpreted as a transgressive or a regressive trend for symmetry reasons (see text). The thin lines connect maxima and minima; the shaded region corresponds to periods of minimal change during which small order fluctuations most likely cause rhythmic sedimentary patterns like marl-clay sequences or clay-black shale alternations.

Fig. 5 illustrates how a sinusoidal "eustatic" signal deforms if a linear trend is added. From the brief analysis we have the following properties of this deformation:

SUBSIDING BASIN	RISING BASIN
duration time	duration time
of transgression > of regression	of transgression < of regression
rate of change	rate of change
of transgression > of regression	of transgression < of regression
absolute depth difference	absolute depth difference
during transgression > during regression	during transgression < during regression

The different properties behave in the same way, i.e. the duration time increases as the 'rate of change' increases. Because of the symmetry between a subsiding and a rising basin Fig. 5 can be interpreted as a regressive or as a transgressive trend. However, it becomes clear that the phase opposite to the general trend vanishes as this trend becomes stronger, the extrema coincide in one point causing a flat 'platform' of relative sea-level stillstand and finally give way to a continuous rise or fall of relative sea-level as the linear trend becomes dominant.

These results can easily be extented to more complex sinusoidal patterns. Assume that a sinusoidal signal with much shorter phase is superimposed onto the transgression curves of Fig. 5. Then the amplitude of these signals is reduced where the slope of the transgression curve is high, while the short signals are not disturbed when the slope of the transgression curve is zero. Thus we can infer the time intervals during which minor cyclic patterns should become visible within a larger transgressive-regressive trend (cf. BAYER et al., this volume). The expected interval is shaded in Fig. 5 and it provides a typical bifurcation pattern with two intervals during which we expect rhythmic patterns if the linear trend is small, a single interval as the linear trend increases and supression of minor signals as the linear trend becomes dominant. If Fig. 5 is considered as a regional model with gradually increasing subsidence rates then the shaded area refers to the spatio-temporal distribution of minor cycles (cf. BAYER et al., this volume) and the point where the extrema vanish corresponds with the "age of conformity" in seismic stratigraphy (VAIL, HARDENBOL & TODD, 1984).

The above discussed deformations of eustatic signals occur not only with different rates of subsidence, but a similar situation arises if short fluctuations are superimposed onto a long term trend (Fig. 6). If the differences between the phases are large, then

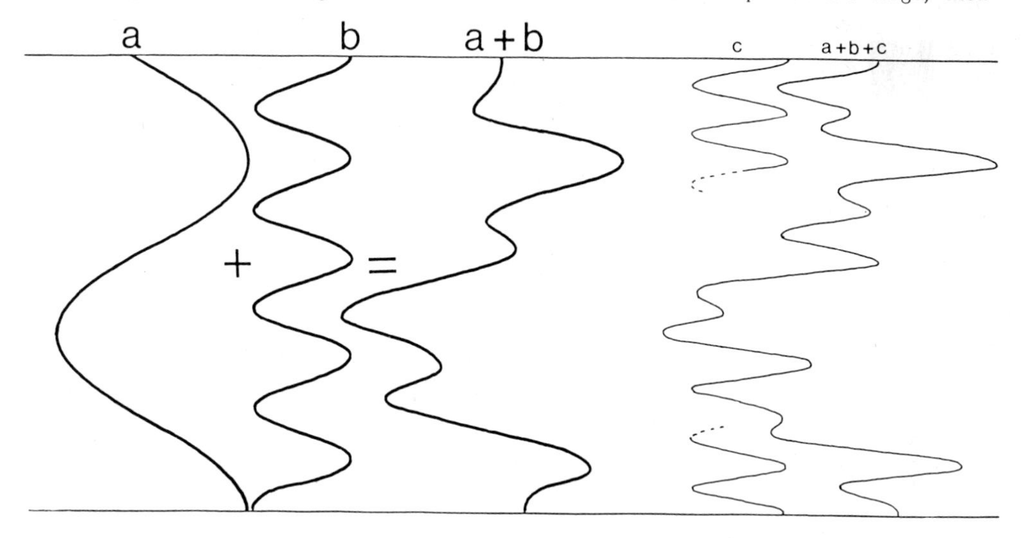

Fig. 6: Superposition of cycles of different phase-length and amplitude (synthetic Fourier diagram).

the linear model applies to this situation as a first approximation. Otherwise we have a non-linear model which, of course, is a synthetic Fourier approximation. However, the synthetic functions of Fig. 6 resemble to some extent the transgression curve of the Late Devonian (Fig. 2) although the theoretical function given in Fig. 6 (at right margin) was only chosen by chance.

CONCLUSIONS

Black shales and iron-stones are often assumed to be characteristic lithologies for the high and low stand of sea-level changes. However, our two examples from the Devonian and Jurassic illustrate that it is not the lithology *per se* but the varying area covered by the typical lithology or "marker horizon" which permits the reconstruction of transgressive-regressive cycles.

The recognition of cyclic sedimentation patterns and their subdivision in "generic sequences" becomes more objective if additional parameters are available in addition to lithology. One additional parameter -- avilable in any stratigraphic column -- is the sedimentation rate, which in the case of the Jurassic provides the most important source for the recognition of cycles.

Simple mathematical modelling allows one to specify the important parameters in more detail which may cause a divergence between 'eustatic changes' and transgressive-regressive cycles. An important result is that the position of rhythmic sequences and of minor cyclic patterns (as far as they have extrinsic causes) can be predicted from rather simple models and thus these sedimentary patterns can potentially be used for the analysis of regressive-transgressive sequences.

Comparison of field data and theoretical data with the chart of "global unconformities" show that the recent data of seismic stratigraphy provide a useful framework for the analysis of cyclic stratigraphic sequences. Of some further interest is that a cyclicity of approximately 4 Myr was found for the Jurassic, and appears likely for the Devonian as well.

Acknowledgement: Some essential ideas expressed in this paper arose during a field trip by one of us (UB) together with A. Hallam and P. Vail in the English Jurassic. The paper was prepared while one of us (GMcG) was supported by the SFB 53 (DFG) during a research stay at Tübingen, and while the other (UB) was Heisenberg Fellow of the DFG.

REFERENCES

Arthur, M.A. & Jenkyns, H.C. 1981: Phosphorites and paleoceanography. In:Berger, W.H., ed.: Ocean geochemical cycles: Oceanologica Acta, Spec. Issue, Proc. 26th Intern. Geol. Congr., 83-96.

Bayer, U. & McGhee, G.R. 1984:Iterative evolution of Middle Jurassic ammonite faunas.- Lethaia, 17.

Byers, C.W. 1977: Biofacies patterns in enxinic basins: a general model. Spec. Publ., Soc. of Econ. Paleontol. & Mineral. 25: 5-17.

Cooper, M.R. 1977: Eustacy during the Cretaceous: its implications and importance. Palaeogeogr., Palaeoclim., Palaeoecol. 22: 1-60.

Einsele, G. & Seilacher, A. eds. 1982: Cyclic and event stratification. Berlin, 536 pp.

Hallam, A. 1961: Cyclothems, transgressions and faunal changes in the Lias of North-West Europe. Transact. Edinburgh Geol. Soc. 18: 124-174.

Hallam, A. 1975: Jurassic environments. (Cambridge Univ. Press) Cambridge, 269 pp.

Hallam, A. 1978: Eustatic cycles in the Jurassic. Paleo^3 23: 1-32.

Hallam, A. 1981: A revised sea-level curve for the early Jurassic.- J. Geol. Soc. London 138: 735-743.

Hallam, A. & Bradshaw, M.J. 1979: Bituminous shales and oolithic ironstones as indicators of transgressions and regressions. Journ. Geol. Soc. London 136: 157-164.

Harland, W.B. & Herod, K.N. 1975: Glaciations through time. In:Wright, A.E. & Moseley, F. eds.:Ice ages: Ancient and modern. Liverpool (Seel House Press), 189-216.

Harland et al. 1982: A geologic time scale (Cambridge Univ. Press) Cambridge.

Kauffmann,E.G., 1977: Cretaceous facies, faunas, and paleoenvironments across the Western Interior Basin. The Mountain Geologist, 14: 75-274.

Klüpfel, W. 1917: Über die Sedimente der Flachsee im Lothringer Jura. Geol. Rdsch. 7: 98-109.

Krebs, W. 1979: Devonian basinal facies. Spec. pap. Paleontology, 23: 125-139.

Mackenzie, F.T. & Pigott, G.P. 1981: Tectonic controls of Phanerozoic sedimentary rock cycling. Geol. Soc. London Journ., 138, 183-196.

McGhee, G.R. Jr. 1982: The Frasnian-Famennian extinction event: a preliminary analysis of Appalachian marine ecosystems. Geol. Soc. Am., Spec. Paper 190: 491-500.

McGhee, G.R. Jr. & Sutton, R.G. 1981: Late Devonian marine ecology and zoogeography of the centrical Appalachian and New York. Lethaia 14: 27-43.

Richard, L.V. 1975: Correlation of the Silurian and Devonian rocks in

New York State: N.Y. State Museum and Science Serv., Map and Chart Series No. 24.

Shanmugam, G. & Moiola, R.J. 1982: Eustatic control of turbidites and winnowed turbidites. Geology, 10, 231-235.

Schlanger, W. et al. 1981: Origin and evolution of marine sedimentary sequences. In: Report of the Conference on scientific Ocean Drilling. (Joides) Washington, 37-72.

Sutton, R.G., Bowen, Z.P. & McAlester, A.L. 1970: Marine shelf environments of the upper Devonian Soyeda Group of New York. Geol. Soc. Am. Bull. 88: 2975-2992.

Vail, P.R. et al. 1977: Seismic stratigraphy and global changes of sea level. In: Seismic stratigraphy - applications to hydrocarbon exploration: AAPG Memoir 26, 49-212.

Vail, P.R. & Hardenbol, J. 1979: Sea-level changes during the Tertiary. Oceanus, 22, 71-79.

Vail, P.R., Hardenbol, J.& Todd, R.G. 1984: Jurassic unconformities, chronostratigraphy and sea-level changes from seismic stratigraphy and biostratigraphy. UCSSEPM Foundation Third Annual Research conference Proc.

SEA-LEVEL CHANGES IN THE UPPER SINEMURIAN AND PLIENSBACHIAN OF SOUTHERN GERMANY

K. Brandt
Tübingen

Abstract: The progressive onlap of marine Lower Jurassic sediments onto the Bohemian and Vindelician Landmasses in SE-Germany indicates a major transgression, which reached its deepest phase during the Toarcian. On this overall deepening trend several minor shallowing/regressive events are superimposed. It is herein shown that correlation of these shallowing-deepening cycles over relatively large areas is possible. This indicates further that most of the shallowing and deepening events in the Upper Sinemurian and Pliensbachian were caused by eustatic changes in sea level, although locally other factors (e.g. tectonics, sediment supply, marine currents) may obliterate eustatic fluctuations.

1. INTRODUCTION

In recent years several authors have recognized that the Lower Jurassic was a time of transgression in many parts of the world, a trend which led to a strong increase in the extension of shallow epicontinental seas. The peak of the Liassic transgression was reached in the Toarcian (HALLAM, 1978, 1981; AGER, 1981, and others), which was characterized in NW-Europe by widespread deposition of bituminous shales.

In Southern Germany, the progressive onlap of marine sediments on the Bohemian and Vindelician Landmasses in the SE also indicates a major transgression. Fig. 2 shows the SE-ward spread of marine facies during the Lower Jurassic. Possible causes for this transgression are eustatic rise in sea level, increasing crustal subsidence, or a decrease in the rate of net deposition accompanied by a constant rate of subsidence. PITMAN (1978) pointed out that, assuming an approximate equilibrium between subsidence and sedimentation rate, a change in the rate of rise or fall in sea level is sufficient to produce transgressions or regressions, respectively. It is also possible that a slowly rising sea level, associated with a rather high rate of net deposition, can cause shifting of the shoreline in seaward direction. In the reverse case, a high rate of erosion may offset the effect of a slowly falling sea level, causing a small transgression. Presumably, these

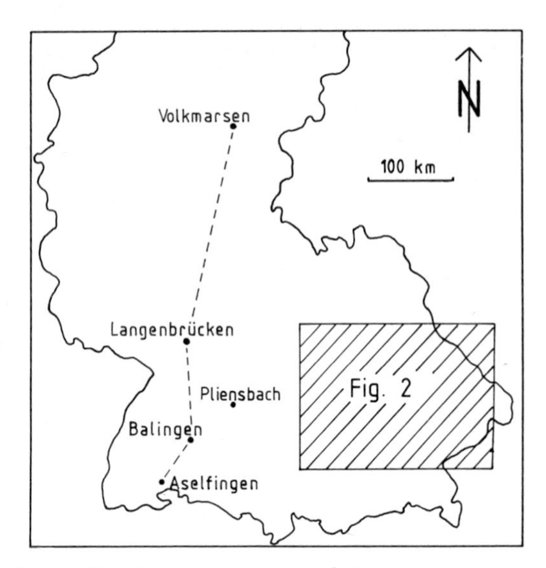

Fig. 1: Location of map Fig. 2 and sections Fig. 3.

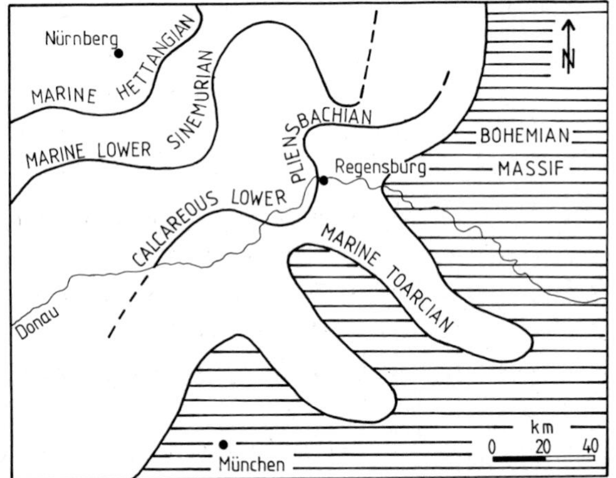

Fig. 2: Approximate SE-ward spreading of marine facies during the Lower Jurassic onto the Bohemian and Vindelician Land (from SCHMIDT-KAHLER, 1979).

mechanisms are important mainly on a local scale. However, since Liassic transgressions are tracable in several parts of the world the more probable cause here seems to be an eustatic rise in sea level. Further, several widespread reworking and condensation layers in the Upper Sinemurian and Pliensbachian of Southern Germany indicate that on the overall deepening trend several minor shallowing events were superimposed.

2. SHALLOWING AND DEEPENING EVENTS

2.1. *obtusum* Zone

In the upper part of the *obtusum* Zone, at the top of the predominantly argillaceous *stellare* Subzone, a 5-30 cm thick limestone bed is exposed which can be traced over large parts of Southern Germany. This so-called "Beta-limestone" bed is characterized by an erosive base, bored clasts, and varying content of iron oolites and bioclasts.

In most areas the sedimentary limestone bed is underlain by calcareous concretions of various sizes. In general, the concretions have flat ellipsoidal shapes and shrinkage cracks, which are mostly filled with calcite. The detrital limestone bed above the concretionary layer contains broken and abraded bioclasts (gryphaeas, belemnites, echinoderms brachiopods). Ammonites are relatively rare. Due to intense bioturbation, primary sedimentary structures are almost totally absent in the limestone bed and in the overlying marls.

A low degree of compaction and reworking indicate that the concretions at the base of the "Beta-limestone" are of early diagenetic origin, and formed at shallow depths. The ellipsoidal shape of most concretions is probably due to a higher permeability parallel to the orientation of clay minerals (HUDSON, 1978). The growth of early diagenetic calcareous concretions was probably favoured or triggered by a low rate of net deposition, resulting from repeated reworking and erosion of fine-grained sediments by storm events.

Erosion in the upper part of the *obtusum* Zone is indicated by bored and broken concretions, pelitic clasts, and gryphaeids which were probably derived from the underlying clays. The varying degree of abrasion of fossils and the the varying sediment fills of tests show that the "Beta-limestone" was not the result of a single storm event.

There are marked differences between clays deposited in a low energy environment (*stellare* Subzone) and predominantly bioclastic sediment deposited above storm-wave base. The "Beta-limestone" was probably deposited below normal wave base because rounding and sorting of bioclasts and lithoclasts is not extensive enough to indicate constant water agitation. The boundary between the limestone and the 10-20 cm thick overlying calcarenitic marls is indistinct, as is the transition between the marl and the overlying clay. This gradual decrease in grain size upsection indicates a new phase of deepening following the deposition of the "Beta-limestone".

This shallowing event in the upper part of the *obtusum* Zone can be correlated

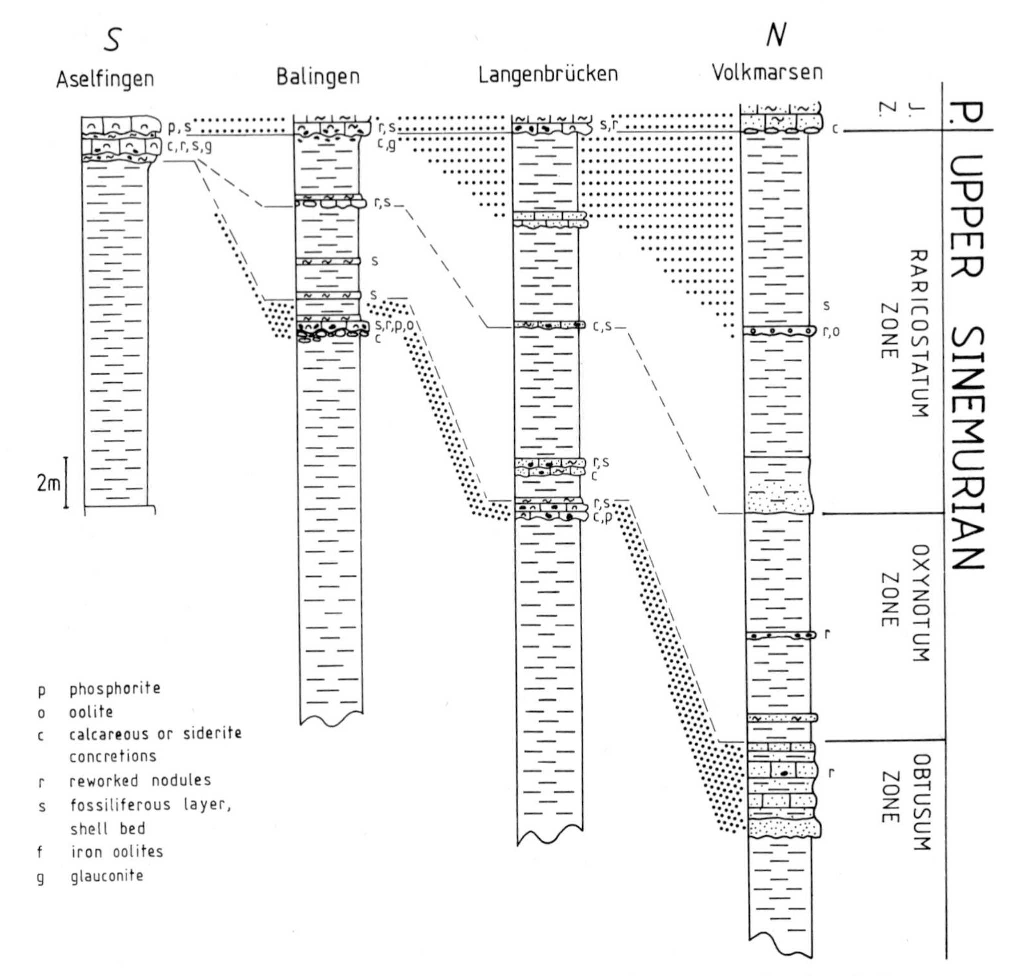

Fig. 3: Correlation of reworking layers in the Upper Sinemurian. In the shallower or more slowly subsiding areas several reworking beds are amalgamated. Sections Langenbrücken and Volkmarsen according to HOFFMANN (1964).

with reworked layers in the Upper Rhine Valley (Fig. 3), Northern Franconia, NW-Germany, Eastern France and England. In the Wutach-area (Aselfingen, in the south-eastern part of the Swabian Alb) the "Beta-limestone" is absent, but it is significant to note that reworked sediments here also lie on top of *obtusum* Zone clays.

In the Upper Rhine Valley of Langenbrücken (Fig. 3), the bioclastic layer at the top of the *obtusum* Zone is up to 3 m thick and consits of shelly, partly marly limestone beds with bored calcareous and phosphoritic concretions (HOFFMANN, 1964). In the upper part of the *obtusum* Zone in NW-Germany an up to 4 m thick calcareous sandstone is intercalated with dark, locally slightly bituminous claystones (BRAND & HOFFMANN, 1963). HALLAM (1969) described a pyritized limestone hardground in Dorset (England), which formes the top of the argillaceous *stellare* Subzone. He interpreted this hardground as being formed in shallower water than the under- and overlying shales. The hardground is overlain by marls and shales of the *raricostatum* Zone; the *oxynotum* Zone and topmost Subzone of the *obtusum* Zone are absent. A marked regression at the base of the *oxynotum* Zone has been previously recorded by DONOVAN (1983) on the northern flank of the London Platform.

2.2. *oxynotum* Zone

The *oxynotum* Zone in the Middle Swabian Alb consits of clays with 1 or 2 thin marl layers. The marls contain abundant bioclasts of bivalves, brachiopods, and echinoderms; belemnites are relatively rare. Ammonites occur mostly uncompacted or only partly compacted, indicating early diagenetic reinforcement of the shells by pyrite linings (SEILACHER et al., 1976). Sedimentary structures in the marls have been obliterated by intense bioturbation. According to SÖLL (1956), the marl beds are reworked layers although erosion and reworking were probably much weaker here than in the "Beta-limestone".

2.3. Lower *raricostatum* Zone

In most areas of the Swabian Alb, the base of the *raricostatum* Zone is formed By a shelly marl layer, similar in lithology (and presumably origin) to the marly horizons of the *oxynotum* Zone. A low sedimentation rate is suggested by

(1) local development of a nodular limestone bed at the base of the marls,

(2) high faunal density,

(3) uncompacted preservation of many ammonites (*Echioceras* sp.), and

(4) more intensive bioturbation as compared with the under- and overlying shales.

In the Franconian Alb this marl bed is not developed, but calcareous and sandy layers at the base of the *raricostatum* Zone have been reported from drillings in the Upper Rhine Valley and some areas in Northern Germany (HOFFMANN, 1964).

2.4. Upper *raricostatum/jamesoni* Zone

A major hiatus exists at the top of the Sinemurian, covering the topmost Subzones (*aplanatum* und *macdonnelli*) of the *raricostatum* Zone in most areas of SW-Germany. Sediments of the *aplanatum* Subzone have been only identified in the deeper parts of the NW-Basin (HOFFMANN, 1964), while in some other areas the whole ***raricostatum*** Zone is absent.

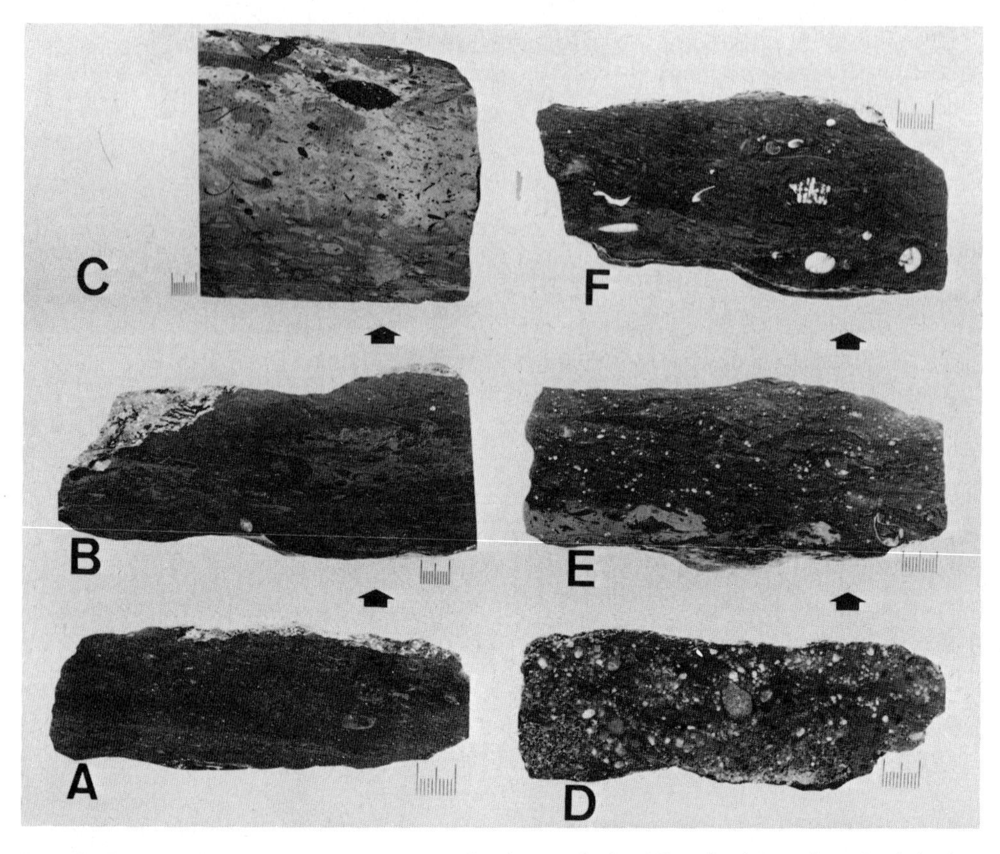

Fig. 4: Two transgressive sequences at the base of the Pliensbachian, deposited in basinal (A-C) and marginal (D-F) settings.

A-C: Fining-upward sequence characterized by a decreasing content of bioclasts, intraclasts, glauconite, and an increasing micrite content. Note preserved low-angle lamination at the base (A) and intense bioturbation in B and C (Pliensbach, Middle Swabian Alb).

D-F: Decreasing content and grainsize of terrigenous sediment (predominantly quartz) and increasing carbonate content (Haimendorf, Franconian Alb).

The base of the overlying Pliensbachian is erosive in all examined sections in Southern Germany. The sediments at the base of the *jamesoni* Zone are characterized by bored and abraded fossils, reworked and bored concretions, phosphorite nodules, glauconite, and pelitic intraclasts. The predominantly angular intraclasts are derived from the erosion of underlying clays and marls of the *raricostatum* Zone - as is indicated by reworked *echioceras* - and show that these sediments were already partially lithified.

Within the first one or two beds of the *jamesoni* Zone a distinctive fining-upward sequence can be recognized in most outcrops (Fig. 4). This gradation is caused mainly by decreasing contents of coarse bioclastic fragments and increasing micrite content. While at the base of the *jamesoni* Zone primary lamination and small-scale cross-bedding are sometimes locally preserved, sedimentary structures in the upper part of the Lower Pliensbachian are obliterated by intense bioturbation.

The observations listed above suggest a major shallowing event at the end of the Sinemurian, followed by a widespread deepening/transgressive event at the base of the Pliensbachian. This shallowing/deepening cycle can be identified not only in SW-Germany, but also in Franconia and Northern Germany. Along the southern margin of the NW-German Basin, the base of the Pliensbachian is formed by iron oolites. As in Southern Germany, there exists a widespread hiatus covering the topmost 1-4 Subzones of the Sinemurian. Transgressive marine sediments of the *jamesoni* Zone from E-Germany, Southern Sweden, and Greenland have also been described.

2.5. *ibex* and *davoei* Zone

In SW-Germany these two Ammonite Zones consist of marls with intercalated micritic limestone or marly limestone layers. They show no major signs of erosion and reworking, except in some shallower areas, suggesting that they were probably deposited below storm-wave base.

2.6. *margaritatus* Zone

Another shallowing event probably occurred in the lower part of the *margaritatus* Zone (*stokesi* and *subnodosus* Subzone). Phosphoritic nodules and ammonites (*Amaltheus* sp.), partially filled with phosphorite are common in the presumably shallower areas. Especially conspicuous are accumulations of bored and corroded belemnites, indicating prolonged exposure on the sea floor. In Weissenburg (Southern Franconian Alb), the *margaritatus* Zone is condensed within a 0.2 m thick bioclastic limestone bed (URLICHS, 1975), containing considerable glauconite and pyrite. Further to the north, this condensed sequence (with accumulations of bored fossils, glauconite, and phosphorite) covers only the lower part of the *margaritatus* Zone. In Northern Franconia a layer of phosphorite nodules has been found within the *subnodosus* Subzone (SCHIRMER, 1974). Thus it can be shown that the period of low sedimentation rate is shorter in more basinal areas than in shallower areas. This is another indicator of a shallowing event, as a slowly falling sea level will cause reworking and winnowing first in the shallower areas and, vice-versa, low energy conditions will be reestablished first in deeper environments when sea level is rising.

The relatively thick argillaceous sediments deposited during the upper *margaritatus* and lower *spinatum* Zone in Southern Germany indicate a widespread deepening/trans-

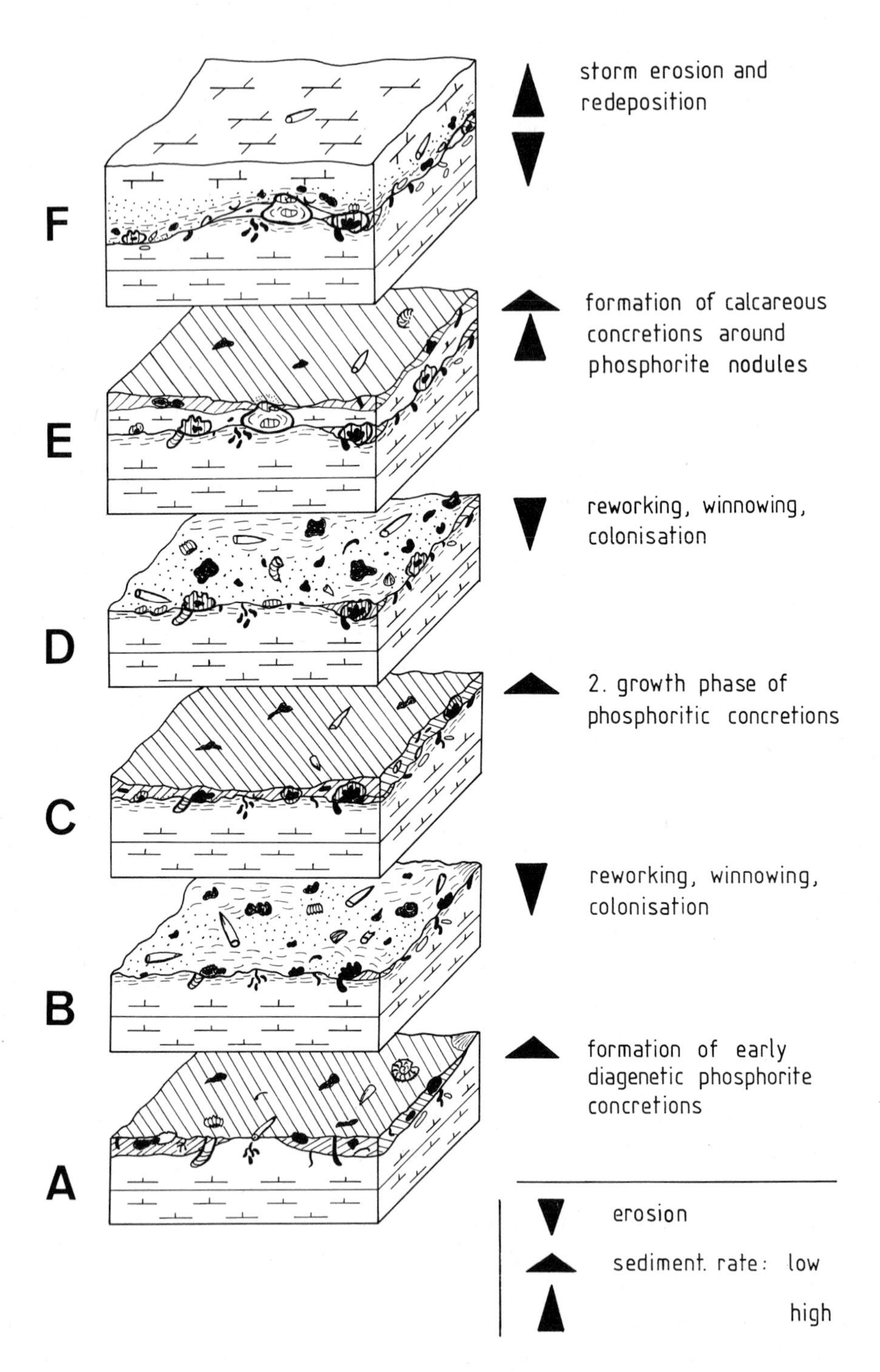

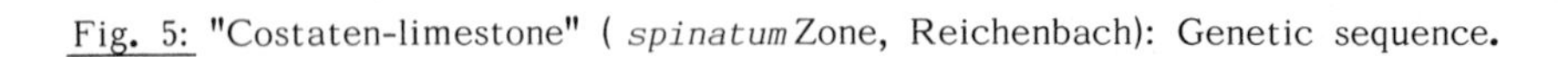

Fig. 5: "Costaten-limestone" (*spinatum* Zone, Reichenbach): Genetic sequence.

gressive event. This assumption is confirmed by transgressive Upper Pliensbachian deposits on parts of the margins of the Armorican Massif in Normandy and the London-Brabant Massif (HALLAM, 1978).

2.7. *spinatum* Zone

In SW-Germany the clays of the lower part of the *spinatum* Zone are overlain by marls and limestones. In the more basinal environments of the Midle Swabian Alb several nodular limestone beds and layers of calcareous concretions are intercalated with the marlstones. Indications of erosion and reworking are rather rare in these areas but higher fossil content and phosphorite nodules suggest a decreasing rate of deposition The nodular micritic limestones and calcareous concretions were lithified relatively early, probably only a short distance below the sediment surface, as they show little or no compaction.

In the NE-Swabian Alb, where only one limestone bed is developed in the upper *spinatum* Zone, phosphoritic concretions, intraclasts, and accumulations of broken and bored bioclasts show that sedimentation rate was probably low in shallower areas, and that several erosion and reworking events occurred. Long intervals with a low supply of terrigenous sediments resulted in the accumulation of phosphate-enriched sediment layers. Only the body chamber and parts of the inner whorls of Ammonites became sediment-filled, suggesting relatively low turbulence levels during these intervals (SEILACHER et al., 1976). Within this phosphate-rich sediment layer, early diagenetic phosphoritized concretions and moulds were formed, thus preventing later compaction. Subsequent erosion and winnowing of fine-grained sediments was probably caused by storm events. These enabled boring organisms to settle on newly formed secondary hardgrounds, provided by exposed phosphorite concretions and bioclasts. At least 3 reworking-deposition events can be identified within the sediments of the upper *spinatum* Zone of Reichenbach (Fig. 5).

In the eastern part of the NW-German Basin up to 200 m of shales of the *margaritatus* and *spinatum* Zone are overlain by fine-grained sandstones (BRAND & HOFFMANN, 1963). In E-Germany fine-grained sandstones and silts of the *spinatum* Zone are capped by coarse-grained quartz sandstones (ERNST, 1967). In parts of England the Pliensbachian is capped by an oolitic ironstone (HALLAM, 1978).

These observations indicate a shallowing phase throughout NW-Europe, reaching its lowest point during the *hawskerense* Subzone. However, compared to the shallowing events in the upper parts of the *obtusum* Zone and *raricostatum* Zone, features of reworking and condensation are less marked and mostly restricted to more marginal areas.

3. DISCUSSION AND CONCLUSIONS

The herein described thin bioclastic reworked and condensed layers -- intercalated with argillaceous or calcilutitic sequences -- were probably caused by long-term increases in turbulence. Changes in the rate and composition of sediment supply were probably of minor importance. A remarkable feature of these bioclastic layers in the Upper Sinemurian and Pliensbachian is their sharp lower boundary. The underlying sediments generally show no signs of gradually increasing water energy. In areas where subsidence is less than sedimentation rate, shallowing is usually indicated by coarsening-upward sequences (EINSELE & SEILACHER, 1982). Although part of the underlying sediments probably have been eroded, the general absence of coarsening-upward sequences below the reworked horizons suggest that sea-level changes may have been responsible for these shallowing-deepening cycles.

Other indications for eustatic control are the widespread geographic distribution of these shallowing-deepening events, and similarities with coastal onlap and offlap patterns(VAIL & TODD, 1981), as determined by seismic stratigraphy.

Due to the wide geographical distribution of the described bioclastic beds, continuous current activity or interruptions of sediment supply by changing current directions (BLOOS, 1982) cannot be the main factor controlling the formation of these beds.

Accumulations of bored and corroded bioclasts, phosphorite nodules, glauconite, hiatus concretions, and calcareous or sideritic concretions at the base of the bioclastic beds indicate long periods with very low rates of deposition, interrupted by storm events with erosion, reworking, and winnowing of fine-grained sediment. Thus, during periods of low sea level, the net rate of deposition tends to be very low. This is only true for shallow, slowly subsiding areas above storm-wave base. In more basinal environments regressions may have the reverse effect.

Small-scale sea-level fluctuations can be detected only within a certain depth range. In basinal environments, far from shallow areas, small-scale sea-level changes will probably leave no alterations in the sedimentary record. In shallow areas with very low rates of subsidence, amalgamation of severals successive events may occur (Fig. 3).

Fig. 6 tentatively shows the relationship between lithology, rates of deposition or erosion, and sea level for the Upper Sinemurian and Pliensbachian in SW-Germany. If one assumes that one Ammonite Zone in the Lias is on the average roughly equivalent to 1 million years, correlation of the lithostratigraphic units with the time-scale (VAN HINTE, 1976) shows that the predominantly argillaceous units were deposited

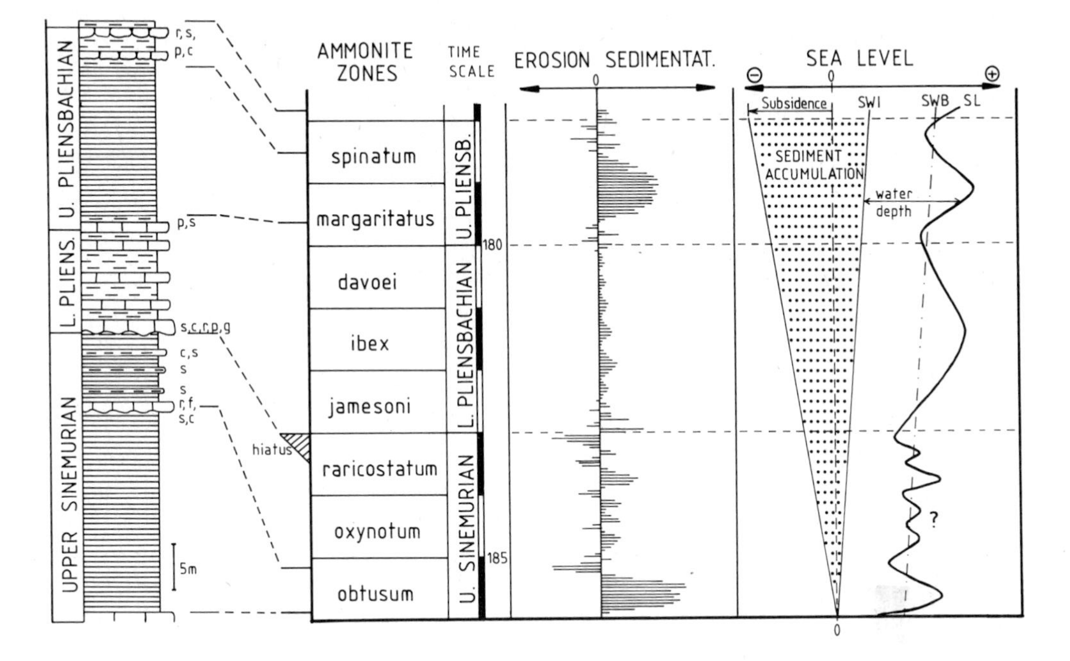

Fig. 6: Schematic section of the Middle Swabian Alb correlated with time scale (VAN HINTE, 1976). Relationship between lithology, rates of deposition, erosion, and proposed sea-level curve. Abbreviations see Fig. 3. SL: sealevel; SWB: storm-wave base; SWI: sediment-water interface.

within relatively short periods. This suggests considerably higher sedimentation rates during these intervals, and correlates well with a generally observed lower faunal density and less intensive bioturbation within the shales. The two argillaceous sequences of the *obtusum* Zone and the upper *margaritatus*/lower *spinatum* Zone were deposited below storm-wave base and presumably mark periods of high sea level. Of course it is not possible to correlate carbonate and clay content or sedimentation rate directly with sea-level, as may be deduced from Fig. 6. Thus, the marl-limestone sequence of the Lower Pliensbachian was deposited during a period of high sea level.

For the sea-level curve given in Fig. 6 it is assumed that water depth was controlled mainly by changes in sea level. Therefore, only an average rate of sediment accumulation is given in the diagram. Since no major tectonic events occured during Sinemurian and Pliensbachian time in Southern Germany and adjacent areas (ZIEGLER, 1982) it is assumed that subsidence was caused by isostatic adjustment of the crust to its loading by sediments (WATTS & RYAN, 1976).

4. COMPARISON WITH OTHER SEA-LEVEL CURVES

A comparison of the sea-level curve developed here (Fig. 6) with the eustatic curves proposed by HALLAM (1978, 1981) shows relatively good correlation of transgressive and regressive peaks, whereas a comparison with the curve proposed by VAIL & TODD (1981) shows only broad similarities. This may be due to the different methods employed by the authors. The curves proposed by HALLAM (1978, 1981) and this report are principally based on the identification of widespread shallowing-deepening events whereas the eustatic curve established by VAIL & TODD (1981) is based on the determination of coastal onlap and offlap by seismic stratigraphy.

A rather conspicuous feature of the eustatic curves of HALLAM (1981) and VAIL & TODD (1981) is the very short duration of the periods of low sea level, as compared with the periods of high sea level, an assumption which cannot be confirmed in this study.

Other contradictions between the HALLAM curve (1981) and the curve proposed in this paper concern small-scale oscillations, as well as time and scale of several transgressive-regressive peaks. The shallowing-deepening event of the upper *raricostatum/jamesoni* Zone, for example, is much more marked in Germany than is indicated in the Hallam curve. Local tectonic overprint may be the chief reason for these differences.

In conclusion, it can be stated that the close correlation of sea-level curves determined by indentification of shallowing-deepening cycles supports a eustatic interpretation, although it is essential to further correlate them with shallowing-deepening cycles from other continents.

Acknowledgements

I thank Prof. G. EINSELE for suggesting this study and for stimulating discussions. He, Dr. U. BAYER, and Prof. G. McGHEE reviewed the manuscript. This is a preliminary report of my thesis and was financially supported by the SFB 53.

REFERENCES

Ager, D.V. 1981: Major marine cycles in the Mesozoic.- J. geol. Soc. Lond., 138: 159-166.

Aigner, T. 1982: Calcareous Tempestites: Storm-dominated Stratification in Upper Muschelkalk Limestones (Middle Trias, SW-Germany).- In: Einsele, G. & Seilacher, A. (eds.): Cyclic and Event Stratification. Springer, Berlin, Heidelberg, New York: 180-198.

Bloos, G. 1982: Shell Beds in the Lower Lias of South Germany.- Facies and Origin.- In: Einsele, G. & Seilacher, A. (eds.): Cyclic and Event Stratification. Springer, Berlin, Heidelberg, New York: 221-239.

Brand, E. & Hoffmann, K. 1963: Stratigraphie und Fazies des nordwestdeutschen Jura und Bildungsbedingungen seiner Erdöllagerstätten.- Erdöl und Kohle, 16/6: 437-450.

Donovan, D.T., Horton, A. & Ivimey-Cook, H.C. 1979: The transgression of the Lower Lias over the northern flank of the London Platform.- J. geol. Soc. Lond., 136: 165-173.

Einsele, G. & Seilacher, A. 1982: Paleogeographic Significance of Tempestites and Periodites.- In: Einsele, G. & Seilacher, A. (eds.): Cyclic and Event Stratification. Springer, Berlin, Heidelberg, New York: 531-536.

Ernst, W. 1967: Die Liastongrube Grimmen. Sediment, Makrofauna und Stratigraphie.- Geologie, 16: 550-569.

Hallam, A. 1969: A pyritized limestone hardground in the Lower Jurassic of Dorset (England).- Sedimentology, 12, 231-240.

Hallam, A. 1978: Eustatic cycles in the Jurassic.- Palaeogeog., Palaeoclimatol., Palaeoecol., 23: 1-32.

Hallam, A. 1981: A revised sea-level curve for the Early Jurassic.- J. geol. Soc. Lond., 138: 735-743.

Hoffmann, K. 1964: Die Stufe des Lotharingien (Lotharingium) im Unterlias Deutschlands und allgemeine Betrachtungen über das "Lotharingien".- In: Colloque du Jurassique, Luxembourg 1962, 135-160.

Hudson, J.D. 1978: Concretions, isotopes, and the diagenetic history of the Oxford Clay (Jurassic) of central England.- Sedimentology, 25: 339-370.

Pitman, W.C. 1978: Relationship between eustacy and stratigraphic sequences of passive margins.- Geol. Soc. Am. Bull., 89: 1389-1403.

Schirmer, W. 1974: Übersicht über die Lias-Gliederung im nördlichen Vorland der Frankenalb.- Z. Deutsch. Geol. Ges., 125: 173-182.

Schmidt-Kaler, H. 1979: Geol. Karte Altmühltal, Südliche Frankenalb; Kurzerläuterungen.- Bayer. Geol. Landesamt, München.

Seilacher, A., Andalib, F., Dietl, G. & Gocht, H. 1976: Preservational history of compressed Jurassic ammonites from Southern Germany.- N. Jb. Geol. Paläont. Abh., 152/3: 307-356.

Söll, H. 1956: Stratigraphie und Ammonitenfauna des mittleren und oberen Lias beta (Lotharingien) in Mittel-Württemberg.- Geol. Jb., 72: 376-434.

Urlichs, M. 1975: Über einen Kondensationshorizont im Pliensbachian (Lias) von Franken.- Geol. Bl. NE-Bayern, 25: 29-38.

Urlichs, M. 1977: The Lower Jurassic in Southwestern Germany.- Stuttg. Beitr. Naturk., Ser. B, 24.

Vail, P.R. & Todd, R.G. 1981: Northern North Sea Jurassic unconformities, chronostratigraphy and sea-level changes from seismic stratigraphy.- In: Illing, L.V. & Hobson, G.D. (eds.): Petroleum Geology of the Continental Shelf of North-west Europe.- Heyden, London: 216-235.

Van Hinte, J.E. 1976: A Jurassic time scale.- Amer. Assoc. Petr. Geol. Bull., 60/4: 489-497.

Watts, A.B. & Ryan, W.B.F. 1976: Flexure of the Lithosphere and continental Margin Basins.- Tectonophysics, 36: 25-44.

Ziegler, P.A. 1982: Geological Atlas of Western and Central Europe.- Shell Internationale Petroleum Maatschappij B.V., The Hague, 130 pp.,40 figs.

EPICONTINENTAL MARL-LIMESTONE ALTERNATIONS: EVENT DEPOSITION AND DIAGENETIC BEDDING (UPPER JURASSIC, SOUTHWEST GERMANY)

W. Ricken
Tübingen

Abstract: Oxfordian to Kimmeridgian marl-limestone alternations of southern Germany result from diagenetic exaggeration of minor fluctuations in the primary sediment by stratiform carbonate redistribution during burial diagenesis (diagenetic bedding).

In regressive phases widespread sea-floor erosion caused channeling and submarine fan deposition (tens of km) from suspension clouds. Transgressive periods are characterized by calcilutitic turbidites shed from northern shoals. Bioturbation destroyed most of the primary sedimentary structures and diminished compositional differences.

Beds which originally had a slightly increased carbonate content became preferentially cemented to form limestone layers suffered approximately 80% compaction, providing the source for the cement by carbonate dissolution. Carbonate redistribution strongly enhanced the primary bedding rhythm, whereas weathering intensified or reduced rhythmicity.

1. INTRODUCTION

Rhythmic bedding, a conspicuous characteristic of many pelagic carbonate-marl sequences, usually results from oscillating carbonate content (SCHOLLE et al., 1983). The rhythmicity is commonly explained as being due to climatic oscillations (e.g. SEIBOLD, 1952; FISCHER & ARTHUR, 1977; FISCHER, 1980; EINSELE,1982; SCHWARZACHER & FISCHER, 1982), but diagenetic explanations have also been discussed (e.g. SUJKOWSKI, 1958; HALLAM, 1964; CAMPOS & HALLAM, 1979; EDER, 1982; WALTHER; 1983; RICKEN, i.pr.). The Upper Jurassic epicontinental marl-limestone sequences in southern Germany provide an excellent opportunity to study the phenomenon because the sequences are not tectonically disturbed, they are well exposed, and the rhythmic bedding is extensive and extraordinarily distinct. Since SEIBOLD's (1952) classic investigations the Upper Jurassic alternations have been attributed to climatic cycles (e.g. WEILER, 1957; GYGI, 1966; FREYBERG, 1966; KÖHLER, 1971; GWINNER, 1976; BAUSCH et al., 1982; EINSELE, 1982). However, new observations are not in accordance with earlier theories and make the cyclic nature questionable.

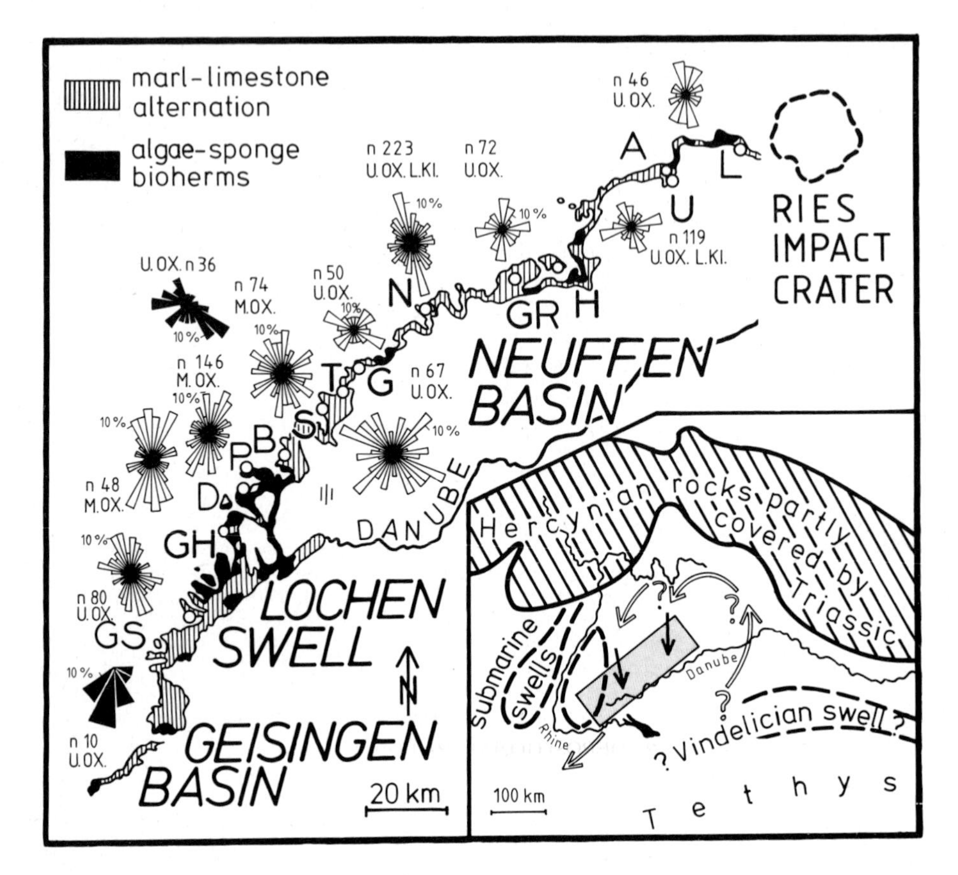

Fig. 1: Investigated area of Oxfordian to Kimmeridgian marl-limestone sequences of the Swabian Alb. Current roses show the bidirectional orientation of belemnites (white roses) and the forsetdip of turbidic ripple marks (black roses). Indicated main outcrops: H = Hausen Landslide, N = Neuffen Quarry, G = Genkingen Quarry, T = Talheim Quarry, S = Schlatt Quarry, P = Plettenberg Quarry, GH = Gosheim Quarry. Inset map shows the paleogeographic situation of the investigated area (shaded). Flow regime marked by arrows.

In this study evidence is provided for a three-step origin of marl-limestone alternations:

1.) Primary sedimentary processes produced weak compositional variations.

2.) Burial diagenesis caused stratiform redistribution of carbonate and strongly enforced the original rhythmic pattern.

3.) Differential weathering separated the rock into distinct beds of marl and limestone.

Although carbonate redistribution is found to be the most important process in generating rhythmic carbonate oscillations ("diagenetic bedding"), this study will also focus on another important factor, primary sedimentary processes.

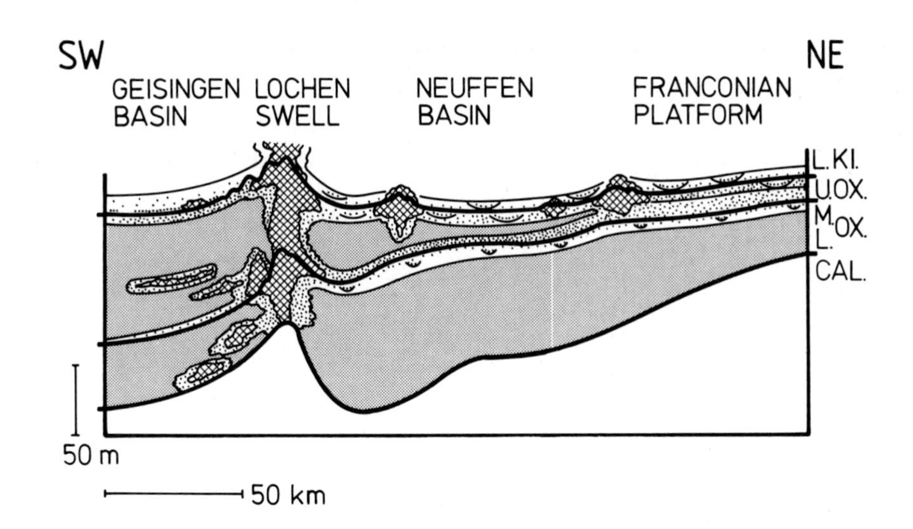

Fig. 2: Schematic cross section through early Upper Jurassic sediments (cf. Fig. 1) along outcrop line, showing sponge — algae buildups (checked), bioclastic marl-limestone alternations (stippled, with positions of channels) and turbidic marl-limestone alternations (shaded).

2. DEPOSITIONAL PATTERNS

The Oxfordian to Lower Kimmeridgian carbonates of southern Germany, which are 50 to 150 m thick, were deposited in a shallow marginal sea of the northern Tethys ocean. Submarine swells were covered with biostromes and reefs of cyanophycean algae and siliceous sponges (FLÜGEL & STEIGER, 1981), while nannoplanktonic and bioclastic mud was deposited in depressions (HILLER, 1964; GWINNER, 1976; ZEISS, 1977; ZIEGLER, 1977; Fig. 1). On the Swabian Alb, two sedimentary basins can be distinguished during the Oxfordian and Kimmeridgian ages, the northern Neuffen Basin and the southern Geisingen Basin, which were separated by a biostrome swell (Lochen Swell) which rose about 50 m over the average level of the sea bottom (Fig. 2).

With the exception of biostromes and related bioclastic taluses the basin sediments are rhythmic marl-limestone sequences. In Lower and Middle Oxfordian sequences (Malm α) the alternation is dominated by marls, whereas the Upper Oxfordian (Malm β) rocks show a masonry-like sequence of limestone beds with a uniform thicknes of 20 to 30 cm (Fig. 3A, B). Within these alternations, belemnite orientations, channels, and ripple marks imply sediment transport from north to south across the ancient shelf to the deeper Tethys (Fig. 1). Therefore, maximum facies differences can be expected in a north-south direction, but the outcrop line of the Upper Jurassic is actually perpendicular to this. Hence, the alternations seem to be more monotonous in an east-west direction than they would be in a north-south orientation. In addition, marl-limestone alternations are usually completely bioturbated by *Planolites*, *Chondrites*, *Thalassinoides*, and *Teichichnus*. Original sedimentary structures are only locally preserved due to high sedimentation rates and reduced diagenetic alteration (such as limestone beds).

In order to gather information about the depositional processes involved, investigations were conducted along three lines:

1.) Search for primary sedimentary structures.

2.) Evaluation of the amount of bioclastic grains.

3.) Bed-by-bed correlation within the entire outcrop area over a distance of 150 km.

In the Oxfordian sequence two bioclastic - lutitic cycles are present. The bioclastic phases represent sea-floor erosion during regressive trends, while the lutites were deposited during transgressions and reveal patterns of turbidic sedimentation.

2.1. Phases of Sea-Floor Erosion

Two coarsening-upward cycles (BAYER et al., this vol.) can be recognized in the early Upper Jurassic marl-limestone alternations, reaching from the Callovian/Oxfordian boundary to the end of the Middle Oxfordian, and from the Upper Oxfordian to the Oxfordian/Kimmeridgian boundary (EINSELE; this vol.). Shallowing is documented by channeling, a decreasing amount of carbonate, and increasing bioclastic content. At the boundary between the Middle and Upper Oxfordian and especially between the Upper Oxfordian and the Lower Kimmeridgian, mud pebbles and shell assemblages are seen throughout the basins, covering the entire South German Platform. These horizons are the so called "ammonite breccias" of the early stratigraphers (e.g. SCHMIDT-KALER, 1962; FREYBERG, 1966; for a discussion see RICKEN; i.pr.). The two cycles are probably related to worldwide regressive trends or at least sea level stillstands (Upper Oxfordian/Lower Kimmeridgian: HALLAM, 1977; VAIL & TODD, 1981; Middle/Upper Oxfordian: HALLAM, 1977).

Fig. 3: A, B: Marl-limestone alternations and position of sections (Figs. 12, 2O) and channels (a - h) in the Neuffen Quarry. Channel (f) is transsected by a fault system. Channels are restricted to the boundaries of the Middle to Upper Oxfordian and Upper Oxfordian to Lower Kimmeridgian deposits. C: Channel system (d, e) of early Lower Kimmeridgian age. (d) Point-bar bedding from right to left (W to E). Flow direction vertical into the plane of the photograph. Neuffen Quarry. D: Lateral view of channel system (a - c) (see Figs. 4, 6). The channel began in the late Upper Oxfordian (a) and shifted continuously eastward to (c), where only the right channel side is visible. Eastward shifting was disrupted by slumping of the left channel wall at (b). Flow direction was into the picture plane. The Oxfordian/Kimmeridgian boundary lies in the vegetation covered floor between (a) and (b). Neuffen Quarry.

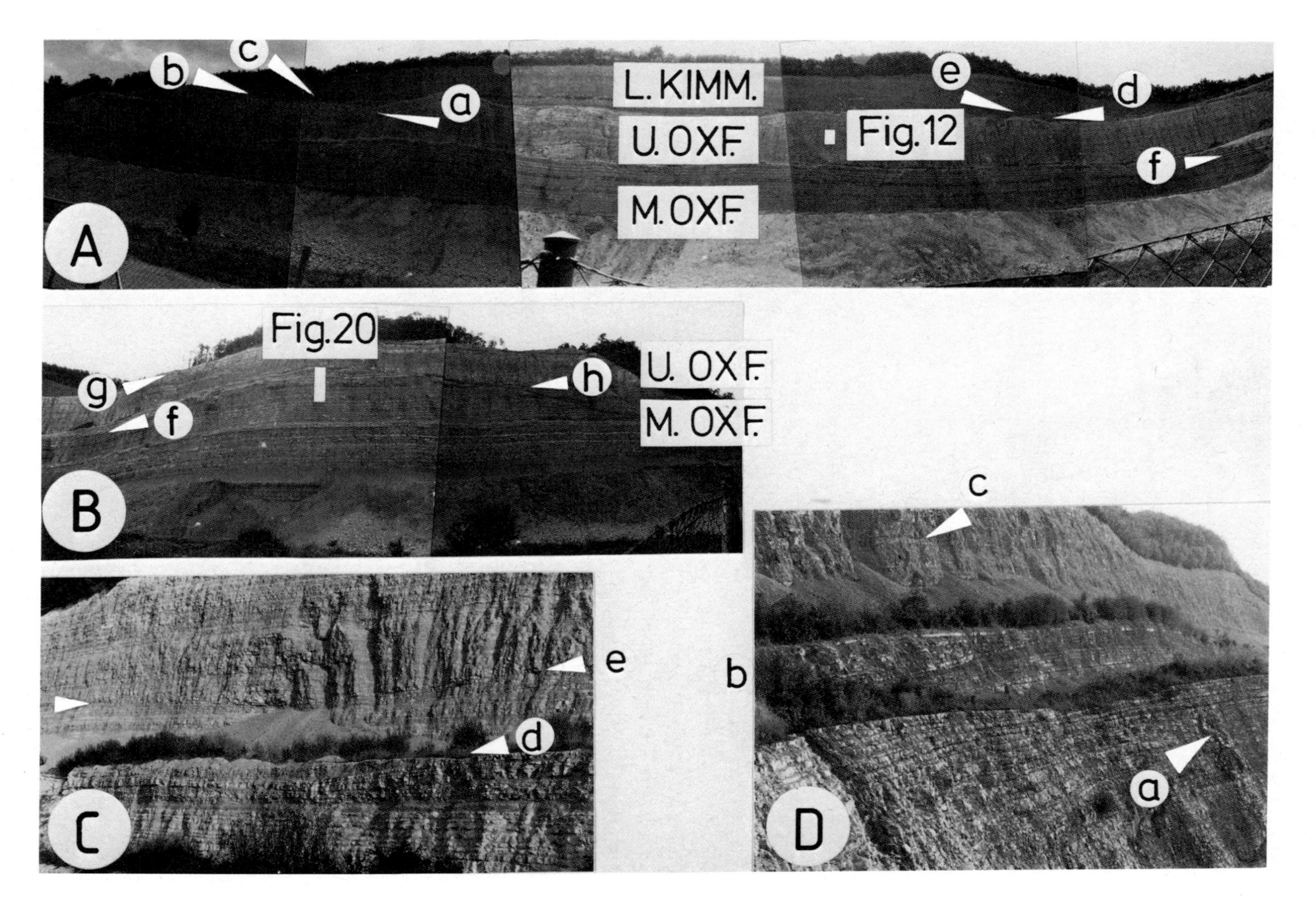
A
b
c
a
L. KIMM.
U. OXF.
M. OXF.
Fig. 12
e
d
f
B
Fig.20
g
f
h
U. OXF.
M. OXF.
C
e
d
b
D
c
a

During regressive periods bioclastic detritus was spread over the basins of the Swabian Alb through a time interval of several 10^5 to 10^6 years (Fig. 2) and persisting channels became eroded, particularly in the Neuffen Basin. Some of the channels were rather small and reached only one meter in width; most of them, however, were 0.5 to several meters deep and over 50 m wide. The distribution of channels is estimated to be about two per kilometer (Fig. 3A, B).

Larger channels commonly show a complicated internal structure due to multiple erosional and sedimentary events (Fig. 3C, D). During the Late Upper Oxfordian and early Lower Kimmeridgian ages the channel systems a - c and e - d (Neuffen Quarry, Fig. 3A) shifted several tens of meters to the east, while deposition was still going on. KENNEDY & ODIN (1982) assume that the Oxfordian age lastet 10 million years (150 m) and that the Kimmeridgian lasted 5 million years (200 m). Therefore, channel system a - c must have existed for 2.3 million years (10 m during the Upper Oxfordian and 25 m during the Lower Kimmeridgian). During this time interval the amount of lateral shifting was about 100 m, while the amount of sediment (compacted) accumulated to a thickness of 35 m (Fig. 4). Lateral shifting of these Upper Jurassic channels was slower by a factor 10^5 to 10^6 (since they were located in much deeper water) than the migration of some present subtidal channels, which can shift some tens of meters per year (REINECK, 1980). Therefore, unlike the rapidly migrating channels of the modern shallow sea, for example North Sea, the point-bars of the Upper Jurassic channels grew slowly upwards during deposition and lateral shifting. This caused a progradational pattern of superposed point-bar layers (Fig. 5), which also appears in marl-limestone alternations. During unidirectional shifting, the erosive channel side had sometimes become so steep that instability caused slumping (Fig. 3D; Fig. 4). Throughout the investigated interval (Middle Oxfordian to Lower Kimmeridgian age) no significant change in belemnite orientation can be recognized in the Neuffen Quarry, indicating a constant flow direction. Therefore, the continuous lateral shift of the channels may be due to a slow migration of meanders, which seems to be a common pattern with slightly incised submarine channels (DAMUTH et al., 1983).

Channel fills consist of muddy intraclasts and debris (lag-deposits). Channels of the Middle and Upper Oxfordian ages chiefly contain articulated crinoid fragments, whereas during the Lower Kimmeridgian age cephalopods were the main bioclastic component. This change in bioclastic content may occur within one channel system (e.g. channel a - e, Fig. 3), implying biological changes, rather than fluctuations in flow conditions. Composition and structure of the lag deposits differ slightly from bed to bed. However, biostromal detritus was rarely found, suggesting that the channels of the central Neuffen Basin only occasionally had source areas in biostromes. Channels served mainly as a transport path of winnowed basin sediments.

Fig. 4: Detailed view of the bedding-pattern of the channel system (a - b) (Neuffen Quarry, Fig. 3 D). The channel shows continuous eastward shifting. Slumping of the eastern side of the channel caused a short period of filling. Locations of the various shifting channel centers are marked with a white circle. Note that most of the erosional contacts are marl beds (white) and not limestone layers (black).

The sedimentation pattern of the slowly migrating Upper Jurassic channels (see Fig. 5B) clearly indicates a relatively deep water origin, while the general environment implies shelf conditions (shelf-channels, SHEPARD & DILL, 1966). As discussed later, water depth may have been of the order of 100 m. Geologic examples of shelf channels are rare, because geologists usually tend to pay more attention to phenomena which are more spectacular than slightly inclined unconformities. The studies of KENNEDY & JUIGNET (1974), JEDLETZKY (1975), and HYDEN (1980), however, indicate the general importance of these weakly incised shelf channels for submarine sediment transport (EINSELE, this vol.).

Within the channel facies, marl-limestone alternations can be clearly identified as representing events. However, the channel deposits grade laterally into normal marl-limestone alternations, which show perfectly rhythmic bedding. The interfingering of channel and interchannel areas can be used to establish the depositional process of the rhythmic marl-limestone alternations. This was done by tracing single beds of the channel systems a - c, d, and f (Fig. 3) from the channel to the interchannel domains using a rope ladder, with the following results (Fig. 6, 7, 8):

1.) Most marly layers in the channel itself represent erosional events, as indicated by the presence of erosional detritus and sedimentary discordances within the channels. Limestone layers are often truncated by bioclastic marl beds (Fig. 4, Fig. 6). Erosional detritus is brought by bioturbation into underlaying micritic limestone layers, causing an irregular pattern of burrows filled with clastic grains (Fig. 8A). The present marl layers are depleted of carbonate by dissolution, and compaction has been prevalent during burial diagenesis. Therefore, delicate sedimentary structures are not preserved.

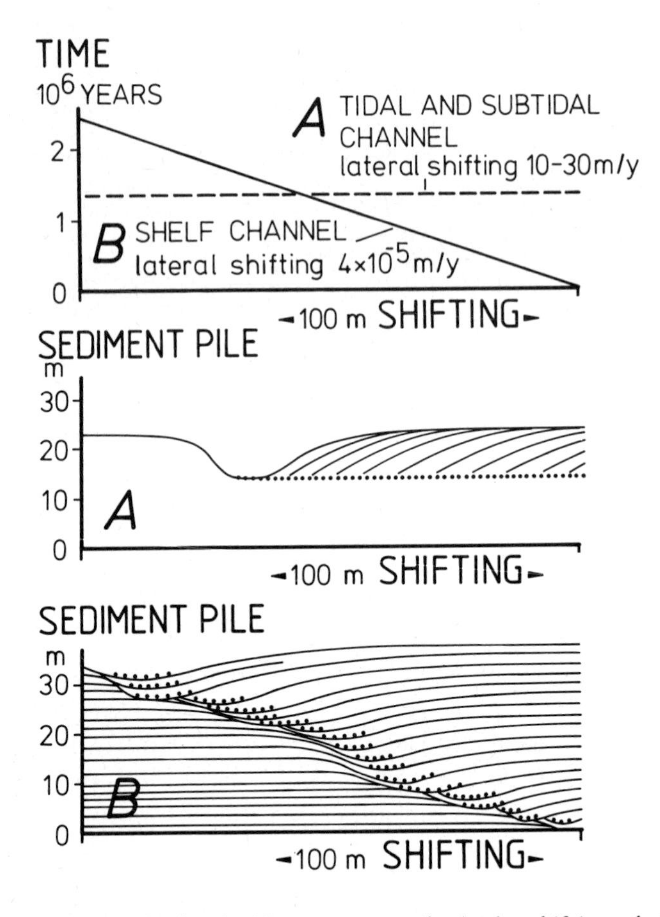

Fig. 5: Interpretation of the bedding pattern of slowly shifting shelf channels (B), which can be derived from the bedding pattern of rapidly migrating subtidal channels (A). The upper diagram shows the relationship between shifting and time of coastal (A) and shelf channels (B). In the lower diagrams time is converted to a sediment column (assuming a constant sedimentation rate). Rapidly migrating channels (A) show the familiar situation of lateral, prograding point-bar bedding with lag deposits (dots) parallel to the sediment surface. In contrast, very slowly migrating channels (B) grow upwards during shifting and generate a bedding pattern of laterally and vertically prograding point-bars.

2.) weak erosion within the channels produced slightly graded, marly lag deposits of broken shells and irregular muddy intraclasts, which imply erosion of relatively soft sediment ("ammonite breccia") (Fig. 8B). In the interchannel domain, lag deposits are replaced by normal marl layers, containing thinly scattered detritus, articulated crinoid stem fragments, belemnites, and other shell fragments (Fig. 7, 8C).

3.) In highly deepened channels true lag sediments are often missing, although the channel marl layers do contain a little bioclastic detritus. However, this is practical-

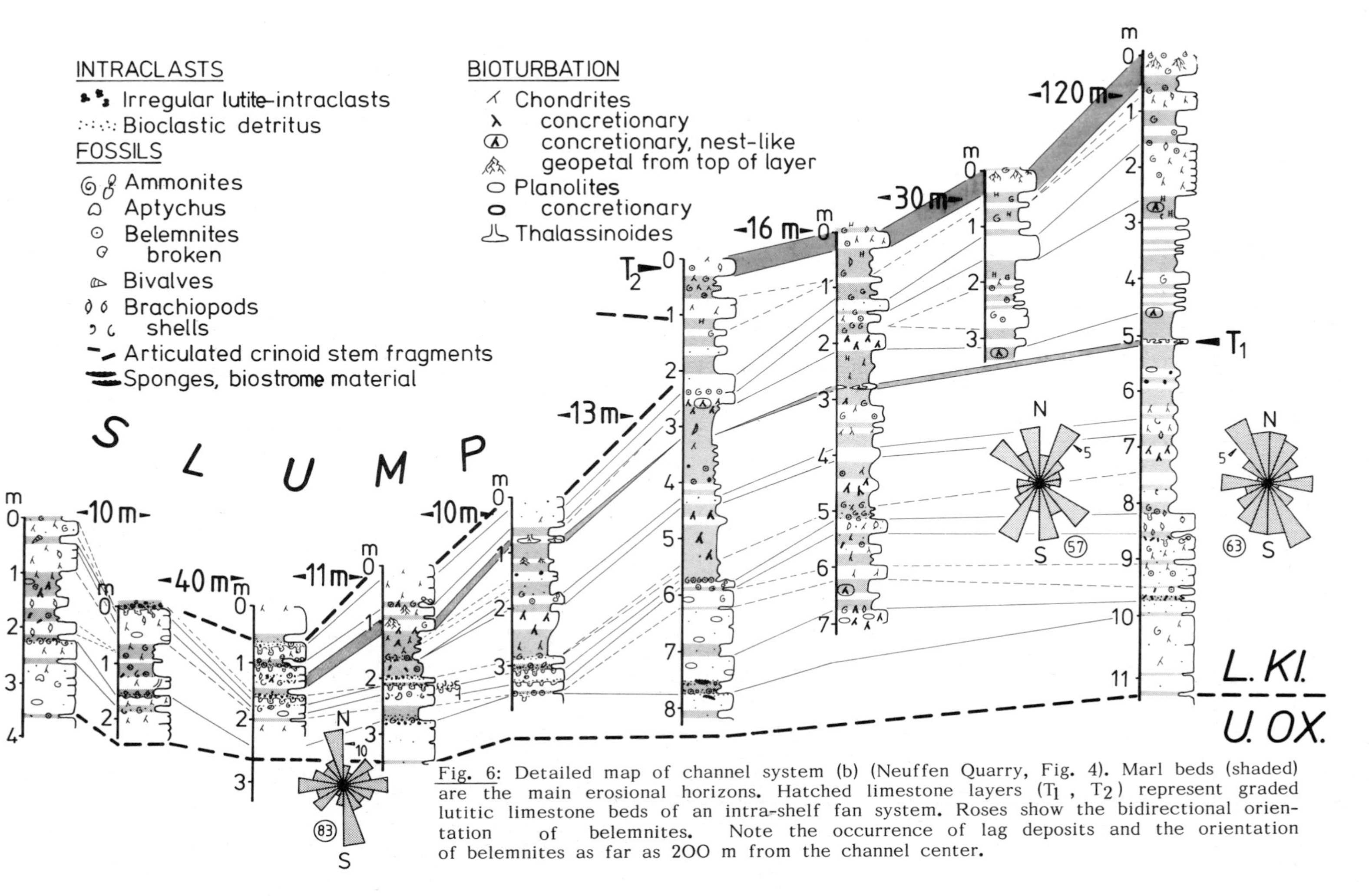

Fig. 6: Detailed map of channel system (b) (Neuffen Quarry, Fig. 4). Marl beds (shaded) are the main erosional horizons. Hatched limestone layers (T_1, T_2) represent graded lutitic limestone beds of an intra-shelf fan system. Roses show the bidirectional orientation of belemnites. Note the occurrence of lag deposits and the orientation of belemnites as far as 200 m from the channel center.

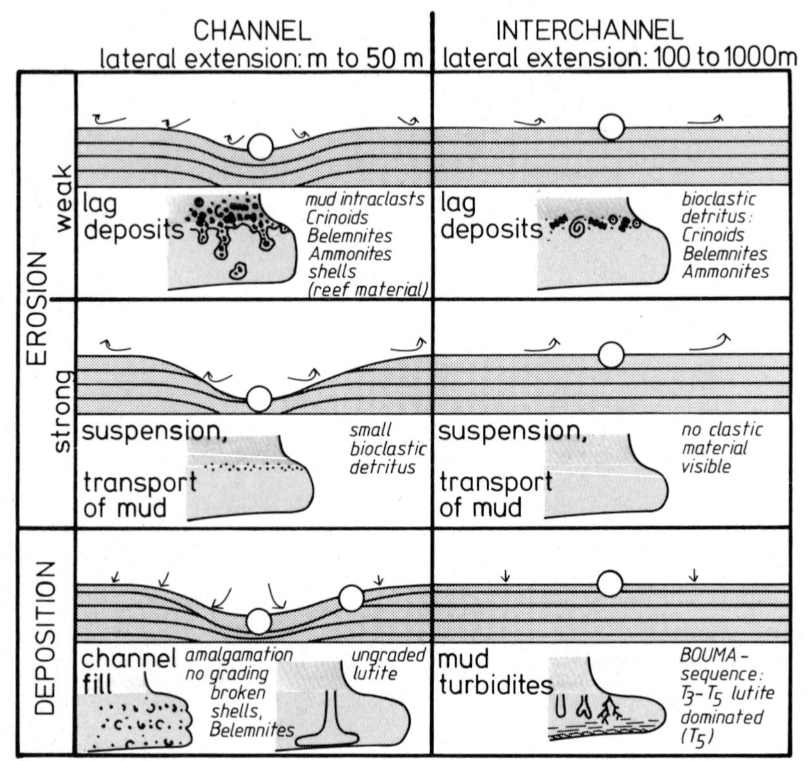

Fig. 7: Simplified model of sedimentary processes and related structures of the channel and the interchannel facies during conditions of weak erosion, strong erosion, and deposition.

Fig. 8: A) Limestone layer of channel system (f) shows erosional contact of a marl bed at the layer top. Bioclastic detritus is brought into the limestone layer by bioturbation. Neuffen Quarry, Middle Oxfordian. B) Detailed view of the same marl--lime - stone contact shown in Fig.8A. Intraclasts are irregular and belemnite shells are broken by erosional event(s). Neuffen Quarry, Middle Oxfordian. C) Upper Oxfordian channel fill (Neuffen Quarry, channel a) which consists of crinoid fragments and some sponges (arrows). Crinoid stem fragments are often articulated. Marl seams in the upper part are due to pressure solution. D) *Thalassinoid* burrow of the graded limestone layer (T_1) of Fig. 6 (comp. Fig. 8 E, F, G). Hausen Landslide, Lower Kimmeridge. E, F, G) Graded lutitic limestone layer (T_1) of Fig. 6 showing BOUMA-PIPER subdivisions D (laminated silt), E_1 (laminated calcareous mud) and E_2 (graded calcareous mud). Fig. 8 E shows the situation at the channel center (Neuffen Quarry), where grading is poorly developed. Fig. 8 F and G are from interchannel areas from the Neuffen Quarry (F) and from the northeastward lying Hausen Landslide (G), showing the same bed 32 km away. Sedimentary structures of G are a little more proximal compared to F. Lower Kimmeridgian.

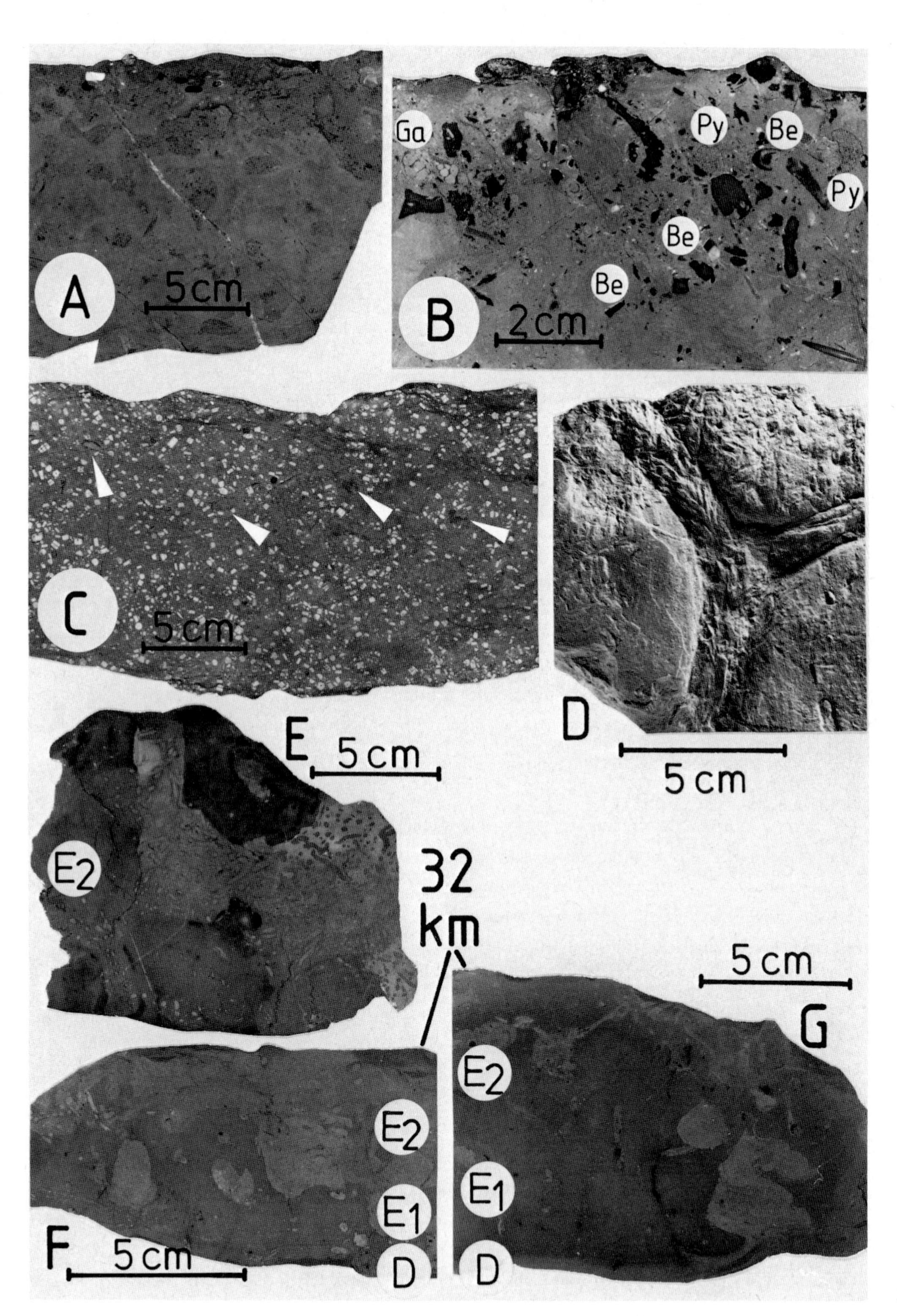
A
5 cm
Ga
Py
Be
Py
Be
Be
B
2 cm
C
5 cm
D
5 cm
E
5 cm
E2
32
km
5 cm
G
E2
E1
D
E2
E1
D
F
5 cm

ly nonexistent in the marl layers of the normal alternations. Hence, erosion was so strong that calcareous mud was brought into suspension and removed (Fig. 7). 4.) Occasionally, layers composed of slightly graded lutite overlie the channel and interchannel areas (Fig. 6, 7). The layers thicken in the channel areas and reveal geopetal bioturbation in the top of the layers (Fig. 8D -G). The layers show little to no grading within the channels, but poorly bioturbated layers exhibit the BOUMA-PIPER subdivisions D, E_1, and E_2 in the interchannel domain (laminated calcareous silt, laminated calcareous mud, graded calcareous mud). They are interpreted as having been deposited from suspension clouds. The thin basal sublayer D consists of bioclastic grains and little quartz, whereas sublayers $E_{1,2}$ reveal nannoplanktonic mud, pellets and other small grains.

The graded beds do not show any reworking by wave action, therefore the suspension clouds must have been transported below the storm wave base. This suspended material reached a considerable amount. In the early Lower Kimmeridgian, a single, graded limestone layer can be accurately traced over a minimum distance of 32 km. This layer contains 750 x 10^6 m^3 of eroded material, assuming a circular distribution and an average thickness of 7 cm (Fig. 8F, G). Comparison of the channel areas indicates that marl layers of the regressive phases are erosional in origin. Single bioclastic marl layers in the early Upper Oxfordian can be traced about 100 km within the Neuffen Basin. The occurrence of articulated crinoid stem fragments together with broken shells, especially belemnites, indicate that erosion was rapid but short term. Storm processes were very likely dominant in generating the sedimentary patterns discussed above.

Larger biostrome complexes were significantly affected by turbulence. Growth of algae-sponge reefs at the Lochen Swell was frequently interrupted by erosion; the resulting erosional detritus was later transformed into stylolitic marl layers. On both sides of the Lochen Swell, a thick bioclastic talus developed. Fig. 9 schematically shows the situation during the regressive phases at a water depth of roughly 100 m. The storm wave base touched a reef complex and larger areas of the basin floor. The eroded material was transported in sinuous and bifurcating channels below the storm wave base, leading distally to huge fan-like depositions of graded lutitic beds. Some of the channels had source areas originating in biostrome buildups, while others located within the basin.

2.2. Phases of Turbidic Deposition

The lower and middle sections of the coarsening-upward cycles mentioned above, reveal relict features of turbidity fan sedimentation, especially in the Gei-

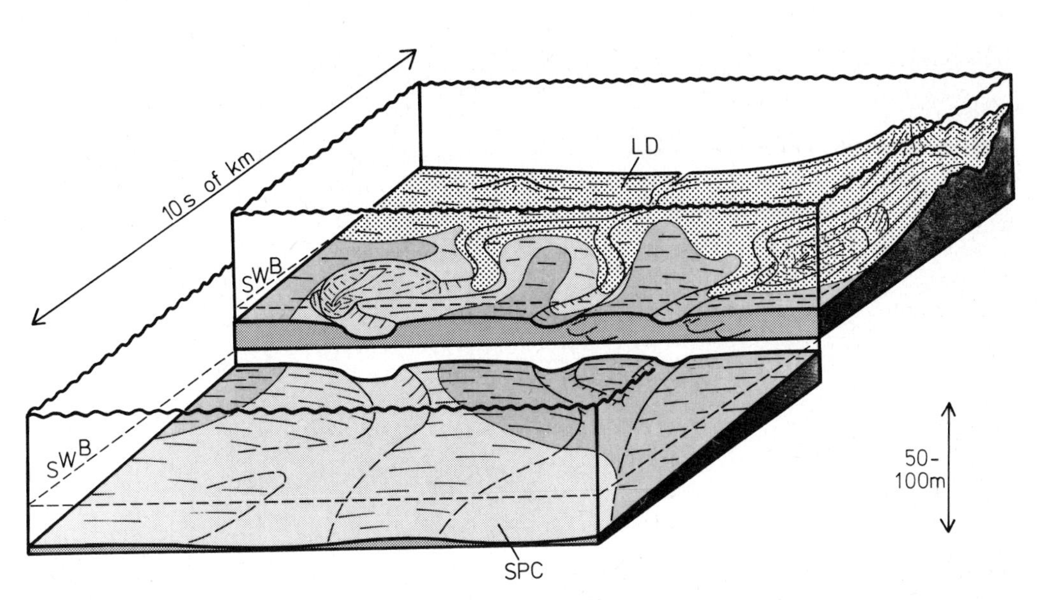

Fig. 9: Sedimentary processes during regressive phases (schematic). The storm wave base (SWB) touches reef bodies as well as the basin floor, generating lag deposits (LD, stippled background) and lobes of suspension clouds (SPC, lightly shaded), which were transported below the storm wave base in the sinuous channels.

singen Basin. The term "turbidite" is used in this paper as a simple descriptive expression for graded lutitic beds. Genetic interpretations are avoided, because in nature all transitions can exist between storm generated suspension clouds (tempestites), which are transported below the storm wave base, and "real" turbidites caused by submarine slumping (AIGNER, 1982; SEILACHER, 1982; DOTT, 1983).

Turbidite sequences are developed as typical marl-limestone alternations, which closely resemble the rhythmic alternations of the regressive phases. Turbidic marl-limestone alternations often overlay submerged reefs and biostromes and their bioclastic taluses, showing the general transgressive character of the turbidic marl-limestone sequences (Fig. 10 B, C). Frequently, turbidites are found in the lowermost parts of those alternations, where bioturbation has been low, presumably due to rapid deposition. Due to increasing bioturbation, turbidic depositional structures tend to disappear gradually towards the higher parts of the alternation. Sometimes turbidic marl-limestone alternations show low angle laminations (Fig. 10 D) and channeling (Fig. 10 E). In the Plettenberg Quarry, depressions between biostromal buildups of the Lochen Swell were filled with turbidites, which strongly leveled the preexisting topography of the reefs (Fig. 10 C). Measurements of the dips of the ripples at the Lochen Swell and in the Geisingen Basin show sediment transport from north to south (Fig. 1). This coincides with some Oxfordian siliclastic turbidites from eastern Switzerland (GYGI, 1969). Turbidites consist of calcareous lutite and small calcareous grains. Therefore, as in

A
20 cm
T
R
B
TURBIDITES
REEF TALUS
C
D
20cm
E

other finegrained turbidites (e.g. KELTS & ARTHUER, 1981), sedimentary structures were originally scarce. The calcareous turbidites can be described by the BOUMA-PIPER (1962; 1978) subdivisions B, C, D, E_1, E_2, and E_3. Relatively proximal turbidites overlying reefs and biostromes show the typical BOUMA - sequence (B - E). Lutite interval E comprises only 1/3 of the turbidite (Fig. 11 A); however, over 2/3 of the turbidite commonly consists of interval E, showing slight changes from laminated (E_1) to graded (E_2) and to non-graded mud (E_3) (Fig. 11 B, C). Distal layers consist entirely of weakly graded lutite, and their turbidic origin can be inferred only from the bioturbation of the top of the layer (Fig. 11 D).

Several layers (up to three) may be comprised in a single limestone layer 20 to 30 cm thick. As will be discussed below, the original carbonate distribution within the turbidites is strongly altered due to carbonate dissolution and cementation processes (see Fig. 16). The base and top of the turbidites and of the interturbidic units were favored sites of carbonate dissolution during burial diagenesis. Therefore, turbidic limestone beds usually show diffuse contacts. Sole marks, like groove casts, are quite rare (MEISCHNER, 1964), although not completely absent (Fig. 10 A).

Bioturbation largely destroys the inconspicuous primary sedimentary structures of the lutitic turbidites, which originally consisted of only an indistinct grading or fine lamination. The elimination of the primary structures and simultaneous formation of "normal" marl-limestone alternations can be demonstrated in all stages (Fig. 11 D, F, G). This may be the reason why only DAVAUD & LOMBARD (1975) recognized the turbidic character of Oxfordian marl-limestone alternations in the French Jurassic.

Turbidites occured roughly every 1000 - 3000 years. They may be partly caused by redeposition of mud by sea floor erosion during storm events. It is possible that during transgressions the erosional facies retreated northward towards the coast, providing the source of calcareous mud of the turbidites. This scenario however, cannot be verified since Upper Jurassic coastal sediments are now completely eroded. Another source of turbidites is mass movement from taluses of biostrome swells. Thus, GWINNER (1962) reported slumping and turbidic sediment mobilization from several Oxfordian and especially Portlandian biostromes of southwest Germany.

Fig. 10: A) Base of a calcilutitic turbidite reveals groove cast and belemnite orientation (Plettenberg Quarry, Upper Oxfordian). B) Turbidites (T) located within the Geisingen Basin were shed over flat biostromes and related bioclastic facies (R). Only in the lower 10 m of (T) are primary sedimentary structures preserved. Geisingen Quarry, Upper Oxfordian. C) Reef talus of the Lochen Swell is overlain by a turbidic marl-limestone alternation, which filled smoothed over the reef topography. Plettenberg Quarry, Upper Oxfordian. D) Weathering in a turbidic marl-limestone alternation shows slightly inclined laminations. Plettenberg Quarry, Upper Oxfordian. E) Channeling within turbidic marl-limestone alternations. Plettenberg Quarry, Upper Oxfordian.

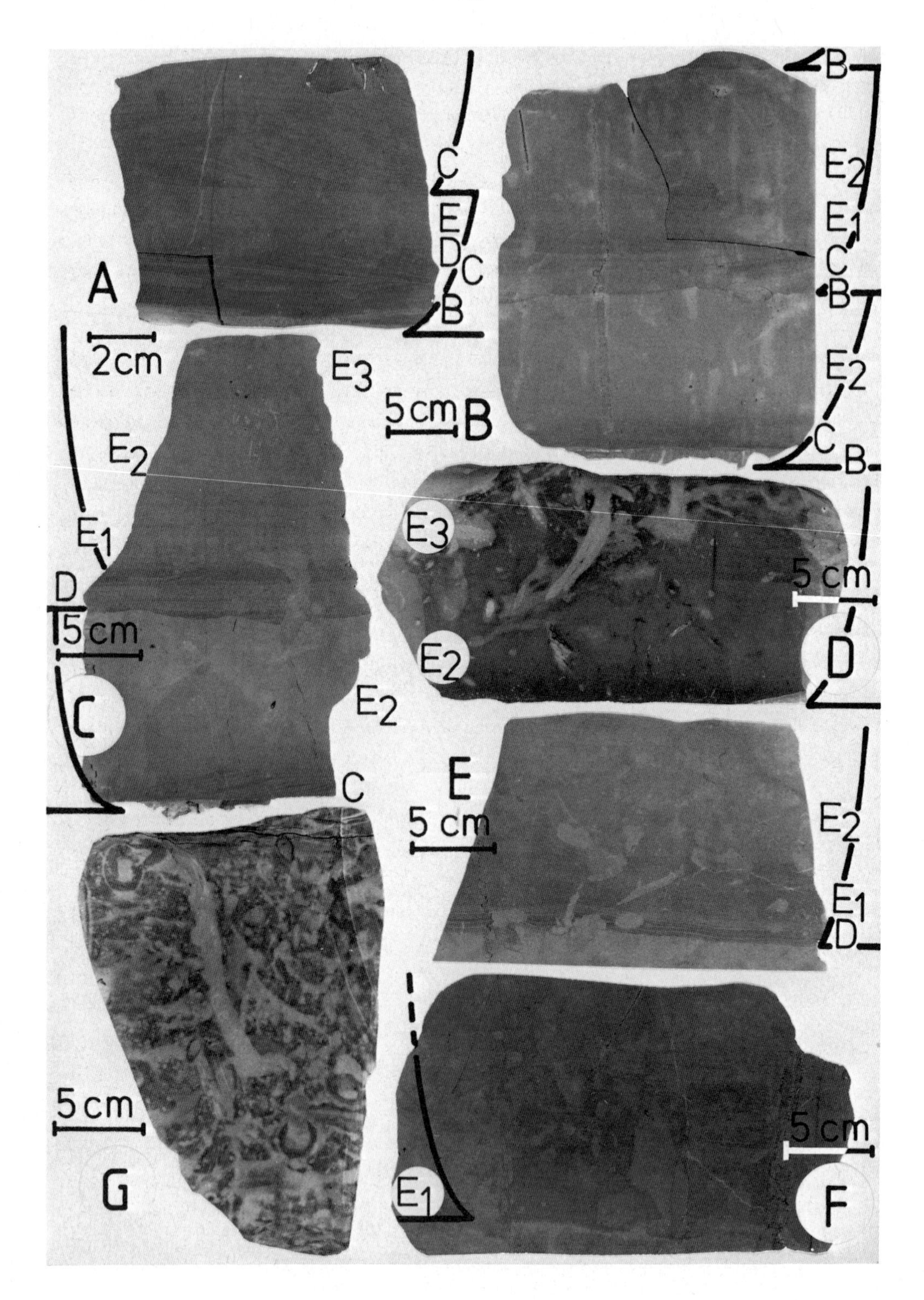
A
2cm
C
E
D
C
B
B
E2
E1
C
B
E2
C
B
5cm
E3
E2
E1
D
5 cm
C
E2
C
E3
E2
5 cm
D
E
5 cm
E2
E1
D
5 cm
G
E1
5 cm
F

2.3. Basin Development

During the Upper Jurassic sea level rise (HALAM, 1977; VAIL & TODD, 1981) the sea expanded from the south onto an island that separated the Tethys from the boreal regions (Fig. 1). South of this island GWINNER (1976) suggest the existence of a reef platform, which shifted southward during later regressions of the Middle and Upper Kimmeridgian, leading to the formation of huge biostromes all over the Swabian Alb. The Lochen Swell supposedly belonged to a southward-striking branch of this platform. During minor sea level rises in the Lower and Middle Oxfordian, turbidites were shed southward from the northern platform, preferably into the deeper Geisingen Basin. In this basin turbidite sequences are interrupted several times by small algae-sponge buildups, which may have been caused by short term shallowing due to rapid depositional filling. In the Neuffen Basin a thick, uniform sedimentary sequence was deposited during the Middle Oxfordian, but not enough primary structures have been found to allow a detailed environmental interpretation. The deposition of carbonate lutites came to an end with the regressive phases of the Middle to Upper Oxfordian, where bioclastic grains were admixed and sea–floor erosion caused channeling. A second sea level rise during the Upper Oxfordian again caused turbidic deposition, which is best documented in the Geisingen Basin. Again, the sequence was interrupted by a horizon of biostromes spreading within the Geisingen Basin (Niveau Knollenschicht, GYGI, 1969). Sedimentary infill and levelling of the topography continued during the late Upper Oxfordian and the early Lower Kimmeridgian, when widespread erosion occurred throughout the South German Basin. Especially in the Neuffen Basin a channel fan system developed during this regressive phase and caused widespread erosion and redeposition.

Detailed bed by bed correlations (RICKEN, i.pr.) show that each basin of the lower Upper Jurassic of the Swabian Alb has its own bedding rhythm. This coincides with the well surveyed marl-limestone alternations of Franconia, where the Oxfordian sequences of the "Südalb" and "Feuerstein" alternation are different from the "Hart-

Fig. 11: A-C) Proximal - distal trends in calcilutitic turbidites of the Upper Oxfordian sequence. Fig. 11 A shows relatively proximal turbidites, which are composed of nearly complete BOUMA - sequences and reveal that most deposition occured in the rippled sublayer C (Geisingen Quarry). Fig. 11 B and Fig. 11 C reveal more common features of Oxfordian turbidites of the Geisingen Basin. The poorly graded lutitic interval E is the dominant layer within the turbidites. Limestone beds are composed of several turbidites (Geisingen Quarry). D) Limestone layer of a distal turbidite consisting entirely of graded lutite (E_2) and ungraded lutite (E_3). Neuffen Quarry, Upper Oxfordian. E, F, G) Progradational destruction of primary sedimentary structures of calcilutitic turbidites (Fig. 11 E, Plettenberg Quarry, Upper Oxfordian; Fig. 11 F, Schlatt Landslide, Middle Oxfordian); and the bioturbation pattern of a typical unweathered, blue-gray Upper Oxfordian limestone layer (Neuffen Quarry).

mannshofer" alternation (FREYBERG, 1966). Though bedding is mainly a result of diagenetic processes (3.3.), the ability to trace single beds laterally clearly reflects primary sedimentary stratification. In the Geisingen Basin stratification changes comparatively quickly over distances of about 10 km, whereas the Neuffen Basin reveals a relatively constant bedding rhythm, with rhythmograms being constant over a minimum of 80 km during the Middle Oxfordian.

Bedding rhythms also remain constant in the bioclastic zone of the early Upper Oxfordian over a minimum of 100 km. However, within the turbidic sedimentation of the middle Upper Oxfordian and during the regression phase at the Oxfordian/Kimmeridgian transition, beds pinch out against the Franconian Platform in the northeast, so that some beds can be traced only 20 to 30 km, while others show a wider occurrence.

Water depth is difficult to estimate from biological indicators. FLÜGEL & STEIGER (1981) emphasize that comparison with recent siliceous sponge colonies fails, because they show a distribution from a few to as much as 1000 meters below sea level. Moreover, not all groups of cyanophycean algae, which build up the mounds together with siliceous sponges, are photosynthesizing. The minimum water depth of the basins must have been of the order of 50 m, because the Lochen Swell reached a relative elevation of this amount. In addition, the Lochen Swell and the basin floor were affected by storms; therefore, the basin cannot have exceeded 200 m in depth. Thus, roughly 100 m seems to be a good estimate for the water depth of the Upper Jurassic basins of the Swabian Alb.

3. DIAGENESIS

In spite of their noncyclic origin, marl-limestone alternations reveal a relatively constant bedding rhythm, which is developed as well in the turbidic and erosional facies.

The Upper Oxfordian alternations are especially masonry-like with a nearly constant thickness of limestone beds. It is obvious a mechanism other than deposition must account for these periodic carbonate variations. This process is summarized here under the term "diagenetic bedding", which results from stratiform carbonate redistribution during diagenesis. The idea of diagenetic bedding was first introduced by WEPFER (1926, "Auslaugungsdiagenese"); KENT (1936) and SUJKOWSKI (1958, "rhythmic unmixing"), and is more recently discussed by BARRET (1964), HALLAM (1964), CAMPOS & HALLAM (1979), EDER (1982), EINSELE (1982, Fig. 11), WALTHER (1983) and RICKEN (i.pr.). All of these authors assume that carbonate is dissolved by pressure solution from single beds, which therefore become enriched in clay (marl layers, "donor

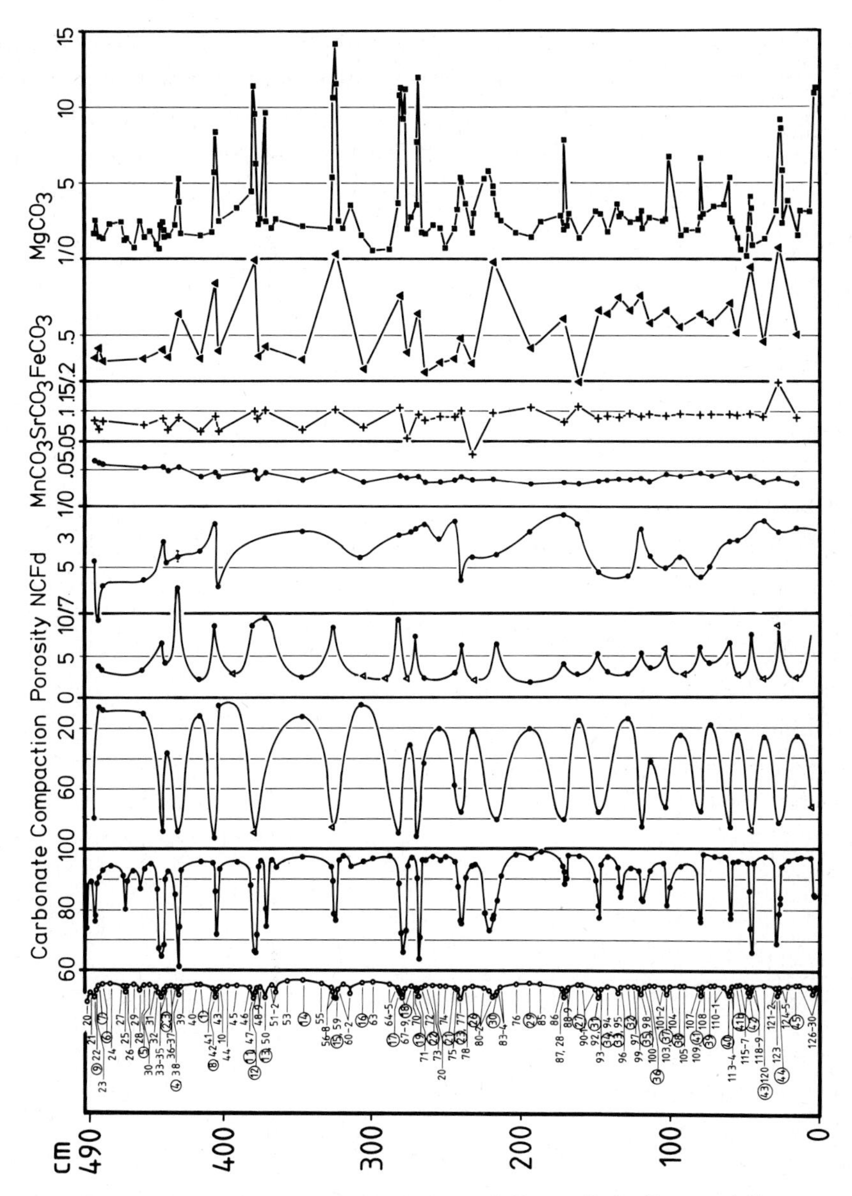

Fig. 12: Mass physical and geochemical data of Upper Oxfordian marl-limestone alternation (Neuffen Quarry, see Fig. 3 A). Columns from left to right: 1) Weathering profile and positions of samples, 2) carbonate curve, 3) compaction, 4) porosity, 5) calculated absolute amount of the noncarbonate fraction (NCFd) of the original decompacted sediment, 6) distribution of trace elements expressed as a percentage of the total carbonate fraction.

limestones", HUDSON (1975). The carbonate is then reprecipitated as pore-cement in other beds, which become enriched in carbonate (limestone layers, "receptor limestones", HUDSON (1975).

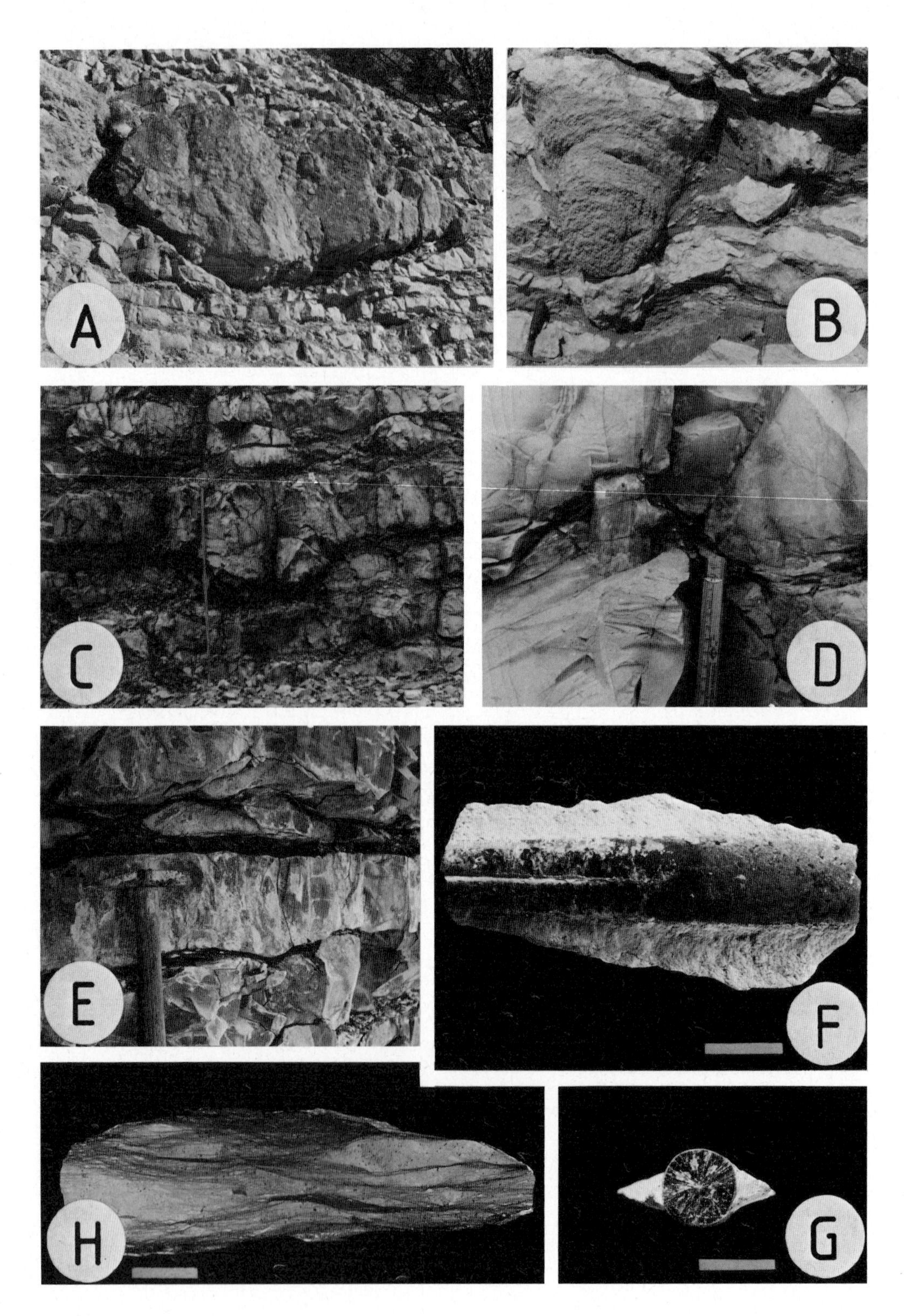
A
B
C
D
E
F
H
G

3.1. Observations

Qualitative evidence of the existence of diagenetic bedding is given by the following observations from the Upper Jurassic marl-limestone alternations:

1.) Calculations from deformed *Planolites* show that limestone beds are only compacted by 20 to 30% of the original sediment thickness (Fig. 12, 20). However, in spite of this small reduction in thickness the porosity of the present rock is only 2 to 5%. Therefore, these limestones must contain a considerable amount of cement (HUDSON, 1975; BATHURST, 1976; MEYERS & HILL, 1983). The calcareous cement is presumably derived from dissolution of adjacent marl layers. This is evident from nearby zones of high $CaCO_3$- content and zones of intense leaching (Fig. 13 D, Fig. 14), implying diffusional transport both with and against the compaction flow. Diffusion plays an important role in ionic transport during diagenesis (EINSELE, 1977; BERNER, 1980; PINGITORE, 1981).

2.) Within the marl layers highly compressed *Planolites*-burrows reveal intense compaction by about 80% (Fig. 12, 20). Weathered marl layers usually appear homogeneous; however, unweathered marl layers contain flasering of secondary marl seams due to microstylolitic carbonate dissolution or chemical compaction (Fig. 13 H) (BARRET, 1964; WANLESS, 1979; GARRISON & KENNEDY, 1977; RICKEN & HEMLEBEN, 1982). Marl seams are connected by small joints, which compensate for vertical movements due to differential compaction (RICKEN & HEMLEBEN, Fig. 2 E, F). Some specific marl layers can be clearly identified as secondary features. They appear at both the bottom and top of resistant nuclei, such as patch reefs, coarse grained channel fills, and provided carbonate for cementation (Fig. 13A-C, Fig. 14).

3.) Major elements and trace elements of the carbonate fraction of the Upper Oxfordian marl-limestone alternation show diagenetic redistribution. Magnesium,

---

Fig. 13: A) Biostrome core showing development of a marly dissolution rim due to differential compaction. Gosheim Quarry, Middle Oxfordian. B) Base of a algae-sponge biostrome. Early cemented sponges acted as stylolites during burial diagenesis. Gosheim Quarry, Middle Oxfordian. C) Diagenetic marl seams below a channel fill consisting of skeletal calcareous material see Fig. 15. Neuffen Quarry, Upper Oxfordian. D) Stylolite due to differential compaction and cementation during burial diagenesis, see Fig. 14. Neuffen Quarry, Upper Oxfordian. E) Marl layers containing relict limestone lenses from carbonate dissolution (see Fig. 13 H). Genkingen Quarry, Upper Oxfordian. F, G) Carbonate aggregation (pressure shadow structure) along the longitudinal sides of a belemnite shell within a marl layer. Cross section in Fig. 13 G. Scale is 1 cm. Talheim Quarry, Upper Oxfordian. H) Typical flasery solution seams (marl enrichment) of a clastic marl bed. Talheim Quarry, early Upper Oxfordian.

Table 1: Minor element composition of the carbonate fraction (Upper Oxfordian, Neuffen Quarry), determined by AAS (means of 61 samples).

	limestone layer		marl bed		differences	
total carbonate content	94.0 %		77.2 %		16.8 %	
	ppm	% $MeCO_3$	ppm	% $MeCO_3$	ppm	% $MeCO_3$
Mg	5898	2.048	10767	3.726	4869	1.678
Fe	2355	0.487	3374	0.694	1019	0.207
Mn	165	0.035	215	0.045	50	0.010
Sr	529	0.090	551	0.094	22	0.004

iron, manganese and minor amounts of strontium are enriched in the marl layers. They correlate inversely with the total carbonate content and increase with progressive compaction (Fig. 12, Table 1).

The greatest amounts of magnesium and iron occur in some marl layers at 40 000 ppm Mg, and 5 000 ppm Fe, that is, about 14 % $MgCO_3$ and 1 % $FeCO_3$, or about 30 % dolomite. Magnesium enrichment is also found in the marl layers of some other marl-limestone alternations of France and Italy (RICKEN, i.pr.). WANLESS (1979), WALTHER (1983), and JØRGENSEN (1983) show that dolomitization may be connected with zones of carbonate dissolution. Depletion may be attributed to dissolution of carbonate in the marl layers, while enrichment is indicative of net precipitation of relatively clean $CaCO_3$-cement within the limestone beds. Distribution of carbonate trace elements in the marl layers can be described mathematically as a process of compaction and concentration in a partly closed system (RICKEN, i.pr.).

Alternating zones of dissolution and cementation within marl-limestone alternations depend more on differential stress than on the presence of some special substances and pore waters in the marl and limestone beds. Precipitation of carbonate occurs even within the marl beds, if pore space remains open and stress is locally reduced. In marl layers carbonate is precipitated along the longitudinal sides of large shells, forming typical pressure shadow structures (Fig. 13 F, G; RICKEN & HEMLEBEN, 1982, Fig. 3).

There was little primary variation in carbonate content before the sediment was subjected to diagenesis. This can be drawn from the following observations:

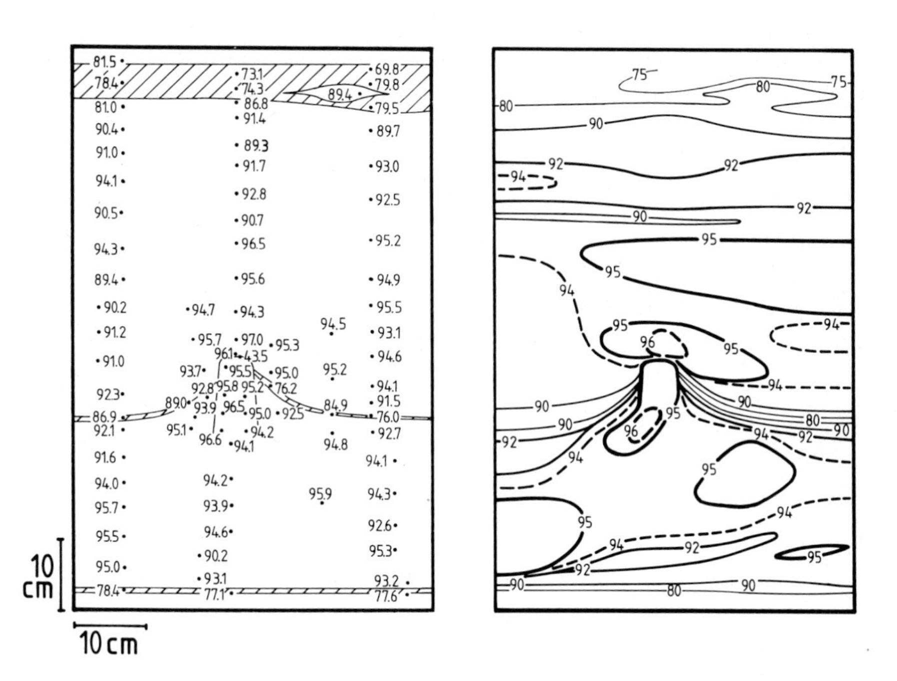

Fig. 14: Carbonate distribution in the neighborhood of a diagenetic stylolite (see Fig. 13 D). Highest carbonate values of more than 96 % are related to zones closely above and below the stylolite. Neuffen Quarry, Upper Oxfordian.

1.) Carbonate content is nearly constant within any particular limestone layer, even if material was locally brought in by bioturbation from overlying layers, which are now marl beds.

2.) The carbonate content of calcareous turbidites varies from 10 to 35 %, depending on whether the bases and tops of the turbidites were subjected to carbonate dissolusion (marl layers). However, carbonate content varies only about 3% if the turbidites were cemented in the center of the limestone beds (Fig. 16).

3.) In the marl layers, the carbonate content of pressure shadow structures resembles the carbonate content of adjacent limestone beds.

3.2. Quantifying the Diagenesis

It is beyond the scope of this paper to give a comprehensive treatment in quantifying carbonate diagenesis. A detailed study is presented in RICKEN (i.pr.). The quantification of carbonate diagenesis depends essentially on gathering rock compaction data. Several methods have been previously proposed, for instance measurement of

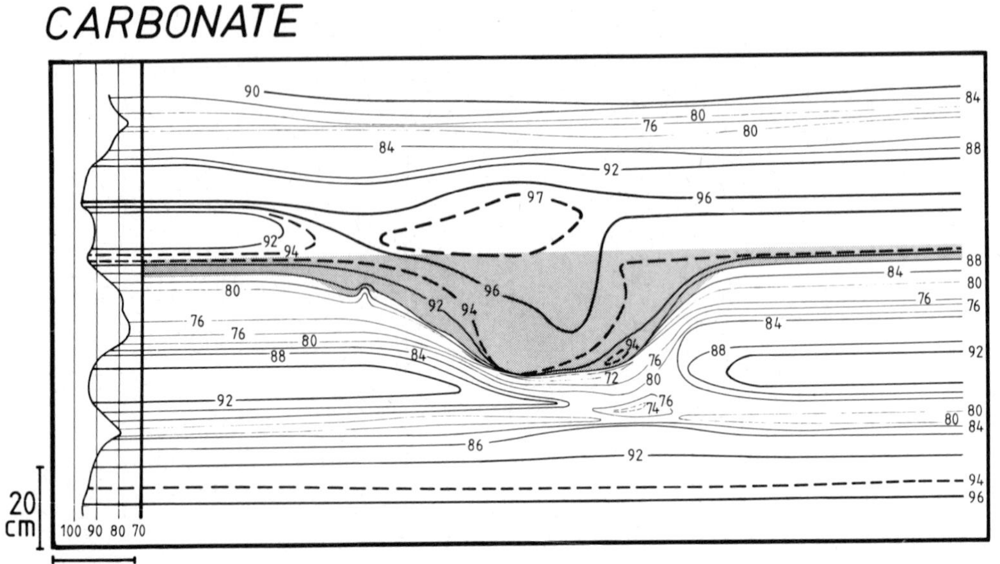

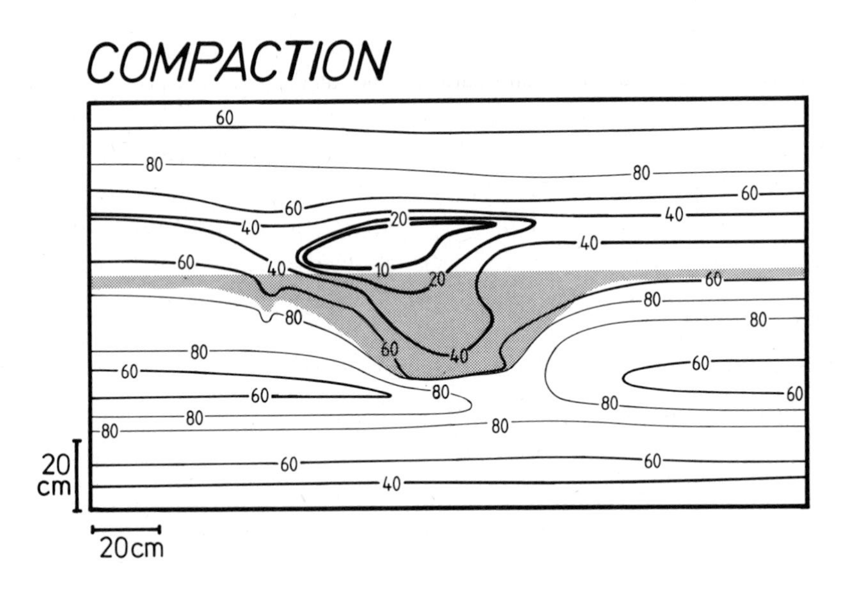

Fig. 15: Carbonate values and calculated compaction data of a channel fill consisting of skeletal material (shaded), see Fig. 13 C. The channel fill and the overlying limestone bed show minor compaction and higher carbonate values than the surrounding alternations. Minor compaction within the channel was compensated by a zone of strong compaction and dissolution, which converted the underlying limestone layer into marl. Neuffen Quarry, Upper Oxfordian.

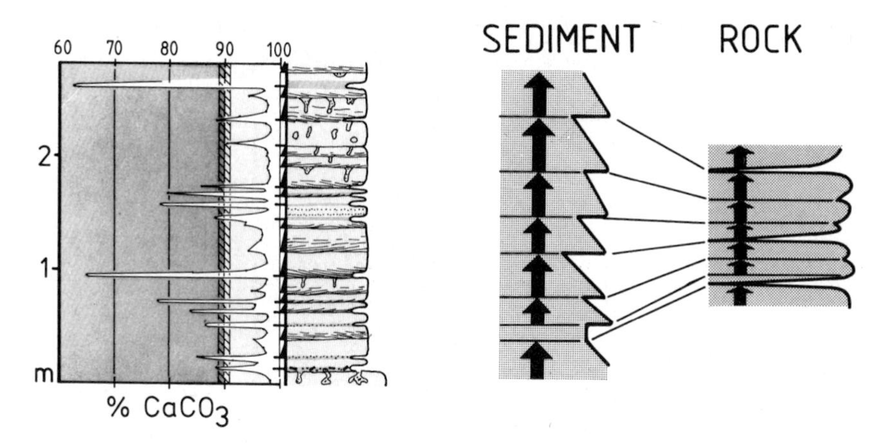

Fig. 16: (A) Turbidic marl-limestone alternation. Detailed carbonate curve shows a weathering limit between marls and limestones at about 90 % $CaCO_3$. (B) Alteration of the assumed primary carbonate curve (decreasing carbonate content upwards within single turbidites) to a diagenetic marl-limestone sequence. Geisingen Quarry, Upper Oxfordian.

the packing density of oolitic grains (COOGAN, 1970) or the microscopic determination of the amocunt of cement in skeletal limestones (MEYERS & HILLS, 1983). However, in micritic marl-limestone alternations, the amount of compaction is best ascertained from deformed bioturbational patterns. The axes of elliptical burrows, which were originally circular, can be measured if the burrow tube is parallel to the bedding and can be three dimensionally exposed. From this data fairly exact compaction values can be determined. The evaluation of the physical-chemical properties of the initial sediment and the calculation of the net carbonate redistribution is based on two fundamental assumptions:

(1) Burial diagenesis of carbonates should occur in a closed system within a column of several meters (MATTER, 1974 "autolithification"; SCHLANGER & DOUGLAS, 1974; SCHOLLE, 1977; GARRISON, 1981; BAKER et al., 1982; RICKEN, i.pr.).

(2) Differences in porosity should be minute between the dissolved and the cemented part of the initial sediment.

HAMILTON (1976) shows that medium deep sea calcareous and terrigenous sediments possess 72 % porosity, however, porosity of clays is about 81 %. The evaluation is carried out by measuring the carbonate content, porosity and compaction in a large number of samples for a rock column some meters thick (Fig. 12, 20). The section is separated into small segments with specific properties, which have to be decompacted by calculation (RICKEN, i.pr.). If no net carbonate redistribution took place, the decompacted rock must resemble the original values of clay, carbonate, and water.

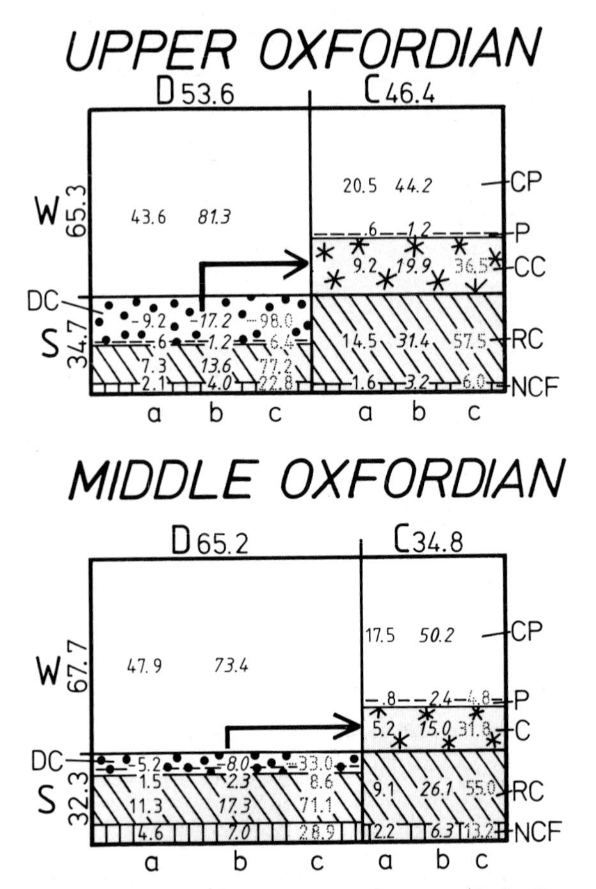

Fig. 17: Box model of calculated primary sediments (Oxfordian, outer box frame) and their alteration to marls and limestones (shaded). Numbers in percent of volume. W = original water content, S = primary content of solids, D = amount of the dissolved zones (representing mainly marl layers),C = amount of the cemented zones (representing mainly limestone layers), a = values relative to the total primary sediment, b = values relative to the cemented or dissolved zones, c = values relative to present rock of the cemented or dissolved zones. Inner values from top to bottom: 1) Dissolved zones: compaction (CP), dissolved carbonate (DC), rock porosity (P), relict carbonate (RC), noncarbonate fraction (NCF), 2) Cemented zones: compaction (CP), rock porosity (P), cemented carbonate (CC), relict carbonate (RC), noncarbonate fraction (NCF).

However, reconstructed limestone sediment yields too much carbonate, due to cementation, but reconstructed marl sediment contains too little carbonate, because of dissolution. Assuming there are no greater variations of original porosity, the compacted values of dissolved and reprecipitated carbonate can be evaluated by mass balance calculation (Table 2).

Calculations show that primary sediments of the Middle and Upper Oxfordian were calcareous muds containing 78 to 91 % carbonate and a porosity of 65 to 68% (Fig. 17). The average variation in the amount of carbonate was only about 2 %

if the original porosity remained constant in the primary sediment. Assuming a maximum porosity variation of 5 %, primary carbonate variations would be about 3 to 4 %. Thus, primary variations were enhanced by a factor 3 to 7 and 4 to 8. The layers which have been preferably cemented were more calcareous, whereas dissolution tended to affect zones of slightly lower carbonate content (Table 2). Considerable amounts of dissolved or cemented carbonate can be calculated for the centers of the beds. The maximum carbonate redistribution (Upper Oxfordian) is about +60 % or -400 % of the present carbonate fraction. This may enhance carbonate variations from 60 to 96 % $CaCO_3$.

Table 2: Calculated $CaCO_3$ - content of present alternations and their original sediments. Relative amounts of dissolved/reprecipitated carbonate in brackets in % of total rock carbonate.

		dissolved domain		cemented domain		differences
Upper	primary sediment	88.5		90.5		2.0
Oxfordian	rock	77.2	[-126.5]	94.0	[38.8]	16.8
Middle	primary sediment	78.3		80.6		2.3
Oxfordian	rock	71,1	[-46.2]	86.8	[36.6]	15.7

The overburden at the beginning of the segregation process can be estimated. The degree of compaction in the center of a limestone layer is considered as that stage of diagenesis, where grain contacts have become so pronounced that further compaction is prevented. The depth at this stage can be estimated by using porosity - overburden formulas from HAMILTON (1976). Porosity at a certain depth below the sediment-water interface can be easily converted into compaction, if the initial porosity is known (Fig. 17). Minimum compaction of the limestone beds is about 30 % in the Middle Oxfordian and about 20 % in the Upper Oxfordian. This corresponds a 55 to 57 % porosity and an overburden of 140 m in the Middle Oxfordian (varying between 50 and 350 m for different beds) and 80 m in the Upper Oxfordian (varying between 40 and 160 m for different beds). Similar data are reported for incipient diagenesis in pelagic carbonates (SCHLANGER & DOUGLAS, 1974; NEUGEBAUER, 1974; MATTER, 1974; SCHOLLE, 1977; GARRISON, 1981; SCHOLLE et al., 1983; RICKEN, i.pr.), where chemical diagenesis starts much later than in shallow seas.

3.3. Rhythmicity

As previously indicated, the carbonate rhythms are mainly a result of carbonate redistribution during burial, which overprinted a slight primary bedding. Though redistribution was stratiform and the bedding rhythm differs between the Geisingen and the Neuffen Basin (see 2.3.), the thicknesses of Oxfordian and Lower Kimmeridgian limestone layers are rather uniform. Fig. 18 shows a high correlation between the amount of bedding planes (marl joints) and the total thickness of the measured Upper Oxfordian.

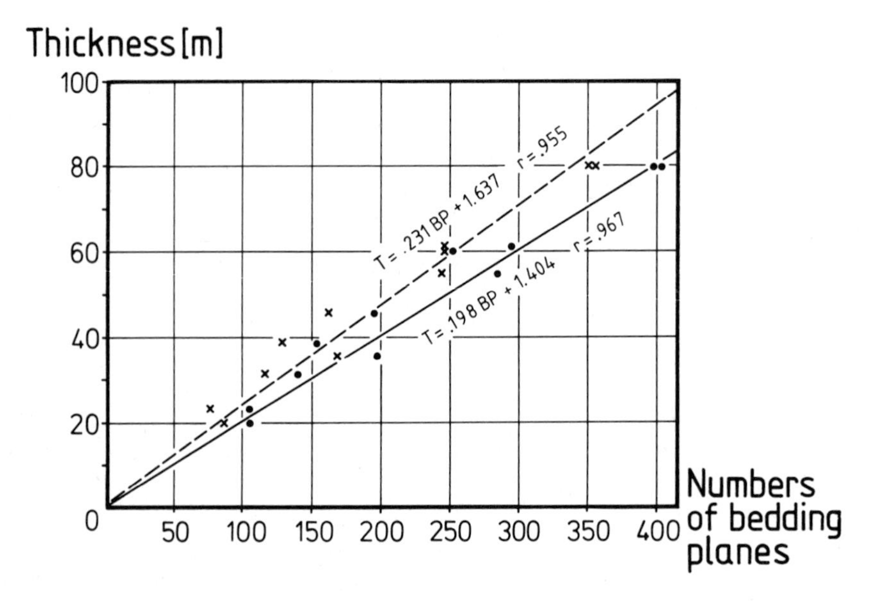

Fig. 18: Relationship between total thickness and numbers of bedding planes (marl joints) of the Upper Oxfordian zone in several locations in the Neuffen and Geisingen Basin. Crosses: Major bedding planes. Dots: Major and minor bedding planes.

Fig. 18 clearly shows that the alternation is discyclic but that the layers are uniform in thickness and are highly rhythmic. These phenomena should be explained by the theory of diagenetic bedding. Redistribution of carbonate enhances the original indistinct sedimentary rhythm in three ways:

1.) If the burial depth of the onset of the formation of carbonate aggregation and the physical-chemical gradients vary within a certain range (see compaction curves Fig. 12, 20), carbonate peaks reveal a similar thickness and a characteristic carbonate distribution profile (RICKEN, i.pr.). Therefore, limestone beds show only small variations in thickness. Aggregation occurs within those layers which contain a somewhat higher carbonate content, or have other conditions, which favor cementation (e.g. relatively high porosity).

2.) Centers of aggregation which originally lay close together, can coincide within one carbonate peak (see carbonate curves of Fig. 12, 20). However,

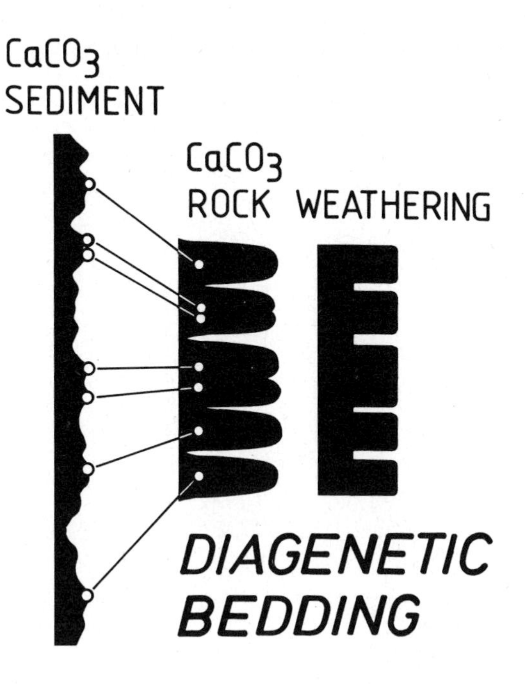

Fig. 19: Schematic illustration of diagenetic bedding process. Weak $CaCO_3$-peaks of the primary sediment are used during burial diagenesis for carbonate aggregation. Differential compaction between dissolved marls and cemented limestone beds greatly enhance the rock rhythmicity.

other smaller carbonate variations between the centers of aggregation are converted to marl layers (see residual carbonate "layers" within marl beds in Fig. 13 E).

3.) Differential compaction by a factor of 3 to 5 between the centers of marl and limestone layers was due to alternating zones of net cementation and dissolution which led to a strong reduction of the intervals between the limestone layers. In the Upper Oxfordian, the thickness of marls has been reduced so much that the sequence has become dominated by limestone layers, which tend to be roughly equal in thickness. Therefore, the alternation resembles a regular masonry-pattern, in which marl layers have been diminished to small marl joints.

It is clear from the above arguments that "diagenetic bedding" caused a "diagenetic rhythmicity". The amplitudes of the diagenetic carbonate rhythmicity are much greater than the primary bedding rhythm, but the frequency of the oscillations is lower. Weathering alter the rhythmic appearance of marl-limestone alternations. The carbonate oscillations of the rock were separated by weathering processes into distinct marl and limestone layers depending on the level of carbonate present in the rock (EINSELE,

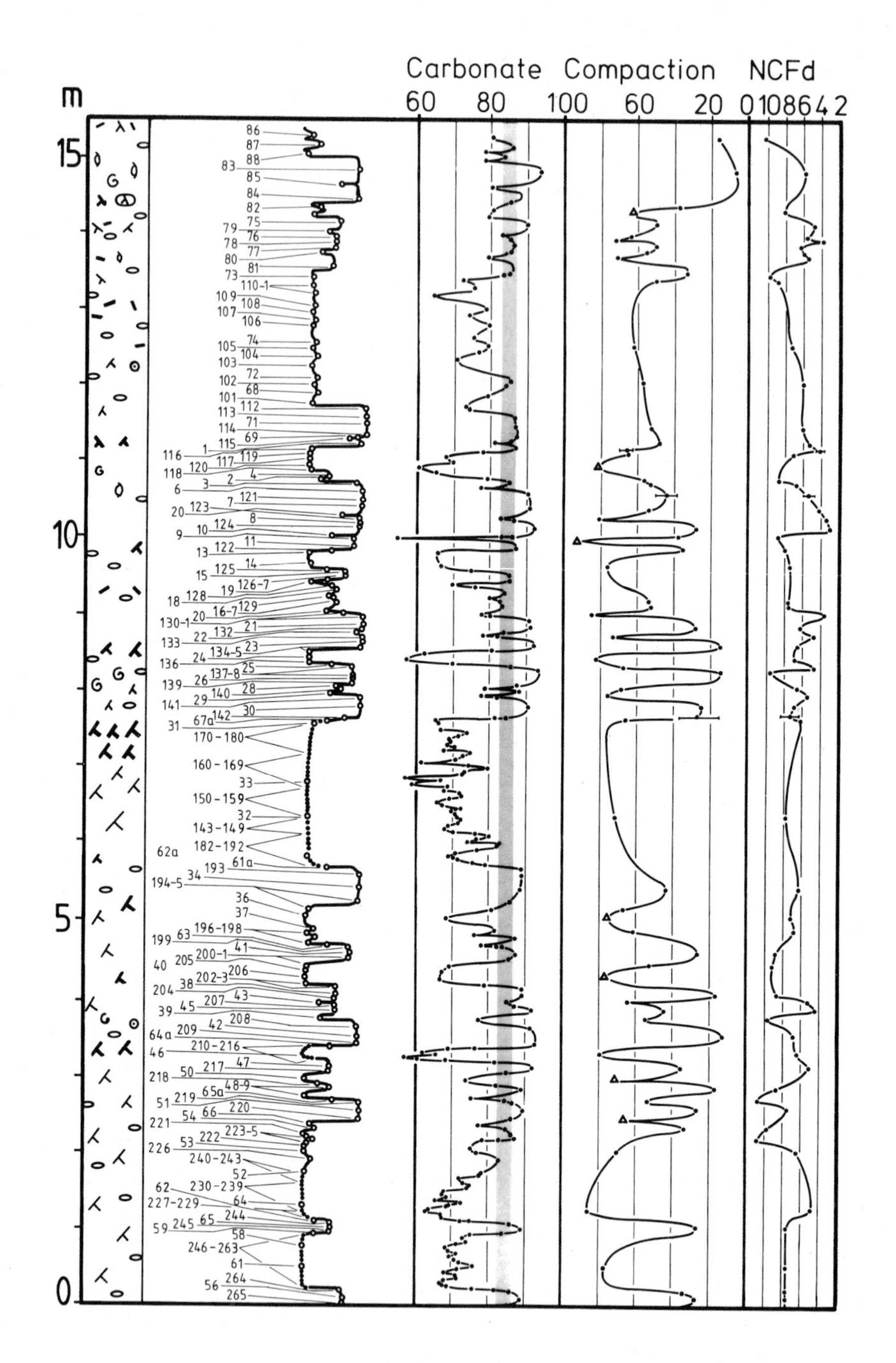

Fig. 20: Middle Oxfordian marl-limestone alternation (Neuffen Quarry, see Fig. 3B) from left to right: 1) Weathering section with sample positions, 2) carbonate curve with the weathering boundary between marl and limestone (85 % $CaCO_3$, shaded), 3) compaction, 4) calculated non-carbonate-fraction of the total decompacted sediment.

1982). In the South German Upper Jurassic the "weathering limit" is in the range of 85 to 90 % carbonate. Weathering may enhance rhythmicity, because insignificant carbonate peaks may be obscured and appear like homogeneous marl. On the other hand, weathering can also diminish rhythmicity, if several adjacent carbonate peaks of an alternation weather as marl. Fig. 20 shows a detailed carbonate curve of the upper Middle Oxfordian (see site of section in Fig. 3 A). Although the weathering profile reveals three marl beds (2 m thick each) between three groups of limestone layers, the carbonate curve oscillates entirely throughout the section. The appearence of the thick marl beds is not due to a lack of carbonate rhythmicity, but is caused by the shifting of the oscillating carbonate curve totally below the "weathering limit" of about 85 %.

3.4. Reinterpretation of the SEIBOLD - Model

SEIBOLD (1952) was the first to develop a quantitative model of the Upper Jurassic marl-limestone rhythmicity. In order to explain the depositional processes, he compared marl and limestone layers by calculating the absolute content of carbonate and clay for each b e d of the alternation (marls and limestones). Marl and limestone beds reveal similar absolute values of clay, but various amounts of carbonate. Because he neglected cementation and compaction, SEIBOLD concluded that constant deposition of clay was superimposed by rapid, rhythmic carbonate precipitation.

In the light of the new data presented here, SEIBOLD's results must be reinterpreted. In the Upper Oxfordian the weathering limit between marls and limestones is identical with the boundary between the dissolved and the cemented domain of the alternation. This is a special feature of the Upper Oxfordian and not the usual case in all marl-limestone alternations (RICKEN, i.pr.). Thus, SEIBOLD compared the d dissolved and highly compacted parts of the alternation (marls) with the cemented and little compacted parts (limestones). From Fig. 17 it is obvious that the original clay content of the two parts was nearly equal. Therefore, during carbonate redistribution and differential compaction between the present marl and limestone layers the a b s o - l u t e clay content of each b e d must remain constant. However, the absolute carbonate content of each bed and the bed thickness must differ considerably. Therefore, SEIBOLD's assumption of a depositional superimposition of carbonate over a constant clay background sedimentation has to be reinterpreted as a diagenetic redistribution of carbonate within a sediment of a nearly uniform composition.

4. CONCLUSION

Facies analyses, bed by bed correlation and geochemical investigation reveal a three-stage origin of Upper Jurassic, epicontinental, marl-limestone alternations in South Germany:

1.) An indistinct primary bedding due to small variations of the physical-chemical composition was caused by deposition of calcareous mud turbidites and the occurrence of widespread sea floor erosion. Bioturbation destroyed primary sedimentary structures and further diminished compositional variations. The slight, primary stratification pattern differs between basins; nevertheless, after diagenesis, it can be traced laterally over distances of 10 to 100 km within a single basin.

2.) Several million years after deposition and with an overburden of about 100 to 150 m, a stratiform redistribution of carbonate began and overprinted the original slight stratification pattern by the development of diagenetic bedding. Layers, which were already more calcareous were preferentially subjected to cementation, whereas the present marl layers provided carbonate by dissolution. Redistribution continued until the porosity of the present limestone layers was reduced to 2 - 5 %. Alternating net cementation and dissolution was the main cause of carbonate variations in the present rock and of its bedding rhythmicity.

3.) Postdiagenetic carbonate variations in the rock are not fully reflected in the weathered section. Above a certain critical carbonate level (85 to 90 %), the rock weathers to a limestone, whereas below this limit the rock weathers to a marlstone. Therefore, weathering may enhance or reduce rhythmicity.

ACKNOWLEDGEMENTS

I acknowledge valuable discussions with several colleagues, particularly G. Einsele. U. Bayer, G. Einsele, B. Haskell, L. Hobert, A. Seilacher, and H. Winder critically read this manuscript. The research was supported by the Sonderforschungsbereich 53, Palökologie.

REFERENCES

Aigner, T. 1982: Calcareous Tempestites: Storm-dominated stratification in Upper Muschelkalk. In: Einsele, G. & Seilacher, A. (eds.): Cyclic and event stratification, 180-198 (Springer, Berlin).

Baker, P.A., Gieskes, J.M. & Elderfield, H. 1982: Diagenesis of carbonates in deep-sea sediments - evidence from Sr/Ca ratios and interstitial dissolved Sr data.- Journal Sediment. Petrol. vol. 52, 71-82.

Barret, P.J. 1964: Residual seams and cementation in Oligocene shell calcarenites, Te Kuiti Group.- Journal Sediment. Petrol. vol. 34, 524-531.

Bathurst, R.G.C. 1976: Carbonate sediments and their diagenesis.- Developments in Sedimentology vol. 12, 658 pp. (Elsevier, Amsterdam).

Bausch, W.M., Fatschel, J. & Hofmann, D. 1982: Observations on well-bedded Upper Jurassic limestones. In: Einsele, G. & Seilacher, A. (eds.): Cyclic and event stratification, 54-62 (Springer, Berlin.

Berner, R.A. 1980: Early diagenesis. 241 pp. (Princeton University Press).

Bouma, A.H. 1962: Sedimentology of some flysch deposits, 168 pp., (Elsevier).

Campos, H.S. & Hallam, A. 1979: Diagenesis of English Lower Jurassic limestones as inferred from oxygen and carbon isotope analysis.- Earth and Planetary Sci. Lett. vol. 45, 23-31.

Coogan, A.H. 1970: Measurements of compaction in oolitic grainstone.- Journal Sediment. Petrol. vol. 40, 921-929.

Damuth, J.E., Flood, R.D., Kowsmann, R.O., Monteiro, M.C., Gorini, M. A., Palma, J.J.C. & Belderson, R.H. 1983: Distributary channel meandering and bifurcation patterns on the Amazon deep-sea fan as revealed by long-range side-scan sonar (Gloria).- Geology vol. 11, 94-98.

Davaud, E. & Lombard, A. 1975: Statistical approach to the problem of alternating beds of limestone and marl (Upper Oxfordian of the French Jura).- Eclogae Helv. vol. 68, 491-509.

Dott, R.H. 1983: Episodic sedimentation - how normal is average? How rare is rare? Does it matter? - Journal Sediment. Petrol. vol. 53, 5-23.

Eder, W. 1982: Diagenetic redistribution of carbonate, a process in forming limestone - marl alternations (Devonian and Carboniferous, Rheinisches Schiefergebirge, W. Germany). In: Einsele, G. & Seilacher, A. (eds.): Cyclic and event stratification, 98-112.

Einsele, G. 1977: Range, velocity and material flux of compaction flow in growing sedimentary sequences.- Sedimentology vol. 24, 639-655.

Einsele, G. 1982: Limestone - marl cycles (Periodites): Diagnosis, significance, causes - a review. In: Einsele, G. & Seilacher, A. (eds.): Cyclic and event stratification, 8-53 (Springer, Berlin).

Fischer, A.G. 1980: Gilbert - bedding rhythms and geochronology. In: Yochelson, E.L. (eds.): The scientific ideas of G.K. Gilbert.- Geological Soc. Am. Spec. Paper vol. 183, 93-104.

Fischer, A.G. & Arthur, M.A. 1977: Secular variations in the pelagic realm. In: Cook, H.E. & Enos, P. (eds.): Deep-water carbonate environments.- Soc. Econ. Paleont. Mineral. Spec. Publ. vol. 25, 19-50.

Freyberg, B.v. 1966: Der Faziesverband im unteren Malm Frankens.- Erlanger geologische Abhandlungen vol. 62, 3-92.

Flügel, E. & Steiger, T. 1981: An Upper Jurassic spongealgal buildup from the northern Frankenalb, West Germany.- Soc. Econ. Paleont. Mineral. Spec. Publ. vol. 30, 371-397.

Garrison, R.E. 1981: Diagenesis of oceanic carbonate sediments: A review of the DSDP perspective.- Soc. Econ. Paleont. Mineral. Spec. Publ. vol. 32, 181-207.

Garrison, R.E. & Kennedy, W.J. 1977: Origin of solution seams and flaser structure in the Upper Cretaceous chalks of southern England.- Sediment. Geology vol. 19, 107-137.

Gwinner, M.P. 1962: Subaquatische Gleitungen und resedimentäre Breccien im Weißen Jura der Schwäbischen Alb (Württemberg).- Zeitschrift deutsche geologische Gesell. vol. 113, 571-590.

Gwinner, M.P. 1976: Origin of Upper Jurassic limestones of the Swabian Alb (Southwest Germany).- Contrib. to Sedimentology vol. 5, 75 pp.

Gygi, R. 1969: Zur Stratigraphie der Oxford-Stufe der Nordschweiz und des süddeutschen Grenzgebietes.- Beiträge geol. Karte Schweiz, N. F. vol. 136, 123 pp.

Hallam, A. 1964: Origin of the limestone-shale rhythms in the Blue Lias of England: A composite theory.- Journal of Geology vol. 72, 157-168.

Hallam, A. 1977: Eustatic cycles in the Jurassic.- Palaeogeogr., Palaeoclimat., Palaeoecol. vol. 23, 1-32.

Hamilton, E.L. 1976: Variations of density and porosity with depth in deep-sea sediments.- Journ. Sediment. Petrol. vol. 46, 280-300.

Hiller, K. 1964: Über die Bank- und Schwammfazies des Weißen Jura der Schwäbischen Alb (Württemberg).- Arbeiten geologisch paläontologisches Institut T.H. Stuttgart, N. F. vol. 40, 190 pp.

Hudson, J.D. 1975: Carbon isotopes and limestone cement.- Geology vol. 3, 19-22.

Hyden, F.M. 1980: Mass flow deposits on a mid Tertiary carbonate shelf, southern New Zealand. - Geological Magaz. vol. 117, 409-516.

Jeletzky, J.A. 1975: Hesquiat Formation (new): A neritic channel and interchannel deposit of Oligocene age, western Vancouver Island, British Columbia.- Geol. Surv. Canada Paper vol. 75, 32 pp.

Jørgensen, N.O. 1983: Dolomitization in chalk from the North Sea central graben.- Journal Sediment. Petrol. vol. 53, 557-564.

Kelts, K. & Arthur, M.A. 1981: Turbidites after ten years of deep-sea drilling - wiringing out the mop? - Soc. Econ. Paleont. Mineral. Spec. Publ. vol. 32, 91-127.

Kennedy, W.J. & Juignet, P. 1974: Carbonate banks and slump beds in the Upper Cretaceous (Upper Turonian - Santonian) of Haute Normandie, France.- Sedimentology vol. 21, 1-42.

Kent, P.E. 1936: The formation of the hydraulic limestones of the Lower Lias.- Geol. Magazine vol. 73, 476-478.

Köhler, K.E. 1971: Zur Sedimentologie der Grenzschichten Dogger/Malm in Südwestwürttemberg.- Arbeiten geol. paläontol. Institut T.H. Stuttgart, N.F. vol. 64, 90 pp.

Matter, A. 1974: Burial diagenesis of pelitic and carbonate deepsea sediments from the Arabian Sea.- Initial Reports DSDP vol. 23, 421-470.

Meischner, K.D. 1964: Allodapische Kalke, Turbidite in Riffnahen Sedimentationsbecken.- Developm. Sedimentology vol. 3, 156-191.

Meyers, W.J. & Hill, B.E. 1983: Quantitative studies of compaction in Mississippian skeletal limestones, New Mexico.- Journ. Sediment. Petrol. vol. 53, 231-242.

Neugebauer, J. 1974: Some aspects of cementation in chalk. In: Hsü, K.J. & Jenkins, H.C. (eds.): Pelagic sediments: on land and under the sea.- Intern. Assoc. Sed. Spec. Pub. No 1, 149-176.

Pingitore, N.E. 1982: The role of diffusion during carbonate diagenesis.- Journal Sediment. Petrol. vol. 52, 27-39.

Piper, D.J.W. 1978: Turbidite muds and silts on deepsea fans and abyssal plains. In: Stanley, D.J. & Kelling, G. (eds.): Sedimentation in submarine canyons, fans, and trenches (Stroudsburg, Downden, Hutchinson & Ross), 163-176.

Ricken, W. 1984: Diagenetische Bankung in Kalk-Mergel-Wechselfolgen. Zementbilanz, Geochemie, Fazies.- Unpublished thesis, Universität Tübingen.

Ricken, W. & Hemleben, C. 1982: Origin of marl-limestone alternation (Oxford 2) in southwest Germany. In: Einsele, G. & Seilacher, A. (eds.): Cyclic and event stratification, 83-71.

Reineck, H.-E. & Singh, I.B. 1980: Depositional sedimentary environments, 549 pp., (Springer, Berlin).

Schlanger, S.O. & Douglas, R.G. 1974: Pelagic ooze - chalk - limestone transition and its implications for marine stratigraphy. In: Hsü, K.J. (eds.): Pelagic sediments: on land and under the sea.- Intern. Assoc. Sedimentology Spec. Publ. vol. 1, 117-148.

Schmidt-Kaler, H. 1962: Stratigraphische und tektonische Untersuchungen im Malm des nordöstlichen Ries-Rahmens.- Erlanger geol. Abhandlungen, vol. 44, 51 pp.

Scholle, P.A. 1977: Chalk diagenesis and its relation to petroleum exploration: Oil from chalks, a modern miracle? AAPG Bull. vol. 61, 982-1009.

Scholle, P.A., Arthur, M.A. & Ekdale, A.A. 1983: Pelagic sediments. In: Scholle, P.A., Bebout, P.G. & Moore, C.H. (eds.): Carbonate depositional environments.- AAPG Memoir vol. 33, 638-691.

Schwarzacher, W. & Fischer, A.G. 1982: Limestone-shale bedding. In: Einsele, G. & Seilacher, A. (eds.): Cyclic and event stratification, 72-95 (Springer, Berlin).

Seibold, E. 1952: Chemische Untersuchungen zur Bankung im unteren Malm Schwabens.- Neues Jahrbuch Geol. Paläontol. Abhandlungen vol. 95, 337-370.

Seilacher, A. 1982: Distinctive features of snandy tempestites. In: Einsele, G. & Seilacher, A. (eds.): Cyclic and event stratification, 333-349 (Springer, Berlin).

Shepard, F.P. & Dill, R.F. 1966: Submarine canyons and other sea valleys, 381 pp., Chicago (Rand Mc Nally).

Sujkowski, Z.L. 1958: Diagenesis.- AAPG Bull. vol. 42, 2692-2717.

Vail, P.R. & Todd, R.G. 1981: Northern North Sea Jurassic unconformities, chronostratigraphy and sea-level changes from seismic stratigraphy. In: Petroleum Geology of the continental shelf of Northern Europe, 216-235 (Institute of petroleum, London).

Wanless, H.R. 1979: Limestone response to stress: pressure solution and dolomitization. J. Sediment. Petrol. 49, 437-462.

Walther, M. 1983: Diagenese gebankter Karbonate im Unterkarbon Nordwest-Irlands.- Thesis, Univers. Göttingen, 76 pp., unpublished.

Weiler, H. 1957: Untersuchungen zur Frage der Kalk-Mergel-Sedimentation im Jura Schwabens.- Thesis, Univers. Tübingen, 57 pp., unpublished.

Wepfer, E. 1926: Die Auslaugungsdiagenese, ihre Wirkung auf Gesteine und Fossilinhalt.- Neues Jahrbuch Mineralogie Beilagen Bd. vol. 54 B, 17-94.

Ziegler, B. 1977: The "White" (Upper) Jurassic in southern Germany.- Stuttgarter Beiträge Naturkunde, B 26, 79 pp.

PART 3

EVOLUTIONARY AND ECOLOGICAL REPLACEMENTS IN MARGINAL EPICONTINENTAL SEAS

Evolutionary processes are long term phenomena (as pre-Pleistocene sea-level changes are), and marginal basins are potential evolutionary centers because of their varying degree of isolation and their continuously and repeatedly changing environmental conditions in the course of global sea-level changes. However, the relation between evolutionary and environmental trends is problematic in two ways:

(1) In the course of an environmental trend a taxon may follow the trend by selection; however, if related species have the possibility and capability to immigrate into the basin, a similar trend may result from purely ecological replacements, which -- on the basis of the local record -- cannot be distinguished from an evolutionary trend.

The second problem results from the fact that sea-level changes may alter environmental conditions at various levels and in different ways. They influence e.g. the degree of isolation, the stratification and illumination of the water column, the substrate conditions etc.

(2) The factors causing an ecological or evolutionary trend, are commonly hard to determine because the conclusion "the particular rock containing a fossil allows to infer its ecology" usually fails: In the fifties of this century a classical correlation between annual birthrate and the frequency of storks was observed in Germany. In the following 20 years both, the birth rate and the frequency of storks declined. Given a comparable paleontological example one would probably construct a causal relationship, although the correlation reflects only a general 'environmental' trend involving different factors for the two correlating processes.

To study the causality of evolutionary and ecological replacements requires the repetition of 'experiments' under slightly differing boundary conditions, allowing to eliminate minor environmental factors. Such repeated or iterative evolutionary and ecological trends are discussed in all three papers.

** *Iterated morphological trends of ammonites result from endemic evolution and from immigration as Bayer & McGhee and Urlichs & Mundlos show. On both levels convergences are observed in morphology and especially in the evolution of suture lines.*

** *The Echinoderms discussed by Hagdorn lack endemic evolution; however, the repeated ecological successions can be traced in much more detail -- a pattern which can be related to their special ecological disposition.*

The studies indicate that the step towards a causal analysis of evolutionary patterns in paleontology requires the integrated study of sedimentary and faunal evolution under well defined boundary conditions such as basin configuration and sea-level fluctuations, and, perhaps, we should view our biostratigraphic boundaries as ecological events rather than as evolutionary ones.

EVOLUTION IN MARGINAL EPICONTINENTAL BASINS: THE ROLE OF PHYLOGENETIC AND ECOLOGICAL FACTORS

AMMONITE REPLACEMENTS IN THE GERMAN LOWER AND MIDDLE JURASSIC

Ulf Bayer & George R. McGhee Jr.

Tübingen / New Brunswick

Abstract: Organisms are bounded within an ecological framework, which in part depends on the physical environment as reflected by the sedimentological characteristics of the strata within a certain region. Of special interest are cyclic sedimentation patterns ('Klüpfel cycles'). These allow the study of evolution of taxa under repeated similar changes of the physical environment. In fact, some groups (e.g. ammonites) show iterative morphological cycles clearly related to the sedimentological ones.

Three aspects of patterns in Jurassic ammonite successions are discussed in detail: rapid faunal replacements, faunal fluctuations, and iterated morphological cycles and their relation to environmental changes.

It turns out that phylogeny, reconstructed from a regionally localized record, is a peculiar mixture of faunal migrations and in situ evolution. The interpretation of phylogenetic patterns, therefore, hinges on one's point of view. Here, ecological models are proposed to explain the punctuated stratigraphic record, thus linking the concept of 'punctuated equilibrium' with ecological replacements and paleogeographic migration events.

INTRODUCTION

The concept of biological evolution is based upon two major observations: the change of species during geologic time as documented in the stratigraphic record, and the adaptation of species within specific ecological constraints. In principle, evolutionary theory is founded on both observations equally, but in actual practice the relative weighting often becomes quite different. In ecology, often even in paleoecology, evolutionary changes of species are ignored -- everything is considered well adapted within a given ecological time section. On the other hand, in phylogenetic and biostratigraphic studies, the sequence of species changes during the course of geologic time heavily overdominates any consideration of adaptational aspects. Biostratigraphy, in an extreme chronostratigraphic approach, requires a biological clock which is independent of external factors, and where the occurrence of new species becomes the ticking of the evolutionary clock.

It is, therefore, not surprising that earlier non-darwinian evolutionary models were based, in large part, upon groups such as ammonites -- groups which were (and are) important for stratigraphic correlation (SCHINDEWOLF, 1950). Outside of the stratigraphic aspects of ammonite successions, adaptational or ecological aspects were traced only in terms of the "facies dependence of ammonites" (KLÜPFEL, 1917; FREBOLD, 1925; HALLAM, 1961); again usually with little relation to evolutionary changes.

Both aspects of evolutionary theory are traditionally viewed as processes which act gradually -- gradually changing species which gradually adapt to specific environments -- although this picture is contradicted by the punctuational nature of the stratigraphic record (SCHINDEWOLF, 1950; ELDREDGE & GOULD, 1972). This discrepancy between theory and observation has classically formed the basis for earlier non-darwinian evolutionary models such as "Typostrophism" (BEURLEN, 1937; SCHINDEWOLF, 1950 -- for a discussion see REIF, 1983), and more recently for non-gradualistic punctuational models in paleontology (ELDREDGE & GOULD, 1972; STANLEY, 1975; GOULD & ELDREDGE, 1977; STANLEY, 1979; VRBA, 1980) and in genetics (PETRY, 1982).

Unfortunately, most of our phylogenetic reconstructions today are still based on monographs of faunas and taxonomic groups from small geographic areas. In order to reconstruct the temporal sequence of evolutionary lineages from such local studies it is still the practice to superimpose a chronostratigraphic framework based upon the empirical fact that species change with time -- a practice which in and of itself is dangerously circular. In addition, however, the ecological context of each small geographic area -- and the consequent local adaptations of the species inhabiting that area -- are commonly not considered in the construction of chronostratigraphic range charts.

The purpose of the present study is to examine the relationship between ecological changes with time and the reconstructed phylogenetic changes in such a local fauna within a small geographic area -- the southern German depositional basin. In Fig. 1 is given the global and regional paleogeographic context of the southern German basin during Lower and Middle Jurassic times.

Throughout the paper we shall argue:

1. that organisms are constrained within an ecological framework,

2. that at least some aspects of physical paleoenvironment are reflected by the sedimentological characteristics of the strata within a given region,

3. and that "phylogenies" reconstructed from regionally localized records represent a mixture of faunal migrations and actual in situ evolution.

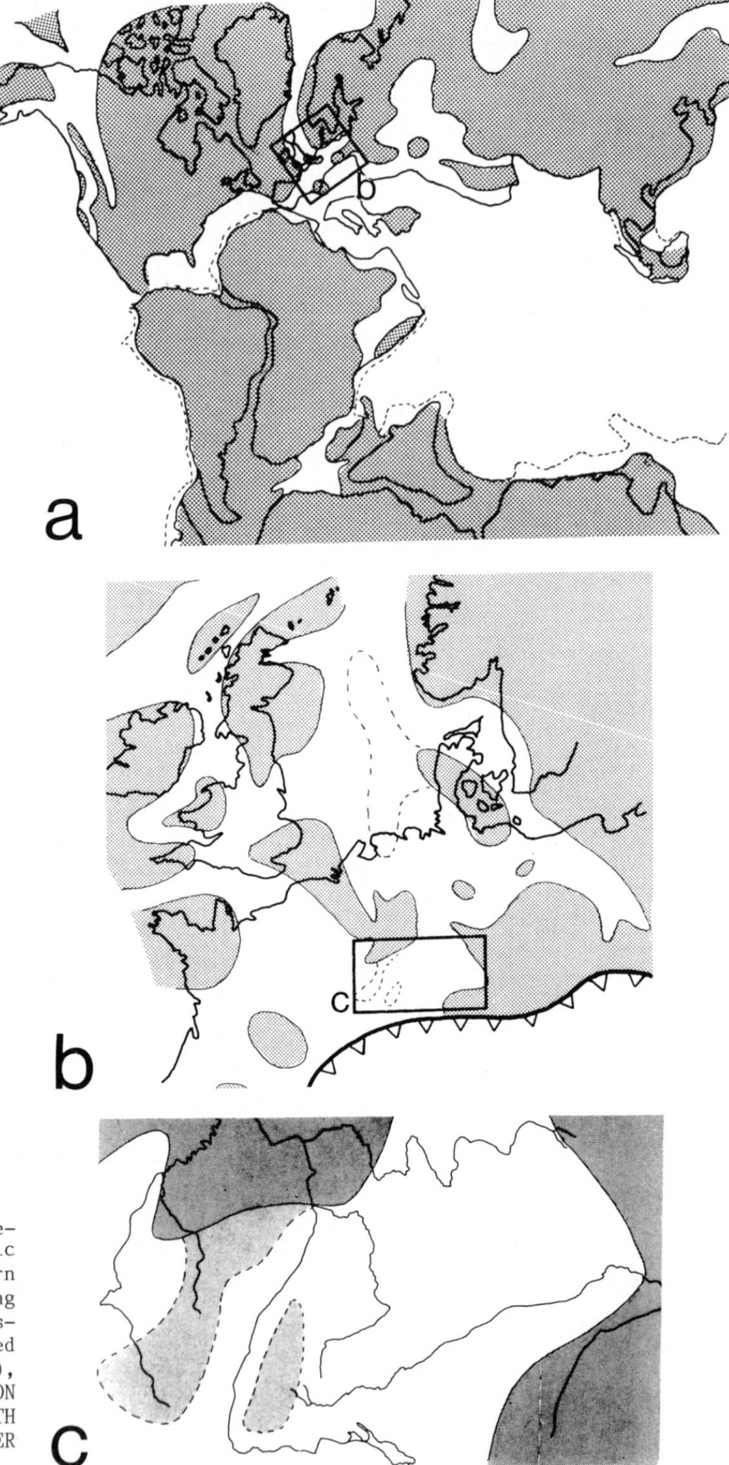

Fig. 1: Global and regional paleogeographic context of the southern German basin during Lower and Middle Jurassic times. Maps modified after SCHRÖDER (1962), ZIEGLER (1978), BARRON et al. (1981), SMITH et al. (1982), BAYER & McGHEE (1984).

The material is presented in three chapters:

In the first chapter we discuss classical stratigraphic sequences of faunal replacements on the level of local duration times and with respect to lithological sequences. We point out that there are two independent explanatory modes for faunal replacements, dependent upon whether the phylogenetic or ecological aspect of evolution is taken as the a priori point of view.

In the second chapter we introduce morphology as an additional evolutionary parameter and focus on a particular example of the phenomenon of iterative morphological evolution within Aalenian ammonites. Iterative morphological pathways, their relationship with ontogenetic pathways, temporal trends in morphological variability, and asymmetric paleoenvironmental cycles within depositional basins are discussed.

In the third and final section we then present a series of models for environmental and ecological changes within a marginal epicontinental basin which are sufficient to explain the local phylogenetic record and evolutionary phenomena discussed in the previous two chapters. These ecological models are outlined in a series of increasingly open systems:

*** an isolated basin model with in situ speciation;*

*** an open basin model with immigration from a constant external source;*

*** environmentally coupled basins with increasing complexity in faunal and ecological interactions.*

1. STRATIGRAPHIC PATTERNS OF FAUNAL REPLACEMENTS

In this chapter we will examine some of the patterns of faunal and species replacements in Jurassic ammonite successions. Our major objective is to elucidate, with some classical examples, how a priori ideas influence the interpretation of empirical stratigraphic data.

1.1 Liassic Replacements in Southern German Ammonites

A correlation between ammonite successions (subfamilial/familial level) and sedimentological changes becomes immediately visible in the Hettangian to lowermost Pliensbachian section (Fig. 2) if lineage durations are plotted versus the lithological column. Every major lithological change -- claystones, sandstones, carbonates -- is mirrored by a faunal replacement on the subfamilial level. These parallel relationships between fauna and lithologic

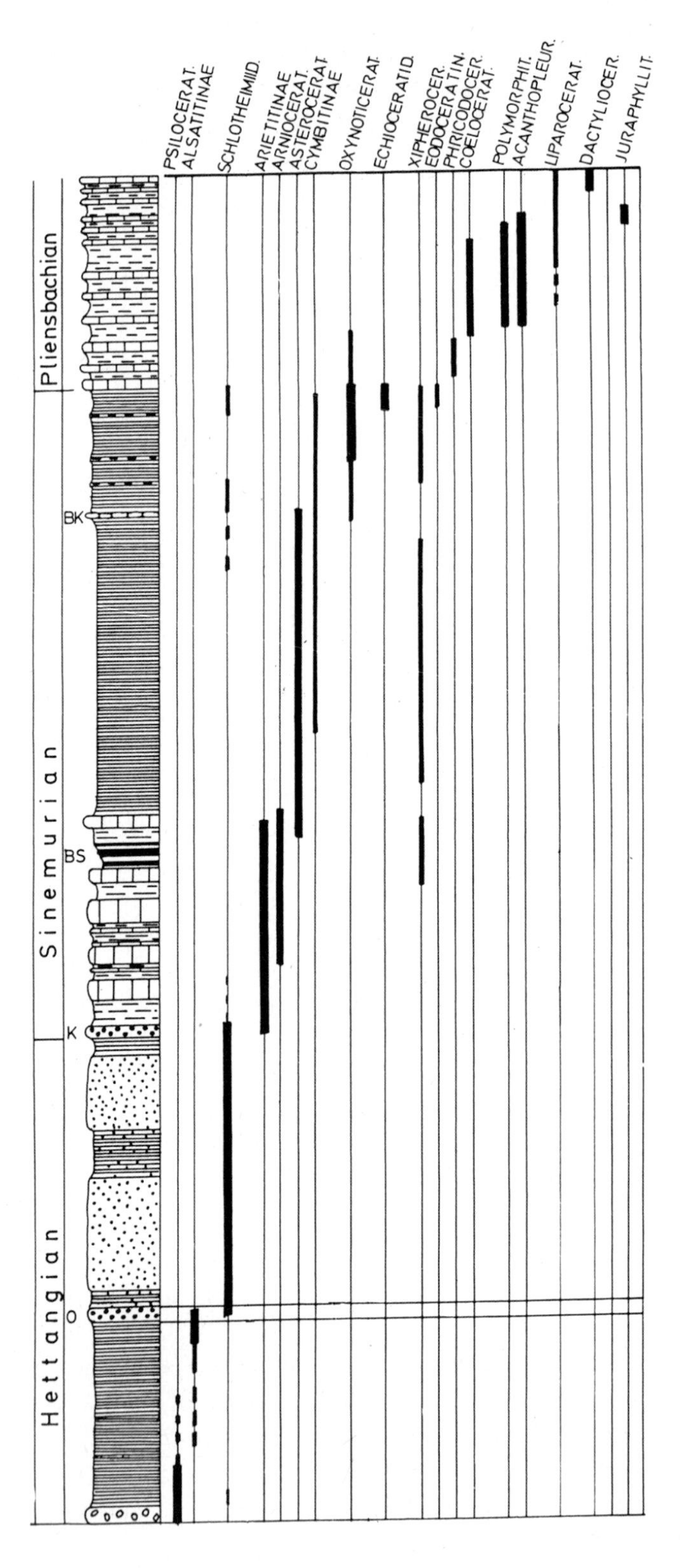

Fig. 2: Ranges of ammonite subfamilies and lithological column of the lower and middle Lias of southern Germany. After HOFFMANN (1936/38), SÖLL (1956), WALLISER (1956), GEYER & GWINNER (1962), SCHLOZ (1972) and others.

changes in the Lias have been widely discussed in the literature (KLÜPFEL, 1917; FREBOLD, 1924, 1925; HEIDORN, 1928; ARKELL, 1957; HALLAM, 1961; and others).

On the other hand, if one destroys the visual match between lineage duration and lithologic breaks by covering up the lithologic column, the faunal transitions appear to overlap more or less sharply. In addition, nice "phylogenetic" sequences within the families and subfamilies are visible, e.g. the succesion:

Arietitinae -- Arnioceratinae -- Asteroceratinae
or
Xipheroceratinae -- Eodoceratinae -- Phricodoceratinae
etc.

These sequences appear to be good temporal sequences of subtaxa within higher taxa. Furthermore, similar temporal sequences are found throughout Europe, thus reinforcing the view that they are genuine phylogenetic lineages. However, the sequence of lithologic changes can also be very similar throughout large parts of Europe. Such widespread changes resemble major Klüpfel-cycles (KLÜPFEL, 1917; HALLAM, 1961), which appear to correspond very well with global sea level fluctuations (VAIL et al., 1977; HALLAM, 1978,1981).

If we examine the lithologic and faunal pattern in detail, a further correlation frequently appears. A faunal break is often rather sharp whenever there is an abrupt change in lithology, i.e. when the transition is characterized by a typical "marker-bed". Such marker-beds in Fig. 2 are the "Oolithenbank" (O), the "Kupferfelsbank" (K), and the "Beta--Kalkbank" (BK) -- mature condensed iron-oolitic carbonates. In contrast, a faunal replacement appears more gradual if the lithological transition itself is gradual (e.g. the local black shales in the "Arietenkalk" = BS, and the lower Pliensbachian faunal replacement). Two of these boundaries have been studied in considerable detail: the "Oolithenbank" by SCHLOZ (1972) and the uppermost Sinemurian (above the "Beta-Kalkbank") by SÖLL (1954). In the next sections these studies will be used to examine in more detail the two types of faunal/lithologic transitions.

1.2 Faunal Punctuations

Within the "Oolithenbank" a sudden and final change from the Psiloceratidae (Alsatitinae) to the Schlotheimiidae occurs (Fig. 2). The detailed paleogeographic pattern of this transition and its relation to lithology is illustrated in Fig. 3. In the northeastern region of southern Germany the "Oolithenbank" is dominated by sand and pebbles which are reworked from the bed material. In the central region the iron-ooid

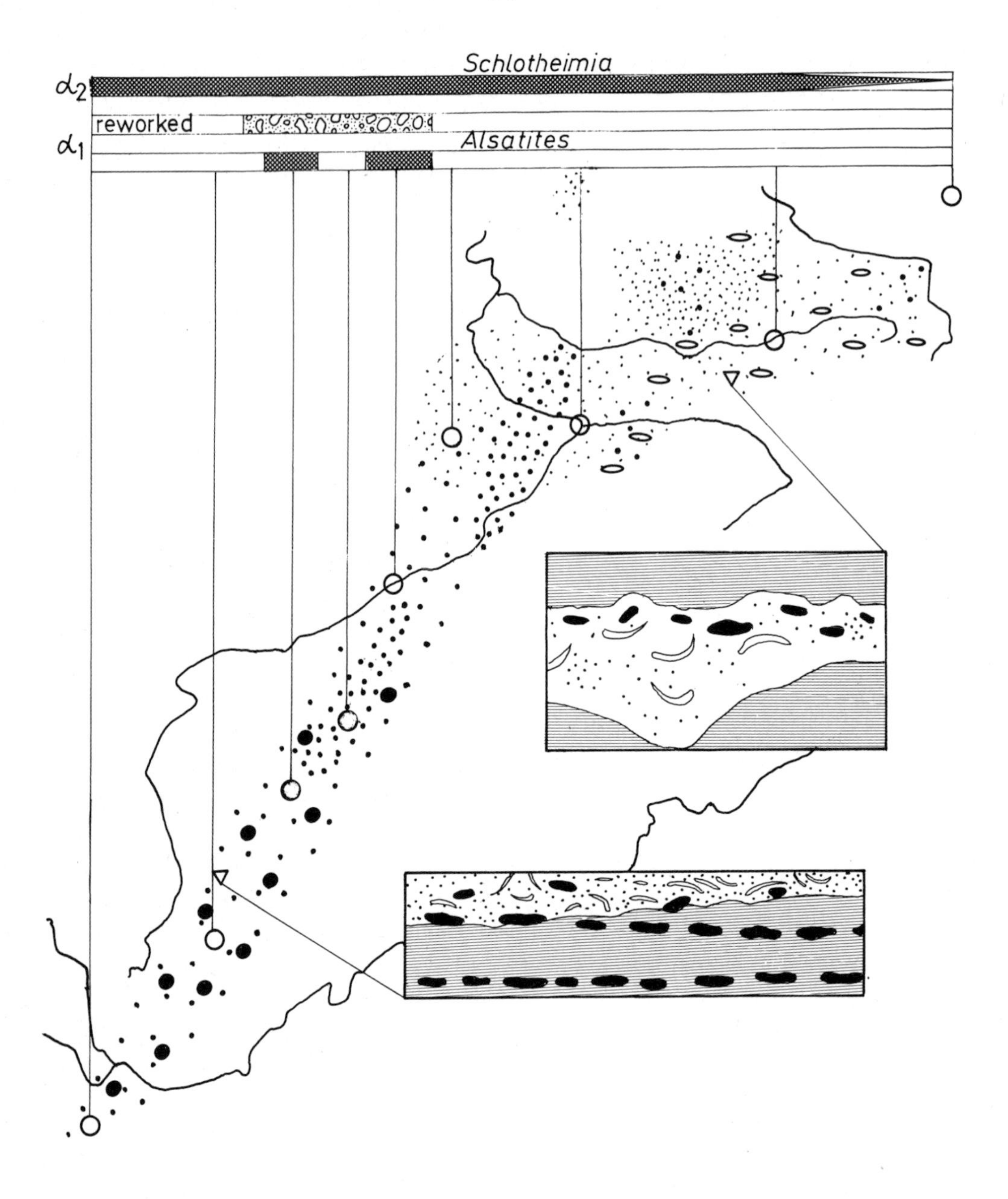

Fig. 3: Ammonite fauna and lithology of the "Oolithenbank" of southern Germany (modified from SCHLOZ, 1972). Sand and reworked pebbles from the bed itself (white ellipses) dominate in the north-east; iron-ooids and reworked carbonate concretions from the clays beneath the bed (black ellipses) are major constitutes in the south. The base of the bed is usually erosional, sometimes with channel structures (cross-sections).

content increases; and toward the southwestern region reworked concretions from the clays below become a major constituent of the bed. The base of the bed is usually erosional, sometimes with channel-like patterns (illustrated in the cross-sections).

As SCHLOZ (1972) has previously pointed out, the faunal transition is not continuously found throughout the outcrop area. The faunal transition within the "Oolithenbank" itself is documented only in the central region where in situ and reworked *Alsatites* appear together with *Schlotheimia*. Sedimentologically, the "Oolithenbank" has all of the features of a classical condensation horizon, with a somewhat asynchronous base, geographically. Thus a sedimentological gap has distorted the phylogenetic record in this region of the stratigraphic column. An additional problem is, however, why do we find condensed beds at zonal boundaries three times within the Lower Jurassic profile (Fig. 2: "Oolithenbank", "Kupferfelsbank", "Beta-Kalkbank") and not within interzonal portions of the stratigraphic profile?

Let us examine the basic observations now from a different conceptual viewpoint, that of "event condensation" (EINSELE & SEILACHER, Eds., 1982). Within this concept we expect bed formation to be the result of long term spatio-temporal processes with the formation of erosional channels filled with coarse relict sediments, and the build-up of shell grounds, hardgrounds, and eventually oolitic sediments. At any given period in the process of bed formation these different sediments occurred in spatially adjacent regions, providing different ecological habitats. However, repeated reworking and winnowing -- together with early diagenetic formation of concretions and increasing shell production -- caused finally the bed to spread over large areas (BAYER et al., this volume; BRANDT, this volume). If the trophic chain of ammonites was based at the benthos, the ecological context the ammonites experienced changed as the event-accumulation proceeded. As the sedimentological process generated a suitable ecological base, appropriately adapted forms entered the basin. Already POMPECKJ (1914) interpreted the discontinuous distribution of the Schlotheimiidae as a response to the environmental conditions. Thus, the faunal overturn may not have been abrupt as it appears in the stratigraphic record, but rather a gradual ecological process. To determine the speed of replacement we would have to have some control or record of absolute time. Lastly, with respect to within-basin correlation, the lithological marker of the condensed bed appears now as useful as the biostratigraphic control of the ammonites contained within the bed -- as the two are now seen as interdependent.

The second Liassic example, taken from SÖLL (1954), illustrates the opposite viewpoint. On the subfamilial level (Fig. 2), the Upper Sinemurian appears as a time of gradual faunal change. At the species level, however, Söll observed several sudden replacements of ammonite faunas which are correlated with small erosional or condensation events (Fig. 4). Thus, Söll's observations fit excellently with the environmental and event concept previously outlined. As one decreases the scale of event intensity and duration, one simultanuously reduces the taxonomic level at which faunal

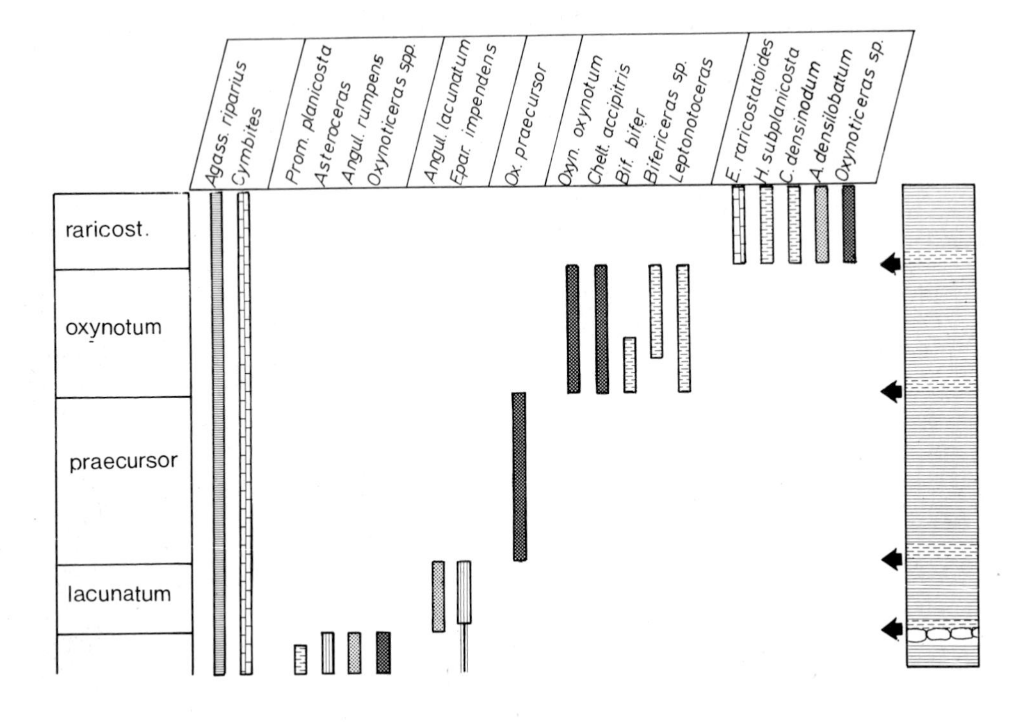

Fig. 4: SÖLL's (1954) representation of faunal events in the uppermost Sinemurian of South Germany. Small erosional or condensational events initiate a faunal replacement.

reactions are visible, finally ending up with purely ecologically determined post-event faunas with duration times on the order of years (EINSELE & SEILACHER, Eds., 1982).

Examination of Söll's original, and often republished figure (e.g. HÖLDER, 1964), shows that Söll arranged the ammonites not in a taxonomic order, as is usually done, but in "event"-faunal groups. Thus, in preparation of the figure he utilized a priori the idea of faunal breaks at observed lithological boundaries. If, however, the ammonite range durations are rearranged in a taxonomic manner (subfamilial/familial) then the pattern changes immediately (Fig. 5). At least one good "phylogenetic" sequence of species appears within the genus *Oxynoticeras*. Furthermore, the gaps within the Schlotheimiidae and Eodoceratidae could be attributable now to shifts of relative abundance -- a gradualist could argue that the missing species links will be found if sufficiently large samples within the lower diversity region of the column are taken. In total, the overall pattern could be explained in terms of a more or less gradual change of faunal diversity (cf. Figs. 4 & 5).

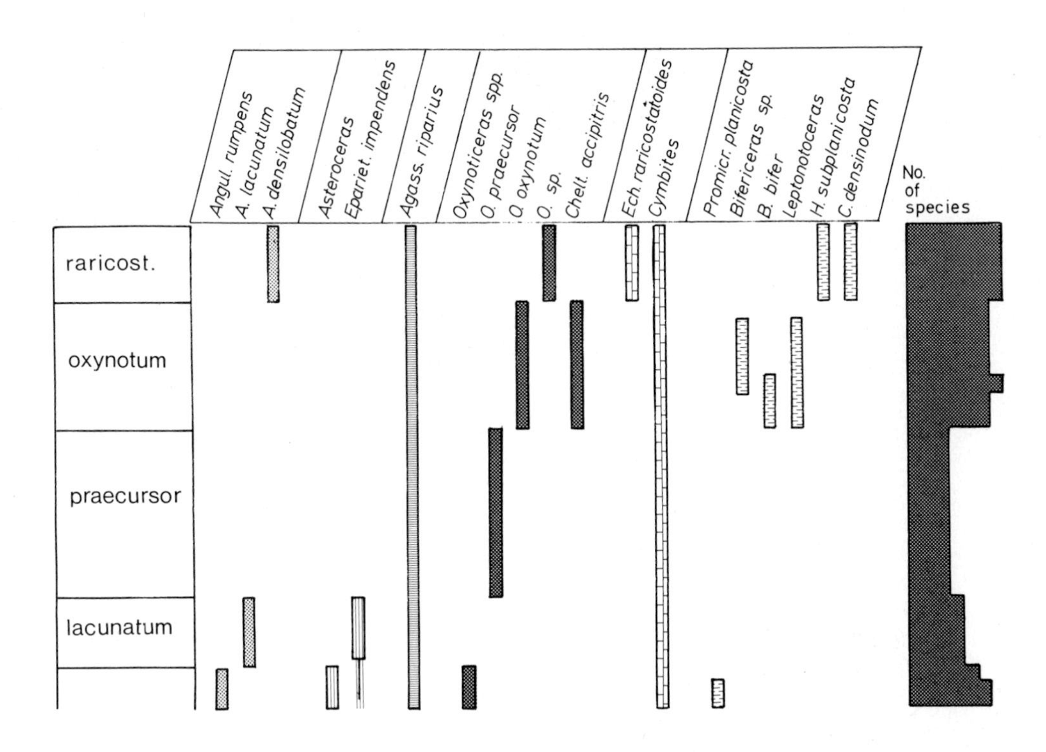

Fig. 5: SÖLL's data (from Fig. 4) reordered in a classical taxonomic manner. A nearly gradual change of diversity is now sufficient to explain the distribution pattern of ammonites. Within the higher taxa "phylogenetic" patterns become visible.

In summary, the phylogenetic record can be artificially punctuated by sedimentological processes but also by our interpretations. Our interpretation depends on the expectations of our initial theory, whether we stress the ecological/adaptational or the phylogenetic/stratigraphic aspect of the record. This viewpoint has been stressed by ELDREDGE & GOULD (1972) and should be taken as a word to the wise whenever empirical studies are taken uncritically at face value.

1.3 Phylogenetic Punctuations Versus Paleogeographic Events

The examples cited so far have been concerned with stratigraphic sequences within a single basin. Now, we shall consider the comparative aspect of stratigraphic information from geographically separated regions. To accomplish this, we will move to the lower Middle Jurassic and consider an example from G.E.G. Westermann's impressive work on ammonites, both here and in the following chapter.

Fig. 6: Possible global pathways for faunal migrations in Middle Jurassic times; partially after WESTERMANN (1969) and HALLAM (1983) (circles: south Alaska and Europe).

In his monograph on the Bajocian ammonites of southern Alaska, WESTERMANN (1969) compared Alaskan ammonite successions with the European one. Fig. 6 gives the principal paleogeographic setting of the world during Middle Jurassic time and possible migration paths for the ammonites. Westermann chose, without any doubt, the best possible b i o s t r a t i g r a p h i c correlation between Alaska and Europe. His correlation is based on the stratigraphic fit of five genera of different families, each with distinct stratigraphic range (Fig. 7). The resulting biostratigraphic pattern is surprising. Five other genera now range into younger strata in south Alaska than in Europe, and with

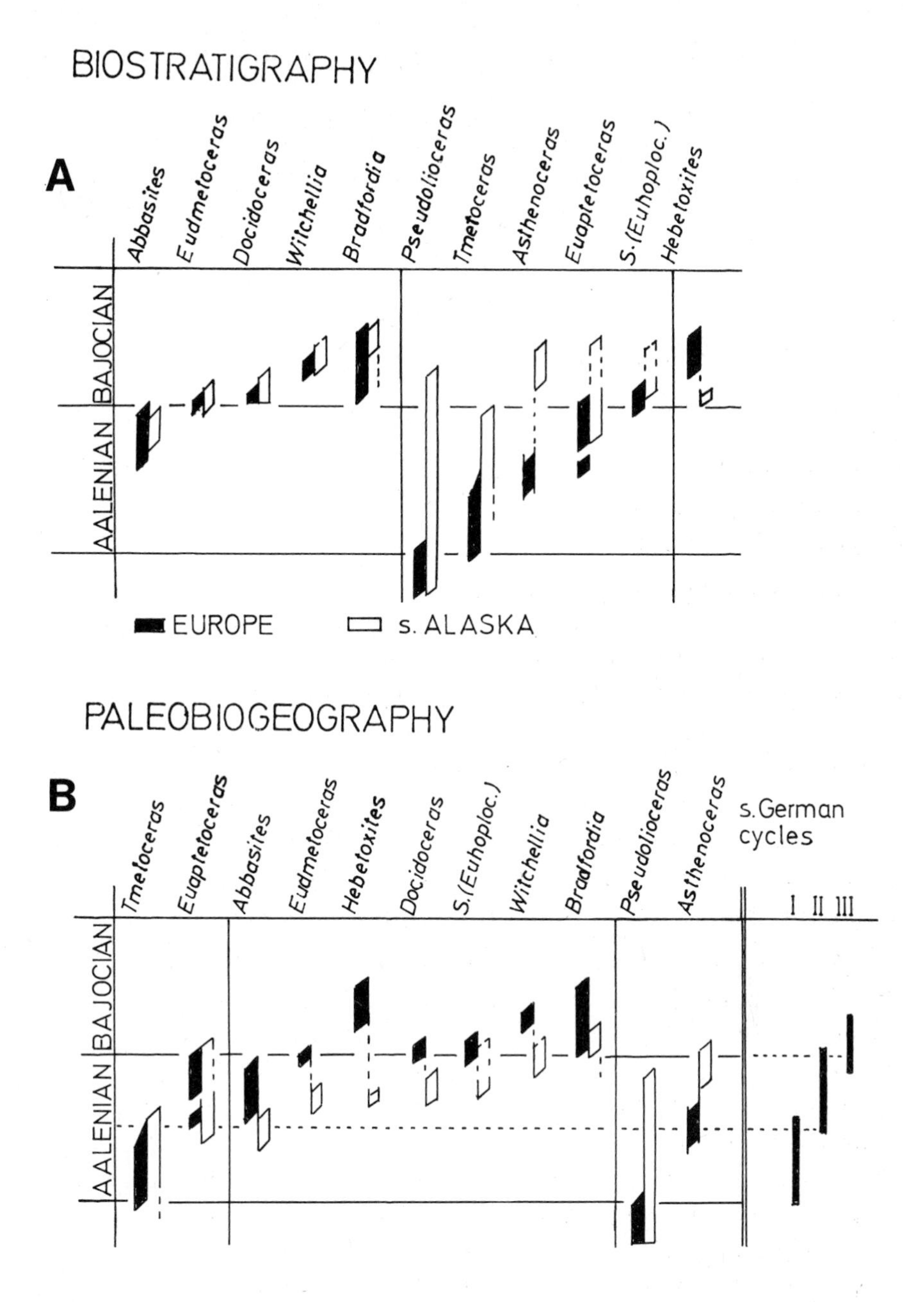

Fig. 7: Two possible correlations of the identical stratigraphic records from Europe and south Alaska. A, the best biostratigraphic fit after WESTERMANN (1969) versus B, a paleobiogeographical event model. For explanation see text.

only one genus is the inverse relationship observed. This pattern (Fig. 7A) has two implications:

(1) most species originated in Europe and later migrated to Alaska, and

(2) a migration route existed from the Tethys to the Pacific throughout the span of the study interval, probably through a connection in Central America (WESTERMANN, 1969, 1983; HALLAM, 1982, 1983).

There are, however, some reasons to consider other correlation possibilities. First, the implications noted above, and secondly the rather sudden widespread appearance of the genus *Sonninia* in Europe. This faunal event is so marked that it is classically used to define the Aalenian/Bajocian boundary. The reputed extremely rare occurrence of *Sonninia* in the *concava*-zone of the uppermost Aalenian does not affect this pattern of sudden change (RIEBER, 1963; BAYER, 1969 a,b).

There are no plausible ancestors for *Sonninia* known from the entire Tethyan realm. In addition, the Stephanoceratidae appear in Europe at the same time (BAYER, 1968; DIETL & HAAG, 1980); again without a sufficient previous phylogenetic record in the European area. WESTERMANN & RICCARDI (1982, this volume) have been able to demonstrate that it is more likely that the sonniniids evolved in the southeasternn Pacific and then expanded into the Tethys, rather than vice versa. HALLAM (1982) also postulates migration of the bivalve *Gryphaea* between the Pacific and the Tethys within the Aalenian/Bajocian, and he assumes an "episode of relatively free intermigration in the Toarcian to early Bajocian time interval" via a shallow epicontinental central Atlantic seaway "with some kind or kinds of physical or ecological restriction" (HALLAM, 1983).

Assume, as an alternative scenario, that the European faunal replacement at the end of the Aalenian reflects a paleogeographic event -- the sudden immigration of Pacific faunas into the Tethys; a possibility which is consistent with the paleoenvironment of the Mexican Gulf region (WESTERMANN, 1983). In this case, the time ranges of genera which occur near the critical boundary in Europe are now no longer useful for global stratigraphic correlation, as they occur first in the Pacific and later in time in the Tethys. There is one genus left in the original list that appears useful for global stratigraphic correlation: the genus *Euaptetoceras*, which is known to have a rather cosmopolitian distribution. Correlation by this genus alone changes the respective European and Alaskan range pattern totally. A second genus, *Tmetoceras*, appears now to have identical range durations in both areas (Fig. 7). From the original five genera with higher ranges in Alaska, only two remain. All other genera now appear earlier in the Pacific Realm than in the Tethys. The Aalenian/Bajocian boundary now appears as a paleogeographic event rather than a phylogenetic event (the sudden appearance globally of new ammonite stocks). Further, in the alternative scenario, there exist some minor faunal patterns which coincide well with several faunal cycles that have been observed in southern Germany (BAYER & McGHEE, 1984). These cycles will be discussed in greater detail in the next chapter.

Thus, on the global level we again come up with two alternative models:

*** *an expansion model (perhaps phylogenetically punctuated) based on biostratigraphic concepts, which require open migration pathways between the Tethys and the Pacific throughout the Aalenian and Lower Bajocian;*

*** *an event-model which postulates only occasionally open pathways, with resultant episodic migration events from realm to realm, and with faunal "revolutions" during the migration phases.*

There is no way to choose definitively between these two alternatives on the basis of ammonite biostratigraphy alone. The two alternative correlations given in Fig. 7 can only be tested using an independent, non-biological, method of time correlation between the two regions. On the other hand, the two different correlations do not cause time differences in excess of one (or two) ammonite-Zones. Therefore, the intercontinental correlation is not really affected by the alternative approach while the interpretation of the replacement pattern changes drastically.

The alternative correlation (Fig. 7B) takes, of course, an extreme position. The true correlation may be somewhere between the two models. The extreme case was choosen to elucidate that "faunal revolutions" at (some) stage boundaries or within debated stages may well express the breakdown of global paleogeographic barriers (compare the historic discussion of the Aalenian -- e.g. MAUBEUGE, 1963; MORTON, 1971, 1974 -- or of the Berriasian -- e.g. WIEDMANN, 1973). Cuvier had good arguments for his (regional) "revolutions" when '... he debated Saint-Hilair at the Academy in 1830 and scored such an oratorical victory that little more was heard of evolution in France for another generation' (SIMPSON, 1953).

2. ITERATED MORPHOLOGICAL CYCLES

Any unique event is in general of little interest other than of pure description. In the search for general evolutionary rules or trends in the fossil record one needs multiple occurrences of similar situations -- that is, repetitions of the natural experiment. To ammonite workers the phenomenon of heterochronous homeomorphy is well known (ARKELL, 1957; KENNEDY & COBBAN, 1977). The repeated re-evolution of similar morphotypes in the stratigraphic record provides a basis for the analysis of underlying ecological causes -- the interaction of environment and morphology over time ("constructional morphology"; SEILACHER, 1970, 1972, 1973; SEILACHER et al., Eds., 1982).

Of still further interest, however -- in terms of "repeated natural experiments" -- are those cases where we have not only isolated homeomorphic morphologies but repeated temporal sequences of similar morphologies. This phenomenon has been variously labelled as *'iterative repetition of identical courses'* (HAAS, 1942) in the literature. The existence of iterated morphological cycles makes it possible to compare multiple time-series of morphological adaptations with changes in the physical environment. Information can be gathered on two different levels -- the process of morphological transformation within a single lineage and the repetition of homeomorphs in multiple lineages. Here, hopefully, one can go beyond the singular and unique nature of the examples discussed in the last chapter.

Two examples will be discussed: the apparent simple fluctuation of a series of morphotypes within a stratigraphic sequence, and the special case of iterated morphological cycles. We demonstrate that there exists a definite relationship between the succession of ammonite morphologies and the sedimentological characteristics of the strata within a given geographical region and time period. But, it will also become clear that in order to understand and relate the stratigraphic and sedimentological record, we need to consider further the kinematics of the local sedimentary basin.

2.1 Faunal Fluctuations

An example of fluctuating ammonite faunas has been described by WESTERMANN (1954) from northern Germany. He studied in great detail two Bajocian profiles and found a repeated replacement of morphotypes which, in this case, belong to two different superfamilies (Fig. 8). His observations were verified by HUF in 1968. Examining other features within the profile, it turned out that the sequence of these fluctuations show no visible relationship to the lithology (Fig. 8; WESTERMANN 1954 p.44: "Den Biofazies entsprechen nicht bestimmte Petrofazies"). Westermann realized some weak correlations with other faunal components from which he concluded that the stephanoid facies had better oxygenated bottom waters than the sonniniid facies. The lithology, however, is extremely monotonous throughout the profile -- consisting mostly of claystones with slightly varying silt content and some shelly layers. One correspondance is visible in Fig. 8, however. The faunal fluctuations are of similar magnitude as the biostratigraphic divisions. Indeed, following WESTERMANN (1954), the faunal fluctuations can provide the same time resolution for correlation of the two profiles he studied as usage of the biozones. Thus, the faunal "reversals" provide (at least locally) an additional stratigraphic tool.

The two apparently fluctuating ammonite faunas not only belong to two different superfamilies, they are of totally different morphological type. Therefore, Westermann concluded, it is most likely that they were adapted to distinctly different environments.

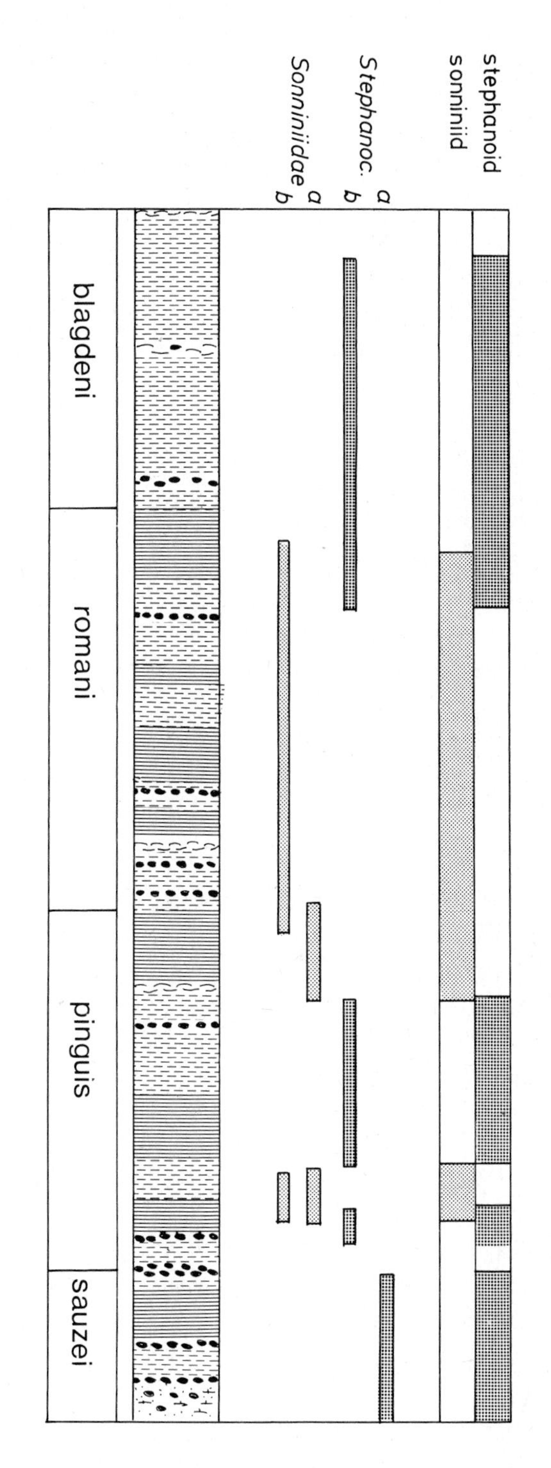

Fig. 8: Lithological section (clays, silty clays, concretions and shelly beds) of the middle Bajocian at Gerzen, North Germany, and faunal fluctuations between sonniniids (a: *Poecilomorphus*; b: *Dorsetensia*) and stephanoceratids (a: Otoitidae; b: Stephanoceratidae). After WESTERMANN (1954) and HUF (1968).

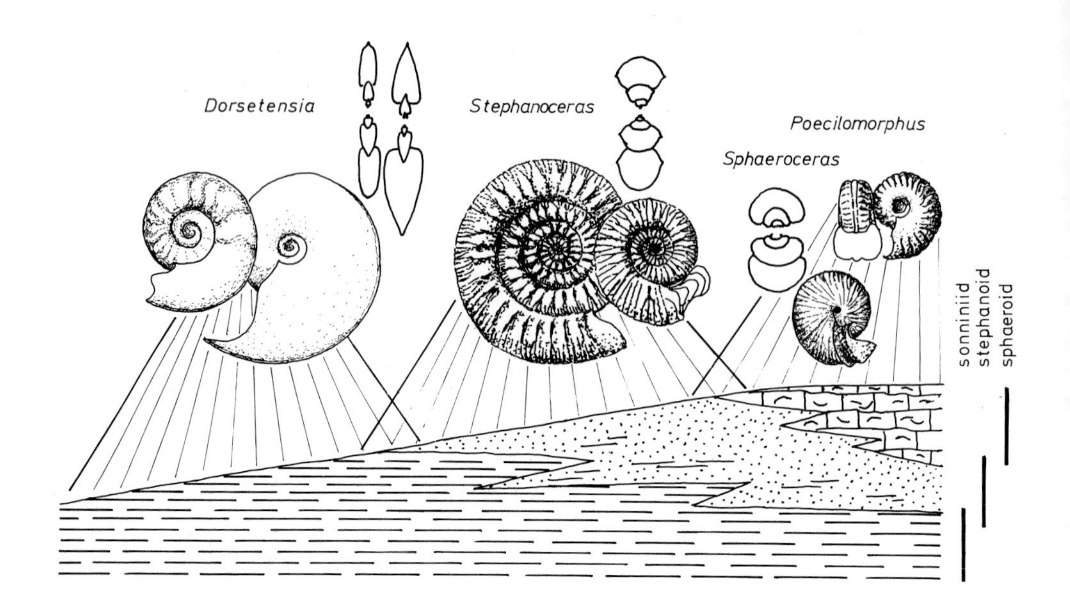

Fig. 9: Facies-relationships of morphotypes of Sonniniids (*Dorsetensia*, *Poecilomorphus*) and Stephanoceratids (*Stephanoceras*, *Chondroceras* + *Sphaeroceras*) in the middle Bajocian of southern Germany. Simplified after observations of STAHLECKER (1934), BAYER (1970) and unpublished field observations.

In the profiles given, however, no marked changes in sedimentary facies are seen, suggesting that the controlling environmental factor was not facies-linked. On the other hand, there are several reports in the literature which demonstrate a facies dependency for these ammonites (for a discussion see WESTERMANN, 1954). Fig. 9 illustrates the facies dependence of the morphotypes under question, as reconstructed from the available information in WESTERMANN (1954), STAHLECKER (1934), KUMM (1952), BAYER (1970). The smooth involute sonniniids are found in the clay facies, the evolute and sculptured stephanoceratids occur preferably in oolitic limestones while involute sphaeroid forms of both superfamilies are abundant in condensed carbonates. The latter show additional morphological features perhaps related to shallow water existence -- a tricarinate embedded keel in *Poecilomorphus* and specially stiffened apertures in the sphaeroceratids (BAYER, 1970 b).

The debate on functional morphology and on the mode of life in ammonites appears endless (ARKELL, 1957). Let us assume a nektic life for the ammonites under discussion, however, with a food chain starting within the plankton. Then, the observed faunal fluctuations reflect some change within the water column (changes of temperature, salinity, oxygen contents etc.). On the other hand, if one follows the model of ARKELL (1957):

> *"Most probably the majority of ammonites hovered and soared through the water not far above the bottom, resting near it without touching it, as does Nautilus",*

then we would expect to find some correspondance with sedimentary facies as Fig. 9 implies. The only way to explain the lack of such correspondance in the profiles (Fig. 8) is to assume that the profiles preserve pure taphocoenoses -- that the ammonites have been deposited by postmortem drift and/or sedimentological transport. Indeed, facies types other than the ones observed do occur within the north German basin, but they are only known from boreholes. The exception is the occurrence of iron-oolitic limestones near the Harz; and they contain, as would be expected from Fig. 9, only very rare sonniniids (KUMM, 1952). Without detailed knowledge of the basin configuration, its facies distribution and the kinematics of sedimentation within the basin, such isolated profiles cannot be used to argue against an elsewhere recognized facies dependency -- as, on the other hand, a diverse fauna in a condensed bed needs not to disprove facies relationships.

2.2 Iterative Morphological Cycles

As the complexity of a system increases, more and more control parameters are needed to describe it, and especially to analyse its relationships to other systems. Obviously, the sedimentological record of a few stratigraphic profiles does not provide enough control even for the simple case of fluctuating faunas. A greater degree of morphological parameters is available in the special case of "iterative repetition of identical morphological cycles", especially if such sequences are themselves iteratively repeated in groups separated in time and taxonomic affinity (BRINKMANN, 1929; SCHINDEWOLF, 1940, 1950; ZOCH, 1940; HAAS, 1942; BAYER & McGHEE, 1984). Repetitions of morphological cycles in space, time, and different taxonomic entities provide an exceptional frame for the analysis of autecological relationships at the morphological level. In order to have sufficient environmental control parameters on the other hand, we need sufficient knowledge of both basin geometry and evolution.

a) Temporal patterns of morphological iterations

A well known example of such repeated morphological cycles is found in the Aalenian to Lower Bajocian of Southern Germany. RIEBER (1963) observed two morphologically convergent lineages which terminate with discoidal and highly involute forms (Staufenia staufensis / Hyperlioceras discites). Because the two lineages reach their similar morphologies at different stratigraphic levels, he concluded that the parallel evolutionary lineages cannot be explained by repeated extrinsic physical factors but that they are due to internal factors which originated in the common ancestral form (RIEBER, 1963

p. 71):

"Dieselbe Entwicklungsstufe wird von beiden Entwicklungsreihen zu verschiedenen Zeiten erreicht, deshalb dürfen meines Erachtens nicht äußere Bedingungen für die Parallelentwicklung verantwortlich gemacht werden, sondern es muß an gleichartige Potenzen gedacht werden, die in den Tieren selbst liegen und die wohl auf die gemeinsamen Ahnen (Artengruppe des Leioceras comptum) zurückgehen."

These evolutionary sequences, with some extensions, have recently been reinvestigated (BAYER & McGHEE, 1984), and it turns out that a rather clear correlation does exist between repetitive regressive-transgressive environmental cycles and the morphological trends within lineages which occur in the southern German depositional basin during the Aalenian and lowermost Bajocian.

The major ammonite taxa which inhabited the basin during the study interval are listed in Fig. 10. Three faunistic cycles occur on the familial/subfamilial taxonomic level. The first cycle is dominated by the Leioceratinae, the second cycle includes the Graphoceratinae and the Hammatoceratidae, and the third cycle is precipitated by the "faunal revolution" at the Aalenian/Bajocian boundary, involving the Sonniniidae. The subfamilies appear and terminate at different stratigraphic levels, with the exception of the Hammatoceratidae which run in parallel with the Graphoceratinae.

The temporal span of the faunal cycles correspond very well with basin-wide regression-transgression cycles (BAYER & McGHEE, 1984). The replacement of one cycle fauna with another occurs at the beginning of the maximum regression. During these times the sediments within the basin are dominated by sands and iron-oolites. At the end of a regressive cycle, when hardgrounds are widespread in the basin, the faunal replacement has completely taken place. These typical marker horizons correspond with the *comptum*--Zone, the *bradfordensis-gigantea*-subzone, the *discites*-Zone and the *stephani*-subzone (Fig. 10). Superimposed onto these smaller regressive-transgressive cycles is an overall regression trend with its lowest water-stand in the *laeviscula*-Zone (WEBER, 1967; BAYER & McGHEE, 1984).

Therefore, with regard to basin-wide environmental parameters, RIEBER's (1963) observation of repeated morphological convergence in the Leioceratinae and Graphoceratinae can be directly related to repeated similar facies changes (though not within every single profile). Furthermore, the parallel morphological trends are not restricted to these two subfamilies -- they also occur within the Hammatoceratidae and the Sonniniidae (Fig. 11), providing an additional iterated cycle and a parallel line. Every morphological cycle (Fig. 11) or subfamily (Fig. 10) begins with evolute, globose, and well sculptured forms (high D and S values) during the period of low water stand and overlaps the termi-

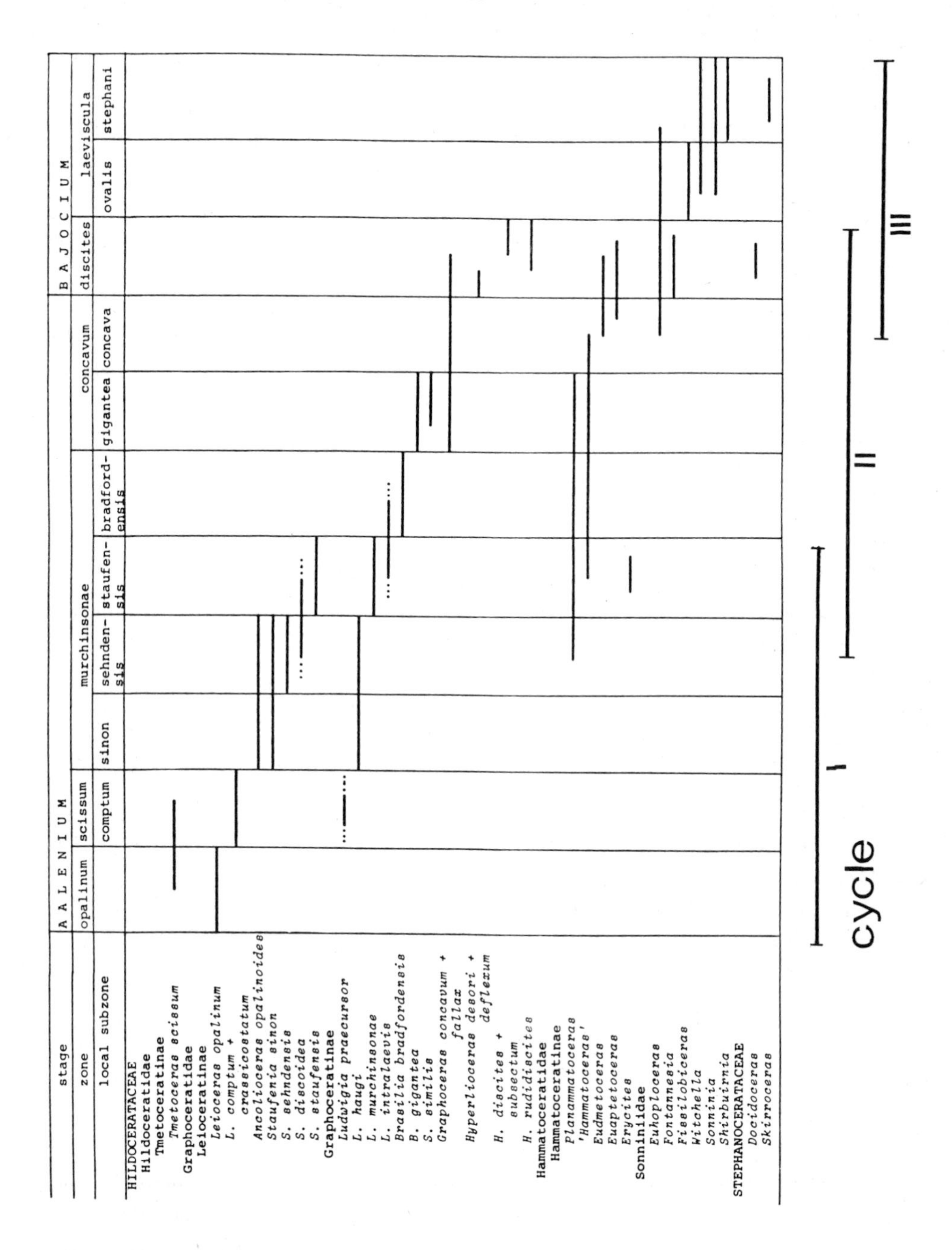

Fig. 10: The major ammonite taxa, and their temporal distribution, which inhabited the South German basin during Aalenian times. Modified after BAYER & McGHEE (1984) -- macroconchs only.

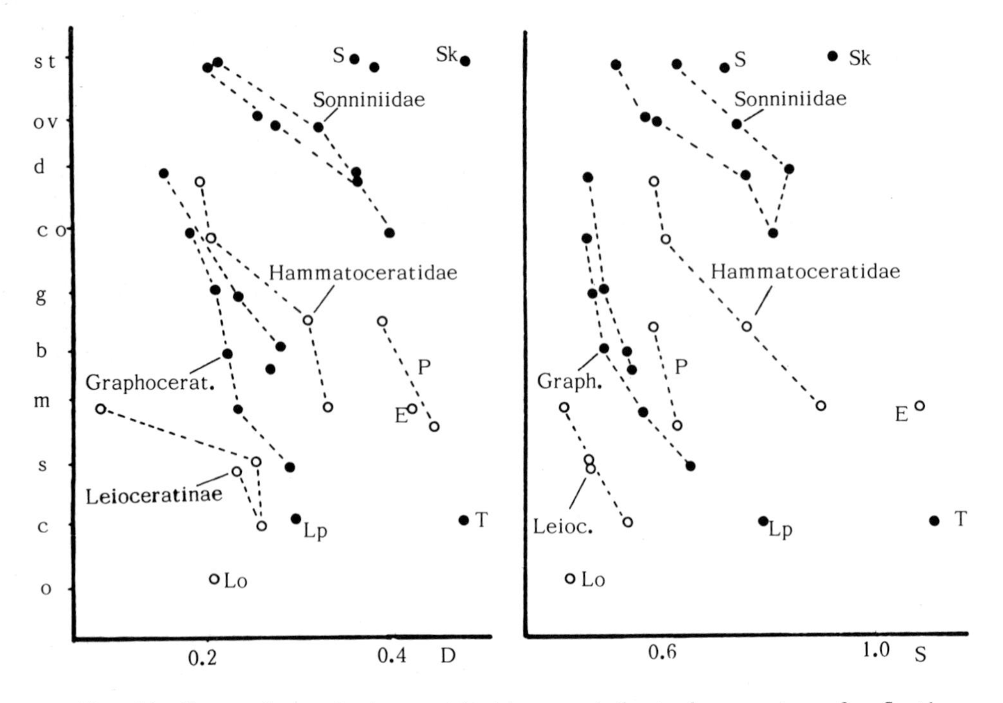

Fig. 11: Temporal trends in quantitative morphological parameters for South German ammonites during Aalenian and lowermost Bajocian times. D and S are RAUP's (1966, 1967) morphological parameters (D: distance from the coiling axis of the generating curve; S: shape of the aperture). A high S-value means rounded whorls, a high D-value characterizes evolute forms and vice versa (for details see BAYER & McGHEE, 1984). Sporadic introductions of rare forms are indicated by capital letters: T, *Tmetoceras*; E, *Ericites*; P,*Planammatoceras*; S: *Sonninia (Papilliceras)*; Sk, *Skirroceras*, Lb, *Ludwigia praecursor*; Lo, *Leioceras opalinum* (lower Aalenian).Ammonite Zones (vertical scale) indicated by letters: *o*, *opalinum*; *c*, *comptum*; *s*, *sinon*; *m*, *staufensis (murchensonae s. str.)*; *b*, *bradfordensis*; *g*, *gigantea*, *co*, *concavum*; *d*, *discites*; *ov*, *ovalis*; *st*, *stephani* (see fig. 10).

nal members of the previous morphological cycle which are extremely involute, compressed, and smooth (low D and S values). These final and extreme morphological forms are

Staufenia staufensis, *Hyperlioceras discites*,

Euaptetoceras amplectens, *Shirbuirnia stephani*

for each cycle or parallel line, respectively. The unusual environmental condition present during low water stand periods is further underlined by the occasional rare occurrence of extremely evolute forms with rounded whorls:

Tmetoceras, *Erycites*, *Stephanoceras (Skirroceras)*

(see Figs. 7 & 11) -- which otherwise are missing within the basin during the study interval. In addition to the correlation between morphological cycles and regressive-transgressive cycles there exists an overall morphological correlation to the overall regressive

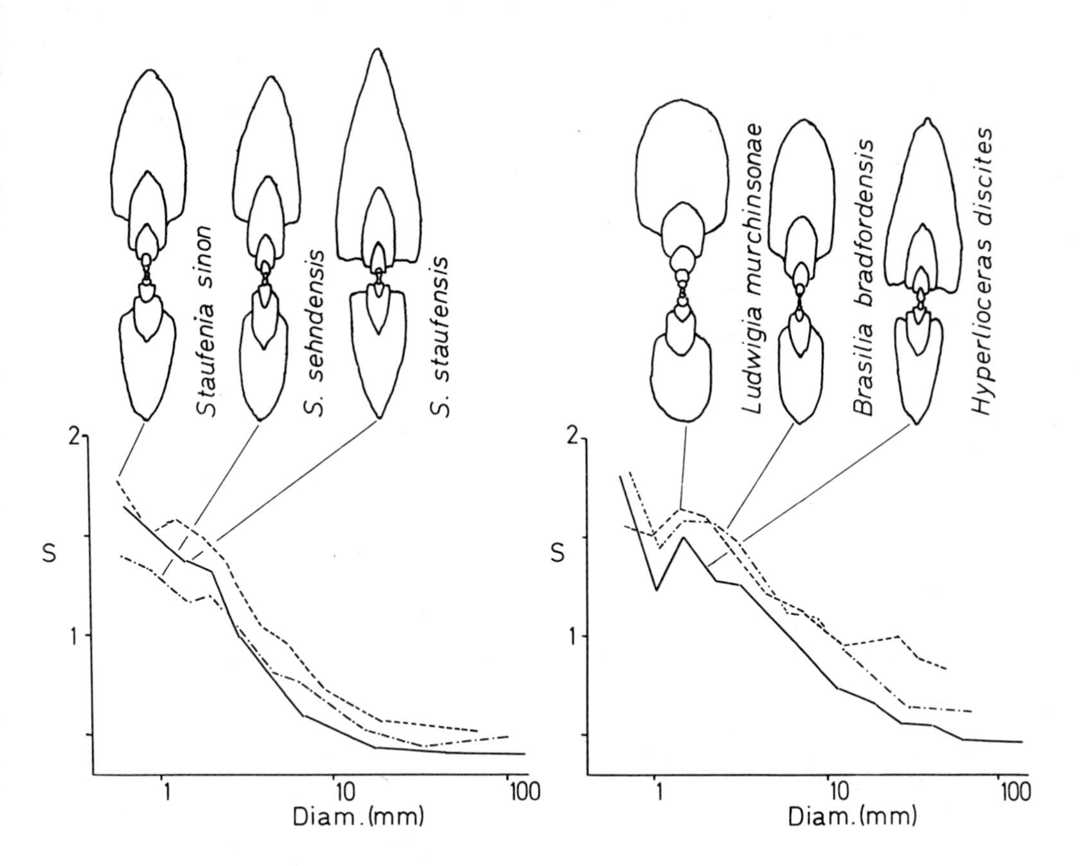

Fig. 12: Developmental acceleration in the Leioceratinae and the Grapho-ceratinae as observed by RIEBER (1963). The parameter S measures the shape of the aperture and is plotted against the diameter of the shell (modified after RIEBER, 1963).

trend: every morphological cycle starts with somewhat more evolute and more intensively sculptured forms (Fig. 11); and, each cycle terminates with somewhat less involute forms -- the extreme case of *Staufenia staufensis* in the first cycle is not fully reached in the second cycle, and so on. On the other hand, the morphological cycles overlap to some extend, thus there is no simple replacement but the same pattern as was observed for the "Oolithenbank". We shall later return (section 3) to this replacement pattern.

b) Ontogeny and phylogeny

In addition to convergent morphological trends, the ammonites under consideration have another interesting feature in common. They also have similar ontogenetic trends: starting with evolute rounded whorls in the juvenile and ending with compressed and more involute whorls in the adult (RIEBER, 1963; BAYER 1969a, 1972). Thus, the ontogenetic

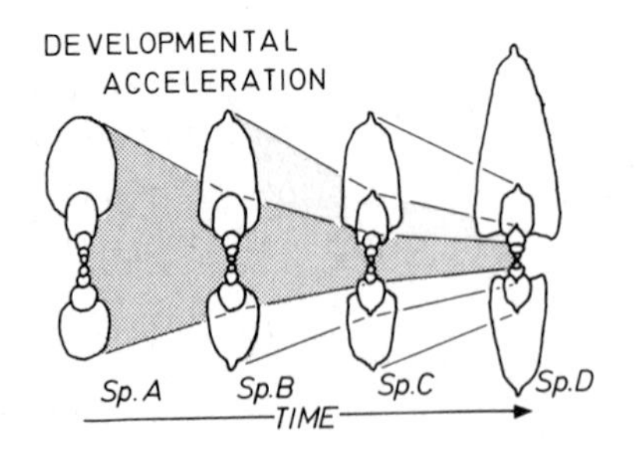

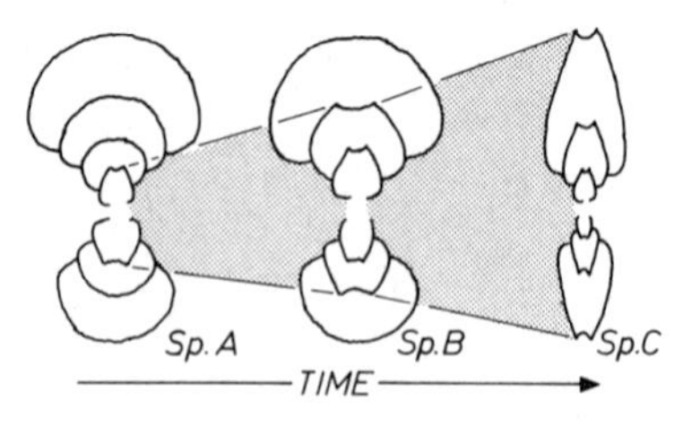

Fig. 13: The classical view of developmental acceleration versus neoteny ("Proterogenesis", SCHINDEWOLF, 1950). In the case of developmental acceleration the ontogeny of the ancestors is later repeated in the early ontogeny of species B to D. In the neotenic case the early states of species A become the adult features of species C.

development of the individual mirrors the observed phylogenetic sequences of species within all groups under consideration. Fig. 12 illustrates some individual ontogenetic pathways for the Leioceratinae and the Graphoceratinae, which have been studied in detail by RIEBER (1963). The morphological changes seen in each iterative cycle thus appear to have been produced by "developmental acceleration", as defined and discussed by GOULD (1977) and ALBERCH et al. (1979). Within a given cycle, each successive species becomes more compressed and evolute -- i.e., the normal developmental trend is carried further in the ontogenetic history of each successive species. Fig. 12 illustrates also that developmental acceleration in these ammonites does not lead to a simple condensed repetition of the ontogeny of previous species in the sense of Haeckel, as illustrated in Fig. 13. The entire ontogenetic pathway is altered to some extent, as far as the precision of the measurements can show. As RIEBER (1963) demonstrated, one can use the early ontogenetic stages of individuals in the Graphoceratidae for classification at the subfamily level.

On the other hand, one has to be careful not to automatically interpret this apparent pattern of developmental acceleration between species as evidence for evolutionary descendency of species. The Hammatoceratidae did not evolve within the basin -- they entered, occasionally, from the Tethys where a more complete phylogenetic record of the family can be found (GECZY, 1966; ELMI, 1963). As discussed in the last chapter, there is also little doubt that the Sonniniidae did not evolve in Europe, and within the Graphoceratinae phylogenetic relationships are still rather unclear. Neither *Brasilia*, *Graphoceras* nor *Hyperlioceras* can be traced to ancestors either within the basin or in middle Europe. The genus *Staufenia* is the only one for which in situ evolution within the German basin is highly probable. *Staufenia staufensis* and the less well documented *Staufenia - discoidea* are true endemic forms, and have never been found outside the German basin (see Fig. 1).

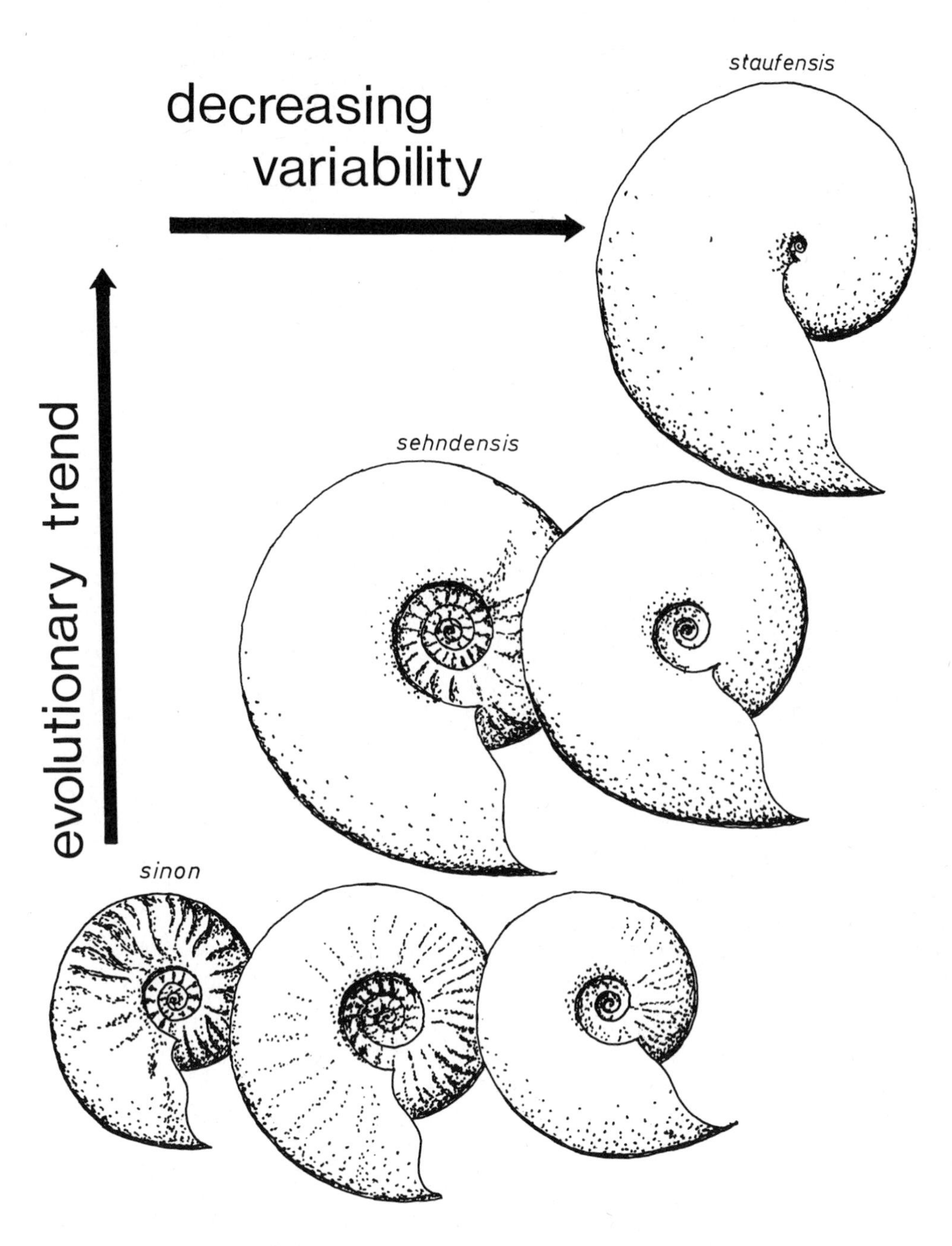

Fig. 14: Qualitative evolutionary pattern of the genus *Staufenia*. The morphological evolutionary trend is accompanied by a size-increase and by a decrease of intraspecific variability.

The early species of this genus, *Staufenia sinon*, is rather variable, particularly with respect to sculpture (RIEBER, 1963). In the evolution of the later species this variability is reduced, size is increased and, by developmental acceleration, the adult features are more hypermorphic, as the shell becomes more involute and smooth. Fig. 14 illustrates these patterns in a qualitative fashion, the result of looking through various collections and monographs.

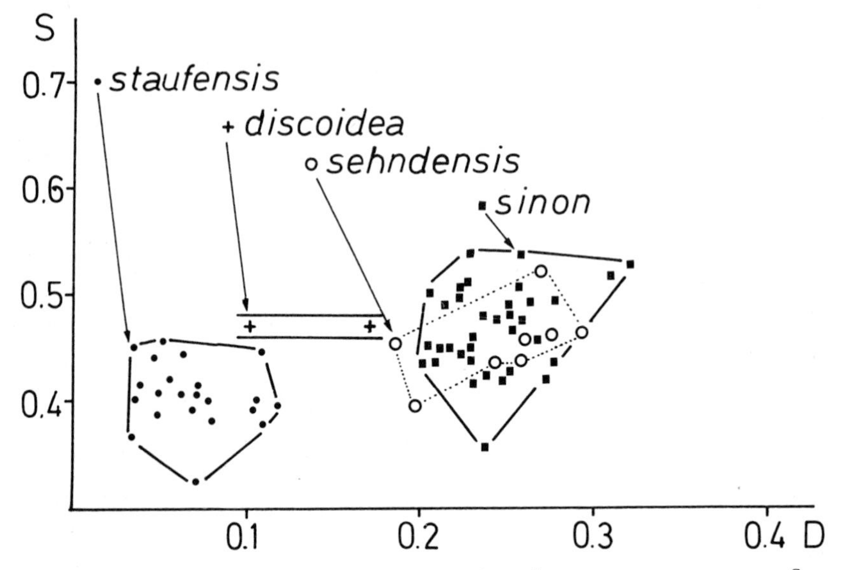

Fig. 15: The endemic lineage of *Staufenia* in the parameter space S and D (see fig. 11). The descendant *Staufenia staufensis* is widely separated from its ancestors (mainly in terms of evoluteness) while *Staufenia sinon* and *Staufenia sehndensis* overlapp in considerably. The two specimens of *Staufenia discoidea* may not be representative.

However, if one tries to q u a n t i f y these trends by measurements of evolutness (D) and cross-section (S) the trend become much less clear. It appears (Fig. 15) that *Staufenia staufensis* is widely separated morphologically from its ancestors. The intermediate form, *Staufenia discoidea*, is poorly documented and the identity of the only two individuals for which we have data is doubtful. The morphological divergence between *Staufenia staufensis* and its ancestors becomes still more pronounced if evoluteness (D) is plotted versus time (Fig. 16). From a statistical viewpoint it appears that there is no evolutionary trend detectable within *Staufenia sinon*, for which data from two subzones are available (Fig. 16). In addition, *Staufenia sehndensis* does not differ morphologically from *Staufenia sinon* (totally overlapping standard deviations). In taking all morphological features into account (Fig. 14) it seems likely that this species represents merely larger individuals of *Staufenia sinon*, i.e. the only detectable evolutionary trend through two subzones is a slight size increase. Thus, even with these clearly endemic species we find rather long periods of evolutionary stasis, which are then punctuated by a rapid species replacement.

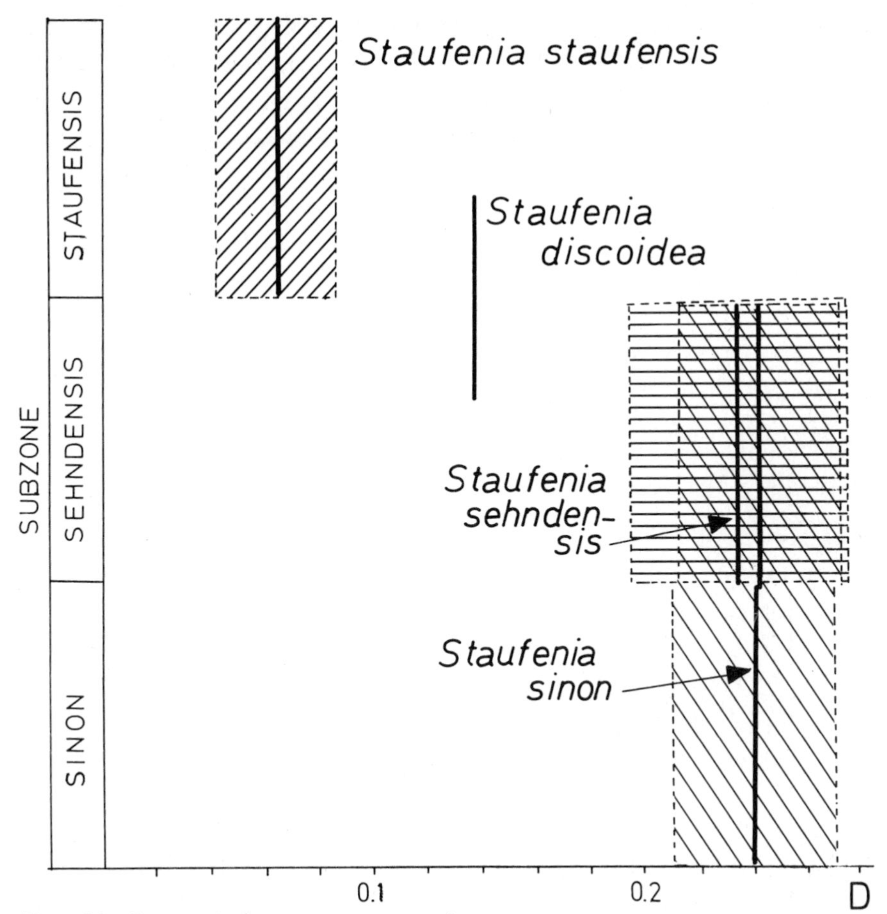

Fig. 16: Temporal (ammonite subzones) evolution of the quantitative parameter of evoluteness (D; mean and standard deviations are plotted). *Staufenia sinon* and *Staufenia sehndensis* cover the same morphological field; they are statistically equivalent because of the total overlap of standard deviations. The data for *Staufenia sehndensis* are from two independent samples from the two subzones.

From these observations it appears highly likely that the iterative morphological cycles consist entirely of successions of discrete species with no gradual evolutionary transformation. In addition, we can be confident of *in situ* evolution in only a very few species. For most of the species an outside source is much more likely, an external source from which they periodically migrated into the basin. Let us now consider the pattern of variability of the successive immigrants within the basin.

c) Interspecific and intraspecific variability

It was observed by RIEBER (1963) for *Staufenia sinon* and the early Graphoceratinae, and by BAYER (1972) for *Leioceras opalinum*, that these species at the onset of the iterative morphological cycles are extremely variable. A similar observation for the early Sonniniidae was made by WESTERMANN (1969) and carefully analysed (Fig. 17). The result was that Westermann lumped the originally described 80 species of *Sonninia (Euhoploceras)* into a single species on the basis of an observed intercorrelation of morphological features (Fig. 17). Strongly ornamented forms are usually evolute with rounded cross-section, while smooth forms are usually involute and compressed. By plotting size (end dia-

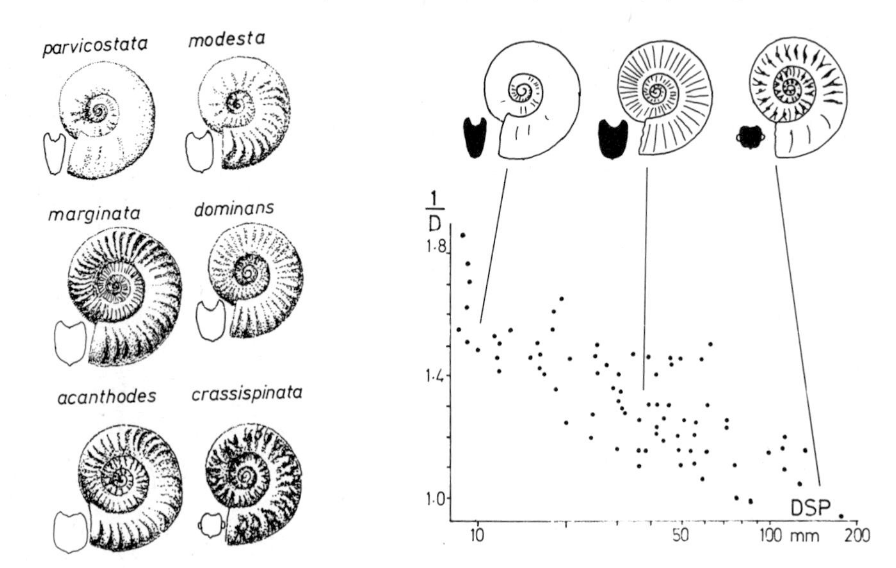

Fig. 17: a) Some representatives of the species *Sonninia adicra*.
b) Covariation of ornament and cross-section of *Sonninia adicra* (WAAGEN), modified from WESTERMANN (1966). The scattergram shows that the morphotypes cover a continuous area in the parameter space; D: distance from coiling axes, DSP: end diameter of the spinous stage.

meter of spinous stage) against evoluteness (D) he showed that it is likely that these highly diverse forms belong to a single species, as the statistical continuum is rather homogeneous (Fig. 17). A similar type of morphological covariation was later found to hold also for *Leioceras opalinum* (BAYER, 1972) and it is likely for the other mentioned forms (Figs. 14 & 15). Now, it is here less important whether the observed variability is viewed as very high i n t r a s p e c i f i c variation in a single species, or if it is viewed as true i n t e r s p e c i f i c variability among diverse separate species. The point is that the high variability or diversity of the individuals at the onset of a faunal event show a morphological covariation which correlates closely with the observed morphological changes that occur within the morphological cycle. Further, it also corresponds with the observed ontogenetic trend in morphological characters. This initial variability indicates that there has been either very low selection pressure on these particular morphological features, or a rather diverse habitat structure was inhabited by this "species complex".

The variability of forms at the onset of a faunal replacement, therefore, might indicate that at this particular time a period of very high habitat diversity existed, which was not occupied by the forms currently within the basin; i.e. the ecological system was not in equilibrium at the initial phase of the cycle, as diversity is commonly reduced in systems approaching equilibrium (AUCLAIR & GOFF, 1971; DAYTON, 1971; CONNELL, 1975). The pattern of morphological variability of the immigrants supports the assumption that this "initial habitat diversity" includes the total spectrum of habitats which are later successively reduced and replaced as expressed in the consequent iterative morphological cycles when the system did approach equilibrium. In the next chapter we shall try to synthesize and explain these observations in a series of models of ecological change and phylogenetic response.

3. ECOLOGICAL MODELS FOR EVOLUTIONARY PATTERNS

New concepts often do not result from new empirical facts but simply from viewing well known data from a new perspective. Even the alternative perspectives themselves have usually already been conceived or prognosed previously. As ELDREDGE & GOULD (1972) pointed out:

> *"The expectations of theory color perception to such a degree that new notions seldom arise from facts collected under the influence of old pictures of the world".*

In the two preceding chapters it was our goal to present alternative views of data which were originally collected in a stratigraphic perspective. We will now attempt to summarize the result in a simplified and unified picture, which will clarify the principal structures of the previously observed patterns. We shall assume three principles which are common to all paleontological thinking:

** *species change with time -- a stratigraphic observation*

** *organisms are adapted to the environment -- an actualistic ecological observation*

** *environments change with time -- a geological and sedimentological observation.*

In using any of these principles in isolation one arrives at the extrema of chronostratigraphy, time-independent paleoecology and functional morphology, and pure lithostratigraphy.

If we consider that the local environment is itself embedded in larger environments (Fig. 1), then its relative condition changes as either its external or internal context is altered. Viewing environmental change from an ecological point of view we expect faunal or species substitutions, while from the alternative evolutionary viewpoint we expect faunal or species adaptation. The local system, therefore, can become highly complicated with respect to the different types of environmental change and faunal response to be expected, and probably it can rarely be analysed in truely dynamic terms.

Sedimentation is related to dynamic processes in the water column, evolution involves rather complex population dynamics, and the ecological system itself represents a dynamic equilibrium of rather fearsome complexity. These aspects of dynamic systems can never truly be preserved in the paleontological record. There is simply no way to measure the various parameters and forces which were involved in the ancient dynamic system. What we observe in the fossil record is the outcome of ancient dynamic processes, the topologic picture they have produced -- the geometry of the local phylogenetic lines and faunal associations, the geometry of facies types and their spatial arrangement, the geometry of the stratigraphic record.

The possible description available is a kinematic one. What we shall do in the following sections is to develop kinematic models which are capable of elucidating the principles -- but not to explain the causes -- of the previously described faunal replacements and morphological cycles in ammonites. In the first section we shall discuss the sedimentological and geographic framework of the southern German Aalenian basin in terms of changing paleoenvironments. In the second section the ecological and evolutionary aspects of these environmental changes will then be formalized in a series of models. We shall begin with the specific case of the Aalenian ammonite cycles, and their relationships to environmental changes as a starting point from which more abstract concepts are then derived.

3.1 Asymmetric Environmental Changes

Two physical aspects are here of importance: the geographic and sedimentological framework of the basin under consideration, and the evolution of this framework through the course of time. Here we summarize these aspects in detail for the South German basin in Aalenian times to provide the background for the explanation of the observed iterated morphological cycles in ammonites. We shall further point out that the dynamic causes of environmental changes are interesting but not necessarily essential for the understanding of the evolutionary pattern produced. We stress this point because in the past relevant observations, like those of "Klüpfel-cycles", have commonly been discredited because they did not agree with current explanations that were themselves later disproved.

a) South Germany at Aalenian times

A generalized paleogeographic map of middle Europe, and simplified facies distributions in Aalenian time, is given in Fig. 18. The area was divided into two major sub-basins with different facies types. In the western basin carbonates dominate (which are usually condensed), while in the eastern basin a clastic belt is developed, reaching from the North Sea to southern Germany. In the south the front of the Alps bounds the area and our knowledge of Aalenian paleogeography ends here. Within the

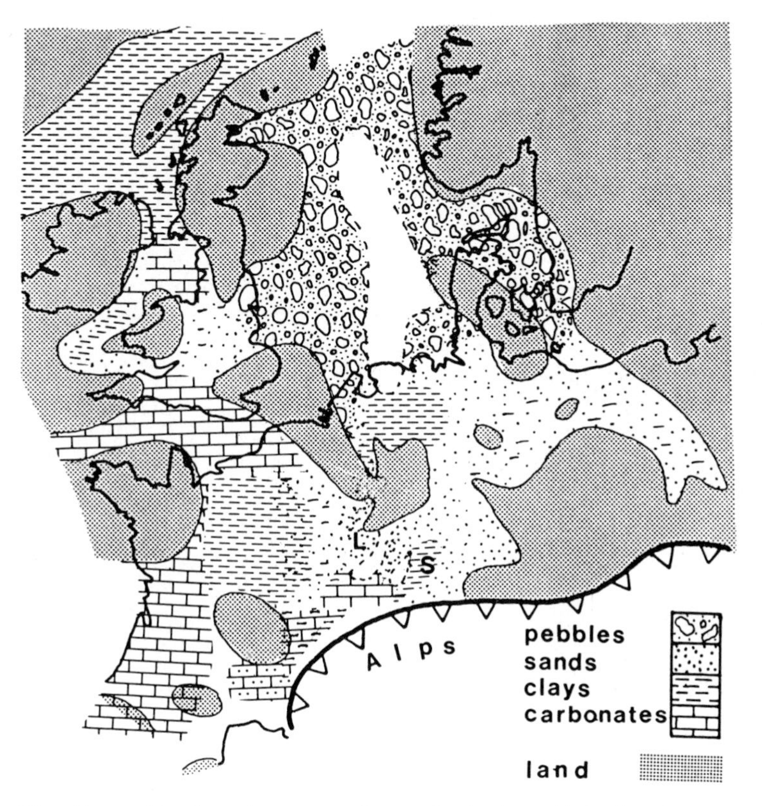

Fig. 18: Generalized paleogeography and facies distribution in middle Europe during Aalenian times. L: Lorraine, S: South Germany. After ZNOSKO (1959), BUBENICEK (1970), RAT (1974), POMEROL (1978), ZIEGLER (1978), ANDERTON et al. (1981), NAYLOR & SHANNON (1982).

clastic belt relatively coarse grained sediments occur in the North Sea area, which were deposited partially under brackish conditions. Towards the south the sands become extremely fine and well sorted. They are distributed mainly along the eastern part of the belt. At the western boundary clays locally dominate the sediment, chiefly in two clusters, the depositional centers of the North German and South German basins.

The South German basin (S) is located at the southern end of this clastic belt and near the major entrance to the Tethys. Fig. 19 illustrates in still more

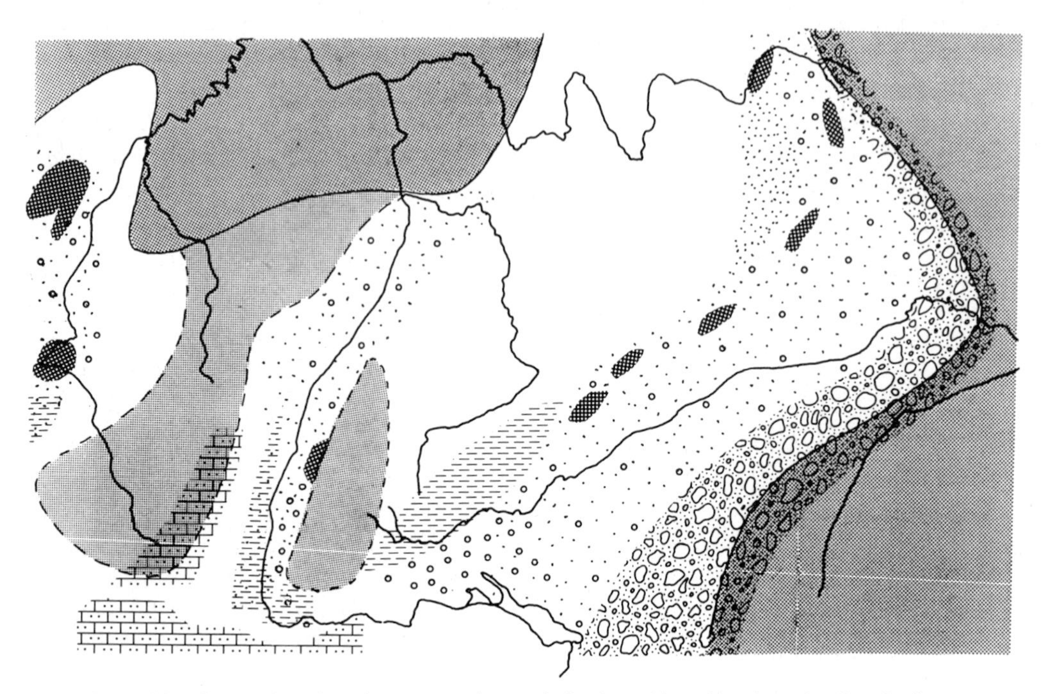

Fig. 19: Generalized paleogeography and facies distribution in South Germany during Aalenian time.

detail the facies distribution within the South German basin in Aalenian times (a similar picture holds for the lower Liassic, ALDINGER,1968; BLOOS, 1976). At its northern end the basin is connected with the North German basin via the 'Hessian strait'; to the west it is bounded by the 'Vogesian swell' which separates it from the "Minette" deposits in Lorraine which in parts are lithologically and temporally equivalent to the South German Aalenian; to the south and southwest the basin is bounded by a facies transition to carbonates and iron-oolitic condensed beds (GENSER, 1966).

The sedimentological pattern which dominates the scene are sand bodies and iron—oolite bodies which parallel the coastline. These cross-bedded offshore bars indicate a coast-parallel transport of sands which passed through the 'Hessian strait'. The analysis of cross-bedding (WERNER, 1959; WEBER, 1964) and grain size distributions (ROGOWSKI, 1971) indicate wave action from the northwest to the southeast, while the general sand transport was longshore (northeast to southwest) in offshore bars. Within the deepest parts of the basin, coarse-grained beds occasionally interrupt the monotonous clay deposits. These generally thin beds contain commonly reworked material from shallower areas -- sands and rounded sandstone-pebbles in the northern region, cemented carbonate fragments with (iron-) ooids in the southern region (BAYER et al., this volume). Graded examples of such beds indicate event (tempestite) deposition.

A synthesis of the available data indicates a longshore transport of sands which entered the basin via the 'Hessian strait'. A similar sediment supply by way of the independent 'Eifel depression' is likely for the sediments of the 'Minette' in Lorraine (west of the 'Vogesian swell'). The temporal differences of siliciclastic sedimentation in these two southernmost clastic basins (mainly Toarcian in Lorraine, Aalenian in South Germany) can be related to local epeirogenetic movements which opened and closed the straits. Synsedimentary epeirogenetic movements are well documented for the 'Eifel depression' (LUCIUS, 1940; MULLER et al., 1976) -- faults separate several local basins with different temporal sedimentary sequences.

Various sedimentary models have been proposed for the Lower and Middle Jurassic siliciclastic sediments in Lorraine and in South Germany (LUCIUS, 1945, 1949; ALDINGER, 1957, 1965, 1968; MULLER, 1967; BUBENICEK, 1970; THEIN, 1975; MULLER et al., 1976; BLOOS, 1976 and others). However, it turns out that there are two principal geographic-geological factors which affect the facies pattern: Local tectonics which causes specific facies zones, and the hydrodynamic conditons within the straits which control overall sediment supply, e.g. regressive-transgressive cycles affected the sediment transport through the straits.

b) Asymmetric sedimentary cycles

Within the South German basin classic asymmetric sedimentary cycles (coarsening—upward sequences) are developed within specific geographic zones (BAYER et al., this volume). Due to a general regressive trend (documented by the offlapping pattern of sandstones, WEBER, 1967; BAYER & McGHEE, 1984) these zones of 'minor Klüpfel cycles' shift towards the deeper parts of the basin during the course of time. Towards the coast these cycles grade into condensed beds with faunal mixing. Towards deeper parts of the basin the cycles separate into series of small beds (BAYER & McGHEE, 1984) which in parts have the characteristics of single event beds. Within this trend two extreme regression stands can be recognized during the *bradfordensis-gigantea* subzones and the "*sowerbyi*-Zone" (= *laeviscula*-Zone). These extreme regression peaks are not only of regional importance but e.g. are also found in the Lorraine basin and, therefore, are likely to reflect true sea-level fluctuations. Especially in the "*sowerbyi*-Zone" a reduced general supply of siliciclastic sediments is likely and can be related to a partial closure of the 'Hessian strait'. During the "*sowerby*-Zone" hardgrounds and condensed beds spread widely over the basin which were accompanied by the occurrence of patched reef corals -- the corals' pollution sensitivity underlining the reduced siliciclastic sedimentation.

The spatial patterns portray a specific temporal sequence within the ideal asymmetric cycle (Fig. 20). Starting with clay sedimentation with occasional nodule layers, small sandy and shelly beds become interbedded and then condense into cross-bedded sands; overlain by shelly beds with erosional contacts; eventually mature carbonates with iron-oolites follow. The cycle is terminated by an erosional horizon which is usually developed as a hard ground. In a time-facies diagram (Fig. 20) such a cycle contains many erosional and non-depositional gaps, with a decreasing frequency of gaps towards the deeper parts of the basin -- but with increasing frequency in the time sequence.

The asymmetric nature of such cycles has been widely discussed. KLÜPFEL's (1917) original interpretation relied upon local tectonics. Later (ALDINGER, 1957), periodic fluctuations in the strength of bottom currents became the favoured explanation, followed by sea level changes (HALLAM, 1961, 1981; ALDINGER, 1965; but also BRANDES, 1912). Climatic changes have always been another single factor to explain cyclic and rhythmic patterns (e.g. POMPECKJ, 1914, 1916; FISCHER, 1981, 1982). These factors have already been controversially discussed by KLÜPFEL (1917). In this volume an event model is presented which works either with sea level changes or long-time climatic fluctuations. All these causal explanations have one aspect in common: by some mechanism the near-bottom flow power is increased. Thus, the various genetic models can be subsumed under the single relevant parameter 'velocity of near-bottom flow' which again may be either permanent or event-dominated. The causal mechanisms for cyclic sedimentation can only be analysed on regional and global scales, i.e. by basin analysis, and several factors may locally interfere in the formation of cyclic patterns as it is likely for the South German basin. The major point, in any case, is to explain the asymmetric nature of sediment supply -- for which the paleogeography of the basin provides the necessary framework.

c) Sedimentation models

The longshore transport of sand and the build-up of offshore bars is a rather slow process especially if the system is wind-dominated as assumed for the South German basin (Fig. 21). Thus, any change in the dynamic system which increases redeposition of sands in the deeper parts of the basin relative to overall sediment supply (sea level fluctuations, decreasing subsidence, climatic changes) will alter the local sedimentary pattern in the same way as decreasing overall sediment supply relative to sediment removal by redeposition (closure of the straits, changing weathering conditions, altered relief of the source land etc.). Relevant only is the relative equilibrium condition between overall sediment supply and sediment redeposition towards deeper areas. This applies both to the spatial facies distribution at a given time and to the temporal pattern of change at a given location. Thus we can formulate:

Through the course of time, at a fixed locality, the pattern of sedimentation depends on relative balance between sedimentation rate and erosion rate.

In this context times of n o n - d e p o s i t i o n would represent states near

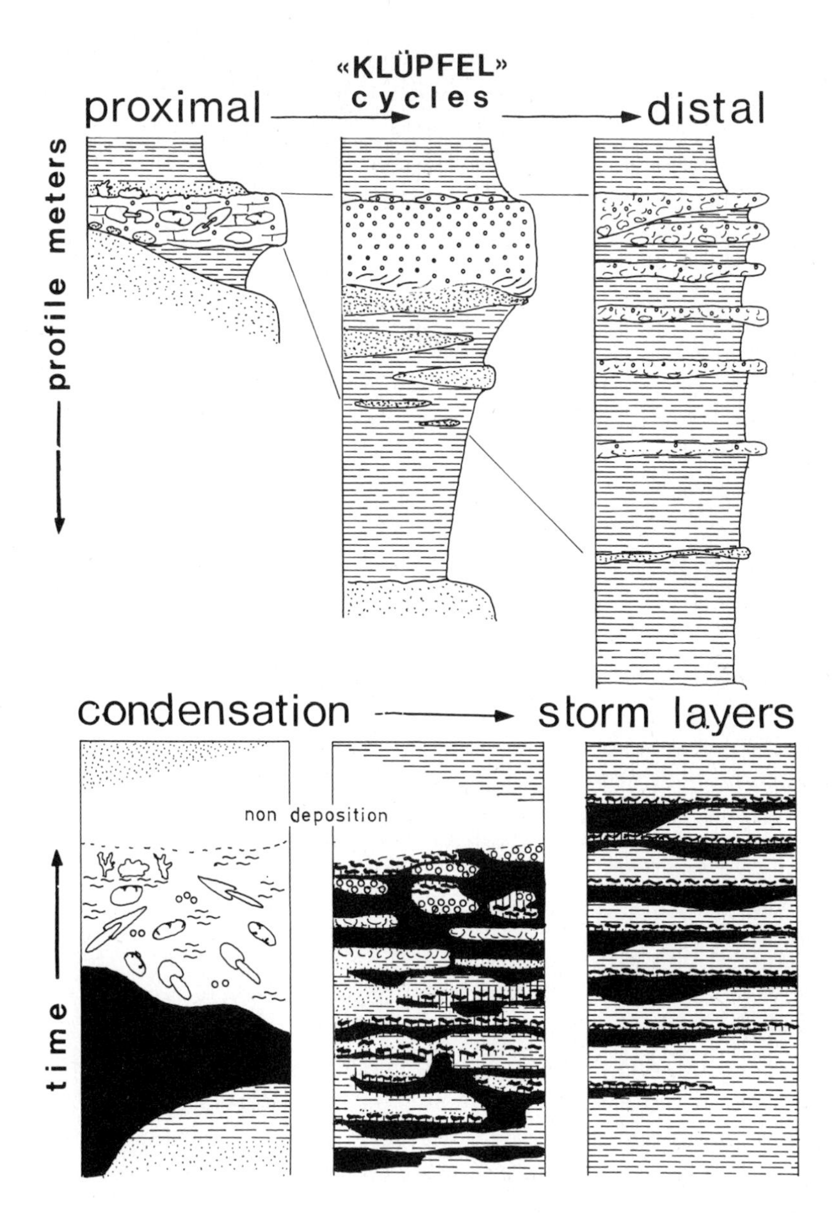

Fig. 20: Proximality trend in idealized asymmetric sedimentary cycles and their image on the time scale. Black areas denote erosion, white areas reworking and non-deposition. The ideal "Klüpfel cycle" grades into condensed beds towards the coast and splits into a "thickening-upward" sequence of event beds in the deeper areas.

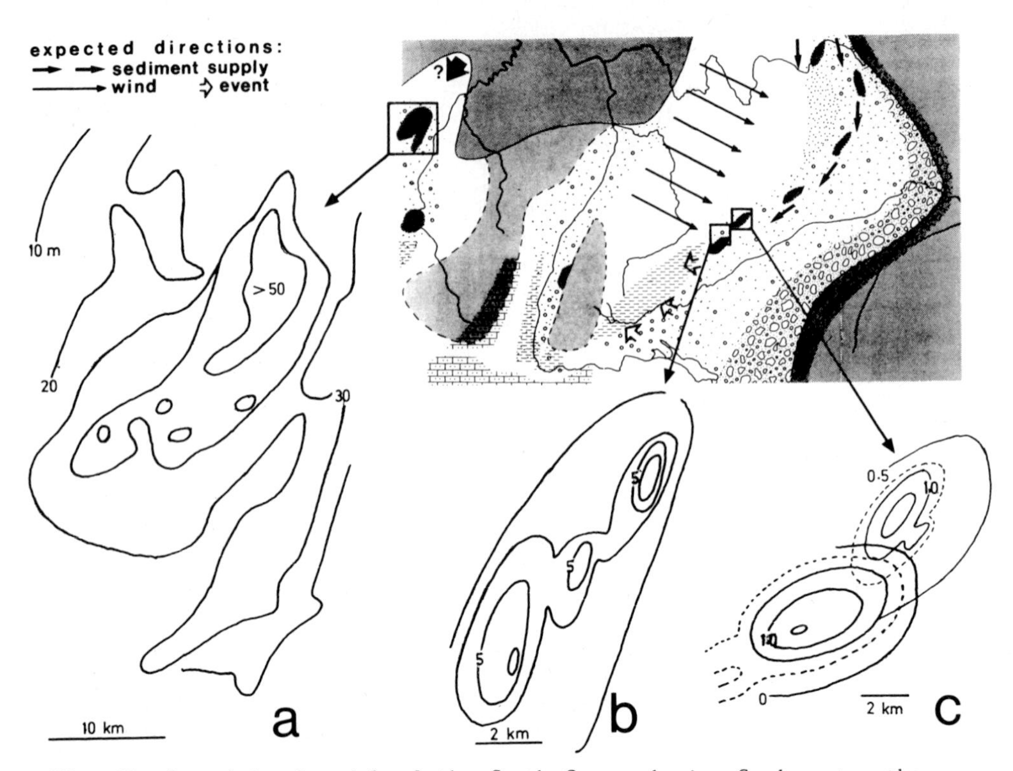

Fig. 21: Depositional model of the South German basin. Sands enter the basin through the 'Hessian strait' and are transported longshore by wave action, which is induced by a dominating northwest to southeast wind direction. The sands accumulate in offshore bars from where they are redeposited in deeper parts of the basin under storm conditions. A similar sedimentation pattern is likely for the Lorraine basin via the independent 'Eifel depression'. The contour lines represent sand and iron-oolite bars, a: the "Minette" after LUCIUS (1945, 1949), b: a sand bar and c: two subsequent iron-oolite bars after WILD (1950).

the actual equilibrium between sedimentation rate and erosion rate. The earlier discussed "Oolithenbank" is an example for deposition near this equilibrium; some 'roof beds' of the asymmetric cycles are of this type, and an actualistic example of non-deposition by fluctuating sedimentation and erosion is given by SEILACHER (this volume). Asymmetric cycles are to be expected whenever the restoration time for local sediment supply is slower than the speed of local erosion or redeposition, i.e. if one has a system with time delay in relative sediment supply, or a "hysteresis".

This view of a hysteresis in sediment supply with respect to the depositional center of a basin is summarized in Fig. 22 in terms of a regression-transgression model (other causes, however, should not be excluded). The sands and oolites which have been normally accumulated in offshore bars are now redeposited in deeper parts of the basin during the regression. Subsequently they are replaced in the nearshore region by hardgrounds, which

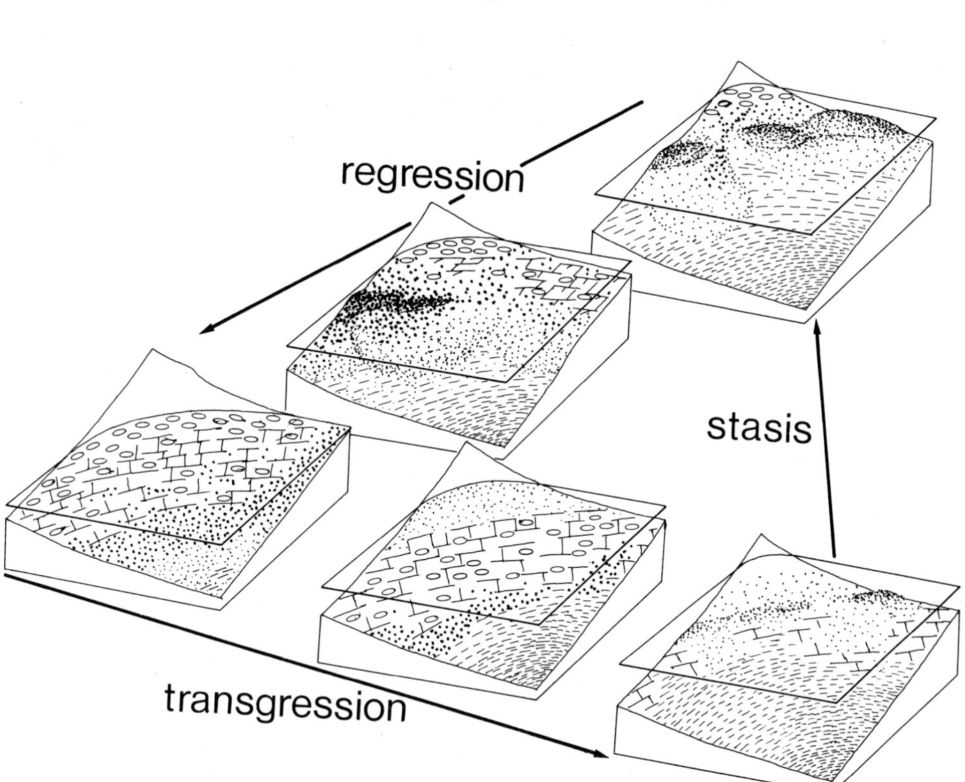

Fig. 22: A 'hysteresis' cycle in sediment supply during a regression-transgression phase. During the regression sediment supply decreases due to closure of the supplying strait and subsequently erosional events dominate the sedimentary pattern. During the transgression the coarse material is initially retained near the coast line due to a time delay in sediment supply, and clays spread over the basin.

finally spread widely over the basin as the clastic supply rate is now reduced to a steady state input below the reworking rate. During the following transgression the coarser material is initially retained near the coastline, where it begins to accumulate again in offshore bars while clays are deposited in the deeper areas. After an initial time of stasis the offshore bars are restored and form a resource of coarser sediment which occasionally can once again be transported into the deeper parts of the basin.

Independent of the special dynamics, we find that any change in the balance between deposition rate and erosion rate affects the facies distribution within the basin. Further, within a sedimentary cycle two monotonous states occur -- clays at its beginning and hardgrounds (shell beds, oolites) at its end. These uniform states are not only locally developed but are widespread within the basin. Between these two uniform states a maximal facies diversity and patchiness is passed. Because of the increasing condensation trend within a cycle this state of maximal facies diversity occurs in the upper part of the stratigraphic column of a cycle, within the "roof bed".

There are other interpretations possible -- e.g. condensation at the base of cyclothems (HALLAM, 1978). However, the iron-stones which form usually the "roof-beds" under discussion point to the highest energy level (HALLAM & BRADSHAW, 1979). Although the sedimentary patterns are clearly asymmetric, a symmetrically oscillating forcing factor is possible (HALLAM, 1978; BAYER et al., this volume), e.g. symmetric sea-level changes will affect the sensitive longshore-transport through the 'Hessian strait' and will cause a depositional "hysteresis".

d) Environmental changes and faunal response

It was pointed out by THOMAS & FOIN (1982) that species diversity is a product of both biological interaction and physical-geographic factors. Biological interaction cannot be observed and hardly reconstructed from the paleontological record -- the factors we can control to some extend are the physical-geographic determinants:

"A model which eliminates biological interactions as a determinant of diversity can be used to assess the potential impact of physical and geographical factors operating alone on temporal trends in diversity (succession) and on spatial differences along latitudinal or environmental gradients"

(THOMAS & FOIN, 1982, p. 45). The South German Aalenian provides an example of repeated temporal trends in physical and geographical factors where regional environmental gradients have been heavily disturbed by sedimentological processes and, therefore, cannot be recognized within the basin with respect to ammonite faunas.

As a direct consequence of the previous discussions one can see that benthic environmental conditions change drastically during the asymmetric sedimentary cycle throughout the entire basin. The changing pattern of sedimentation affects directly the local pattern of environmental diversity. The cycle is bounded by the two monotonous states of clay sedimentation at its beginning and hardground formation at its termination. Between these boundaries a state of increased evironmental diversity is achieved and passed and it is not unreasonable to assume that during this period the local patchiness of environments also increases.

These considerations are summarized in Fig. 23 with reference to the previously discussed temporal cycles in ammonite morphology. It turns out that the observed facies relationship of ammonite morphologies is a rather indirect one. The morphological changes are not directly related to a particular static facies, but rather to the asynchronous spatio-temporal pattern of facies shifts and to the diversity of facies types within the basin available at any given time interval. Thus, the observed faunal changes correspond to a simple ecological substitution of species as the basin framework produces new or

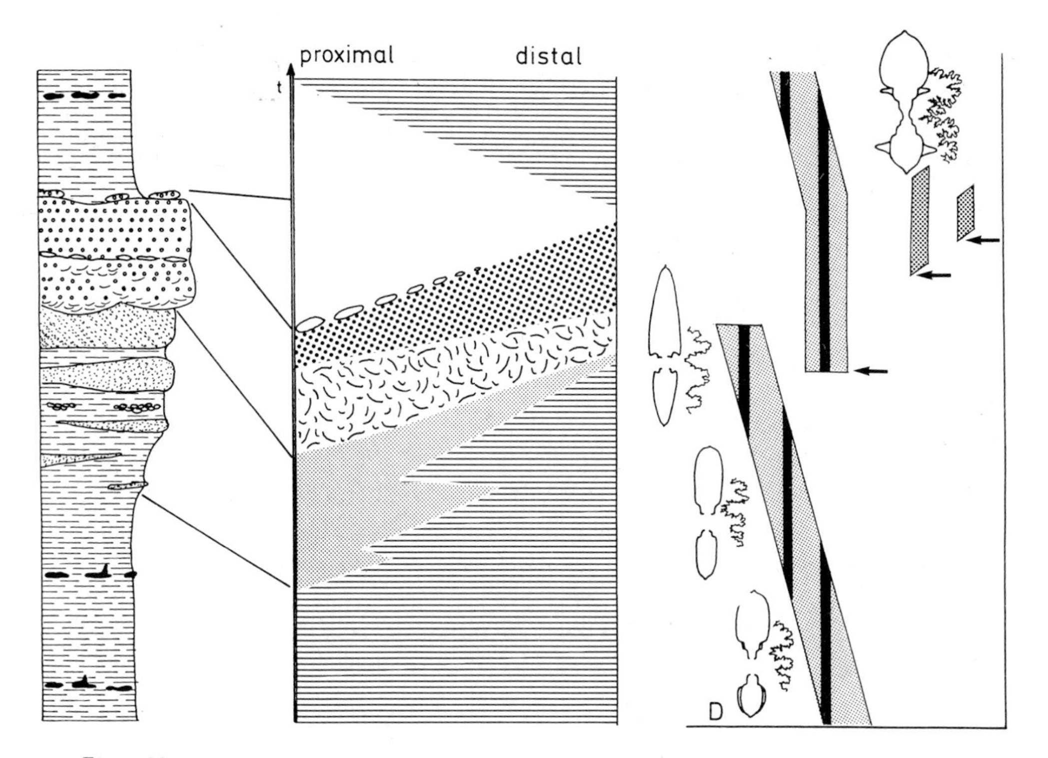

Fig. 23: An ideal Klüpfel cycle and the response of the ammonite fauna. During the early phase the habitat of the *in situ* ammonites is reduced and this causes an increase in directed selection pressure. In the same course, the total habitat diversity increases and already adapted forms enter the basin from outside. At the end of the cycle habitat diversity decreases again and selection for another morphotype begins, from which a similar cycle may result.

deletes old environments. Under this aspect the observed overlap of morphological cycles is easily understandable (section 2.2a, Figs. 10 & 11). It corresponds to the decreasing size and the increasing patchiness of the original habitat, while at the same time totally new habitats are appearing which are normally found only outside the basin - such as the condensed carbonate regions of the western sedimentary belt (Fig. 18). We would expect that these new habitats would be occupied by the already adapted forms (via migration) from the nearby extrabasinal regions -- clearly, at least the hammatoceratids and the occasionally occurring extremely evolute and rounded forms verify this consideration. As the environmental variability within the basin during this period of time is relatively large, we would expect either a high species diversity or a high intraspecific variability within the immigrants, an expectation which also agrees with the observed variability of species. HALLAM (1961) observed a similar relationship between Liassic ammonites and sedimentary cycles:

"A notable point ... is the slight overlap of genera immediately below the top

of cyclothems. This asymmetry seems to be genuine, for the tops of the cyclothems definitely mark the disappearance of given genera."

The Liassic examples of the first chapter illustrate such replacement patterns which were summarized by HALLAM (1978b):

"Times of low sea level are characterised by comparatively unstable high stress environments, high extinction rates, restricted marine communications promoting high endemism, and r-selection as the dominant adaptive strategy".

There are many parallels to the eustatic speciation model proposed by HALLAM (1978b) -- the iterated morphological cycles are another example of the punctuated euquilibria model of species replacements "with the exception of phyletic size increase, which appears within the limits of the data to be gradualistic" (HALLAM, 1978b) as the *Staufenia*-lineage illustrates. Adding the ecological dimension, the local successive contractions and expansions of environments and especially their changing patchiness, may well lead to a deeper understanding of the punctual fossil record as HALLAM (1978b) pointed out.

KLÜPFEL (1917) was, perhaps, the first to recognize the close relationship between sedimentary cycles and faunal evolution. Indeed, one should expect a close relationship between facies and benthic organisms, but a much stronger relationship has been found with ammonites, as in our example, which usually are considered to be nektic (KLÜPFEL, 1917; FREBOLD 1924, 1925; HEIDORN, 1928) and HALLAM (1961) pointed out:

"An intimate correlation exists between cyclothems and evolutionary and migrational changes in the ammonites. Other faunal groups were less influenced by the environmental changes expressed in the sediment".

Similar observations are common in the literature and were much discussed at Klüpfel's time; thus his ideas were not isolated during the first decades of this century. POMPECKJ (1914) summarized the ecological, evolutionary, and migrational aspects of ammonite faunas which were under discussion. He pointed out that the rather confused (local) patterns of ammonite faunal evolution are clarified under the paradigm of migrations and ecological substitutions:

"Solche Einwanderungen (von Ammoniten) schufen mit das in so vielen Teilen als zusammenhangslos sich darstellende Faunenbild, welches die Gesamtheit der Lebewelten des schwäbischen Jurameeres uns enthüllt."

POMPECKJ (1914) -- and previously to him NEUMAYR (1878) -- also depicted the Tethys as the origination area of the colonists, because there are several phylogenetic lines much better documented in this region than in middle Europe. Pompeckj illustrated the sporadic introduction of ammonites into the South German Jurassic basin by the phylloceratids and the lytoceratids -- taxa with an excellent record in the Tethyan realm. The temporal distribution of these rare introductions is redrawn in Fig. 24 and compared with

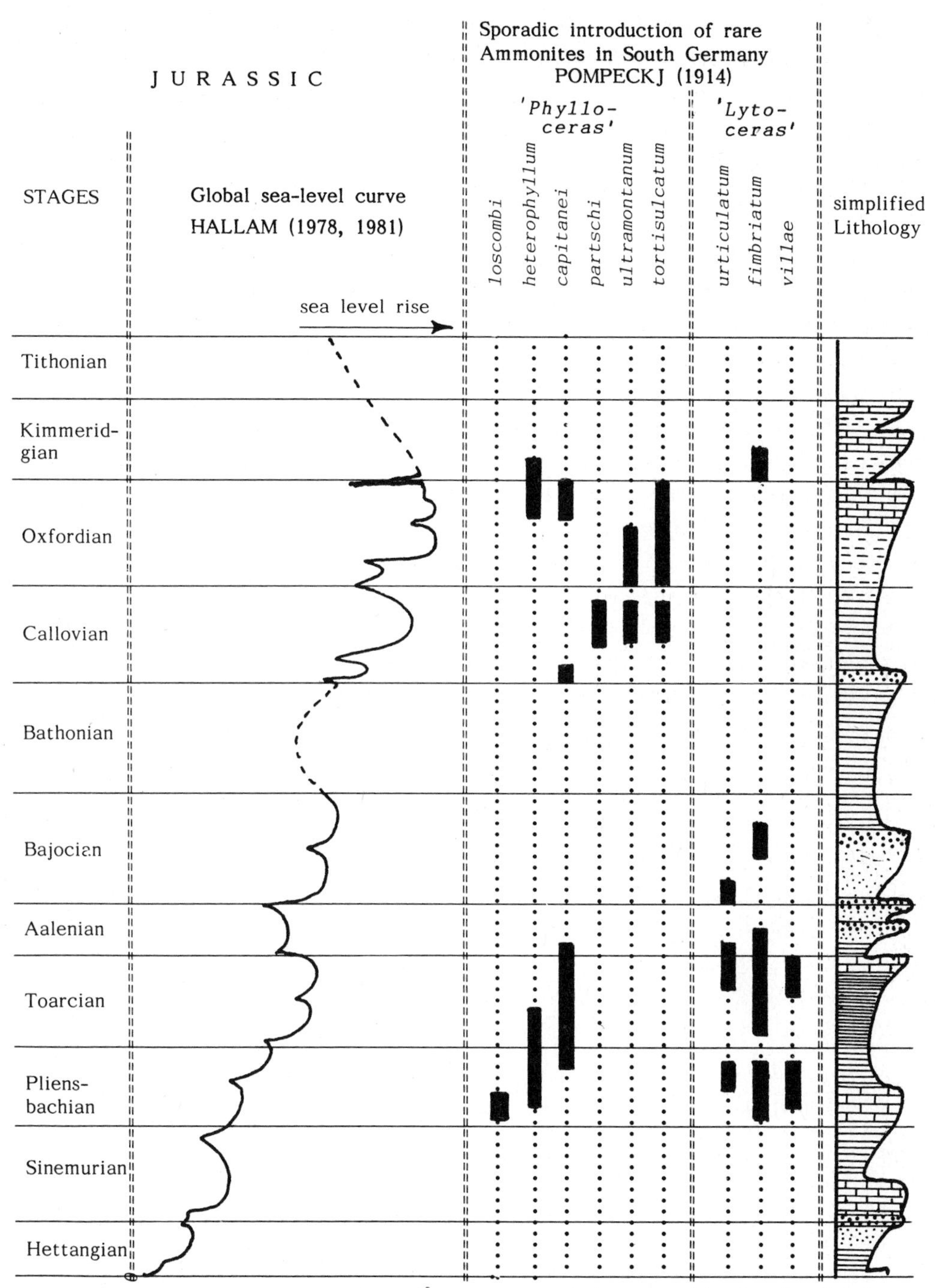

Fig. 24: The occurrence of phylloceratids and lytoceratids in the South German basin during the Jurassic. The faunal data are modified after POMPECKJ (1914) with his original taxonomy. For comparison the global sea--level curve after HALLAM (1978, 1981) and a simplified lithological profile are given.

HALLAM's (1978, 1981) global sea level curve as well as with the generalized local lithological section. Clearly two clusters appear in the faunal list which correspond roughly with the high transgression stands while no correlation with lithology is visible. This pattern one would relate to nektic forms and it clearly contradicts the earlier discussed facies dependency of ammonites -- if and only if one assumes a single common mode of life for all ammonites and one general ecological factor which affects ammonite communities.

The phylloceratids and the lytoceratids are the most typical forms of the Tethyan realm -- they are likely true nektic forms (e.g. GEYER, 1971). In addition, during the Jurassic, they are the morphologically most conservative forms, they are the evolutionary "K-strategists" within the ammonites. In contrast, the ammonites in a narrow sense, the forms useful for stratigraphy, behave like evolutionary and ecological "r-strategists", and it is this group where usually facies relationships are found. Of course, this notion is somewhat oversimplified. However, the evolutionary trends analysed here are very close to HALLAM's (1978b) eustatic speciation model with r-selection during the regression and K-selection during the transgression.

A relationship of ammonites and ammonite morphologies with transgressive-regressive cycles is also pointed out by DONOVAN (this volume). Transgressions and regressions change the environment in several ways, the general bottom conditions as well as the water column. Transgressive-regressive cycles, therefore, give merely some overall measurement of the environment, e.g. some measurement of environmental complexity as discussed above. Changes of environmental diversity due to physical factors may, of course, be the main mechanism which produced the various temporal faunal patterns.

The use of transgressive-regressive cycles as a "control-parameter" for evolutionary processes may even cause confusion. "Radiations" of ammonite faunas during transgressions and extinction events during regressions are well established empirical facts (COOPER, 1977; GINSBURG, 1965; MOORE, 1954; NEWELL, 1952, 1956, 1967; SCHOPF, 1974; WIEDMANN, 1969, 1973). Especially WIEDMANN's (1973) model of ammonoid evolution in relation to changes of sea level seems to contradict our result in two ways:

** The speciation events occur during the transgressions

** the surviving ammonites are constantly oxycones.

However, Wiedmann's examples are related to major sea-level changes like the global Permian or Upper Cretaceous regressions, which affect the entire shelf areas. Our examples are related to minor sea-level changes which cause different degrees of isolation of and within epicontinental basins. The effect of scale becomes an important point which should be taken into account whenever such a generalized control variable like sea-level fluctuations is used. The same argument holds for ammonite morphology -- there is a difference whether such relationships are discussed on the species level or on the level of families. An interesting point with the oxycones is that the terminal forms in the Aalenian have all simplified sutures while the conservative forms cited by Wiedmann have rather complex suture lines.

3.2 Ecology and Evolutionary Pattern

Evolutionary processes, as they are recorded in local phylogenies, are the result of complex spatio-temporal processes rather than simple time successions. At any given time the environment determines habitat complexity and, therefore, the local ecological system. Changes of the habitat cause primarily ecological substitutions rather than directed evolution by gradual adaptation. The latter can probably only be seen when a simple one-to-one relationship exists between a species and an external selective force. On the level of ecological systems, interactions are much too complicated to allow for the recognition of simple evolutionary equilibria between external forces and biological adaptations.

Ecology is based upon actualistic observations; it does not consider *a priori* evolutionary adaptations, because it usually deals with time intervals much too small for the detection of evolutionary responses. On the other hand, it is primarily concerned with faunal and species substitutions, as in the case of "island biogeography" (McARTHUR & WILSON, 1967) or the case of changing seasonal environments (LEVINS, 1968; FRETWELL, 1972). On the ecological level the much discussed problem of evolutionary 'preadaptation' loses its importance. No one doubts that a local environmental change will select those species from an available pool which are best "preadapted" to the altered conditions -- or, those forms will be chosen from the available pool which have the highest fitness of the various species in competition under the new conditions.

In terms of geological time, however, environments and ecosystems are not stable and, by observation, we see the result that species themselves change. As paleontologists we can analyse faunal changes from two viewpoints:

1) We can examine how the entire fauna reacts to some environmental disturbance, i.e. we have to study the geological succession of faunal communities. If we hold environmental conditions fixed by examining equivalent environments through time, we are examining the fauna for patterns of community evolution.

2) The other viewpoint is to confine our attention to a certain taxon and to trace its evolutionary changes through time. This, of course, is the classic phylogenetic approach.

These analytic viewpoints differ only in what we keep fixed or hold constant; in a theoretical sense they are both descriptions of the same phenomena within two different coordinate systems. Therefore, models developed within one framework should be transferable into the other, as they are only sections through the same system where different parameters have been held constant. To understand the processes themselves one has to view both parameters as changing.

We shall now consider this last aspect of complex systems with special regard to our earlier discussions. There are two major controlling factors at the local basin level, geographical or ecological isolation of the fauna and the environmental changes which the faunas experience. With regard to the environment we can further distinguish between three factors:

1) the size or extension of a specific environment changes relative to others: we are especially interested in environments which decrease in area with time.

2) The continuity of the environment is altered, i.e. the patchiness of the overall environment changes: here we are specifically interested in increasing patchiness with time.

3) The quality of the environment is affected, i.e. soft bottom conditions turn into hard grounds in the specific example of Aalenian asymmetric sedimentary cycles.

These are the major environmental changes which we would expect within a certain basinal environment and a certain region. The primary control for modelling ecological evolutionary patterns in basinal systems shall be the relative isolation of the system; that is, the availability or non-availability of an external species pool as a potential source of species for an ecological substitution.

a) The isolated basin

Starting with the simplest system, we consider a totally closed system which can be visualized as an isolated basin (Fig. 25) -- of course, other isolation mechanisms could be chosen but they are less simply visualized. For simplicity we shall consider only two species which belong to the same higher taxon, and which occur in two distinct facies types. During a regression the environment of the basin central species "a" decreases in area. At the same time the patchiness of its environment increases, producing fragmented and increasingly isolated populations. With increasing patchiness and declining geographic range of the population isolates it is likely that the quality of the environment is affected too. The net result is a series of allopatric originations of new species from the adapted populations of the parent species. The expected phylogeny of the central "a" species is a punctuated one (a to a' to a"), with more or less well documented intermediates. Increasing patchiness of the habitat, decreasing size of the patches, and, therefore, of populations, and increasing geographical/ecological distance between the patches provides the

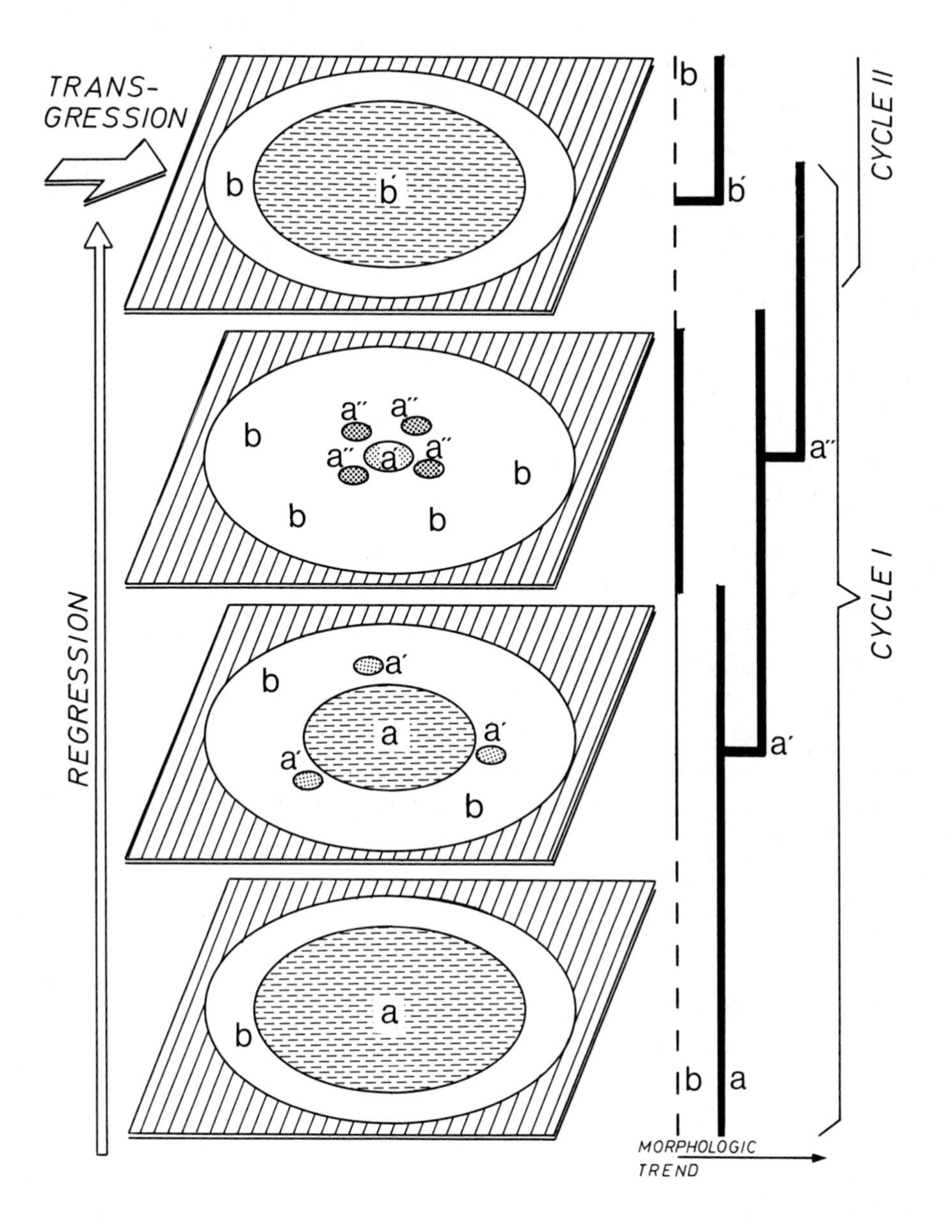

Fig. 25: Evolution within a closed basin during a regression phase. The environment (facies) of species (a) reduces in size and splits into (isolated) patches. A punctuated sequence (a) to (a") evolves. During the same time the facies of species (b) spreads over the basin, and (b) adapts to the central area after the extinction of the (a)-line and the restoration of this facies during the subsequent transgression. As the environmental trend is directed, a directed morphological trend is expected.

basis for intrabasinal allopatric speciation and, of course, punctuation as illustrated by *Staufenia*.

The habitat patches of the "a" species lineages continue to shrink and change while during the same interval of time the second facies spreads over the basin and populations of the species "b" become very abundant. In the most simple case this process would produce a fluctuating faunal record in the central parts of the basin -- as in the example of repeated sonniniid and stephanoid faunas in the central clay facies of the North German Bajocian.

On the other hand, if the "a" lineage becomes extinct then an "empty environment" evolves (an ecological system far from equilibrium) within our closed system when the original situation is restored by renewed transgression. In this case we would expect subpopulations of "b" to adapt to and invade this new central environment. If morphology is linked to environment, then we would expect the origination of a new form (b') morphologically convergent on the previous "a" form, but one which originated in the "b" lineage. This convergence in morphology can be expected if the constructional/historical constraints of "b" allow it to parallel "a" -- but these conditions have already been met by assuming that "a" and "b" are species members of the same higher taxon (subfamily or family). Thus, we use the same "internal factors" as RIEBER (1963) to explain the morphologically convergent lineages, the phylogenetic (historic) constraints, but now there is no mysterious evolutionary process (orthogenesis) necessary but simple adaptation to a slowly changing environment.

b) "Ecological islands"

Next we consider a more open system, in that the basin is open to an external pool of various species (x, y, z). By changing the ecological conditions within the basin in the same way as outlined above, we arrive at a totally different pattern (Fig. 26). Nothing has changed in terms of the relative frequency and distribution of the marginal species "b"; it appears in increasing numbers and spreads throughout the basin during the same time intervals. However, the observed morphological trend in basin-center species consists of a series of immigrants, and not true ancestor-descendant lineages (a, a', a" of Fig. 25 is replaced by x, y, z in Fig. 26). Each time the quality of the environment is changed, a new, already adapted but extrabasinal form successfully invades from the external species pool. Again, if morphology is linked to environment then we would expect a morphological cycle of forms which parallels the true evolutionary one of the closed basin model. In the case that the successive species belong to the same higher taxon, we would observe a local "phylogenetic lineage", which is not a true evolutionary lineage but the result simply of a series of ecological substitutions. This, of course, corresponds with the observed and earlier outlined patterns in the South German hammatoceratid "lineage" and, per-

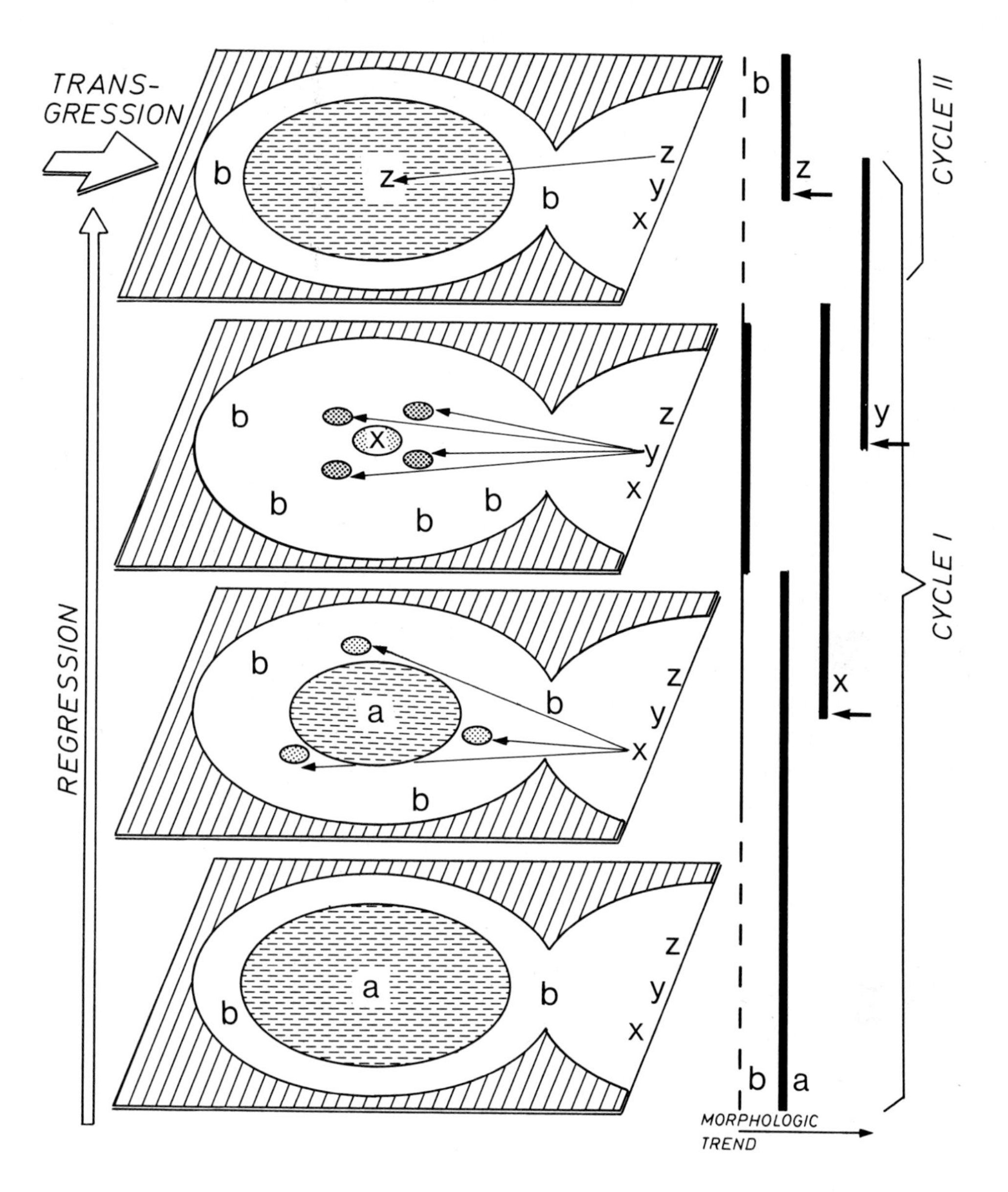

Fig. 26: The identical sequence as in Fig. 25 but this time an external (static) source pool of species is available. The best "preadapted" species from the source pool invade as soon as new habitats evolve through the regression. In the event the immigrants belong to the same higher taxon, a local "phylogenetic" line results which is not a true evolutionary lineage.

haps, the Sonninniidae which immigrate more or less continuously and parallel morphologically the Leioceratinae and the Graphoceratinae.

The model here reduces to "island biogeography" with an island (the basin) which is not stable but changes its environment through time. Each time the "adaptive" equilibrium with the mainland (the external species pool) is altered, the faunal composition of the basin is restored to the "adaptive" equilibrium conditions by extinctions and successive new immigrants.

A generalization from these two models follows immediately: in situ evolution of new species is most likely in isolated basins, while in open basins evolution via natural selection is always slower than the immigration of already adapted forms into an open basin. Whether we shall have adaptation by in situ evolution or the simple ecological replacement of species in a semi-open system depends on the ratio of:

$$\frac{\text{(velocity of adaptation via natural selection (in situ speciation))}}{\text{(velocity of immigration from external species pools)}}$$

which again represents an equilibrium model as was found earlier for the sedimentological equilibrium deposition/erosion. If we consider a single level of geological time, the model, of course, reduces to classical island biogeography and the classical models as cited in THOMAS & FOIN (1982) provide a corollary to this evolutionary model.

c) Coupled systems

We can develop the previous models further in order to model the discussed iterated morphological cycles within a slightly more realistic and complex framework. The basic parameter which controls the equilibrium condition is the velocity of migration, and this is a function of the degree of geographical (or ecological) isolation of the basin. The effect of changing degrees of geographical isolation in a changing environmental context is illustrated by a regression-transgression model with two coupled basins (Fig. 27). In this case, the geographical and ecological distance between the coupled basins is e.g. a function of changing sea-level. Common to two similar environments within the coupled basins we show again simply one stable facies with one species "b" and one instable facies.

During regression the distance between similar facies types in the two basins increases, the central facies of the marginal basin becomes isolated as it steadily decreases in area. This effect decreases the velocity of possible migration and makes more probable that in situ speciation will occur, as in the case of an isolated basin system.

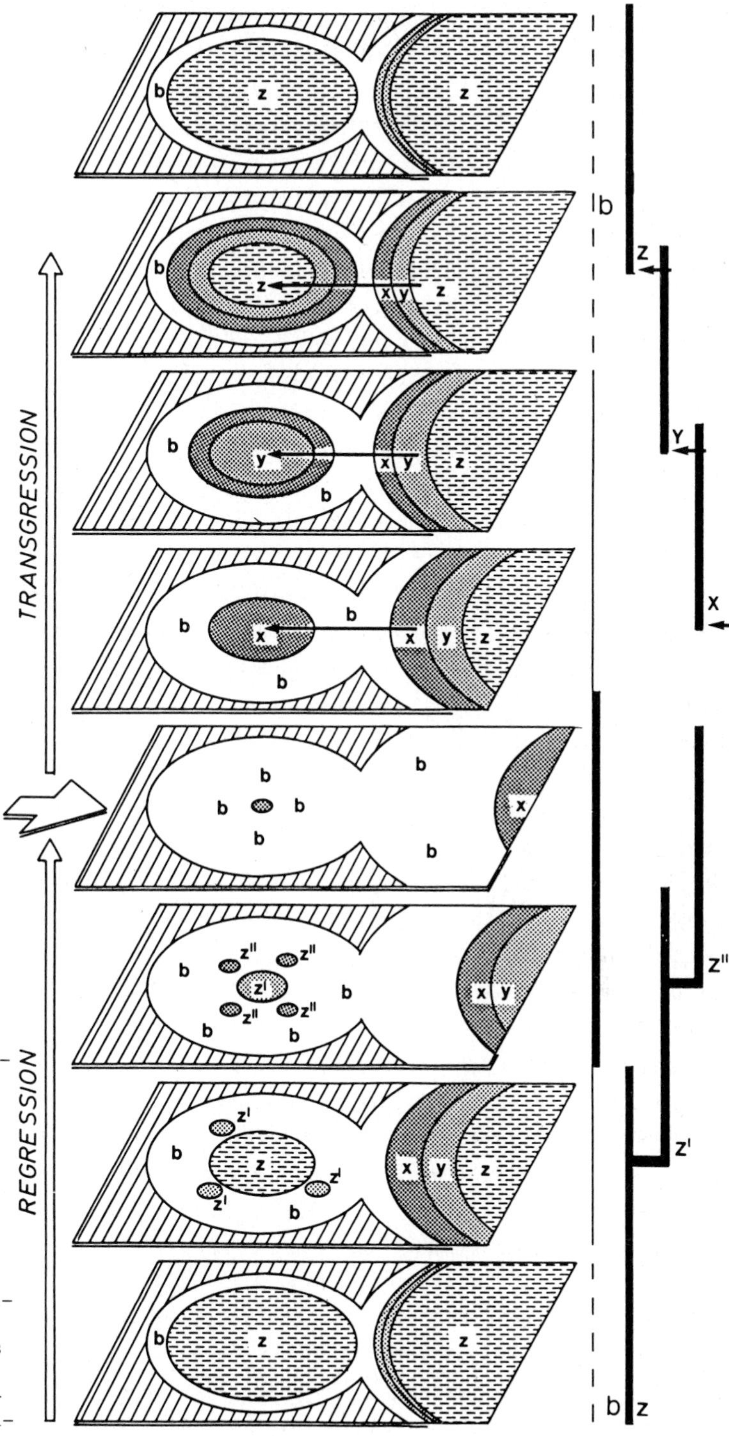

Fig. 27: In coupled basins the geographical and environmental distances change during regressions and transgressions. During a regression phase isolation increases and causes true *in situ* speciation. During transgression isolation is reduced and immigrations dominate. The resulting local "phylogenetic" pattern is a mixture of true *in situ* speciations and of multiple migrations.

During transgression, in contrast, the geographic distance between similar facies types within the basins reduces as the original facies distribution is restored and, therefore, ecologically adapted forms are more likely to migrate into the marginal basin and to occupy the appropriate environments before any *in situ* evolutionary response can occur. Providing that this changing equilibrium relationship between rates of speciation and migration affects forms within the same higher taxon (subfamily, family) then the resulting local "phylogenetic" pattern is a peculiar mixture out of true evolutionary ancestor-descendant relationships and of sequential ecological replacements. Even if the *in situ* evolution phases were "gradual", our "phylogenetic pattern" would still consist in large part of "punctuated" ecological replacements of species -- and even if one assumes a continuous sedimentation one cannot expect a continuous record by applying the outlined models.

In addition, Darwin's belief that the stratigraphic record is incomplete is well known and the outlined depositional pattern in the South German Aalenian underlines this viewpoint -- most of the record is condensed or consists of gaps. This viewpoint has been stressed by SADLER (1981) who doubts that evolutionary hypotheses can be tested because of the discontinuity of sedimentation. Punctuation, therefore, is the pattern we have to expect in the general case from sedimentological, ecological and, perhaps, from evolutionary grounds.

The scenario developed thus far is, indeed, an ecologically motivated type of "punctuated equilibrium". Ecology springs from actualistic concepts; it is biased in favour of the stability of species because everything is "well adapted" during the short time intervals under study. The evolutionary concept of 'punctuated equilibria' extends this short-term ecological pattern into the geological time dimension. Small isolated populations (which allow allopatric speciation) increase the ecological range of a species or species complexes, as we are used to thinking of different species inhabiting different habitats. The replacement of such populations through time in evolution corresponds to a simple ecological substitution of species under changing environmental conditions, yet it often cannot be distinguished from a simple migratory ecological replacement. Figs. 25 to 27 illustrate that increasing facies diversity within the basin may resemble the paleogeographic models on a smaller geographical scale. The relationship between isolated facies patches and the remaining central source pool is of the same type as the inter-basin relationship of the regional models. Thus, the biogeographic models project down to habitat patches and, perhaps, even down to fractions of patches. This supports the concept of "punctuated equilibria" from an ecological viewpoint or, respectively, relates "punctualism" to an ecological selection mechanism, the selection of the fittest population or species from a source pool.

We began this chapter with an observation from ELDREDGE & GOULD (1972), and thus shall close it:

"Paleontologists should recognize that much of their thought is conditioned by a peculiar perspective that they must bring to the study of life: they must be retrospective".

The starting point for this retrospection must be ecology; neither population dynamics nor genetics can be transfered into the fossil record. However, to consider the unfolding of ecological time sections through geological time may well be the key to a better understanding of evolutionary patterns and processes.

SUMMARY AND CONCLUSIONS

Darwin was well aware of the empirical fact that discontinuous rapid faunal changes are the usual pattern in the stratigraphic record. In its generality this pattern affected the paradigm of gradual natural selection, a problem which Darwin overcame by the concept of the incompleteness of the fossil record, a concept which was stressed over and over again. The reconstruction and description of phylogenies -- or of the history of life -- dominated paleontology for decades, and the incomplete record gave rise to non-darwinian models which commonly were non-ecological as well. The apparently contradictory concepts of phylogeny and ecology are illustrated here by well known local and global examples, and we stress the idea of migrational events on the entire scale -- from global paleogeography down to environmental patches. Migrations are simply explained by biological, ecological, and paleogeographic configurations and events; a punctuated record is not unexpected on this explanatory level, and natural selection -- selection on the level of individuals -- can still be considered the microdynamic cause for evolution. Even the biostratigraphic concept is not affected -- as migrations are much faster than adaptation by natural selection, a migrational event may even be a better stratigraphic marker than an evolutionary "revolution".

The close relationship between faunal replacements and facies breaks within the stratigraphic record is a well established observation in the literature and has, generally, been explained by a facies dependency of the fossils, e.g. of certain ammonites. Faunal fluctuations within lithologically monotonous profiles and associations of various morphotypes in single beds seem to disprove sometimes the otherwise established facies relationship. The single profile or bed, however, does not provide the necessary control -- e.g. post mortem processes disturb the local record to a high degree. Sufficient control can only be gathered on the regional level by basin analysis.

But, even on the regional level, the paleoecological pattern provides only a static historical picture. To establish faunal changes and trends with respect to environmental changes requires multiple repetitions of the "natural experiment" under similar constraints. "Iterated morphological cycles" provide such repeated natural experiments and the discussed example of South German Aalenian ammonites points clearly to a rather strong facies dependency of various morphotypes. On the other hand, other environmental relationships are well known in the literature and one has to be careful in overemphasizing a specific example. One remains, with the specific example, or even with a collection of examples, at a ninteenth century descriptive approach of single case studies in terms of e.g. ammonite ecology and mode of life. The general results of such studies cannot provide an explanation of all ammonites' mode of life, but they can elucidate that the 'mysterious' phylogenetic patterns need not be explained by intrinsic evolutionary forces (e.g. orthogenesis) but can be related to simple ecological processes.

In addition, such case studies show that local "phylogenetic lineages" should not automatically be seen as direct ancestor-descendant relationships. Immigrations and ecological substitutions are an extremely common pattern which on the level of any regional analysis overshaddows heavily the true *in situ* species evolution. If ecological competition and replacements occur within the same higher taxa (subfamily, family), it becomes difficult do discern true *in situ* species evolutionary lineages from sequential ecological replacements. The result is a punctuated phylogenetic record which -- if not a result of the incompleteness of the stratigraphic record -- can well be related to ecological faunal replacements -- on global, regional and local (population) levels.

Each of the questions discussed -- rapid faunal replacements, faunal fluctuations, iterated morphological cycles, and punctuated phylogenetic records - hinge finally on one's point of view, as Eldredge & Gould have so often stated. The ecological viewpoint of this study allows one to explain the various empirical observations of "typostrophism", "orthogenesis", and "punctuation" in an unique and consistent way, at least for the paleontologist. Although natural selection, the selection of the fittest individual, is lastly the motor for evolution -- the microscopic dynamics, the biological "Brownian movement" -- it is not the process which formed the bulk of the paleontological record. Using the actualistic ecological approach -- the experience from a single time section -- geologically, the replacement or substitution of species and populations will much more likely be documented in the stratigraphic record than any gradual change.

ACKNOWLEDGEMENTS

We thank A. Hallam, Birmingham; W.-E. Reif, A. Seilacher, Tübingen; and G.E.G. Westermann, Hamilton for reviewing this paper and for stimulating remarks. However, the responsibility for the interpretations given herein remains ours alone. We hope that such re-interpretations of evolutionary patterns will lead to fruitful future discussion.

REFERENCES

Alberch, P., Gould, S.J., Oster, G.F., Wake, D.B. 1979: Size and shape ontogeny and phylogeny.- Paleobiology, 5(3): 296-317.

Aldinger, H. 1957: Eisenoolithbildung und rhythmische Schichtung im süddeutschen Jura.- Geol. Jb., 74: 87-96.

Aldinger, H. 1965: Über den Einfluß von Meeresspiegelschwankungen auf Flachwassersedimente im Schwäbischen Jura.- Tschermarks min. u. petrogr. Mitt. 10: 61-68.

Aldinger, H. 1968: Die Palaeogeographie des schwäbischen Jurabeckens.- Eclogae Geol. Helv., 61(1): 167-182.

Anderton, R., Bridges, P.H., Leeder, M.R., Sellwood, B.W. 198 : A dynamic stratigraphy of the British Isles.- London, 301 pp.

Arkell, W.J. 1957: Introduction to Mesozoic Ammonoidea. In: Moore, R.C. (ed.): Treatise on Invertebrate Paleontology, Part L, Mollusca 4, L 81- L 129. Geol. Soc. Amer. and Univ. Kansas Press, Lawrence.

Auclair, A.M. & Goff, F.G. 1971: Diversity relations of upland forests in the western Great Lake area.- Am. Nat. 105: 449-528.

Barron, E.J., Harrison, Ch.G.A., Sloan, J.L., Hay, W.W. 1981: Paleogeography, 180 million years ago to the present.- Eclogae Geol. Helv., 74(2): 443-470.

Bayer, U. 1968: Docidoceras cf. liebi Maubeuge aus dem Unteren Bajocium des Wutachgebietes.- Stuttgarter Beitr. Naturkde., 183: 1-3.

Bayer, U. 1969a: Die Gattung Hyperlioceras Buckman.- Jber. u. Mitt. oberrh. geol. Ver. 51: 31-70.

Bayer, U. 1969b: Euaptetoceras und Eudmetoceras (Ammonoidea, Hammatoceratidae) aus der concava-Zone (Ober Aalenium) Süddeutschlands.- N. Jb. Geol. Paläont. Abh., 133: 211-222.

Bayer, U. 1970a: Das Profil des Erz-Tagebaus Ringsheim (Ober-Aalenium/ Unter Bajocium) N. Jb. Geol. Paläont. Mh., Jg. 1970, 5, 251-269.

Bayer, U. 1970b:Anomalien bei Ammoniten des Aaleniums und Bajociums und ihre Beziehung zur Lebensweise. N. Jb. Geol. Paläont. Abh., 135(1): 19-41.

Bayer, U. 1972: Zur Ontogenie und Variabilität des jurassischen Ammoniten Leioceras opalinum.- N. Jb. Geol. Paläont. Abh., 140(3): 306-327.

Bayer, U. & McGhee, G.R. 1984: Iterative evolution of Middle Jurassic ammonite faunas.- Lethaia, 17.

Bayer, U., Altheimer, E., Deutschle, W. 1985: Environmental evolution in shallow epicontinental seas: Sedimentary cycles and bed formation. This volume.

Beurlen, K. 1937: Die stammesgeschichtlichen Grundlagen der Abstammungslehre.- (Fischer) Jena, 264 pp.

Bloos, G. 1976: Untersuchungen über Bau und Entstehung der feinkörnigen Sandsteine des Schwarzen Jura α (Hettangium u. tiefstes Sinemurium) im schwäbischen Sedimentationsbereich.- Arb. Geol. Pal. Inst. TH Stuttgart, NF 71.

Brandes, Th. 1912: Die faziellen Verhältnisse des Lias zwischen Harz und Egge-Gebirge mit einer Revision seiner Gliederung. N. Jb. Min. etc., Beil. Bd. 33, 325-508.

Brinkmann, R. 1929: Statistisch-biostratigraphische Untersuchungen an mitteljurassischen Ammoniten über Artbegriff und Stammesentwicklung. Abh. Ges. wiss. Göttingen, math. phys. Kl., NF 13: (3) 1-249.

Bubenicek, L. 1970: Geologie des Gisements de fer de Lorain. Inst. Rech. Sidér.Franc. IRSID, 132 pp.

Connell, J.H. 1975: Some mechanisms producing structure in natural communities. 460-490. In: Cody, M.L. & Diamond, J.M. eds.: Ecology and Evolution of Communities.- (Belknap Pr.) Cambridge, Mass.

Cooper, M.R. 1977: Eustacy during the Cretaceous; its implications and importance. Paleo^3, 22: 1-60.

Dayton, P.K. 1971: Competition, disturbance, and community organization in a rocky intertidal community.- Ecol. Monogr. 41: 351-389.

Dietl, G. & Haag, W. 1980: Über die "sowerbyi"-Zone (=laeviscula-Zone, Unter-Bajocium, Mittl. Jura) in einem Profil bei Nenningen (östl. Schwäb. Alb). Stuttgarter Beitr. Naturk. 60: 1-11.

Donovan, D.T. 1984: Ammonite shell form and transgression in the British Lower Jurassic.- This volume.

Einsele, G. & Seilacher, A., eds. 1982: Cyclic and event stratification. (Springer) Berlin, 536 pp.

Eldredge, N. & Gould, S.J. 1972: Punctuated equilibria: an alternative to phyletic gradualism. 82-115. In: Schopf, J.M. ed.: Models in Paleobiology. (Freeman) San Francisco.

Elmi, S. 1963: Les Hammatoceratinae (Ammonitina) dans le Dogger inférieur du bassin Rhodanien.- Trav. Lab. Geol. Fac. sci. Lyon, NS 10.

Fischer, A.G. 1981: Climatic Oscillations in the Biosphere. 103-131. In: Biotic crises in ecological and evolutionary time. (Acad. press).

Fischer, A.G. 1982: Long-Term Climatic Oscillations recorded in stratigraphy. 97-104. In: Studies in Geophysics, Climate in Earth History. (Nat. Aca. Press).

Frebold, H. 1924: Ammonitenzonen und Sedimentationszyklen und ihre Beziehung zueinander. Zbl. Min. Geol. Pal., 313-320.

Frebold, H. 1925: Über cyklische Meeressedimentation. Leipzig.

Fretwell, S. 1972: Seasonal Environment.- Mon. Pop. Biology, 5.

Géczy, B. 1966: Ammonoides Jurassiques de Csernye, Montagne Bakony, Hongrie. Part I (Hammatoceratidae). Geol. Hungar. 34.

Genser, H. 1966: Schichtenfolge und Stratigraphie des Doggers in den drei Faziesbereichen der Umrandung des Südschwarzwaldes.- Oberrhein. geol. Abh. 15: 1-60.

Geyer, O.F. 1971: Zur paläobathymetrischen Zuverlässigkeit von Ammonoideen-Faunen Spektren. Paleo^3 10: 265-272.

Geyer, O.F. & Gwinner, M.P. 1962: Der Schwäbische Jura. Samml. Geol. Führer, 40 (Bornträger) Berlin.

Ginsburg, L. 1965: Les régressions marines et le problème du nouvellement des faunes au cours des temps géologiques. Bull. Soc. géol. France, 6: 13-22.

Gould, S.J. 1977: Ontogeny and Phylogeny. (Harvard Univ. Pr.) Cambridge, Mass.

Gould, S.J. & Eldredge, N. 1977: Punctuated equilibria: the tempo and mode of evolution reconsidered.- Paleobiology, 2: 115-151.

Haas, O. 1942: Recurrence of morphologic types and evolutionary cycles in Mesozoic ammonites.- Journ. Paleont. 16: 643-650.

Hallam, A. 1961: Cyclothems, transgressions and faunal change in the Lias of North-West Europe.- Transact. Edinburgh Geol. Soc. 18(2): 124-174.

Hallam, A. 1978a: Eustatic cycles in the Jurassic.- Paleo[3] 23: 1-32.

Hallam, A. 1978b: How rare is phyletic gradualism and what is its evolutionary significance? Evidence from Jurassic bivalves. Paleobiology, 4: 16-25.

Hallam, A. 1981: A revised sea-level curve for the early Jurassic. J. geol. Soc. London 138: 735-743.

Hallam, A. 1982: Patterns of speciation in Jurassic Gryphaea. Paleobiology, 8(4): 354-366.

Hallam, A. 1983: Early and mid-Jurassic molluscan biogeography and the establishment of the central Atlantic seaway. Paleo[3], 43: 181-193.

Hallam, A. & Bradshaw, M.J. 1979: Bituminous shales and oolithic ironstones as indicators of transgressions and regressions. Journ. Geol. Soc. London, 136: 157-164.

Heidorn, F. 1928: Paläogeographisch-tektonische Untersuchungen im Lias zeta von Nordwestdeutschland. N. Jb. Min. etc., 59B: 117-244.

Hölder, H. 1964: Jura. Handb. Strat. Geol. Bd. IV, (Enke) Stuttgart, 603 pp.

Hofmann, K. 1936/38: Die Ammoniten des Lias Beta der Langenbrückener Senke I-II. Beitr. naturk. Forsch. Südwestdeutschl., 1 (1936), 3 (1938).

Huf, W. 1968: Über Sonninien und Dorsetensien aus dem Bajocium von Nordwestdeutschland.- Beih. Geol. Jahrb., 64: 1-126.

Karrenberg, H. 1942: Paläogeographische Übersicht über die Ablagerungen der Dogger-beta-Zeit in West- und Südwestdeutschland. Arch. Lagerstättenforsch. 75: 78-79.

Kennedy, W.J. & Cobban, W.A. 1977: The role of ammonites in biostratigraphy. In: Kaufmann, E.G. & Hazel, J.E. (eds.): Concepts and Methods of Biostratigraphy, 658 pp. (Dowden, Hutchinson & Ross, Inc.,) Stroudsburg.

Klüpfel, W. 1917: Über die Sedimente der Flachsee im Lothringer Jura. Geol. Rdsch., 7: 98-109.

Kumm, A. 1952: Das Mesozoikum in Niedersachsen (Der Dogger). Veröff. nieders. Amt Landesplanung u, Statistik, A1, 2, 329-509.

Levins, R. 1968: Evolution in changing environments. Monogr. Popul. Biol., 2: 1-120.

Lucius, M. 1940: Der Luxemburger mesozoische Sedimentationsraum und seine Beziehung zu den herzynischen Bauelementen. Veröff. Lux. geol. Landesaufn., II, 41-102.

Lucius, M. 1945: Die Luxemburger Minetteformation und jüngere Eisenerzbildungen unseres Landes. Serv. Carte Géol. Luxembourg.

Lucius, M. 1948: Das Gutland.- Serv. Géol. Luxembourg, V.

MacArthur, R.H. & Wilson, E.O. 1967: The theory of island biogeography. Monogr. Popul. Biol., 1: 1-203.

Maubeuge, P.L. 1963: La position stratigraphique du Gisement ferrifère Lorrain (Le problème de l'Aalénien). Bull. Techn. Synd. Mines Fer de France, 72.

Morton, N. 1971: The definition of standard Jurassic stages. Coll. Jurassique Lux. Mm. B.R.G.M., Fr., 75: 83-93.

Morton, N. 1974: The standard Zones of the Aalenian stage. Ann. Inst. Geol. Publ. Hungarici, LIV, 2: 433-437.

Moore, R.C. 1954: Evolution of late Paleozoic invertebrates in response to major oscillations of shallow seas. Mus. compar. Zool. Bull. 112: 259-286.

Muller, A. 1967: "Die Mergel und Kalke von Strassen". Serv. Géol. Luxembourg, XVII, 1-136.

Muller, A., Preugschat, F., Schreck, H. 1976: Tektonische Richtungen und Faziesverteilung im Mesozoikum von Luxemburg-Lothringen. Jber. u. Mitt. Oberrh. geol. Ver. NF 58, 153-181.

Naylor, D. & Shannon, P.M. (1982): The Geology of offshore Ireland and West Britain. (Graham & Trotman) London, 161 pp.

Neumayr, M. 1878: Über unvermittelt auftretende Cephalopodentypen im Jura Mittel-Europas. Jahrb. k.k. geol. Reichsanst. 28.

Newell, N.D. 1952: Periodicity in invertebrate evolution. J. Paleontol. 26: 371-385.

Newell, N.D. 1956: Catastrophism and the fossil record. Evolution, 10: 97-101.

Newell, N.D. 1962: Paleontological gaps and geochronology. J. Paleontol., 36: 592-610.

Newell,N.D. 1967: Revolutions in the history of life. Geol. Soc. Am. Spec. Pap. 89: 63-91.

Petry, D. 1982: The pattern of phyletic speciation.- Paleobiology, 8 (1): 56-66.

Pomerol, Ch. 1978: Evolution paleogéographique et structurale du Bassin de Paris, au Précambrien a l'actuel, en relation avec les régions avoisinantes. Geologie en Mijnbow, 57(4): 533-543.

Pompeckj, J.F. 1914: Die Bedeutung des Schwäbischen Jura für die Erdgeschichte. (Schweizerbart) Stuttgart, 64 pp.

Pompeckj, J.F. 1916: Über den Einfluß des Klimas auf die Bildung der Sedimente des Schwäbischen Juras.- Jh. vaterl. Naturkde., 72, XXXII-XXXIII.

Rat, P. 1974: Visages de la France entre l'orogenése alpin. In: Debelmas, J.: Géologie de la France. Paris.

Raup, D.M. 1966: Geometric analysis of shell coiling: general problems.- Jour. Paleontology 40: 1178-1190.

Raup, D.M. 1967: Geometric analysis of shell coiling: coiling in ammonoids. Jour. Paleontology 41: 43-65.

Reif, W.-E. 1983: Evolutionary theory in German paleontology. In: Grene, M. ed.: Dimensions of Darwinism. (Cambridge Univ. Press), 173-203.

Rieber, H. 1963: Ammoniten und Stratigraphie des Braunjura der Schwäbischen Alb. Paleontographica 122 (A): 1-89.

Rogowski, E. 1971: Sedimentpetrographische Untersuchungen in den Dogger-beta-Sandsteinen (Oberes Aalenium) der östlichen Schwäbischen Alb. Arb. Geol. Paläont. Inst. TH Stuttgart, NF 65, 117 pp.

Sadler, P.M. 1981: Sediment accumulation rates and the completeness of stratigraphic sections.- J. Geol. 89: 569-584.

Schindewolf, O.H. 1940: "Konvergenzen" bei Korallen und bei Ammoneen.- Fortschr. Geol. Paläont. 12(41): 387-491.

Schindewolf, O.H. 1950: Grundfragen der Paläontologie. 506 p. (Schweizerbart) Stuttgart.

Schloz, W. 1972: Zur Bildungsgeschichte der Oolithenbank (Hettangium) in Baden-Württemberg.- Arb. Inst. Geol. Paläont. Univ. Stuttgart, NF 67: 101-212.

Schopf, T.J.M. 1974: Permo-Triassic extinctions: relation to sea floor spreading. J. Geol., 82: 129-143.

Schopf, T.J.M. 1981: Punctuated equilibrium and evolutionary stasis. Paleobiology, 7: 156-166.

Schröder, B. 1962: Schwermineralführung und Paläogeographie des Doggersandsteins in Nordbayern.- Erlanger geol. Abh. 42: 1-29 p.

Seilacher, A. 1970: Arbeitskonzept zur Konstruktionsmorphologie. Lethaia 3, 393-396.

Seilacher, A. 1972: Divaricate patterns in pelecypod shells. Lethaia, 5: 325-343.

Seilacher, A. 1973: Fabricational noise in adaptive morphology. Syst. Zool., 22: 451-465.

Seilacher, A. 1984: The Jeram shell bed -- a modern example of event condensation. This volume.

Seilacher, A. et al.: Oyster beds in the Upper Jurassic of Poland. This volume.

Seilacher, A., Reif, W.-E., Westphal, F., eds. 1982: Studies in Paleoecology. N. Jb. Geol. Pal., 164: 1-305.

Simpson, G.G. 1953: Life of the past. (Yale Univ. Pr.) New Haven, 198 pp.

Smith, A.G., Hurley, A.M., Briden, J.C. 1982: Paläokontinentale Weltkarten des Phanerozoikums. (Enke) Stuttgart, 102 pp.

Söll, H. 1956: Stratigraphie und Ammonitenfauna des mittleren und oberen Lias-beta (Lothringien) in Mittel-Württemberg. Geol. Jb., 72: 367-434.

Stahlecker, G. 1934: Stratigraphie und Tektonik des Braunen Jura im Gebiet des Stuifen und Rechbergs. Jh. Ver. vaterl. Naturkde. Württ., 90: 59-121.

Stanley, S.M. 1975: A theory of evolution above the species level. Proc. Nat. Acad. Sci., 72, 646-650.

Stanley, S.M. 1979: Macroevolution: Pattern and Process. (Freeman) San Francisco.

Thein, J. 1975: Sedimentologisch-stratigraphische Untersuchungen in der Minette des Differdinger Beckens. Pub. Serv. Géol. Luxembourg, XXIV, 60 pp.

Thomas, W.R. & Foin, Th.C. 1982: Neutral hypothesis and pattern of species diversity: fact or artifact? Paleobiology, 8(1): 45-55.

Vail, P.R., Mitchum, R.M., Thompson, S., Todd, R.G., Sangree, J.B., Widmier, J.M., Bubb, J.N., Hatlelid, W.G. 1977: Seismic stratigraphy and global changes of sea level.- Mem. Am. Assoc. Petrol. Geol. 26: 49-212.

Vrba, E.S. 1980: Evolution, Species and Fossils: How Does Life Evolve. South African Journ. Sci. 76.

Walliser, O.H. 1956: Chronologie des Lias $alpha_3$ zwischen Fildern und Klettgau (Arietenschichten, Südwestdeutschland). N.Jb.Geol.Paläont., Abh., 103.

Walliser, O.H. 1956: Stratigraphie des Lias $alpha_3$ zwischen Fildern und Klettgau (Arietenschichten, SW-Deutschland). N.Jb. Geol. Pal., Abh., 103.

Weber, H.-S. 1964: Zur Stratigraphie und Ammoniten-Fauna des Braunjura (Dogger beta) der östlichen Schwäbischen Alb. Arb. Geol. Pal. Inst. TH Stuttgart, NF 44, 174 pp.

Weber, H.-S. 1967: Zur Westgrenze der ostschwäbisch-fränkischen Facies des Braunjura (Dogger) beta in der Schwäbischen Alb (Württemberg). Jber. Mit. Oberrh. geol.Ver. NF 49, 47-54.

Werner, F. 1959: Zur Kenntnis der Eisenoolithfacies des Braunjura beta von Ostwürttemberg. Arb. Geol. Pal. Inst. TH Stuttgart, NF 23, 169 pp.

Westermann, G.E.G. 1954: Monographie der Otoitidae (Ammonoidea). Beih. Geol. Jb. 15, 364 pp.

Westermann, G.E.G. 1966: Covariation and taxonomy of the Jurassic ammonite Sonninia adicra (WAAGEN). N. Jb. Geol. Paläont. Abh. 124 (3): 289-312.

Westermann, G.E.G. 1969: The ammonite fauna of the Kialagvik Formation at Wide Bay, Alaska Peninsula; Part II. Bull. Am. Paleontology, 57: 1-226.

Westermann, G.E.G. 1983: The Upper Bajocian and Lower Bathonian (Jurassic) Ammonite faunas of Oaxaca, Mexico and West-Tethyan affinities. Paleontologica Mexicana 46: 1-63.

Westermann, G.E.G. & Riccardi, A.C. 1982: Ammonoid fauna from the early Middle Jurassic of Mendoza Province, Argentina. J. Paleont., 56: 11-41.

Wiedmann, J. 1969: The heteromorphs and ammonoid extinction. Biol. Rev., 44: 563-602.

Wiedmann, J. 1973: Evolution or revolution of ammonoids at Mesozoic system boundaries. Biol. Rev., 48: 159-194.

Wild, H. 1951: Zur Bildungsgeschichte der Braunjura-beta-Flöze und ihrer Begleitgesteine in NO-Württemberg. Geol. Jb. 65, 271-298.

Ziegler, P.A. 1978: North-Western Europe: Tectonics and Basin Development. Geolog. Mijnbourow, 57(4): 589-626.

Znosko, J. 1959: Development of the Aalenian and Bajocian Transgression in the Polish Lowland. Kvortalnik Geologic, 3, 529-562.

Zoch, W. 1940: Die stammesgeschichtliche Gestaltung der Doggerbelemniten Schwabens und ein Vergleich mit Lias-und Kreidebelemniten. N. Jb. Beil. Bd. 83, B, 3.

IMMIGRATION OF CEPHALOPODS INTO THE GERMANIC MUSCHELKALK BASIN AND ITS INFLUENCE ON THEIR SUTURE LINE

M. Urlichs & R. Mundlos
Stuttgart, Bad Friedrichshall

Abstract: Only few, probably more tolerant, genera of the stenohaline group of cephalopods have managed to get established in the Germanic Basin. These include *Germanonautilus*, *Serpianites* and *Paraceratites* as well as *Ceratites* which evolved from the latter in this new province. In *Serpianites* (Lower Muschelkalk/Upper Anisian and the early ceratites (Upper Muschelkalk/Upper Anisian) we observe an iterative reduction of sutural complexity. Radiation of ceratites from the ancestral *Paraceratites* (*Progonoceratites*) *atavus atavus* took place during the *atavus* and *pulcher*/*robustus* Zones of the Upper Anisian.

INTRODUCTION

During the last decade major revisions of the Germanic Muschelkalk (AIGNER, 1982; DURINGER, 1982; HAGDORN, 1978; KOZUR, 1974; SCHULZ, 1972; SCHWARZ, 1975 etc.) have provided faciological and stratigraphic background on which the history of the cephalopods appears in a new light. This review builds also on previous taxonomic revision of part of the Muschelkalk cephalopods (URLICHS & MUNDLOS, 1980, MUNDLOS & URLICHS, 1984). We follow the stratigraphic schemes of GEYER & GWINNER (1964) for Southwest Germany and of KOZUR (1974) for the rest of the basin.

LOWER MUSCHELKALK

A. Geographic distributions

In Lower Muschelkalk times (Upper Anisian) the Germanic Basin had its connections with the Tethys in the east (Silesian-Moravian and East Carpathian Gates; see HAGDORN, this volume). The cephalopods reflect this immigration route, but their distributional patterns differ significantly from those of epibenthic immigrants. Thus

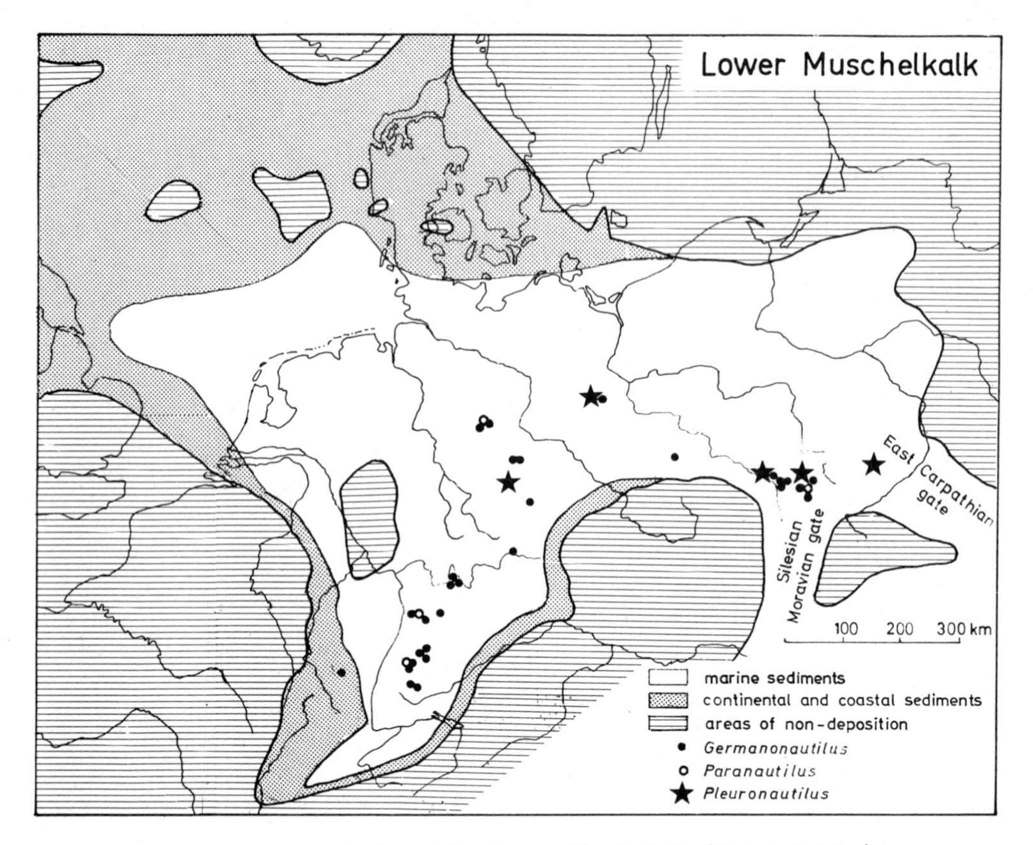

Fig. 1: Range of nautiloids in the Lower Muschelkalk (Upper Anisian) of the Germanic Basin. Paleogeography after ZIEGLER (1982), FISHER (1984) and SCHWARZ (1975). Fauna compiled after ALBERTI 1864), ASSMANN(1937), ECK (1872), FRANZ (1903), FRITSCH (1906), LAUGHIER (1963), MUNDLOS & URLICHS (1984), RASSMUSS (1915), TRAMMER (1972), VOSSMERBÄUMER (1970, 1972), WALTHER (1927).

the nautiloids show high concentrations in the southwest (Fig. 1), which may be due to post-mortem drift in southerly currents of the otherwise poor swimmers (LEHMANN; 1976: 110). It should also be noted that the data base is different for the genera *Paranautilus* and *Pleuronautilus*, which are known only from a few localities, and for *Germanonautilus*, which is common throughout the basin.

The distribution of the ammonoids is similar: occurrences of *Serpianites antecedens* are sparse, but distributed all over the euhaline parts of the Germanic Basin. *Noetlingites strombecki* (see map by KELBER, 1977; Fig. 1) and *Beneckeia buchi* have similar distributions and are therefore not contained in our paleogeographic map (Fig. 2). This map shows, however, the distribution of rarer ammonoids, of which *Acrochordiceras* is known only from Lower Silesia (NOETLING, 1880) and *Beyrichites* only from the Holy Cross Mountains (TRAMMER, 1972), i.e. from areas close to the

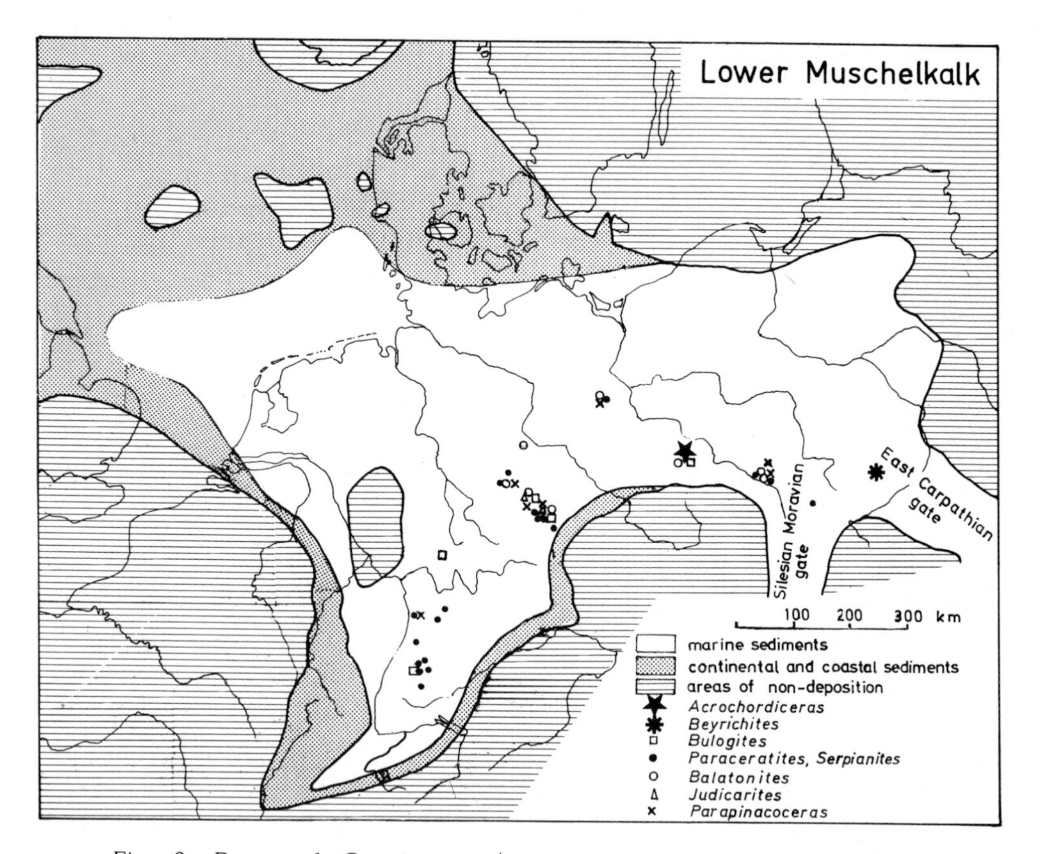

Fig. 2: Range of Ceratitaceae (except *Beneckeia* and *Noetlingites*) and *Parapinacoceras* in the Lower Muschelkalk (Upper Anisian) of the Germanic Basin. Paleogeography after ZIEGLER (1982), FISHER (1984) and SCHULZ (1975). Fauna compiled after ASSMANN (1937), BEYRICH (1854, 1858), CLAUS (1921), ECK (1865, 1891), FRITSCH (1906), GIEBEL (1853), MAYER (1971), NOETLING (1880), PICARD (1889, 1899), RASSMUSS (1915), ROTHE (1951, 1959), SCHMIDT (1907, 1935), TRAMMER (1972), VOSSMERBÄUMER (1970, 1972), WAGNER (1888, 1923), WALTHER (1927), WURM (1914).

gates. Immigration from this direction is also expressed by the higher species diversities in Silesia and Thuringia as compared to Hessen, Franconia and Swabia, where salinities temporarily were unfavourable for stenohaline genera.

B. Stratigraphic ranges

Species of *Balatonites* and *Bulogites* are restricted to the lower and middle parts of the Lower Muschelkalk (Upper Anisian; KOZUR, 1974: 9, 12), while *Judicarites* is found only in its upper part. For some species, stratigraphic ranges differ in different areas. *Beneckeia buchi*, for instance, occurs in Thuringia from the "Myophorien-Schichten" (mu 1α) to the "Schaumkalk" (mu 2α), while in southwest Germany it appears not before the "Mergelige-Schichten" and reaches to the "Spiriferinabank" (mu 2). Similarly, *Serpianites antecedens* ranges in Thuringia from the "Oolithbank " (mu 1ß 00) to the "Schaumkalk" (mu 2x) and the Black Forest only in the *buchi* Horizon (mu 2) and the beds with *Homomya albertii*. *Noetlingites strombecki* is known from the Lower as well as the Upper Gogolin Beds (ECK, 1865: 59, 107), while it is restricted to a narrow interval below the "Oolithbank " (mu 1ß 00) in Thuringia (VOLLRATH, 1924: 134) and occurs somewhat earlier in two dolimitic benches of the "Mergelige-Schichten" (mu 1; SCHMIDT, 1907: 28).

In summary we find that during the Lower Muschelkalk (Upper Anisian) many cephalopod genera have entered the Germanic Basin, but that few of them managed to spread into the western parts of the basin and only during part of their presence in the dispersal center.

This pattern probably reflects regional and temporal fluctuations in salinity, which may have been also responsible for the disappearance of newly immigrated species of ammonoids and nautiloids with the exception of the more tolerant genus *Germanonautilus*. SCHULZ (1972, Fig. 2) has recognized in the Lower Muschelkalk four cycles of increasing salinity. He also noted (p. 166) that "cyclic changes in facies are probably controlled less by changes in water depth than by fluctuating intensities of water exchange with the open sea through the Upper Silesian Gate".

MIDDLE MUSCHELKALK

After closure of the East Carpathian and shallowing of the Silesian-Moravian Gates by the end of the Lower Muschelkalk (SENKOWICZOWA & SZYPERKO-SLIWCZYNSKA, 1975), a new connection was established in the Southwest of the basin (RICOUR, 1963; Fig. 4), as evidenced by the appearance of the first ceratites in the upper part of the Middle Muschelkalk in Lorraine (LAUGHIER, 1963: 51).

UPPER MUSCHELKALK

A. Geographic distributions

By the beginning of the Upper Muschelkalk (Upper Anisian), the Silesian-Moravian Gate was closed and the East Carpathian Gate reopened to a limited extent (SENKOWICZOWA & SZYPERKO-SLIWCZYNSKA, 1975). As a result, faunal immigrations were now limited to the Burgundy Gate.

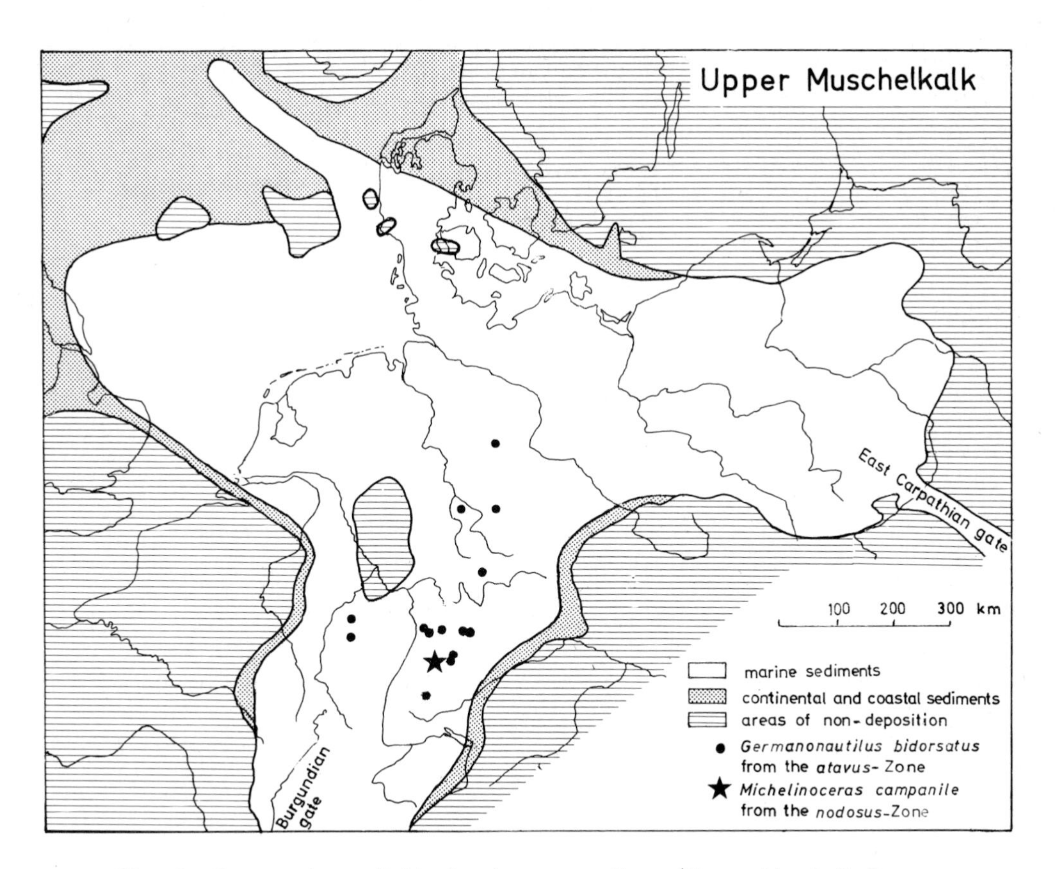

Fig. 3: Range of nautiloids in the *atavus* Zone (Upper Muschelkalk, Upper Anisian) of the Geramnic Basin. Paleogeography after ZIEGLER (1982) and SENKOWICZOWA & SZYPERKO-SLIWCZYNSKA (1975). Fauna compiled after ALBERTI (1964), BUSSE (1954), KELBER (1974), KÖNIG (1920), MUNDLOS & URLICHS (1984), STROMBECK (1849), URLICHS & SCHRÖDER (1980).

The first recorded occurrence of ceratites *Paraceratites* (*Progonoceratites*) *atavus* and *Ceratites (Doloceratites) primitivus* is below the "Hauptencrinitenbank" of Würzburg (GEISLER, 1939); but since two such beds are developed in this area (HOFFMANN, 1967: 22, 24) that GEISLER failed to identify, the exact level of these occurrences is Upper "Hauptencrinitenbank". The oldest well correlated occurrence *Paraceratites (Progonoceratites) flexuosus flexuosus* is in the "Trochitenbank 2" of Northern Baden (KÖNIG, 1920: 27; URLICHS & MUNDLOS, 1980: 3), which HAG-DORN & MUNDLOS (1982: 347) identify with the lower "Hauptencrinitenbank" of Lower Franconia.

Somewhat higher in the section ("Haßmersheimer Mergel 2") two specimens of *P. (Pr.)atavus atavus* have been described (KÖNIG, 1920: 31), which are somewhat older than the fauna of the same species and its relatives recently discovered in the "Haßmersheimer Mergel 3" of Swabia immediately below "Trochitenbank 4" (UR -

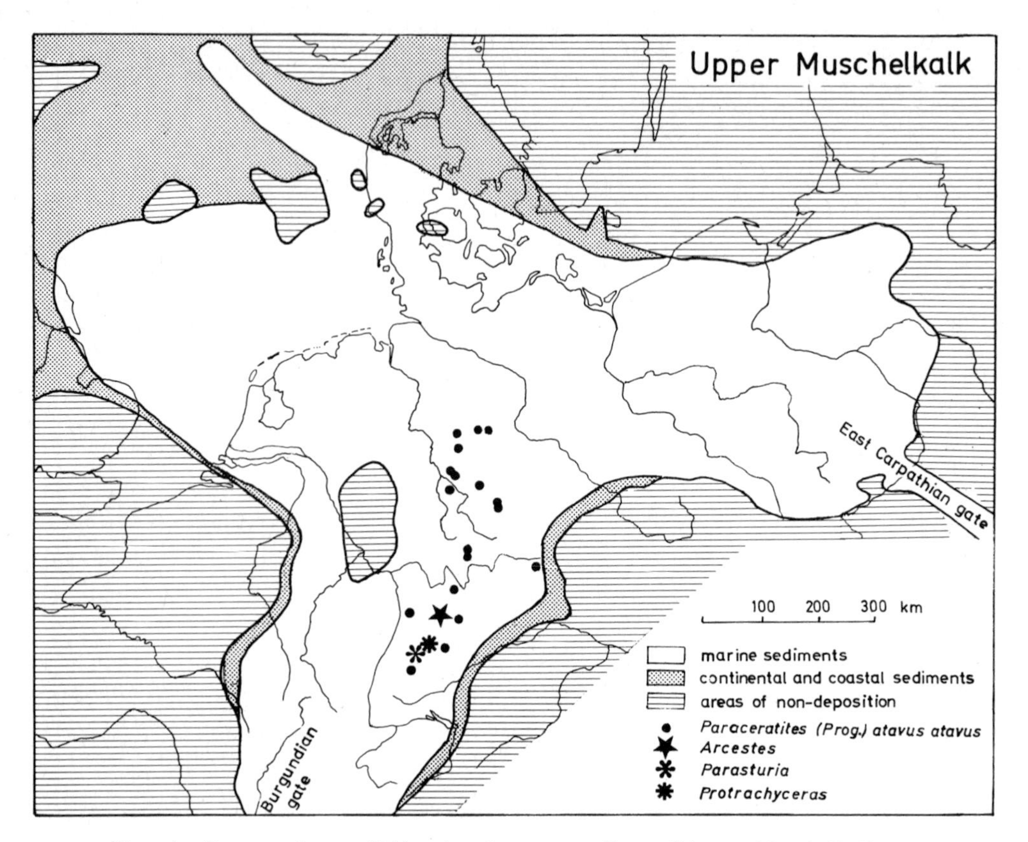

Fig. 4: Range of ceratitids in the *atavus* Zone (Upper Muschelkalk, Upper Anisian) of the Germanic Basin. Paleogeography after ZIEGLER (1982) and SENKOWICZOWA & SZYPERKO-SLIWCZYNSKA (1975). Fauna compiled after PENNDORF (1951), PHILIPPI (1901), RIEDEL (1918), ROTHE (1955), SPATH (1934), URLICHS (1978), URLICHS & MUNDLOS (1980).

LICHS & MUNDLOS, 1980). Other occurrences of *P. (Pr.) atavus atavus* are known from Upper Franconia, Hessen, Thuringia and from the margin of the Harz Mountains (Fig. 3) as well as in Lorraine (MAUBEUGE, 1947); exact locality unknown). Above the "Trochitenbank 4" only *P. (Pr.) flexuosus flexuosus* and *C. (Do.) primitivus* seen to survive.

The fast dispersal of the ceratites from the western gate over the entire western part of the Germanic Basin during the *atavus* Zone was probably facilitated by the establishment of euhaline conditions. During the following *pulcher/robustus* Zone the ceratites reached also the eastern parts of the basin, for which the East Carpathian Gate did not provide a direct immigration route.

The spread of *Germanonautilus* follows a similar pattern. It first appears in south-west Germany in the "Zwergfauna-Schichten" immediately above the base of the

Upper Muschelkalk, i.e. earlier than the first ceratites (MUNDLOS & URLICHS, 1984; Fig. 6), whose range it shares in the *atavus* Zone (Fig. 3) as well as during the geographic expansion in subsequent stages.

Other cephalopods (*Michelinoceras*, Fig. 3; *Arcestes*, *Parasturia* and *Protrachyceras*, Fig. 4) are only known in Swabia, i.e. near the Burgundy Gate, through which they came either as unsuccessful immigrants or as drifted shells.

MODIFICATION OF SUTURE LINES

A. Lower Muschelkalk (Upper Anisian)

The transition from an ammonitic to a ceratitic configuration in the suture line in the earliest ceratites has been described earlier (URLICHS & MUNDLOS, 1980), so that a short review plus the comparison with a similar reduction in ceratite immigrants of the Lower Muschelkalk will suffice in this context.

In the Lower Muschelkalk *Serpianites antecedens* has the largest stratigraphic range. Upon reexamining material from the margin of the Black Forest (M. SCHMIDT 1935; MAYER, 1971) it became evident that it only partly matches the holotype (MB C 436) with respect to cross section, rib pattern and umbilical width. Associated are specimens that are more densely ribbed and have a wider umbilicus, and should properly be assigned to different species (see KOZUR, 1974: 11).

The close relationship of these forms to Alpine species has long been known. "Of all known ceratites, none is closer to *Ceratites antecedens* from the German Lower Muschelkalk than *Ceratites binodosus*. The only profound difference is in the suture line with shallow lobes that are uniformly incised only at the base and with broad, non-incised saddles" (MOJSISOVICS, 1882: 20). In perfectly preserved specimens, however - the one studied by us was affected by neither pressure solution nor deformation - one may see that the incisions climb the sides of the lobe and that the saddles are slightly undulose (Fig. 5b). Except for the undulose saddles, this configuration is so similar (deep, narrow lobes and narrow saddles) to that in *Paraceratites binodosus* (Fig. 5a) that *"Ceratites" antecedens* might be attributed to the same genus. Since juvenile stages, however, habe a week keel as characteristic for *Serpianites*, we tentatively assign *"Ceratites" antecedens* to this genus. In other specimen from the same horizon the lobes are not as deep and incised only at the base (Fig. 5c) suggesting that the suture line was quite variable in these populations. The specimens from the stratigraphically younger "Schaumkalk" (mu x) of Thuringia have "a broad and shallow first auxiliary lobe" (SCHMIDT, 1935: 205), which indicates that there has been a progressive "ceratitization" of the suture line also in this lineage.

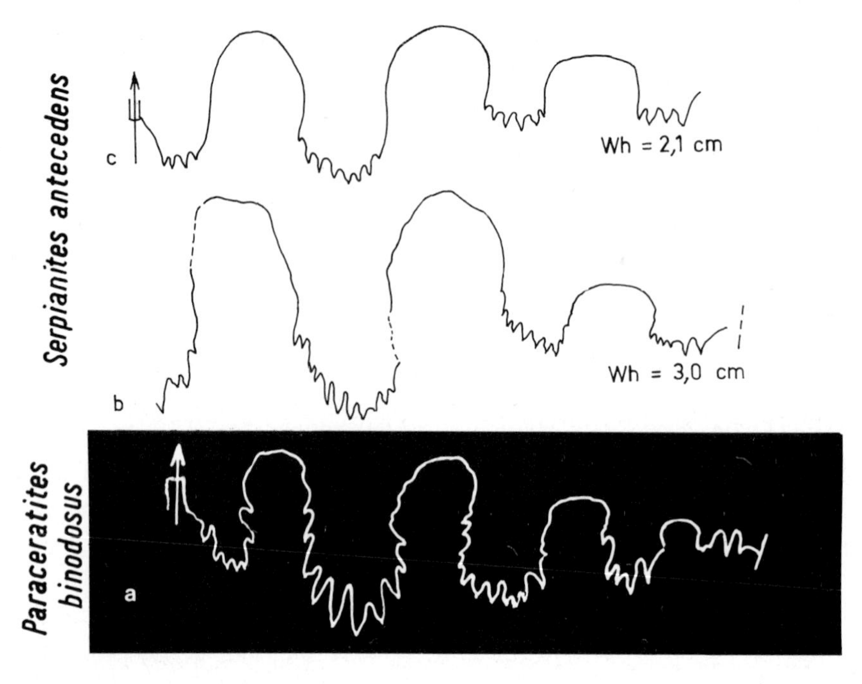

Fig. 5: External sutures of *Paraceratites* and *Serpianites* (drawn reversed)
a) *Paraceratites binodosus* (HAUER). Upper Anisian; Diliskelessi/ Turkey. Orig. ARTHABER 1915, pl. 12, fig. 1; SMNS 12505.-- x 3.
b) *Serpianites antecedens* (BEYRICH). Lower Muschelkalk; beds with *Homomya albertii*; Freudenstadt/Northern Württemberg. Orig. M. SCMIDT 1935, pl. 13, fig. 6; SMNS 12420.-- x 2,2.
c) *Serpianites antecedens* (BEYRICH). Lower Muschelkalk; beds with *Homomya albertii*, Bösingen near Freudenstadt/Northern Württemberg. Orig. M. SCHMIDT 1907, pl. 2, fig. 6; SMNS 15105.-- x 3,0.

A ceratitic suture line is also observed in the Lower Muschelkalk species *Noetlingites strombecki*. But since its close relationship to *Grambergia* (KOZUR, 1974:11) has been challenged by PARNES (1975: 19), it can not be quoted as another case of secondary ceratitization.

A similar statement could be made for *Beneckeia*. In specimens from the Lower Muschelkalk of Thuringia (FRITSCH, 1906: 264; KOZUR, 1974: 10) the lobes are incised at the base. But neither relationships to *Intornites* (KOZUR, 1974) nor to *Beneckeia buchi* (with smooth lobes) have

B. Upper Muschelkalk (Upper Anisian)

The ceratites are one of the best known and representative examples of evolutionary radiation in marginal epicontinental seas. Having the highest similarity with paraceratites of the Tethys, *Paraceratites* (*Progonoceratites*) *atavus atavus* is regarded as the founder species. Its representatives from the Haßmersheim Marls 3 of Neckarrems (near Stuttgart) show a highly variable suture line. Some specimens (Fig. 6b-c) have highly undulose saddles and strong lobe incisions ascending at the flanks, similar to the sutures of Alpine species of *Paraceratites* (Fig. 6a). Other specimens from the same bed, however, have nearly smooth saddles. Ratioes of sutural amplitude to whorl height (A/Wh) are 0,28 - 0,30 in specimens with strongly undulose saddles and 0,25 in individuals in which this undulosity is reduced.

Re-examination of specimens from a probably higher level (Elm, Lower Saxony), in contrast, showed that saddles were invariably smooth, and the incision of the lobes weaker and not ascending (Fig. 6 f-g). The A/Wh ratioes of the shallow sutures are 0.22 - 0,25. This means that these forms already had a typical ceratitic suture line, although their shell geometry does not allow taxonomic separation from *P. (Pr.)atavus atavus*.

Since *P. (Pr.) flexuosus flexuosus* is associated with the previous species in very early assemblages, it is not clear whether we deal with a simultaneous immigrant or a descendant. In *P. (Pr.) flexuosus flexuosus* from the "Haßmersheimer Mergel 3" in northern Swabia we find similarly variable suture lines, ranging from deep and heavily incised lobes and undulose saddles to nearly smooth saddles and shallow lobes (A/Wh = 0,26 - 0,33) in geometrically identical shells. Presumably younger specimens from the Elm, Lower Saxonia, as well as from the "Haßmersheimer Mergel" of Swabia, in contrast, have reduced sutural amplitudes (A/WH = 0,23) and almost to completely smooth saddles (Fig. 7d). Again, shell geometry remains unaffected and does not warrant taxonomic separation even at the subspecies level.

Among the immediate descendants of *Paraceratites (Pr.) atavus atavus*, *P.(Pr.) philippii neolaevis* and *P. (Pr.) philippii philippii* resemble their ancestor in whorl section and number of marginal nodes, but they have a wider umbilicus and smooth sutural saddles (Fig. 6 h-i). Other subspecies, such as *P. (Pr.) atavus discus* and *P. (Pr.) atavus sequens* also have shallow lobes and smooth saddles i.e. the reduction of the suture line to the ceratitic configuration has clearly taken place within the Germanic Basin and within the subgenus *Progonoceratites* , whose end in the *pulcher/ robustus* Zone marks the end of *Paraceratites* in this basin.

P. (Pr.) atavus atavus, however, seems to have given rise to a second and more successful lineage. It is heralded by specimens with a less arched venter, and with dichotomous sculpture and shallow lobes in the phragmocone. These features they share with *Ceratites (Doloceratites) primitivus*, in which the dichotomous sculpture extends

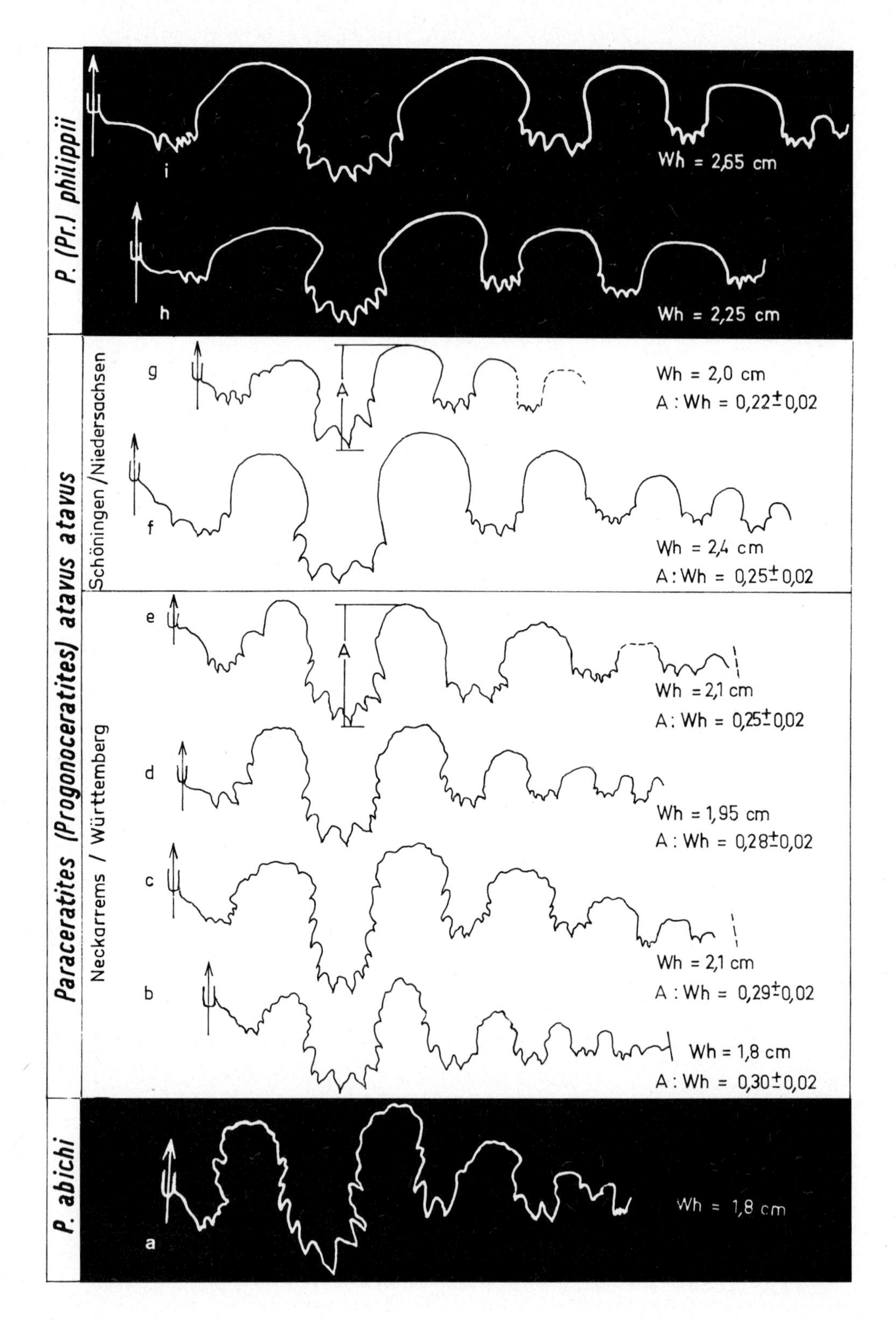
P. (Pr.) philippii
i
Wh = 2,65 cm
h
Wh = 2,25 cm
Paraceratites (Progonoceratites) atavus atavus
Schöningen / Niedersachsen
g
A
Wh = 2,0 cm
A : Wh = 0,22±0,02
f
Wh = 2,4 cm
A : Wh = 0,25±0,02
Neckarrems / Württemberg
e
A
Wh = 2,1 cm
A : Wh = 0,25±0,02
d
Wh = 1,95 cm
A : Wh = 0,28±0,02
c
Wh = 2,1 cm
A : Wh = 0,29±0,02
b
Wh = 1,8 cm
A : Wh = 0,30±0,02
P. abichi
a
Wh = 1,8 cm

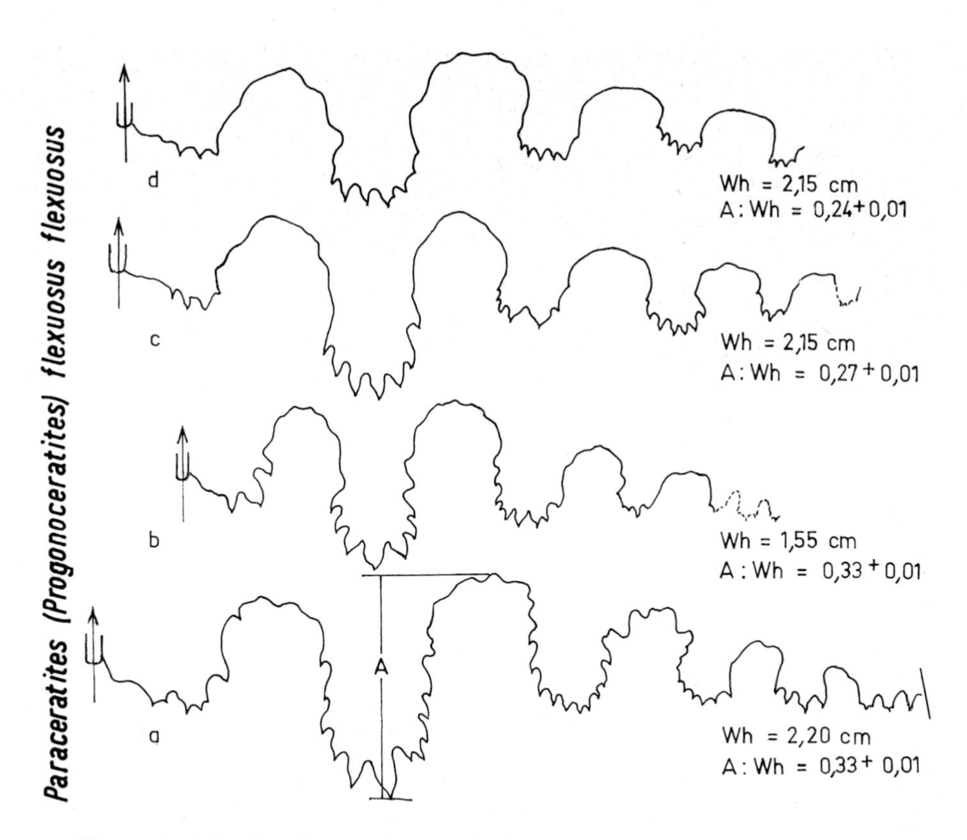

Wh = whorl height. A = amplitude, depth of the lateral lobe.

Fig. 7: Variations of the external sutures of *Paraceratites* (*Progonoceratites*) *flexuosus flexuosus* (PHILIPPI).-- x 3.
a-c) Upper Muschelkalk, *atavus* Zone, Haßmersheimer Mergel 3, Upper Anisian; Neckarrems/Northern Württemberg. SMNS 24570, 24601, 24603.
d) Upper Muschelkalk, *atavus* Zone, Upper Anisian. Schöningen/Niedersachsen; SMNS 23066.
Wh = whorl height, A = amplitude, depth of the lateral lobe.

Fig. 6: Phylogeny of the external sutures of *Paraceratites* (*Progonoceratites*).-- x 3.
a) *Paraceratites abichi* (MOJSISOVICS); Upper Anisian, Schreyeralm/Austria; SMNS 10987-1.
b-e) *Paraceratites* (*Progonoceratites*) *atavus atavus* PHILIPPI); Upper Muschelkalk, *atavus* Zone/Upper Anisian, Neckarrems/Northern Württemberg; SMNS 24506, 24507, 24520, 24528.
f-g) *Paraceratites* (*Progonoceratites*) *atavus atavus* (PHILIPPI), Upper Muschelkalk, *atavus* Zone/Upper Anisian, Schöningen/Niedersachsen; SMNS 24562, 24563.
h) *Paraceratites (Progonoceratites) philippii neolaevis* (PENNDORF); Upper Muschelkalk, *pulcher/robustus* Zone, Schöningen/Niedersachsen; SMNS 14691.
i) *Paraceratites (Progonoceratites) philippii philippii* (RIEDEL); Upper Muschelkalk, Schöningen/Niedersachsen, SMNS 24685.

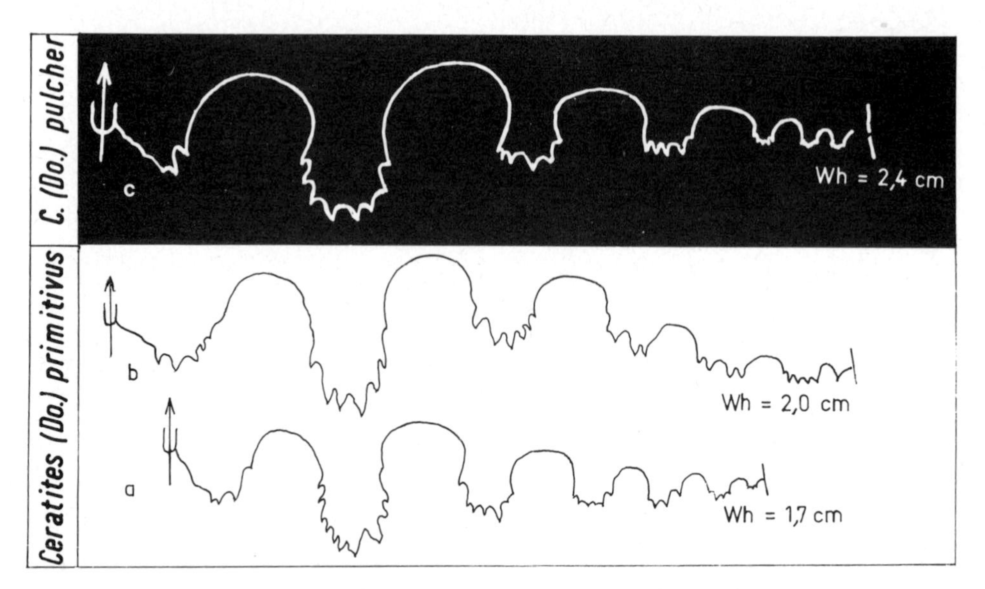

Fig. 8: External sutures of *Ceratites (Doloceratites)*.-- x 3.
a-b) *Ceratites (Doloceratites) primitivus* RIEDEL; Upper Muschelkalk, *atavus* Zone, Upper Anisian; Neckarrems/Northern Württemberg; SMNS 24724, 24709; (b: reversed drawn).
c) *Ceratites (Doloceratites) pulcher* RIEDEL; Upper Muschelkalk, Schöningen/Niedersachsen; SMNS 23068.

to the body chamber even in adult specimens, while the venter is flat or slightly arched and the suture ceratitic with ascending incisions (Fig. 8 a-b). *C. (Do.) primitivus*, in turn gave rise to *C. (Do.) pulcher* with a completely ceratitic suture line (Fig. 8c) and thereon to the radiation of the other ceratitids as outlined by WENGER (1957; Fig. 44).

Although taxonomic procedure requires clear distinction of features at the generic level, some tolerance must be allowed for in the transition fields. Therefore we assign the Germanic species of *Progonoceratites* to the morphologically similar *Paraceratites* of the Alpine province, even though their suture may be ceratitic. Only species in which this suture is combined with truely ceratitic shell geometries and sculptures should be assigned to the descendant genus *Ceratites*.

Since the suture line is variable in the earliest Germanic ceratites, separation of the subfamilies Paraceratitinae SILBERLING and Ceratitinae MOJSISOVICS is not justified. As neither SILBERLING (1962) nor TOZER (1981) provide a diagnosis for the subfamily Paraceratitinae, it is implied that its distinction should be mainly based on the suture.

Outside the Germanic Basin, ceratitic suture lines are known from *Gevanites* and *Israelites* from the Sephardic Province. This emphasizes that we deal with a general trend related to the immigration into marginal epicontinental basins. The multiple occurrence of such evolutions accounts for the renaissance of the ceratitic suture line in the Middle Triassic in a fashion that had been dominant already in the Lower Triassic.

Acknowledgements: We thank A. Seilacher (Tübingen) for encouraging us to write this paper and for translating it. K. Bandel (Erlangen), A. Bartholomä (Neuenstein), H. Hagdorn (Künzelsau), H. Jaeger (Berlin) and H.U. Schwarz (Bochum) made valuable suggestions.

Location of specimens: MB = Paläontol. Museum of the Humboldt University, Berlin; SMNS = Staatliches Museum für Naturkunde, Stuttgart.

REFERENCES

Aigner, T., 1982: Calcareous Tempestites: Storm-dominated Stratification in Upper Muschelkalk Limestone (Middle Trias, SW-Germany).- In: Cyclic and Event Stratification: 180-198, Berlin-Heidelberg-New York (Springer).

Alberti, F.v., 1864: Übersicht über die Trias, mit Berücksichtigung ihres Vorkommens in den Alpen.- 353 pp.,Stuttgart (Cotta).

Arthaber, G.v., 1915: Die Trias von Bithynien (Anatolien).- Beitr.Paläontol. u. Geol. Österreich u. Ungarn, 27: 85-206.

Assmann, P., 1937: Revision der Fauna der Worbellosen der Oberschlesischen Trias.- Abh.preuss.geol.Landesanstalt, N.F., 170: 1-134, Berlin.

Beyrich, E., 1854: Übersicht über die im Muschelkalk zu Rüdersdorf bei Berlin bis jetzt aufgefundenen Ammoniten.- Z.deutsch.geol.Ges., 6: 513-515, Berlin.

Beyrich, E., 1858: Über Ammoniten des unteren Muschelkalks.- Z.deutsch.geol.Ges., 10: 208-214.

Busse, E., 1954: Profil der Unteren und Mittleren Ceratitenschichten vom Eisenberg bei Hessisch-Lichtenau und Wahlburg.- Notizbl.hess.Landesamt Bodenforsch., 82: 152-167, Wiesbaden.

Claus, H., 1921: Über *Ptychites* und *Arnoites* aus dem Schaumkalk von Jena.- Cbl.Miner. Geol.Paläont., 1921: 120-126, Stuttgart.

Claus, H., 1955: Die Kopffüsser des deutschen Muschelkalks.- Neue Brehm-Bücherei 161: 1-60, Wittenberg.

Duringer, P., 1982: Sédimentologie et paléocollogie du Muschelkalk supérieur et de la Lettenkohle (Trias germanique de l'est de la France. Diachronie des facies et reconstitutions des paléoenvironments.- Thèse Univ. Louis Pasteur (3ème cycle) 96pp., Strassbourg.

Eck, H., 1865: Über die Formation des bunten Sandsteins und des Muschelkalks in Oberschlesien und ihre Versteinerungen.- 148pp., Berlin (Friedländer).

Eck, H., 1872: Rüdersdorf und Umgebung. Eine geognostische Monographie.- Abh.geol. Spezialkarte Preussen u. thüring.Staaten, 1/1: 1-183, Berlin.

Eck, H., 1891: *Ceratites antecedens* BEYR. von Wenden in Württemberg.- Z.deutsch.geol. Ges., 43: 734-735, Berlin.

Fisher, M.J., 1984: Triassic. In: K.W. Glennie (Ed.): Introduction to the Petroleum Geology of the North Sea.- pp. 85-101, Oxford-London-Edinburg (Blackwell).

Franz, V., 1903: Ueber "*Nautilus bidorsatus*" und seine Verwandten.- N.Jb.Mineral. Geol. u. Paläont.Beil.-Bd. 17: 486-497, Stuttgart.

Fritsch, K.v., 1906: Beitrag zur Kenntnis der Tierwelt der deutschen Trias.- Abh. naturforsch.Ges.Halle, 24: 219-285, Stuttgart.

Geisler, R., 1939: Zur Stratigraphie des Hauptmuschelkalks in der Umgebung von Würzburg mit besonderer Berücksichtigung der Ceratiten.- Jb.Preuss.geol.Landesanst. 59(1938): 197-248, Berlin.

Geyer, O.F. & Gwinner, M., 1964: Einführung in die Geologie von Baden-Württemberg. 223pp., Stuttgart (Schweizerbart).

Giebel, G.G., 1853: Über *Ammonites dux* aus dem Muschelkalk von Schraplau.- Z.Ges. Naturwiss., 1: 341-345, Halle.

Hagdorn, H., 1978: Muschel/Krinoiden-Bioherme im Oberen Muschelkalk (mo_1, Anis) von Crailsheim und Schwäbisch Hall (Südwestdeutschland).- N.Jb.Geol.Paläont.Abh., 156: 31-86, Stuttgart.

Hagdorn, H. & Mundlos, R., 1982: Autochthonschille im Oberen Muschelkalk (Mitteltrias) Südwestdeutschlands.- N.Jb.Geol.Paläont.Abh., 162: 332-351, Stuttgart.

Hoffmann, U., 1967: Erläuterungen zur Geologischen Karte von Bayern 1:25000, Blatt Nr. 6225 Würzburg-Süd, 134 pp., München.

Kelber, K.P., 1974: Terebratel/Placunopsiden-Riffe im basalen Hauptmuschelkalk Untermuschelkalk Unterfrankens.- Der Aufschluß 25: 643-645, Heidelberg.

Kelber, K.P., 1977: *Hungarites strombecki* GRIEPENKERL aus dem mainfränkischen Wellenkalk.- Der Aufschluß 28: 145-149, Heidelberg.

König, H., 1920: Zur Kenntnis des unteren Trochitenkalkes im nördlichen Kraichgau.- Sitzungsber.Heidelberger Akad.Wiss.,math.-naturwiss.Kl.,Abt. A, 1920:1-47, Heidelberg.

Kozur, H., 1974: Biostratigraphie der germanischen Mitteltrias.- Freiberger Forschungsh., C 280: Part 1, 56 p., Part 2, 72 p., Leipzig.

Laughier, R., 1963: Trias de facies germanique en Lorraine.- Mém.Bur.Rech.géol. min.,15: 39-65, Paris.

Lehmann, U., 1976: Ammoniten - Ihr Leben und ihre Umwelt. 171 pp., Stuttgart (Enke).

Maubeuge, P.L., 1947: Sur les Cératites du Muschelkalk lorrain (Moselle, Meurthe et Moselle, Vosges).- Soc.Géol.France, 1947: 163-164, Paris.

Mayer, G., 1971: Der erste *Ceratites antecedens* BEYRICH aus dem Wellendolomit von Ittersbach (Kreis Karlsruhe) und weitere Vorkommen dieser Art.- Der Aufschluß 22: 126-128, Göttingen.

Mojsisovcs, E.v., 1882: Die Cephalopoden der mediterranen Triasprovinz.- Abh.k.k. geol.Reichsanst., 10: 1-322, Wien.

Mundlos, R. & Urlichs, M., 1984: Revision von *Germanonautilus* aus dem germanischen Muschelkalk (Oberanis - Ladin).- Stuttg.Beitr.Naturk., B 99: 1-43, Stuttgart.

Noetling, F., 1880: Die Entwicklung der Trias in Niederschlesien.- Z.deutsch.geol. Ges., 32: 332-381, Berlin.

Parnes, A., 1975: Middle Triassic ammonite biostratigraphy in Israel.- Geol.Surv.Israel, Bull. 66: 1-25, Jerusalem

Penndorf, H., 1951: Die Ceratiten-Schichten am Meißner in Niederhessen.- Abh.senckenberg.naturforsch.Ges., 484: 1-24, Frankfurt/Main.

Philippi, E., 1901: Die Ceratiten des oberen deutschen Muschelkalks.- Paläont.Ab.,N. F., 4(8): 347-457, Jena.

Picard, K., 1889: Über einige seltenere Petrefacten aus dem Muschelkalk.- Z.deutsch. geol.Ges., 41: 635-640, Berlin.

Picard, K., 1899: Über Cephalopoden aus dem unteren Muschelkalk bei Sondershausen.- Z.deutsch.geol.Ges., 51: 299-309, Berlin.

Rassmuss, H., 1915: Alpine Cephalopoden im niederhessischen Muschelkalk.- Jb.kgl. preuss.geol.Landesanst., 34, Part 2/1913): 283-306, Berlin.

Ricour, J., 1963: Esquisse paléogéographique de la France aux temps triassiques.- Mém.Bur.Rech.géol.Min., 15: 715-734, Paris.

Riedel, A., 1918: Beiträge zur Paläontologie und Stratigraphie der Ceratiten des deutschen Oberen Muschelkalks.- Jb.kgl.preuss.Landesanst., 37(1916): 1-116, Berlin.

Rothe, H.W., 1951: *Ceratites antecedens* BEYR. aus dem Unteren Muschelkalk von Lieskau bei Halle.- Hallesches Jb.mitteldeutsch.Erdgesch., 1: 67-69, Halle.

Rothe, H.W., 1955: Die Ceratiten und die Ceratitenschichten des Oberen Muschelkalks (Trias) im Thüringer Becken.- Beitr.Geol.Thüringen, 8: 255-323, Frankfurt/M.

Rothe, H.W., 1959: Ammonoideen aus dem Unteren Muschelkalk Thüringens.- Der Aufschluß, 10: 66-68, Göttingen.

Schmidt, M., 19o7: Das Wellengebirge der Gegend von Freudenstadt.- Mitt.geol.Abt. kgl.württ.statist.Landesamt, 3: 1-99, Stuttgart.

Schmidt, M., 1934: Über *Ceratites antecedens* BEYRICH und verwandte Formen.- Jb. preuss.geol.Landesanst., 55: 198-213, Berlin.

Schneider, E., 1955: Beiträge zur Kenntnis des Trochitenkalks des Saarlandes und der angrenzenden Gebiete.- Ann.Univ.Saraviensis, 4: 88-92, Saarbrücken.

Schulz, M.-G., 1972: Feinstratigraphie und Zyklenglieder des Unteren Muschelkalks in N-Hessen.- Mitt.Geol.-paläont.Inst.Univ.Hamburg, 41: 133-170, Hamburg.

Schwarz, H.U., 1975: Sedimentary structures and facies analysis of shallow marine Carbonates (Lower Muschelkalk, Middle Triassic, Southwestern Germany).- Contr. Sedimentology, 3: 1-100, Stuttgart.

Senkowiczowa, H. & Szyperko-Sliwczynska, A., 1975: Stratigraphy and Paleogeography of the Trias.- Bull.Geol.Inst.Warszawa, 252: 131-147, Warszawa.

Silberling, N.J., 1962: Stratigraphic distribution of Middle Triassic Ammonites at Foot Hill, Humboldt Range, Nevada.- J.Paleont., 36: 153-160, Tulsa.

Spath, L.F., 1934: The Ammonoidea of the Trias. Catalogue of the fossil Cephalopoda in the British Museum (Natural History), Part IV, 521 pp., London.

Strombeck, A.v., 1849: Beitrag zur Kenntnis der Muschelkalkbildung im nordwestlichen Deutschland.- Z.deutsch.geol.Ges., 1: 115-231, Berlin.

Tozer, E.T., 1981: Triassic ammonoidea: Classification, evolution and relationship with Permian and Jurassic forms.- Systematic Assoc., spec.vol. 18: 65-100, London, New York.

Trammer, J., 1972: *Beyrichtites (Beyrichites)* sp. from the Lower Muschelkalk of the Holy Cross Mts.- Acta geol.Polonica 22: 25-28, Warszawa.

Urlichs, M., 1978: Über zwei alpine Ammoniten aus dem Oberen Muschelkalk SW-Deutschlands.- Stuttg.Beitr.Naturkde. B, 39: 1-13.

Urlichs, M . & Mundlos, R., 1980: Revision der Ceratiten aus der *atavus* Zone (Oberer Muschelkalk, Oberanis) von SW-Deutschland.- Stuttg.Beitr.Naturkde. B,48: 1-48.

Urlichs, M. & Schröder, W., 1980: Erstfund eines orthoceratiden *(Michelinoceras campanile)* im germanischen Muschelkalk.- Stuttg.Beitr.Naturkde. B 59: 1-7.

Vollrath, P., 1924: Beiträge zur Stratigraphie und Paläogeographie des fränkischen Wellengebirges.- N.Jb.Miner.Geol.Paläont.Beil., 50: 120-288, Stuttgart.

Vossmersbäumer, H., 1970: *Germanonautilus* (Cephalopoda, Nautilida) im Würzburger oberen Wellenkalk.- Geol.Bl.NO-Bayern, 20: 46-51, Erlangen.

Vossmersbäumer, H., 1972: Neue Cephalopoden-Funde aus dem Wellenkalk Mainfrankens.- Der Aufschluß 23: 240-252, Göttingen.

Wagner, R., 1888: Über einige Cephalopoden aus dem Röth und unteren Muschelkalk von Jena.- Z.deutsch.geol.Ges., 40: 24-38, Berlin

Wagner, R., 1923: Neue Beobachtungen aus dem Muschelkalk und Röt von Jena.- Jb. preuss.geol.Landesanst., 42(1921): 1-16, Berlin.

Walther, K., 1927: Zwölf Tafeln der verbreitetsten Fossilien aus dem Buntsandstein und Muschelkalk der Umgebung von Jena.- 2nd ed., 48 pp., Jena

Wenger, R., 1957: Die germanischen Ceratiten.- Palaeontographica, A, 108: 57-129, Stuttgart.

Wurm, A., 1914: Über einige neue Funde aus dem Muschelkalk der Umgebung von Heidelberg.- Z.deutsch.geol.Ges., 66: 444-448, Berlin.

Ziegler, P.A., 1982: Geological Atlas of Western and Central Europe.- 130 pp., The Hague (Shell).

IMMIGRATIONS OF CRINOIDS INTO THE GERMAN MUSCHELKALK BASIN

H. Hagdorn

Ingelfingen

Abstract: Since Muschelkalk crinoids were stenohaline and required hard substrates to get attached, they could settle continuously only on paleohighs near the connection to the Tethys. In Upper Muschelkalk times they managed to spread from these centers throughout the western parts of the basin as bottoms became shelly due to regressive conditions. In the Lower Muschelkalk basin continuous settlement was possible only near the eastern gates, from where crinoids spread to the West together with the oolitic Schaumkalk facies. In contrast to the less substrate-controlled ceratites and conodontophorids, crinoids fail to show endemic evolution.

INTRODUCTION

The large diversity of Paleozoic pelmatozoans became drastically reduced by the Permian extinction (Kier 1973). It put an end to the Cystoidea and Blastoidea, but also to the bulk of the crinoids. Only the ancestor group of the Articulata managed to survive into the Triassic.

Unfortunately the fossil record of Lower Triassic crinoids is very scanty and little studied. The record improves in the Middle Triassic, but the majority of described species comes not from the Tethys, but from the Muschelkalk facies, where taphonomic conditions allowed articulated specimens, including the calices, to be preserved. Only a few species are based on isolated columnals that are characteristic enough to be used as index fossils.

In contrast to the Germanic Muschelkalk, the Middle Triassic of the Tethys itself and of other marginal basins has yielded almost exclusively isolated crinoidid ossicles (Kristan-Tollmann & Tollmann 1967; Mostler 1972). Accordingly our knowledge about the actual distribution of Middle Triassic crinoids is very incomplete. Some of the more common Muschelkalk species have never been recorded from other areas. Nevertheless we may assume that the crinoid fauna of the Tethys and its marginal epicontinental seas was rather uniform, and that diversity in the Tethys itself was much higher than appears from the poor fossil record.

From an evolutionary point of view it is very unfortunate that the Germanic Basin became terrestrial in the Upper Triassic. Being deprived of a "fossil trap" analogous to the Muschelkalk, we have to reconstruct the important radiation of crinoids in the Lower Carnian from the Alpine record, where most species have to be based on isolated columnals except for the few calices known from the St. Cassian beds (see plates in Zardini 1976).

ECOLOGICAL "NICHES" OF MUSCHELKALK CRINOIDS

By early Middle Triassic times the crinoid stock that had survived the Permian extinction was as yet little diversified. Its members were still largely attached by a root-callus and thus restricted to adequate hard substrates. Thus substrate condition was the main ecological factor, besides salinity and nutrient supply, that limited the distribution of Muschelkalk crinoids. This is in marked contrast to the diversified Paleozoic pelmatozoans which had been adapted to a variety of substrates, as well as to the situation in the Jurassic, when the evolution of a new mode of attachment had allowed the isocrinids to again invade muddy substrates.

Suitable hard substrates for the attachment of crinoid larvae were only episodically available in the Muschelkalk sea, in contrast to muddy bottoms with an endobenthic fauna that were as yet inaccessible. Such hard substrates are of three types:

* Hardgrounds, i.e. cemented bottoms characterized by borings *(Trypanites)* and incrustations of the small bivalve *Placunopsis*. They are particularly common in the Lower Muschelkalk at the top of meter-thick and commonly oolitic beds of shelly lime-limestone. Crinoids are attached to such surfaces by a discoid callus from which the stem emerges at a right angle. Individuals grow to relatively large sizes.
  A typical example is *Chelocrinus carnalli*, which colonized the top of the "Untere Schaumkalkbank" near Freyburg/Unstrut (DDR) solitarily or in bundles (Biese 1927; Müller 1963, Figs. 503 and 506).
* Shell beds of coarse shell material may be allochthonous or, more commonly, autochthonous with a biostromal or biohermal structure (Hagdorn & Mundlos 1982). The root-calluses of crinoids formed an important element in the frame-building of terquemiid/crinoid bioherms (Hagdorn 1978) by incrusting the basal stem parts of other individuals, the shells of terquemiid "oysters" *(Enantiostreon, Newaagia)* and byssally attached *Myalina*. In this situation the crinoids reach maximum sizes.
  This type of occurrence is particularly common in the "Trochitenkalk" (Lower part of the Upper Muschelkalk), but also occurs in the Karchowice-Beds and the

Diplopora Dolomite in the Lower Muschelkalk of Upper Silesia, where corals and siliceous sponges are additional frame builders (Assmann 1944, Zawidzka 1975).

* Shell islands on soft or firm grounds. Shells of *Germanonautilus* and of large epifaunal bivalves (*Plagiostoma lineatum*, solitary *Newaagia noetlingi*, in the Gogolin Beds also endobyssate bakeveliid bivalves (Wysogorski 1903; Hagdorn, in prep.)), served as anchoring grounds of *Dadocrinus* in the Lower Muschelkalk. But because of the instability of their anchoring grounds, such crinoids grew only to small sizes.

BASIN HISTORY

During Anisian and Ladinian times the epicontinental sea of the Germanic Muschelkalk covered large parts of Central Europe; but it was connected with the Tethys ocean only by narrow straits that changed their positions due to epirogenetic movements. Whether or not the Lower Triassic connection of the basin with the Northern Skandic persisted into Lower Muschelkalk times (Schwarz 1970) is still an open question. At least it had no faunistic consequences. Micropaleontological evidence (Kozur 1973, 1974) suggests that the first marine immigration in the Lower Anisian ("Röt" to "Oolith-Bänke") came through the East Carpathian Strait, while during Middle and Early Upper Anisian times ("Oolith-Bänke" to the base of the Middle Muschelkalk) the main connection was through the Silesian-Moravian Strait. The occurrence of Radiolarians at the level of the "*Spiriferina*-Bank" in the Southern Black Forest (Braun 1983) and special facies developments in this area (Schwarz 1970) suggest that there might also have been temporary connections to the west. The areas near the eastern gates had a more diverse marine fauna than the central and western parts of the basin, where the Wellenkalk facies is claimed to indicate intertidal conditions (Schwarz 1970) and where the waters seem to have been subject to cyclic changes in salinity (Schulz 1972).

The salinity crisis of the Middle Muschelkalk (mid - upper Anisian), expressed by evaporite deposition in large parts of the basin, wiped out the marine invertebrate fauna except near the Silesian gate, where a highly diverse fauna of Tethyan affinities managed to persist for some time (*Diplopora* Dolomite). High salinity is also indicated by stromatolites that locally flourished on oolite beds in SW Germany.

The Upper Muschelkalk transgression entered the basin through the Burgundy Gate in the Southwest and reached its peak in the Lower Ladinian (*spinosus* zone; see Aigner, this volume). Accordingly marine conditions persisted in the southwest during the following regressive phase, while the eastern and northern parts, from the *enodis/laevigatus* zone on, got increasingly dominated by a delta prograding

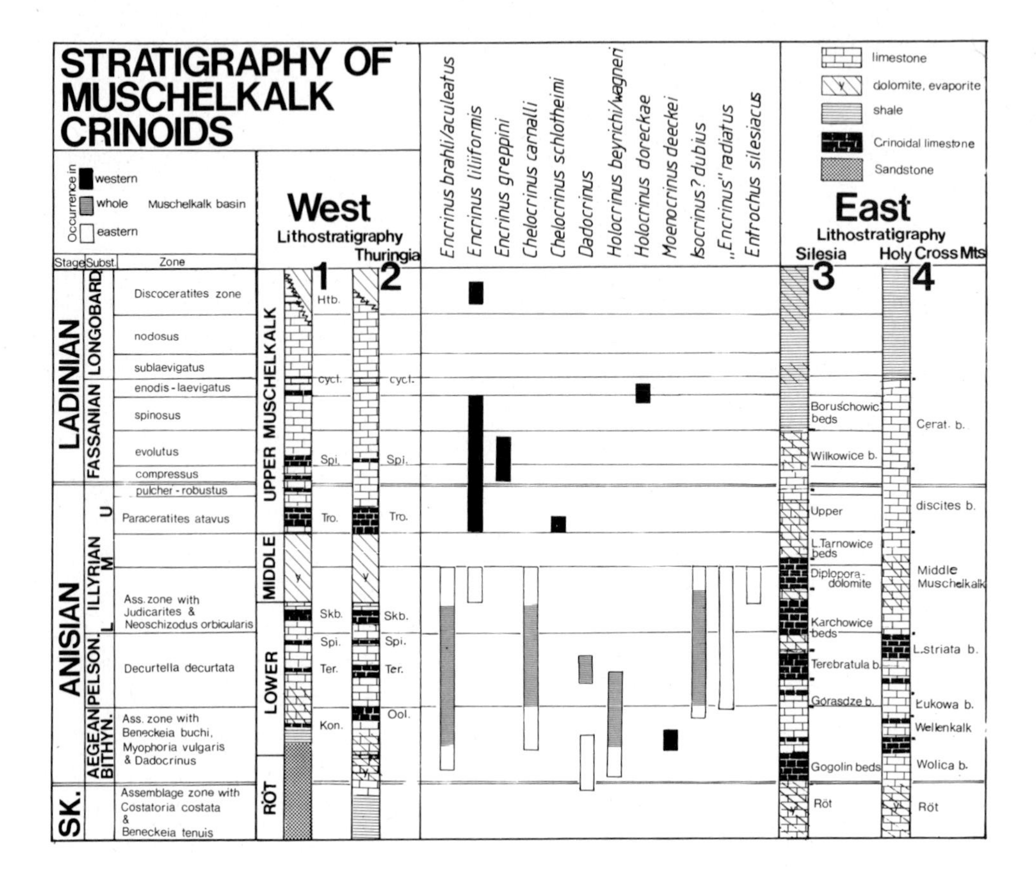

Fig. 1: The selected lithostratigraphic sections (not to scale; for locations see Fig. 4) illustrate major crinoid occurrences with reference to the biostratigraphic zones.
In the Lower Muschelkalk dispersal centers were in the eastern part of the basin (marine facies with *Dadocrinus*), evaporitic and clastic facies dominated in the west. Only some species managed to invade areas farther to the west during the short intervals of shelly marker beds. Restriction to the eastern gate was even more pronounced during the Middle Muschelkalk, in the later part of which a salinity crisis wiped out marine faunas altogether.
In the Upper Muschelkalk dispersal centers had switched to the new gate in the west, but of the immigrant crinoid species only *Encrinus liliiformis* managed to survive for a larger period in the Germanic Basin. (Sections after Geyer & Gwinner 1968, Kozur 1974, Zawidzka 1975, biostratigraphic divisions follow Kozur 1974. HTB = Hauptterebratelbank; cycl. = cycloides-Bank; Spi = Spiriferina-Bed; Tro = Crinoid cohemnal cycl. = *cycloides*-Bank; Spi = *Spiriferina*-Bank; Tro = Crinoid columnal Trochitenbank; SKB = Schaumkalkbänke; Ter = Terebratelbänke; ool = Oolithbänke; Kon = Konglomeratbänke.

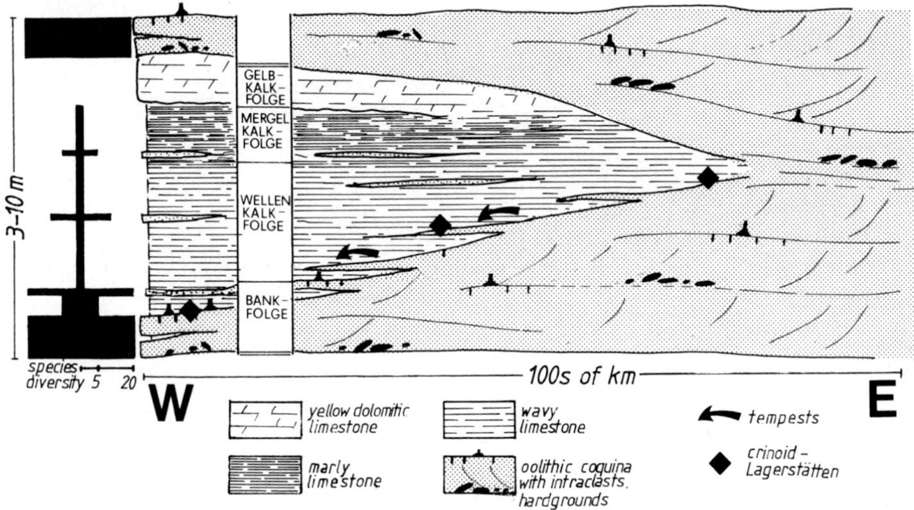

Fig. 2: Facies model of "Schaumkalk"-horizons.
The shallow-marine "Schaumkalk" facies of the Lower Muschelkalk dominant in the east of the basin throughout the Pelnian and Lower Illyrian episodically intertongued with the western "Wellenkalk" facies. These Schaumkalk advances (which would be transgressive if the intertidal origin of the Wellenkalk is correct; Schwarz 1970 allowed the expansion of stenohaline epibenthic communities from their eastern dispersal centers. Minor expansions followed the deposition of individual coarse tempestites, which are more common in the west, where amalgamation has not wiped out the minor events.
(Lithologic cycles after Schulz 1972; for faunal successions in such a cycle see Hagdorn & Simon 1983).

from the north, heralding the facies of the following Letten-Keuper.

PATTERNS OF CRINOID DISTRIBUTION

Lower Muschelkalk

Several sequences of thick-bedded and partly oolitic shelly limestones in Upper Silesia (Poland) with interbedded shaly horizons contain crinoids together with other stenohaline epifauna. Examples are the *Pecten*- and Dadocrinus mestones and Conglomeratic Horizon of the Lower Gogolin Beds, the Gorasdze Beds, the *Terebratula* Beds and the Karchowice Beds of the Lower and the *Diplopora* Dolomite of the Middle Muschelkalk (Assmann 1944; Zawidzka 1975). These occurrences suggest continuous colonization over extended periods. Only the Upper Gogolin Beds have failed to yield crinoids.

More than 70 m of fully marine "Schaumkalke" were deposited during the later part of the Lower Muschelkalk in the Dano - Polonian - Trough, were they outcrop in Rüdersdorf near Berlin (Hardt 1952). They probably extended also along the northern margin of the Bohemian Massive, where the Muschelkalk became subsequently eroded.

The central and western parts of the basin, however, are dominated by the Wellenkalk facies. Here crinoid and brachiopod remains are restricted to the "Oolithbänke", "Terebratelbänke" and "Schaumkalkbänke". These beds can be followed as marker horizons through large parts of the basin, but they become gradually thinner to the West and disappear in the northern part of Baden-Württemberg. Crinoids may also occur, however, in thinner tempestite beds of more limited extension.

Schulz (1972) has subdivided the Lower Muschelkalk into several asymmetric cycles 3 - 11 m in thickness. They typically start with oolitic marker beds that contain intraclasts and commonly have a hard-ground top incrusted by crinoids. Their diverse fauna also includes brachiopods and occasional corals. They are overlain by Wellenkalk beds, in which tempestites of 2 - 10 cm still carry a stenohaline fauna. In the following phase (Wellenkalk and marly limestones), however, the fauna becomes less diverse and contains only euryhaline bivalves and gastropods. Some cycles end with yellowish dolomitic limestones without fossils. Schultz relates these cycles to changes in turbulence and salinity, but their relation to the shallowing- and coarsening-upward cycles observed by Aigner (this volume) in the Upper Muschelkalk remains to be studied.

In a paleogeographic sense, the Terebratelbänke and Schaumkalkbänke may be viewed as episodic westward extensions of the more fully marine facies of the east into the wellenkalk basin. If the Wellenkalk facies is of intertidal origin (Schwarz 1970), they would mark transgressive events. Otherwise they could - like Aigner's coarsening-upward cycles (this volume) - reflect regressions, but perhaps in a sea with a strong vertical salinity gradient.

In any case the crinoids followed these temporal oscillations of salinity and suitable substrates into the western part of the basin.

Upper Muschelkalk

The distributional patterns of crinoids in the Upper Muschelkalk are clearer, because outcrops are plentiful and stratigraphic subdivisions well established in SW Germany, which is also closest to the Tethyan connection during this period. Where thicknesses reach a maximum in this part of the basin (Heilbronn-Kraichgau; see

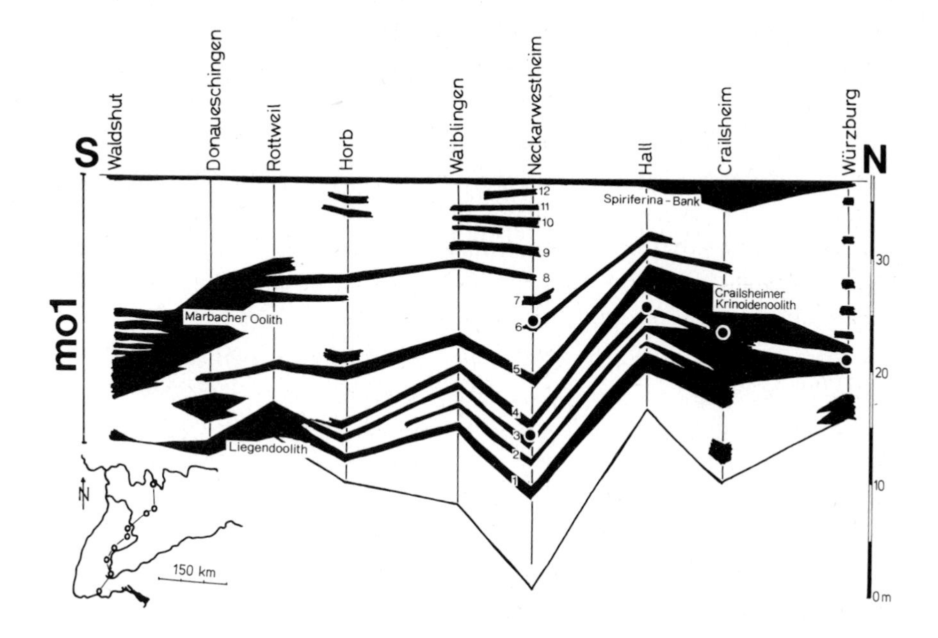

Fig. 3: N/S transsect of the crinoidal Trochitenkalk (mo_1) of SW Germany. An oolitic shoal in the Crailsheim area was continuously inhabited by *Encrinus liliiformis*. Two ammonite zones later, a similar oolite shoal was established 220 km to the south (Marbach Oolite), when a micritic and shaly facies with soft bottom communities took over in the Crailsheim area.
In the area of maximum thickness the Trochitenkalk consists of an alternation of such micritic and shaly beds with individual crinoid horizons ("Trochitenbänke" 1 - 12), which Aigner (this volume) interprets as the coarse tops of minor transgressive/regressive cycles. They also reflect periods, during which epibenthic faunas could spread from the shoals over large parts of the basin before they became smothered by the muddy sediments of the next transgressive cycle. These smothering events resulted in echinoderm conservation deposits (dotted).
During the earlier time (Trochitenbänke 1 - 5) the Crailsheim area, later (Trochitenbänke 6 - 12) the Marbach and possible other oolite shoals served as dispersal centers, while the *Spiriferina* Bank marks an immigration through the Burgundy Gate during a regressive interval.
(Modified from Geyer & Gwinner 1968; Figs. 12 and 13).

Neckarwestheim section in Fig. 3) the common crinoid *Encrinus liliiformis* and its associates (Hagdorn 1978) are restricted to a number of "Trochitenkalk" beds (1 - 12), while the interbedded shales and "Tonplatten" contain a typical soft bottom fauna. Near the margin of the basin and in the area of submarine swells the oolitic crinoidal and shelly limestones may be amalgamated into units more than 10 m in thickness which indicate more or less continuous crinoid populations.

In the coarsening-upward, regressive cycles recognized by Aigner (this volume) crinoids are only found in the coarse top layers, which provided a suitable substrate.

They became smothered by the muds of the next cycle, in which a soft bottom community became again established.

Regional transsects (Fig. 3) show that the "Trochitenbänke" 1-5 of central Württemberg are extensions of the crinoidal oolites of the Crailsheim and of the Liegend-oolith in Southern Württemberg, while the crinoidal horizons 7 - 9 spread from the Marbach Oolite and from crinoidal oolites west of the Rhine at a time when soft bottoms had already taken over in the previous biological spreading center.

The biological spreading center of the encrinids in the "*Spiriferina*-Bank", one of the major marker horizons, remains unknown. Their association with the Alpine species *Punctospirella ("Spiriferina") fragilis* suggests that it was outside the Germanic Basin. Unlike *Encrinus* , this "exotic" brachiopod, as well as the somewhat "younger" scallop *Praechlamys reticulatus* managed to expand their ranges to Eastern Upper Silesia, where the two species coexisted throughout the deposition of several meters of the Wilkowice Conglomerate . Since their migration route via Lower Frankonia and Thuringia is perfectly documented, direct immigragion through an eastern connection can be excluded. This is also consistant with the fact that *Encrinus*, being present in the Ladinian of the Tatra Mountains, could not reach the geographically adjacent Silesian Basin.

In northern Germany the fingering-out of the crinoidal oolites into the basin is not as clear because we lack a modern detailed stratigraphic control. Nevertheless it has been shown (Kleinsorge 1935; Rosenfeld 1978) that the crinoidal facies of the "Oberer Trochitenkalk", in which crinoids are associated with corals, shifted basinward (from South to North)during the *atavus* and *evolutus* zones.

CONSERVATION DEPOSITS FOR CRINOIDS

As has been pointed out, the preservation of complete, or at least articulated crinoids is a relatively common feature in the Germanic Muschelkalk. The same is true for other echinoderms, such as ophiuroids, asteroids and cidaroids. According to Rosenkranz (1973) the most common cause for this mode of preservation in benthic echinoderms is catastrophic smothering (obrution), because sudden mud sedimentation may kill these animals by plugging their ambulacral systems as well as provide an instant burial of the victims. In the Muschelkalk situation, the highest probability for this to happen was in the depth zone between the normal and the storm wave base and at the stratigraphic boundary of cycles, i.e. during the transition of the coarse regressive phase to the following transgressive mud sedimentation.

In fact the best chance to find articulated crinoids is (a) in the central part of the basin at the very top of coarsening-upward cycles (Trochitenbank 6 of Neckarwestheim, see Hagdorn & Simon, in print; Trochitenbank 3) and (b) at the margin of paleohighs where the transgressive phase is reduced to a thin mud separating the massive crinoidal oolite beds (terminal wedge of Mergelschiefer 3 of Hassmersheim Beds near Schwäbisch Hall; *Encrinus*-beds near Crailsheim, see Hagdorn 1978).

In a similar way important obrution deposits of crinoids in the Lower Muschelkalk are found at the tops of the Terebratelbänke and of the Schaumkalkbänke as well as on minor tempestites in the lower parts of the Wellenkalk cycles.

PALEOBIOGEOGRAPHIC MAPS
(Figs. 4 - 13)

The maps are based on Ziegler (1982) without taking account of changes in coast lines during Muschelkalk times. The distribution of crinoid occurrences is in part controlled by the outcrop situation. This is particularly obvious at the northern margin of the Bohemian Massive, which lacks Muschelkalk outcrops. Regional differences in collecting efforts are another factor. Actual plots are based on reliable publications and on own field observations. They do not include occurrences south of the Alpine deformation front, whose palinspastic positions by Middle Triassic times are still questionable.

Fig. 4: Upper Skythian (Röt).
Marine limestones are restricted to Upper Silesia and the Holy Cross Mountains, while red marls, evaporites and eventually sandstones are deposited in the west. With this first transgression marine fauna from the Asiatic Province of the Tethys entered the Germanic Basin through the Eastern Carpathian Gate. It included *Dadocrinus kunischi* and *D. grundeyi* as pioneer crinoid species, which are also known from the Lower Anisian of the Tatra Mountains and the Upper Skythian of the Northern Alps. Their expansion to the West, however, was blocked by a steep salinity gradient.

Fig. 5: Lower Anisian; lower part of the assemblage zone of *Beneckeia buchi*, *Myophoria vulgaris* and *Dadocrinus* (mu 1α).
Marine limestones with *Dadocrinus kunischi* (and occasional *Holocrinus beyrichi* in eastern Upper Silesia) are still restricted to the eastern part of the basin, while dolomitic limestones and evaporites are the dominant facies in the west, except for the sandy zone of the *Voltzia* Sandstone near the western margin.
Faunal connections continue through the East Carpathian Gate (conodonts of Asiatic affinities in the Holy Cross Mountains, but not in Upper Silesia and in the Austroalpine Province that would be connected by a Silesio-Moravian Gate; Kozur 1974). But since *Dadocrinus* also occurs in the Lower Anisian of Upper Silesia, the Austroalpine and the Dinarid region, crinoid data would also allow connections through a Silesian-Moravian Gate.

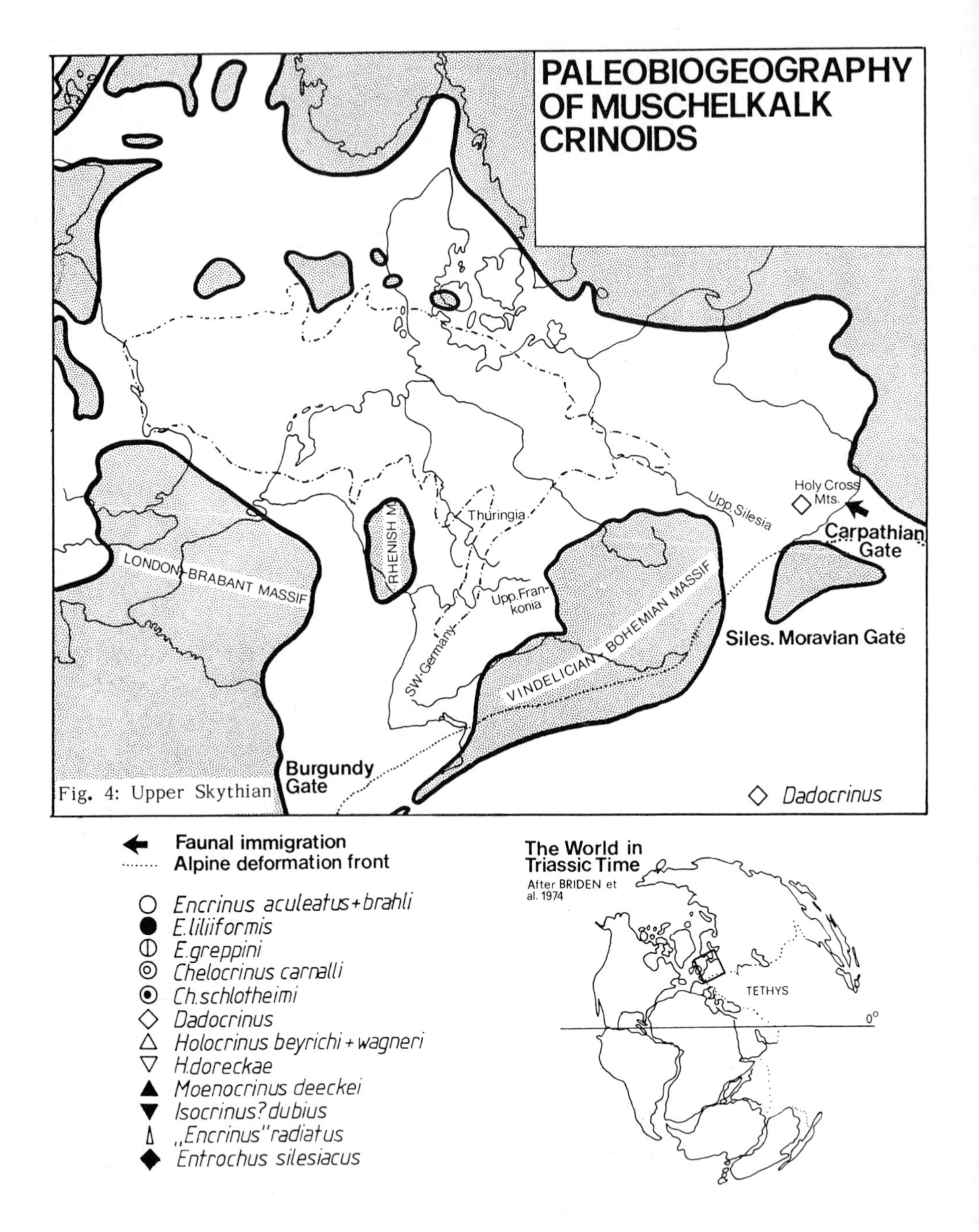

Fig. 4: Upper Skythian

Explanations for Figs. 4 - 13 see pp. 245 and 250 - 252.

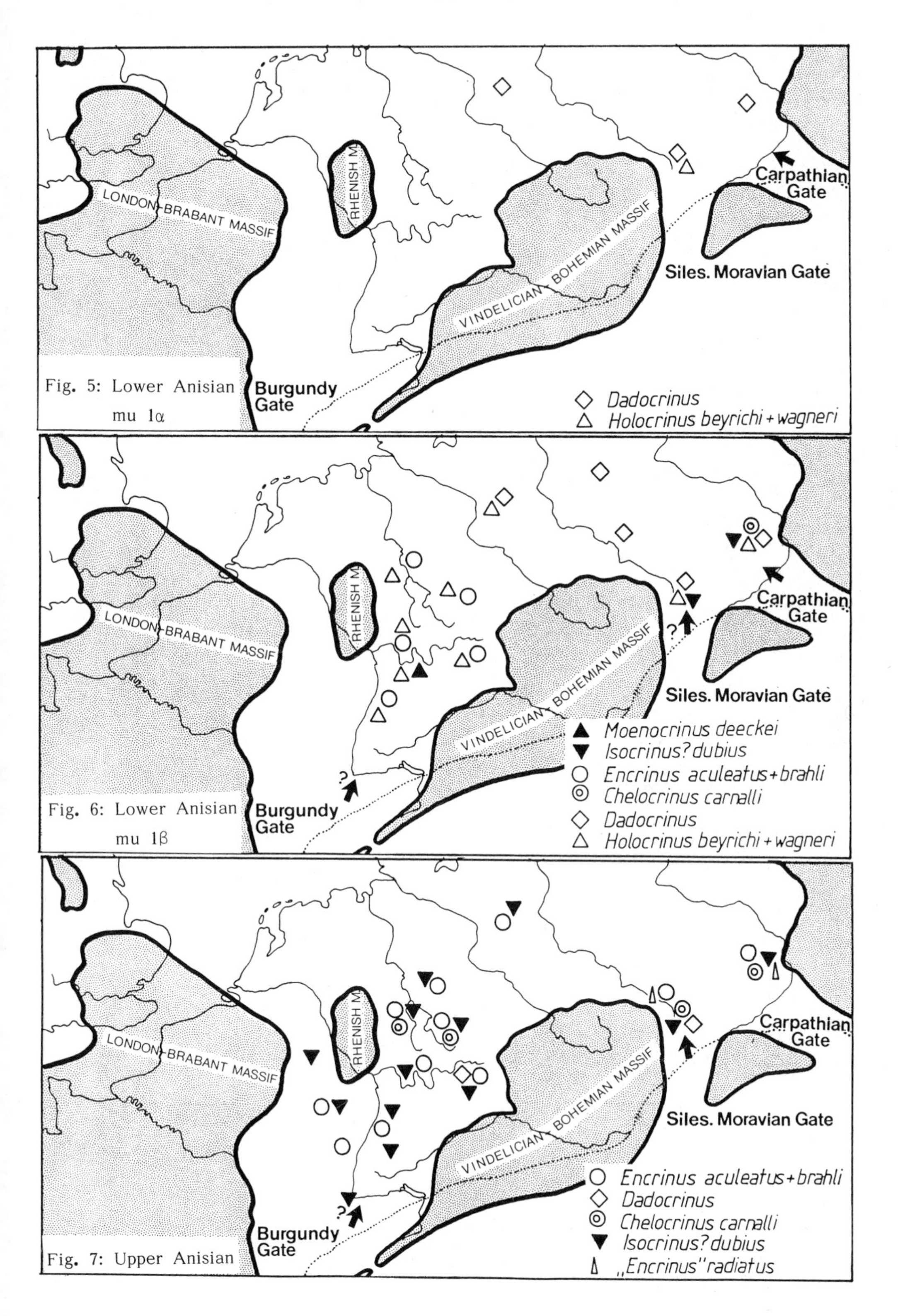

Fig. 5: Lower Anisian mu 1α

Fig. 6: Lower Anisian mu 1β

Fig. 7: Upper Anisian

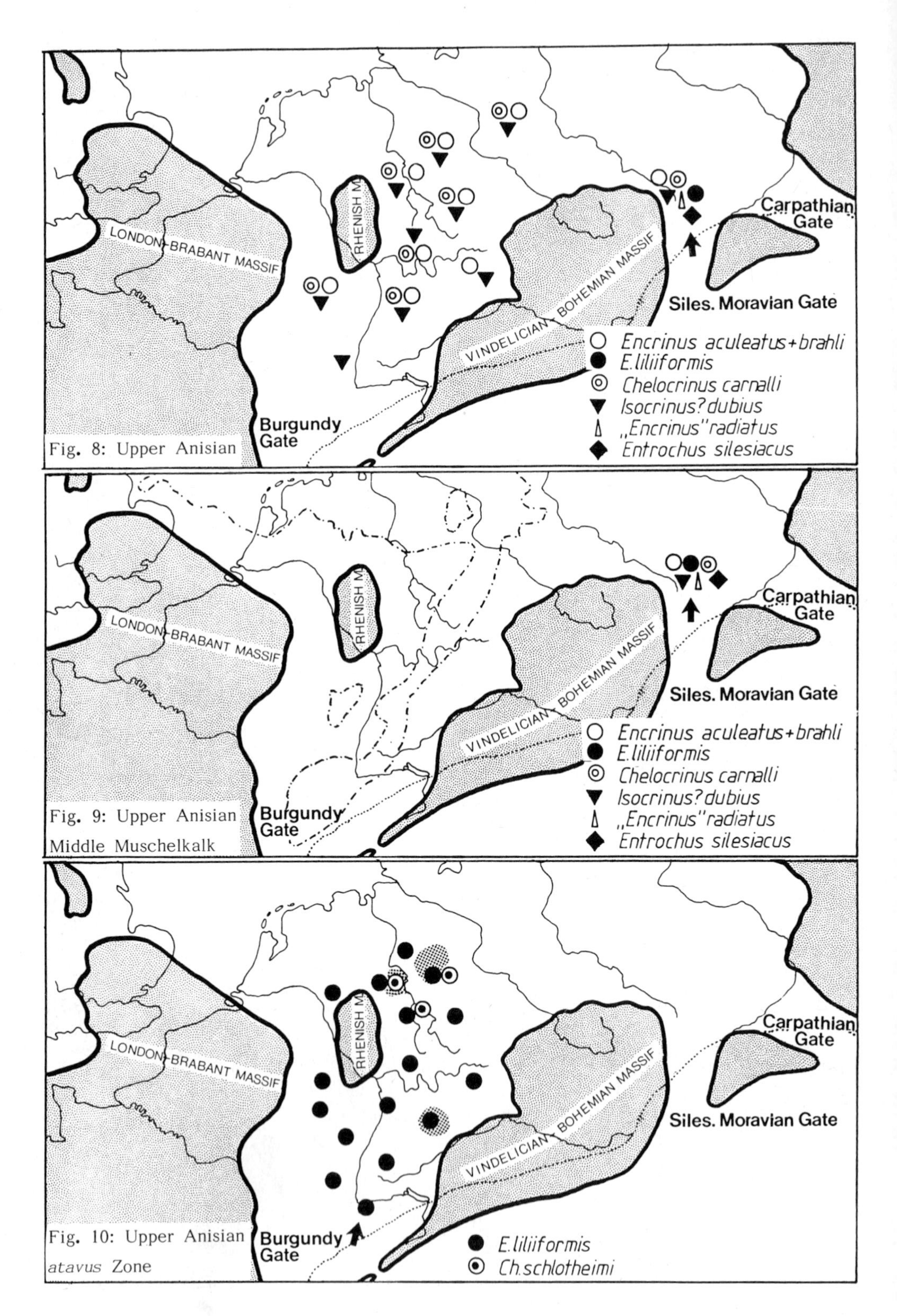

Fig. 8: Upper Anisian

Fig. 9: Upper Anisian Middle Muschelkalk

Fig. 10: Upper Anisian *atavus* Zone

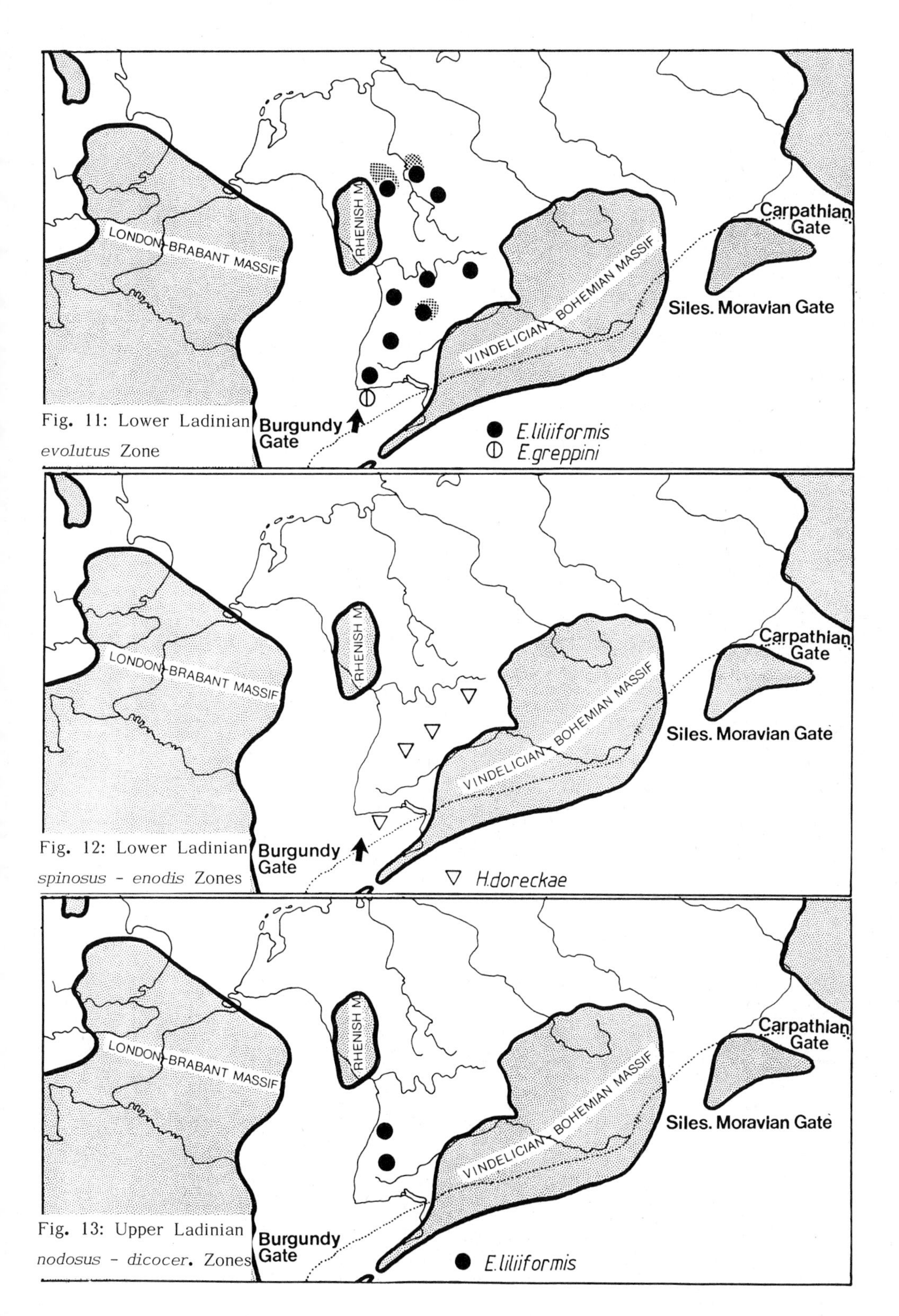

Fig. 11: Lower Ladinian *evolutus* Zone

Fig. 12: Lower Ladinian *spinosus* - *enodis* Zones

Fig. 13: Upper Ladinian *nodosus* - *dicocer.* Zones

Fig. 6: Lower Anisian; upper part of the assemblage zone of *Benneckeia buchi*, *Myophoria vulgaris* and *Dadocrinus* (mu 1ß).
This map is based on the assumption (Kozur 1974) that the Upper Gogolin Beds of Upper Silesia are equivalent to the Lower Wellenkalk including the Oolithbänke i.e. that during this interval large parts of the basin were marine. This assumption, however, meets some difficulties:
1. Why did *Dadocrinus* not extend its range to the west in spite of its ability to invade muddy bottoms by settling on "mud stickers" such as "*Bakevellia*"?
2. Why does *Encrinus* start to appear in the west at the base of the Lower Muschelkalk, but in Upper Silesia only in the Gorasdze Beds, which would be younger in this correlation ?
In contrast, the crinoid data suggest that during *Dadocrinus* times unfavorable salinities continued in the western parts and that marine conditions became established only by the transgression that allowed the immigration of *Encrinus* in Upper Silesia and its immediate dispersal over the basin on the coarse substrate of the conglomerate beds. Therefore the seeming coexistence of *Dadocrinus* in the east and of encrinids and holocrinids in the west as shown in the map is probably misleading. Also the isolated occurrence of *Moenocrinus* is probably a preservational artefact, since this genus can be distinguished only in the vary rare complete calices.
As to the immigration routes during this time (Kozur favors the East Carpathian over the Silesio-Moravian Gate) the crinoids provide no answer, since *Dadocrinus gracilis* is known from the southern as well as the northern Alps and ranged probably through the entire western Tethys.

Fig. 7: Upper Anisian; *decurtata* Zone
In Upper Silesia (Gorasdze Beds, *Terebratula* Beds) and at the southern margin of the Holy Cross Mountains (*Lima striata* Beds) crinoids show a maximum diversity. "*Encrinus*" *radiatus*, an isocrinid widely dispersed in the southern Alps, has failed to invade other parts of the Germanic Basin. Isolated ossicles of an unidentified dadocrinid are common in the *Terebratula* Beds of Upper Silesia and in the Terebratelbank near Kronach (Upper Franconia), where it is associated with the Alpine brachiopod *Aulacothyris angusta ostheimensis* . The probable migration route of these forms along the coast can no more be documented due to the subsequent erosion on the north side of the Bohemian Massive.
Most successful imigrants were the encrinids and and *Isocrinus? dubius*, which followed the transgressive deposition of the *Terebratula* beds to the west and reached as far as Switzerland and the clastic facies of the "Muschelsandstein" in Luxemburg.
Similar patterns are observed among the brachiopods, of which *Decurtella decurtata*, *Tetractinella trigonella* and *Meutzelin mentzeli* remind restricted to Upper Silesia, while *Coenothyris vulgaris*, *Aulacothyris angusta ostheimensis*, *Hirsutella hirsuta* and *Punctospirella fragilis* spread far into the basin during favorable shelly intervals.
Occurrence of radiolarians at Southern Black Forest indicate short term western connection to Tethys.

Fig. 8: Upper Anisian; assemblage zone of *Neoschizodus orbicularis* and *Judicarites*.
Again diversity is highest near the gate in Upper Silesia (Karchowice Beds) and along the Polish-Danish through to Rüdersdorf near Berlin, where *Entrochus silesiacus* has been reported but not corroborated. Among the crinoids that advanced with the Schaumkalk facies to the west, encrinids and *Isocrinus? dubius* were again the most successfull. Others did hardly disperse beyond Upper Silesia, where the fauna by now

has strong Austroalpine affinities (Trammer 1973). In the Karchowitz Beds bioherms contain, besides crinoids, siliceous sponges and corals, but neither these nor the diverse brachiopods managed to disperse into the western part of the basin.

Fig. 9: Upper Anisian; Middle Muschelkalk.
While the western part of the basin is dominated by an evaporitic facies without marine benthos, fully marine conditions persist during the early part of the Middle Muschelkalk in Upper Silesia (*Diplopora* Dolomite) and the southern pre-sudetic monocline (Lower Silesia). There a crinoid fauna of Austroalpine character is associated with diploporan algae, siliceous sponges, corals and with Alpine echinoids unknown from other parts of the basin.

Fig. 10: Upper Anisian; *atavus* Zone.
Encrinus liliiformis now has its maximum extension and occurs throughout the western part of the basin, including northern Germany. It is mostly restricted to individual crinoid beds in a mudstone background facies, but locally (dotted areas) it forms crinoidal oolites several meters in thickness. A profound change in paleogeography is expressed by the fact that *Encrinus liliiformis* failed to reach the northernmost and eastern parts of the basin (Lüneburg, Helgoland, Rüdersdorf, Poland including Silesia). Unfortunately the exact age of the first occurrences of *Encrinus* in the Swiss Jura, near Briancon and in the Provence, i.e. in the area of the new gate, is not well established. Seemingly its immigration was somewhat retarded, because soft bottom conditions hostile for this crinoid continued from the dolomites of the Middle Muschelkalk (dwarfed faunas) into the earliest parts of the Upper Muschelkalk, but a strong facies differentiation may have complicated distribution patterns during this period.
The isolated occurrence of *Chelocrinus schlotheimi* in Northern Germany is difficult to interpret. Possibly this crinoid immigrated episodically through the Burgundy Gate, but did not yet find suitable substrates in the south (Hagdorn 1982). Instead it flourished in the north ("Gelbe Basisschichten"), before it became replaced by the larger *Encrinus liliiformis*.

Fig. 11: Lower Ladinian; *evolutus* Zone.
With the shelly facies of the *Spiriferina* Bed (regressive top of a shallowing-up cycle) *Encrinus liliiformis* could colonize large parts of the basin (SW Germany, Thuringia), accumulating thick crinoidal oolites in nearshore and paleo-high environments ("Upper Trochitenkalk", *Astarte* Bank) at the northwestern margin of the Rhenish Massive (dotted areas). Since the previous *atavus* Zone, this facies has continuously shifted northward ("Ältere Tonplatten"), with local crinoidal oolites reaching east of the Weser river.
Encrinus greppini is an immigrant represented only by few specimens and only in the Southwestern corner (Hochrhein area). However it has not yet been found either in areas to the south.
An isolated specimen of *Encrinus liliiformis* has been found in the "Tonplatten" facies of the *evolutus/spinosus* Zone near Schwäbisch Hall (leg. Seilacher; see Linck 1965).

Fig. 12: Lower Ladinian; transition of the *spinosus* and *enodis/laevigatus* Zones.
The only crinoid known from this interval is *Holocrinus doreckae*, which spread with the shelly facies on top of another regressive cycle (*Holocrinus* Bed) from southwest Germany along the Vindelician-Bohe-

mian massive to Upper Franconia. Only a single columnal of this species has been found in the *cycloides* Bed 1 meter higher up in the section. It may be assumed that its dispersal center was in the Hochrhein area (Middle Oolite). Holocrinids are also known from the Ladinian of Spain (Hagdorn 1983).

Fig. 13: Upper Ladinian; *nodosus* - to discoceratite Zones.
In this interval, we know only isolated columnals from the base of the *Trigonodus* Dolomite (Baden) and a calyx of *Encrinus liliiformis* from the same unit in Rottweil. In view of the availability of suitable substrates (shell beds, *Placunopsis* bioherms, hard grounds, large cephalopod shells) and the presence of other stenohaline elements (starfish, articulate brachiopods, cephalopods, conodonts) the failure of *Encrinus liliiformis* to develop large populations is difficult to explain. It parallels the response of echinoids, which are missing completely in this interval. Ecological competition seems not to have been involved, because the tier of the crinoids above the bottom was not occupied by other organisms. Possible explanations are changes in food supply by changing current patterns and the introduction of detrital material (silts and fine sands) by northern rivers, whose deltas by now prograded into the basin.

SPECIATION PATTERNS

Our knowledge about the spatial and temporal distribution of Triassic crinoid species is as yet too spotty to allow the definition of speciation centers. This is particularly true for the Ladinian. Theoretical considerations suggest that speciation happened mainly in isolated populations such as the ones of the Germanic Muschelkalk basin. Within this basin, the more central parts were colonized by crinoids for too short periods to expect such an effect. But the areas of more continuous colonization near the Upper Silesian Gate would appear as ideal sites, where speciations could occur and where the proximity of soft bottom fostered the evolution of novel adaptational strategies. Thus the evolutionary habitat change of holocrinid and isocrinid lineages as expressed by the development of a preformed, crypto-symplectic fracture line that made them largely independent from hard substrates (Hagdorn 1983 and in print) could well have taken place in this area. But in general epibenthic organisms seem to be less prone - and perhaps too sluggish in an evolutionary sense - to show the kind of short-term endemic evolution that can be observed in more bottom-independent organisms, such as the conodontophorids (Dzik & Trammer, 1980) and the ceratitic ammonoids (Urlichs & Mundlos, this volume) of the Muschelkalk sea.

Annotation

Some taxa whose taxonomic validity is still in doubt, have not been included inin this study. This is true for *Encrinus robustus* and *Encrinus spinosus* from the Karchowice Beds, which seem to fall in the synonymy of *E. brahli* and *E.liliiformis* respectively.

Acknowledgements

Thanks are due to the participants of the 1983 symposium and to Dr. H. Sieverts-Doreck (Stuttgart) for critical discussions. Valuable informations about crinoid occurrences in Upper Silesia were provided by Dr. H. Loewenstam (Pasadena). Material was made available by many museums, institutions and private collectors. T. Aigner assisted me with his ideas during completion of the manuscript which was translated by A. Seilacher into its present form. The Deutsche Forschungsgemeinschaft kindly supported some of the field work.

REFERENCES

Aigner, T. this volume: Dynamic stratigraphy of the Upper Muschelkalk, South-German Basin.- A summary.

Assmann, P., 1944: Die Stratigraphie der oberschlesischen Trias. Teil 2: Der Muschelkalk.- Abh.Reichsamt f.Bodenf. N.F. 208: 1-124, Berlin.

Biese, W., 1927: Über die Encriniten des unteren Muschelkalkes von Mitteldeutschland. - Abh.preuß.geol.L.-Anst. N.F. 103: 1-119, Berlin.

Biese, W., 1934: Crinoidea Triadica. Fossilium Catalogus. I. Animalia, Pars 66: 1-255, Berlin (W. Junk).

Braun, J., 1983: Mikropaläontologische und sedimentologische Untersuchungen an einem Profil im Unteren Muschelkalk in der Wutachschlucht (SE Schwarzwald).- Unveröff. Dipl.Arb., Tübingen.

Dzik, J. & Trammer, J., 1980: Gradual evolution of Conodontophorids in the Polish Triassic.- Acta Palaeont. Polon. 25: 55-89, Warszawa.

Geyer, O.F. & Gwinner, M.P., 1968: Einführung in die Geologie von Baden-Württemberg.- 2. Aufl., Stuttgart (Schweizerbart).

Hagdorn, H., 1978: Muschel/Krinoiden-Bioherme im Oberen Muschelkalk (mol, Anis) von Crailsheim und Schwäbisch Hall (Südwestdeutschland).- N.Jb.Geol.Paläont. Abh. 156: 31-86, Stuttgart.

Hagdorn, H., 1982: *Chelocrinus schlotheimi* (QUENSTEDT) 1835 aus dem Oberen Muschelkalk (mol, Anisium) von Nordwestdeutschland.- Veröff.Naturk.Bielefeld, 4: 5 - 33.

Hagdorn, H., 1983: *Holocrinus doreckae* n.sp. aus dem Oberen Muschelkalk und die Entwicklung von Sollbruchstellen im Stiel der Isocrinida.- N.Jb.Geol.Paläont.Mh. 6: 345-368, Stuttgart.

Hagdorn, H., in print: *Isocrinus? dubius* (GOLDFUSS 1831) aus dem Unteren Muschelkalk (Trias, Anis).- Z.Geol.Wiss., Berlin

Hagdorn, H. & Mundlos, R., 1982: Autochthonschille im Oberen Muschelkalk (Mitteltrias) Südwestdeutschlands.- N.Jb.Geol.Paläont.Abh., 162: 332-351, Stuttgart.

Hagdorn, H. & Simon, T., 1983: Ein Hartgrund im Unteren Muschelkalk von Göttingen.- Der Aufschluß 34: 255-263, Heidelberg.

Hagdorn, H. & Simon, T., (in print): Oberer Muschelkalk (Hauptmuschelkalk, mo).- In: Brunner, H., Erläuterungen zur Geol. Kte. 1:25000 von Baden-Württemberg, Blatt 6921: Großbottwar.

Hardt, H., 1952: Die Rüdersdorfer Kalkberge. Einführung in ihre Geologie.- Berlin (Aufbau).

Klier, P.M., 1973: The Echinoderms and Perminian-Triassic time.- Mem.Canad.Soc.Petr. Geol. 2: 622-629.

Kleinsorge, H., 1935: Paläogeographische Untersuchungen über den Oberen Muschelkalk in Nord- und Mitteldeutschland.- Mitt.Geol.Staatsinst. 15: 57-106, Hamburg.

Kozur, H., 1973: Faunenprovinzen in der Trias und ihre Bedeutung für die Klärung der Paläogeographie.- Geol.Paläont.Mitt.Innsbruck 3: 1-41.

Kozur, H., 1974: Biostratigraphie der Germanischen Mitteltrias. Tle. 1-3.- Freiberger Forschungsh. C 280, 1-56; 1-71, 12 Tab., Leipzig.

Kristan-Tollmann, E. & Tollmann, A., 1967: Crinoiden aus dem Zentralalpinen Anis (Leithagebirge, Thörler Zug und Radstädter Tauern).- Wiss.Arb.Burgenland H.36 (Naturwiss. H.24), Eichstätt.

Linck, O., 1965: Stratinomische, stratigraphische und ökologische Betrachtungen zu *Encrinus liliiformis* LAMARCK.- Jh.geol.Landesamt Bad.-Württ. 7: 123-148.

Mostler, H., 1972: Die stratigraphische Bedeutung von Crinoiden-, Echiniden- und Ophiuren-Skelettelementen in triassischen Karbonatgesteinen.- Mitt.Ges.Geol. Bergbaust. 21: 711-728, Innsbruck.

Müller, A. H., 1963: Lehrbuch der Paläozoologie. Bd. 2 Invertebraten. Tl. 3 Arthropoda - Stomochorda.- Jena (G. Fischer).

Rosenfeld, U., 1978: Beitrag zur Paläogeographie des Mesozoikums in Westfalen.- N.Jb. Geol.Paläont.Abh. 156: 132-155, Stuttgart.

Rosenkranz, D., 1971: Zur Sedimentologie und Ökologie von Echinodermen-Lagerstätten.- N.Jb.Geol.Paläont.Abh. 138: 221-258, Stuttgart.

Seilacher, A., 1981: Towards an evolutionary stratigraphy.- Acta geol.hispan. (Concepts and methods in Paleontology) 16: 39-44, Barcelona.

Schulz, M.G., 1972: Feinstratigraphie und Zyklengliederung des Unteren Muschelkalks in N-Hessen.- Mitt.Geol.Paläont.Inst.Univ. Hamburg 41: 133-170, Hamburg.

Schwarz, H.U., 1970: Zur Sedimentologie und Fazies des Unteren Muschelkalkes in Südwestdeutschland und angrenzenden Gebieten.- Diss. Tübingen.

Sokolowski, S. (ed.), 1976: Geology of Poland. Vol. 1 Stratigraphy. Part 2 Mesozoic.- Warsaw (Publishing House Wydawnictwa Geologiczne).

Trammer, J., 1973: The particular paleogeographical setting of Polish Muschelkalk in the German basin.- N.Jb.Geol.Paläont.Mh. 1973, H.9: 573-575, Stuttgart.

Trammer, J., 1975: Stratigraphy and facies development of the Muschelkalk in the south-western Holy Cross Mts.- acta.geol.Polon. 25: 179-216, Warszawa.

Urlichs, M. & Mundlos, R. (this volume): Verbreitung und Entwicklung von Nautiliden und Ceratiten im Germanischen Muschelkalk (Oberanis bis Ladin).

Wysogorski, J., 1903: Die Trias in Oberschlesien.- In: Frech, F. (ed.). Lethaea geognostica. Handbuch der Erdgeschichte. II. Teil: Das Mesozoikum. 1. Heft Trias, 54-64, Stuttgart.

Zardini, R., 1976: Atlante degli echinodermi cassiani (Trias medio-superiore) della regione dolomitica attorno a Cortina d'Ampezzo.- Cortina d'Ampezzo (Foto Ghedina).

Zawidzka, K., 1975: Conodont stratigraphy and sedimentary environment of the Muschelkalk in Upper Silesia.- acta geol. Polon. 25: 217-256, Warszawa.

Ziegler, P.A., 1982: Geological Atlas of Western and Central Europe.- The Hague.

PART 4

GASTROPOD EVOLUTION IN LAKES:
A PROGRAM

Phylogenetic lineages of Gastropods in lakes are among the earliest paleontological contributions to evolutionary theory. The reinterpretation and re-reexamination even of well known examples has never ended because such localized lineages were always considered as ideal to test a certain evolutionary model. The three papers that follow deal with rather similar objects but illuminate them from different theoretical viewpoints:

(1) The history of scientific research in the case of the classical example of the Steinheim lake is examined by W.-E. Reif. From his neo-Darwinian point of view he concludes that the evolutionary interpretation is hampered because of insufficient knowledge of the ecological evolution of the lake, the unknown taphonomy of the snails, the lacking geographical control, and still insufficient knowledge of the gastropods' genetics.

(2) Another classical example, the freshwater Gastropods of Kos, is reconsidered by R. Willmann. While his taxonomic approach comes close to the cladistic program, he also favors a gradualistic evolution driven by environmental changes, and he provides examples for allopatric adaptational trends with morphological convergences in different genera. As he points out, population size is a primary factor controlling evolutionary rates.

(3) Gorthner & Meier-Brook compare the Steinheim lake with a Recent 'ancient lake'. In their Recent example, they recognize repeated convergent morphological trends and relate them to sympatric speciation by niche partitioning. The more important factor in their interpretation is the 'age of the ecosystem' rather than the environmental condition.)*

The three contributions focus on the same topic, however, from different points of view, and thus provide quite different results. Together they describe a multitude of convergent evolutionary trends. With respect to the diverging results, the contributions may provide a program rather than definitive answers -- although (or because) they are all definitive and consistent themselves: The program is the integrated study of environmental history, ecology and phylogeny of paleontological and recent examples. Similar trends occur in the Cephalopods, as previously discussed; however, the gastropods allow to apply the actualistic principle including genetic studies. Therefore, it should be possible to test correlations between morphology and environment by biological experiments.

**) The contribution of Gorthner & Meier-Brook arrived in the last moment so that other authors had no possibility to respond.*

ENDEMIC EVOLUTION OF *GYRAULUS KLEINI* IN THE STEINHEIM BASIN (PLANORBID SNAILS, MIOCENE, SOUTHERN GERMANY)

Wolf-Ernst Reif

Tübingen

Abstract: Published data on the gastropod evolution in the Steinheim Basin are reanalyzed from a neo-Darwinian point of view. Data from the population genetics of Recent pulmonates are used to interpret the complex relationship of ecophenotypy, genetics, polymorphism and speciation in the Steinheim Lake.

1. INTRODUCTION

The reconstruction of microevolutionary patterns from the fossil record has been an important research program in paleontology since at least the middle of the last century. Since 1859, the date of publication of DARWIN's "On the Origin of Species", such studies were cften regarded as tools to prove the Theory of Descent. The snails from the Miocene Steinheim Basin played a prominent role in such attempts and in the discussion of Darwinism. It took some decades until paleontologists did no longer question the Theory of Descent, but rather started to regard the process of evolution either as a fact, which did not have to be proven, or as an axiom, which need not to be proven, but which provided a wealth of interesting research problems.

The term microevolutionary patterns as used here, does not imply that I accept the dichotomy and the "decoupling" of microevolution and macroevolution as advocated by GOULD (198O), STANLEY (1979) and other authors. Rather, the term implies the study of changes within species and of the origin of new species).

Microevolutionary studies are the most relevant means to gain information about the tempos and modes of evolution. However, most studies are greatly impeded by the fact that what appears to be evolutionary change within one outcrop or within a part of a sedimentary basin might simply have been produced by migrations of clines or by the immigration of new species (see BAYER & MCGHEE, this volume and REIF, in press (a), for discussions and examples). In most studies the real problem is that the biogeographic control, which is the only means to exclude the possibility of migration, is lacking. Two parts of the fossil record may provide a biogeographic control. However, these two parts are only a very small fraction of the fossil record. The first part is the record of pelagic microfossils since the late Miocene. Due to a sampling program of sediment cores during the last decades it is now at least theoretically possible to study lineages of fossilizable planktic organisms from the late Miocene to the Recent on a worldwide basis (PROTHERO & LAZARUS, 1980). The other part of the record are fossil freshwater lakes which are commonly very rich in gastropod shells. It is much easier to demonstrate that evolution in such lakes was endemic than it is in a marine basin. Additionally, within-basin differentiation of populations can be demonstrated if the outcrops are sufficiently large. Among the most famous examples are the Plio-Pleistocene lakes of Kos, Greece and Turkuna Basin, Kenya (WILLMANN, 1981 and this volume; WILLIAMSON, 1981a) and the Miocene Steinheim Basin. Williamson's paper on the Turkuna Basin aroused a very extensive discussion (BOUCOT, 1982; CHARLESWORTH & LANDE, 1982; COHEN & SCHWARZ, 1983; DINGUS & SADLER, 1982; FRYER et al., 1983; GINZBURG & ROST, 1982; JONES, 1981; KAT & DAVIS, 1983; LINDSAY, 1982; MAYR, 1982; SCHINDEL, 1982; VAN VALEN, 1982; WILLIAMSON, 1981b, 1982, 1983). This discussion is of some interest for the present paper, because it demonstrates the possibilities and the limitations of the interpretation of Cenozoic gastropod-bearing freshwater sections.

The study of the freshwater gastropods from the Steinheim Basin has had a long history. It is impossible here to give an overview of the probably more than 100 publications dealing with the snails (other publications deal with the rest of the fauna and with geology and sedimentology). Rather an attempt will be made to review critically the published data and hypotheses from a neo-Darwinian point of view. The purpose is to find out how much of the microevolutionary patterns can be reconstructed in such a locality which is both highly fossiliferous and favourable with regard to the geological setting.

The following problems will be discussed:

* How many gastropod species populated the lake? Was there one or several founder species of the genus *Gyraulus*?"Polymorphism and biogeography of the founding species *Gyraulus kleini* in the South German Miocene".(chapter 3).
* Were the gyraulids in Steinheim reproductively isolated from gyraulids in other lakes?
* What caused the radiation of the gyraulids in Steinheim? (chapter 3).
* Phylogeny of the gyraulids in Steinheim (chapter 4).

* Influence of the geochemical regime on gastropod evolution in Steinheim (chapter 5).

* Evaluation of the ecophenotypic model of Steinheim gyraulids (chapter 6).

* Does ecophenotypy play such a dominant role in Recent gastropods as is claimed by many textbooks? (chapter 7).

* What role did genetic polymorphism, geographic differentiation, dispersal and intralacustrine speciation play in the Steinheim lake? (chapter 8).

* Can evolutionary species be defined objectively in the Steinheim gyraulids? (chapter 8).

* Do *Lymnaea dilatata* and *Pseudamnicola pseudoglobulus* display patterns of evolution in Steinheim? (chapter 9).

* Can the endemic radiation of *Gyraulus* and other Planorbidae in "ancient lakes" serve as a model for the *Gyraulus* radiation in Steinheim?

2. THE STEINHEIM BASIN

There are two Miocene meteorite craters in Southern Germany both of which became lakes. One is the famous Ries with a diameter of 30 km, which separates the Swabian Jurassic Mountains (Schwäbische Alb) from the Franconian Jurassic Mountains (Fränkische Alb) (Fig. 1). The other one is the Steinheim Basin 40 km to the west on the Schwäbische Alb. It has a diameter of only 4 km, is surrounded by marine Jurassic limestones of Portlandian age, and was formed approximately 14.8 million years ago by a fragment of the Ries meteorite (GROSCHOPF & REIFF, 1966; GENTNER, LIPPOLD & SCHAEFFLER, 1963; STORZER & GENTNER, 1970). It was originally 150 m deep.

It should be mentioned in passing that until the advent of the meteorite hypothesis the Ries and Steinheim Basins were regarded as either volcanic or as tectonic structures by various authors and the occurrence of aragonite nodules and algal reefs in Steinheim was interpreted as the product of hot springs (ADAM, 1980; REIF, 1975, for historical overviews).

After the impact both craters filled with water and formed freshwater lakes. The duration of these lakes is not known. JANKOWSKI (1981) calculated that sedimentation in the Ries lake persisted for 0.3 to 2 myr. One can assume that the Steinheim lake did not get older than a few 10^5 years; in any case it is not possible to show a difference in stratigraphic age between lower and upper beds by means of the terrestrial vertebrate fossils which are found in the sediments (MENSINK, 1984, for references). The end of the Steinheim lake was brought about by an erosion of the crater margins tapping the lake. Since that time the sediment fill (consisting largely of lime muds and gastropod coquinas; MENSINK, 1984, for an overview) has been eroded. The most prominent structure of the basin is the central hill, an expansion structure caused by the impact and consisting of lower, middle and upper Jurassic pressed up from below (Fig. 2;

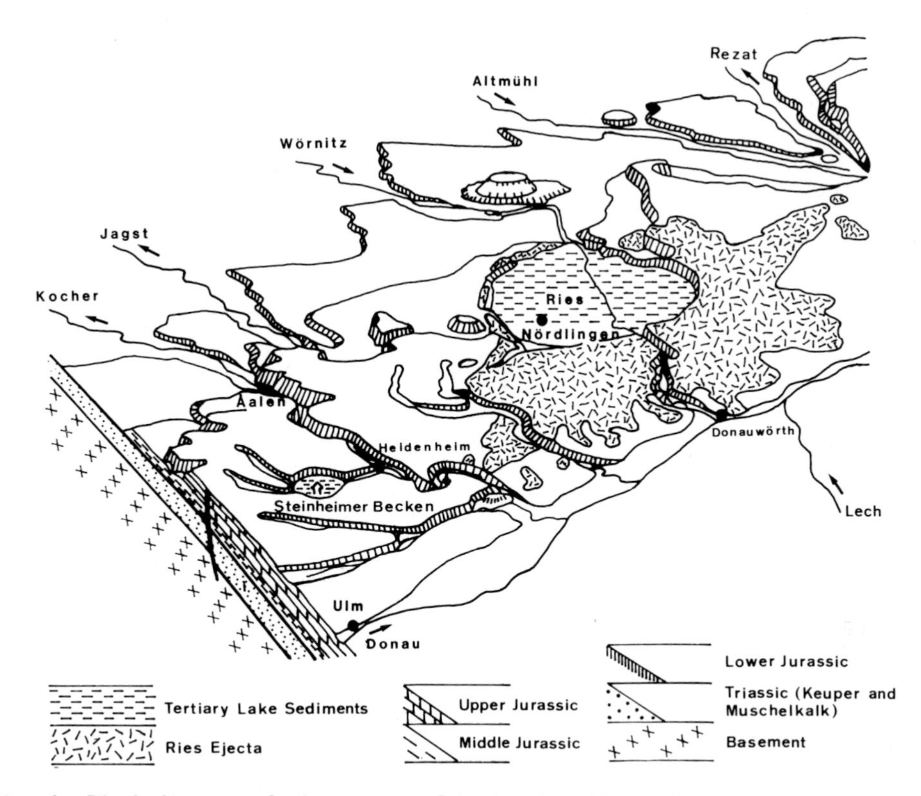

Fig. 1: Block-diagram of the eastern Schwäbische Alb and Ries. The section at the lower left has a N-S-direction. The diameter of the Ries is 30 km (after MENSINK, 1984, redrawn).

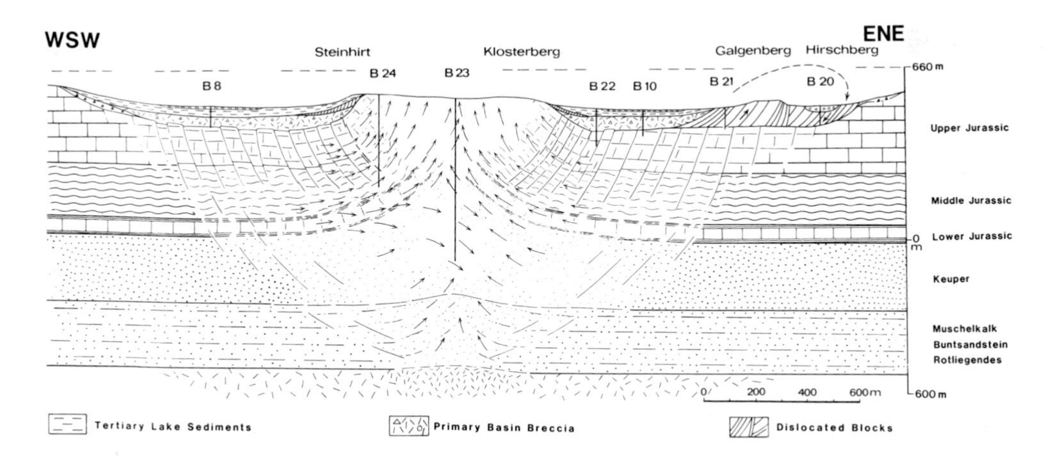

Fig. 2: Cross-section through the Steinheim Basin. B 8 etc. are bore-holes, (after REIFF, 1974, redrawn).

GROSCHOPF & REIFF, 1979). Most paleontological data come from sand-quarries around the central hill, while the lowermost beds of the section (*kleini*-beds) are exposed only at the western margin of the basin. In the marginal facies the complete section is up to 20 m thick. In boreholes between the outer margin and the central hill up to 40 m of section have been found (JANKOWSKI, 1981).

MENSINK's stratigraphic revision (based on different species of *Gyraulus*) distinguishes 7 beds:

7) *Supremus* - Beds

6) *Revertens* - Beds

5) *Oxystoma* - Beds

4) *Trochiformis*- Beds

3) *Sulcatus* - Beds

2) *Steinheimensis* - Beds

1) *Kleini* - Beds

This sequence has been recognized by MENSINK and his coworkers not only in the three sand-quarries at the central hill, but also in numerous bores-holes of varying depths.

Fossil flora and fauna of the lake consist of blue-green algae, green algae, Characeae, diatoms, gastropods, bivalves, ostracods, fishes, frogs, crocodiles, and otters. In addition numerous terrestrial plants, gastropods, reptiles, birds and mammals were washed into the basin (KRANZ, et al., 1924; MENSINK, 1984). The study of the geology and of the very rich vertebrate and invertebrate faunas has had a long history which is outlined by REIF (1975) and ADAM (1980). A complete bibliography on Steinheim can be compiled from KRANZ et al., (1924), KRANZ (1936), ADAM (1980) and MENSINK (1984).

3. *Kleini* -BEDS

The bottom of the Steinheim Basin, which is known only from boreholes, is formed by breccias of material that fell back into the crater after the impact and of coarse material which was washed into the crater some time after the impact. The basal lake deposits are the *kleini* -beds. They occur only in a small area at the western margin of the basin. In all other areas they are either covered by younger sediments or were eroded in post-Miocene times after the lake had disappeared. They are now completely overgrown, so that MENSINK (1984) could study them only in bore-holes and found only a few fossils. However, there must also be fossiliferous parts, because GOTTSCHICK (1925) and earlier authors report a rich fauna from non-permanent exposures.The exact localities of these exposures, however, are no longer known.

The lake was initially populated by a diverse mollusc fauna consisting of 17 gastropod species (with several "subspecies") and two bivalve species (KRANZ et al., 1924). Since there is no recent revision of these taxa, different publications provide different numbers (e.g. GOTTSCHICK, 1919/192O: GOTTSCHICK, 1925). From what can be inferred from the poor outcrops of the *kleini*-beds, no within basin evolution took place during that comparatively short time. By *steinheimensis* -times the bivalves and all gastropod species disappeared except for *Radix (=Lymnaea) dilatata*, *Pseudamnicola pseudoglobulus*, *Gyraulus kleini* and its descendants. The evolutionary history of *G. kleini* is the main theme of this paper.

Gyraulus kleini (a planorbid) is ubiquitous in Miocene freshwater lakes in southern and central German (DEHM et al., 1976; GOTTSCHICK, 1919/20; GOTTSCHICK & WENZ, 1916; SEEMANN, 1941; WENZ, 1923). According to GOTTSCHICK (1919-192O) *Gyraulus kleini* is polymorphic and was originally grouped into three species: *applanatus*, *dealbatus* and *kleini* (Fig. 3). Gottschick regarded them as ecophenotypic morphs

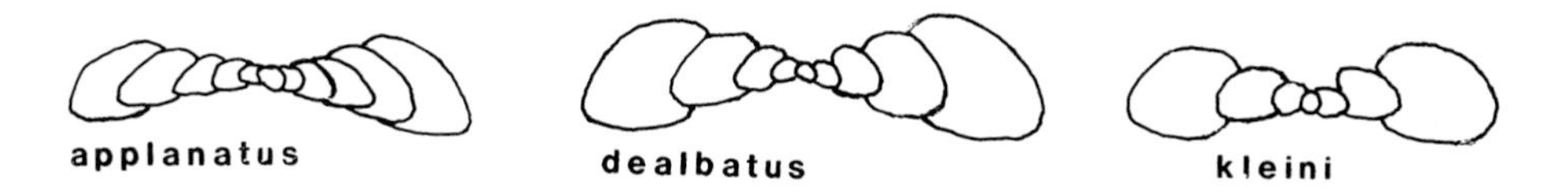

Fig. 3: The three morphs of *Gyraulus kleini*; for explanation see text (after GOTTSCHICK, 1919/20, redrawn).

which grade into each other within one species, *G. kleini* (sensu lato). As will become obvious later, Gottschick, Wenz and other authors before WWII, interpreted all morphs of *Gyraulus* in Steinheim as nonheritable ecophenotypic modifications under the influence of hot springs. Hence Gottschick did not discuss the question whether *applanatus* , *dealbatus* and *kleini* (sensu stricto) could be geographic races or genetically controlled sympatric morphs.

The informations on *G. kleini* provided by GOTTSCHICK (1919/192O) are very complicated and difficult to interpret. As is shown below, he described a very irregular pattern of the occurrence of the various morphs in South and Central German Miocene lakes. In addition to that he found a regular stratigraphic change with two morphs dominating in the Lower Miocene and the third morph in the Upper Miocene if one summarized the information on all lakes. However, individual lakes show completely different trends in the stratigraphic distribution of the morphs. Taken altogether it seems to be impossible at the moment to analyse the factors which controlled the geographic and the stratigraphic distribution of the morphs of *G. kleini*.

GOTTSCHICK (1919/1920, p. 165) summarized the distribution of the morphs in southern and central German localities as follows:

1) *Applanatus* and *dealbatus* occur together (Hochheim-Flörsheim; St. Johann,Rheinhessen)

2) *Applanatus* dominates, but *dealbatus* occurs and transitions to *kleini* can be found (Budenheim near Mainz; Donaurieden)

3) *Dealbatus* dominates; transitions to *kleini* (Gau-Algesheimerkopf, Rheinhessen; *Sylvana*-Kalk Schwaben)

4) Transitional forms between *dealbatus* and *kleini* dominate (Frankfurt-Ginnheim).

This distribution shows that the three morphs can not be geographic races. From the scarce data given by Gottschick, *applanatus* and *dealbatus* occur in the Lower Miocene and *dealbatus* and *kleini* in the Upper Miocene predominantly; *kleini* seems to evolve from *dealbatus* and *applanatus*. In Steinheim the three morphs occur "nebeneinander" (side by side; GOTTSCHICK, 1919/1920, p. 167). *Applanatus* dominates in the lower parts of the *kleini*-beds; *dealbatus* is rare; *kleini* occurs throughout the *kleini*-beds. All these data provided by Gottschick, an amateur fossil-collector, are non-quantitative but based on careful collecting over a long time. Gottschick did not explain why, in contrast to other Upper Miocene localities *applanatus* dominates and not *dealbatus*.

There are no data on the "*G. kleini* problem" (i.e. the question of what factors controlled the geographic and the stratigraphic distribution of the morphs of *G. kleini*) in the more recent literature. The *kleini*-beds are almost inaccessible in Steinheim; MENSINK (1984) had only one sample (number O, Fig. 4). His biometrical data demonstrate a unimodal distribution (see below for Mensink's method) and he identified the sample as *G. kleini* (probably sensu lato) without any specification.

In contrast to the ubiquitous *G. kleini* (sensu lato) no morph of *Gyraulus* which occurs above the *kleini*-beds has ever been found outside the Steinheim Basin. As will be shown below, no immigration of new species can be observed in the *steinheimensis*- to *supremus*-beds. This clearly indicates that *Gyraulus* (but probably also *Radix* and *Pseudamnicola*) formed endemic populations in the Steinheim Basin. It must be stressed that the differences in shell morphology between *G. kleini* and its descendants are the indicator of reproductive isolastion between the typical Steinheim morphs and the synchronous *G. kleini* populations in other lakes; but the differences in shell shape are by no means necessarily the cause of this reproductive isolation. It could very well be that identical shell shapes occurred in different lakes and yet complete reproductive isolation was still be case. Such "crypto-species" (SUDHAUS, 1978) could not be discriminated in the fossil record.

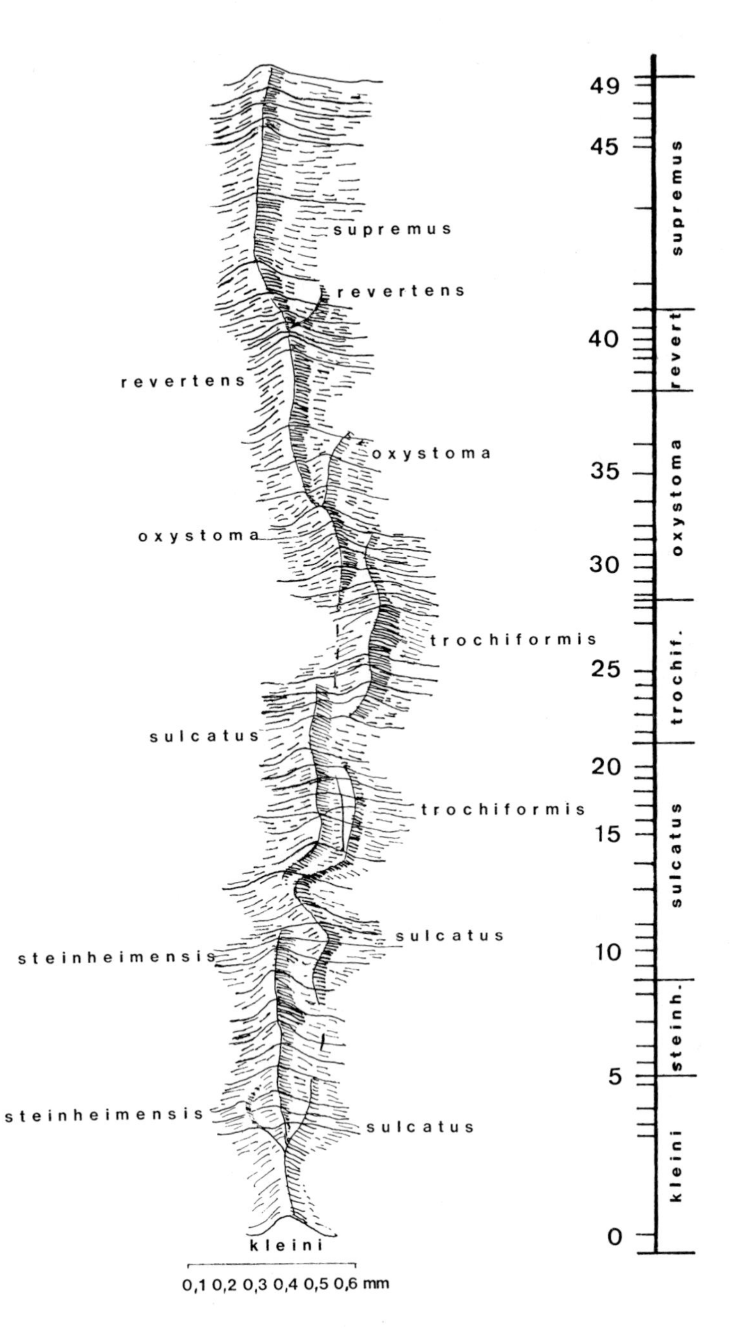

Fig. 4: Evolution of the main lineage (after MENSINK, 1984, combined and redrawn). In the right column: stratigraphy and sample numbers from a combined section of 18 m. Small black dots indicate gaps in the section. On the right: branching and transformation patterns in the main lineage. Thin continuous lines: univariate diagrams for shell width after second whorl for all 50 samples.

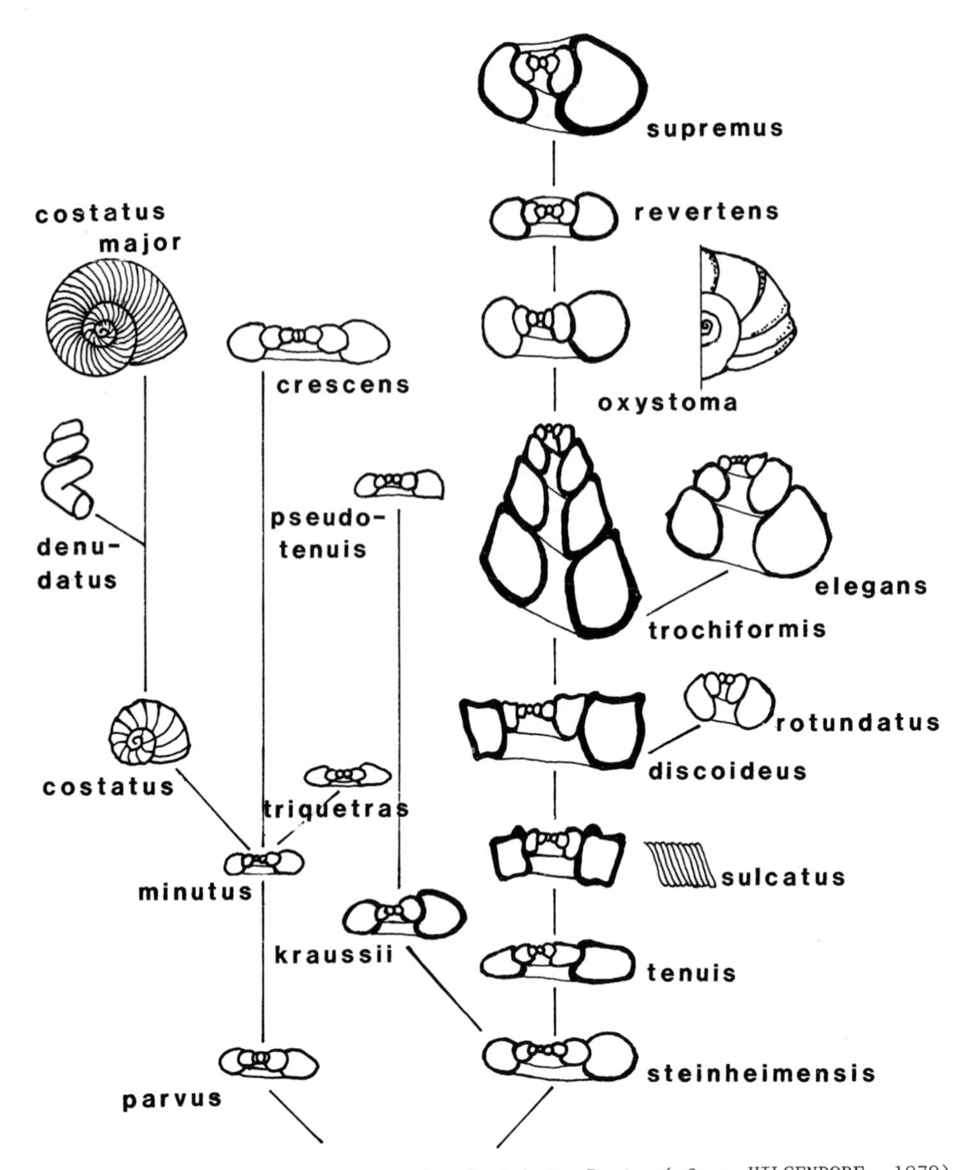

Fig. 5: Evolution of *Gyraulus* in the Steinheim Basin (after HILGENDORF, 1879). Hilgendorf did not indicate the common ancestor of the three lineages. Main lineage on the right; minor branch lineage in the center (*kraussi* – *pseudotenius*); major branch lineage on the left. Correct name for *"triquetras"* is *triquetrus*.

It may also be that reproductive isolation and thus endemism was not directly caused by an interruption of gene flow, i.e. by an interruption of the transport of the snails from one lake to another. Rather, *G. kleini* seems to have had effective means of passive dispersal (by birds). The cause of reproductive isolation seems to have been the adaptation to a special geochemical regime in the Steinheim lake, which was different from the geochemical regimes in other lakes. Thus any new migration of *G. kleini* into the lake, or of Steinheim morphs into other lakes, were bound to be unsuccessful. Transport of the snails was most probably not interrupted but it did not lead to an exchange of genes between the Steinheim population and other lakes during *steinheimensis*- to *supremus*-times.

The special geochemical regime in Steinheim was caused by high evaporation rate in a semiarid climate (BAJOR, 1965; JANKOWSKI, 1981; WOLFF & FÜCHTBAUER, 1976; MENSINK, 1984). The exact hydrography (influx from tributaries and connection with ground water) is not known; it suffices, however, to state that the geochemistry in the Steinheim Basin differs significantly from that in the probably contemporaneous Ries (40 km to the east) and in the Randecker Maar a volcanic crater lake 40 km to the west (JANKOWSKI, 1981). Steinheim Basin and Randecker Maar were alkaline-earth lakes with much aragonite and a high Mg/Ca-ratio; Ries was an alkaline lake with little aragonite, with authigenic silicates and high Na- and K-content. Salinity in the Steinheim Basin was lower than in the Ries. As a consequence, the faunas in the three lakes differ significantly. Gastropods are rare in the Ries and in the Randecker Maar. *G. kleini* was found in both lakes, but morphs of the post-*kleini* lineages, which are typical for the Steinheim Basin, do not occur. The evaporation hypothesis is based on analyses of delta ^{18}O, delta ^{13}C, Sr and Ba and the reconstruction of the Mg/Ca-ratio (see below).

Several authors have assumed that the evaporation rate directly influenced the evolution of the snails, in other words that highspired (trochiform) shells occur during high evaporation and flat (planorbiform) shells occur during low evaporation. This is, however, by no means the case. The correlation between evaporation rate and shell shape is very poor (see below). This is already clearly demonstrated by the fact that the trochiform shells of the main lineage occur together with planorbiform shells of the lateral lineages (Figs. 5 and 6). Additionally, there are no genetic laws that would explain such a parallelism between shell shape and evaporation rate.

The radiation of the descendants of *G. kleini* in the Steinheim Basin has two complementary explanations:

1) The geochemistry of the lake prevented immigration of specimens from outside. It thus led to a reproductive isolation between *G. kleini* in Steinheim and *G. kleini* populations of other lakes. This is very important considering the fact that migration of *G. kleini* in southern and central German "normal" lakes of the Miocene was strong enough to prevent divergent evolution within them (see above). Geochemistry of the lake water may thus have been the main cause for the divergent evolution of *G. kleini* in Steinheim.

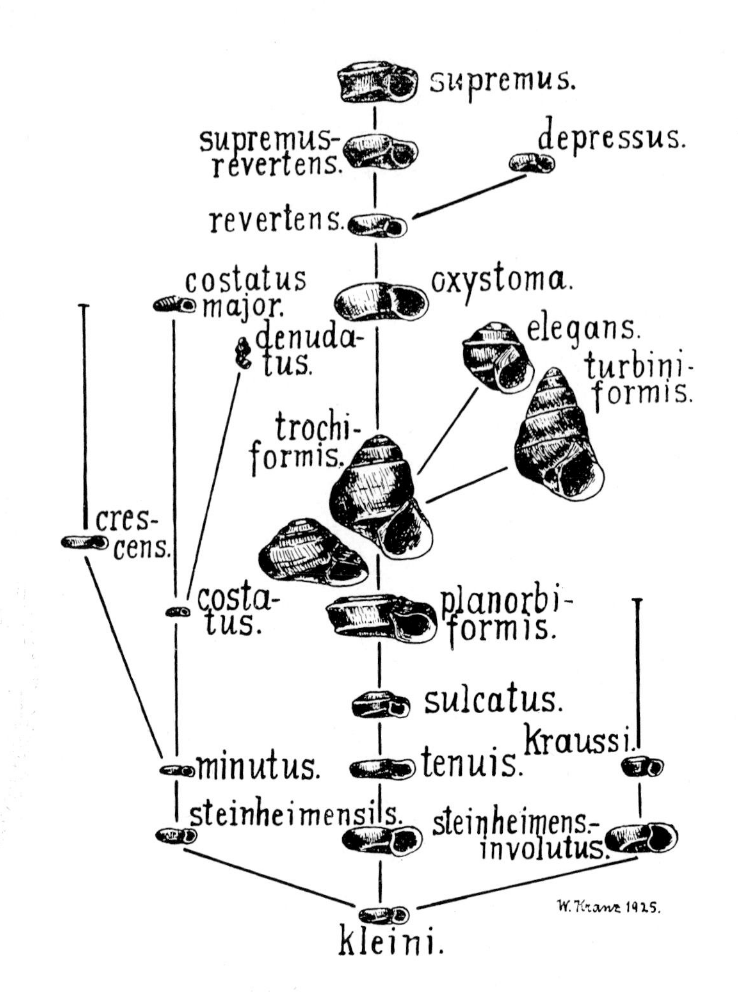

Fig. 6: Phylogenetic tree of *Gyraulus* (after GOTTSCHICK & WENZ; from KRANZ, 1926).

2) 14 gastropod species and two bivalve species which coexisted with *G. kleini*, *Radix dilatata* and *Pseudamnicola pseudoglobulus* during *kleini*-times went extinct in the lake afterwards. This may have led to a considerable reduction of competition for these three species and thus to an increase in ecologic opportunities. The diversification of *G. kleini* (as reflected in the shell shapes) may thus reflect an ecological diversification and a filling of "empty niches", left open by the extinct species. (This conclusion does not mean that empty niches "exist" in any real sense).

4. POST-*Kleini* BEDS

In order to understand the differences between the various phylogenetic trees of the Steinheim planorbids one has to know the methods used and assumptions made by the authors of the trees.

The first phylogenetic tree was constructed by HILGENDORF in 1863 (unpublished Ph.-D. thesis; see REIF, 1983a, b). Hilgendorf's tree was based on a taxonomic study by KLEIN (1846) who had distinguished 4 species of the genus *"Planorbis"* and one species of *"Valvata"*, with five "varieties" (=subspecies). Von Klein did not take into consideration that these species and subspecies were of different stratigraphical age. In bed by bed samples Hilgendorf discovered that the species and morphs occurred in different strata. He was influenced by the theoretical phylogenetic diagram in Darwin's "On the Origin of Species" and by the fact that Darwin strongly emphasized the lack of any essential difference between varieties and species; rather varieties, according to Darwin, grade into species. Hilgendorf reconstructed a phylogenetic tree on the basis of several assumptions and methodological tools:

(1) All species and varieties of *"Planorbis"* and *"Valvata"* form a single evolutionary tree.

(2) They are derived from a single founder species. (In 1863 Hilgendorf was not yet sure, which the founder species was. In 1866 he called it *P.aequeumbilicatus).*

(3) Possible patterns of evolutionary change are transformation (change within lineages) and splitting (a bifurcation of a lineage or a splitting off of a side branch). A fusion of lineages is impossible.

(4) Stratigraphic distribution and morphological similarity are the only available tools to reconstruct a phylogenetic tree. Stratigraphic distribution is established on the basis of a dense sampling program. Hilgendorf studied the stratigraphy of the Steinheim Basin sediments for the first time and distinguished 9 different beds.

Before dealing with Hilgendorf's results it seems necessary to comment briefly on his phylogenetic method which is mentioned under (4). This method had already been developed by pre-Darwinian stratigraphers, especially by those stratigraphers (such as Hilgendorf's professor F.A. Quenstedt; see REIF, in press b), who were interested in variation and temporal change of lower level taxa. (In modern terms one could say that they accepted evolution for lower level taxa, sometimes up to the family level, but not for the organic kingdom as a whole). Hilgendorf followed these stratigraphers by identifying morphs within beds and tracing them through a section. If new morphs occurred, he tried to connect them with the already known morphs on the basis of similarity. This method, which was also used by Hilgendorf's followers, sufficed to erect a phylogenetic tree. The morphs were usually regarded as species. The method did not require a species concept or a concept of how to distinguish between primitive and advanced characters or a model of how intralacustrine speciation could have taken place.

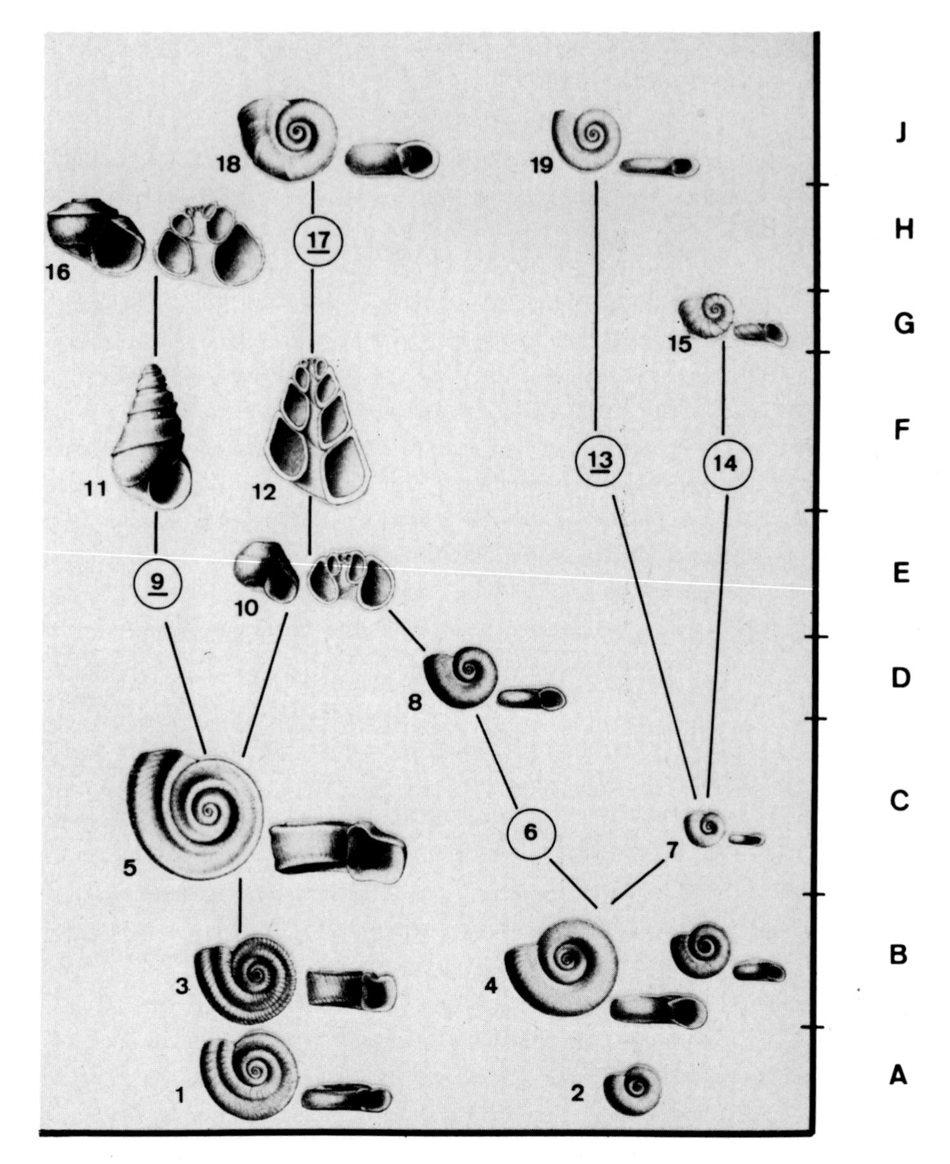

Fig. 7: Phylogenetic tree of *Gyraulus* (after collection of HILGENDORF, 1863, snails glued on card-board; from REIF, 1983a); main lineage on the left; major branch lineage on the right; minor branch lineage (nos. 6 and 8) in the center. Common ancestor of the three lineages is either no. 1 or no. 2.

Hilgendorf's result was a phylogenetic tree (Fig. 7) which he deposited in the museum of the Tübingen department without including it as a diagram in the text of his thesis. In the text, however, he discussed details of the tree. The tree consists only of branching and of transformation. In the diagram Hilgendorf indicated tacitly that morph no. 8 fuses with morph no. 10. In the text, however, he clearly emphasized that such a result would contradict the Darwinian paradigm, In his publications (1867, 1879 etc.) HILGENDORF never suggested fusion of lineages. His first evolutionary tree (of 1863) already shows the pattern of his later trees (of 1867 and of 1879; see REIF, 1983a and Fig. 5): There is one main lineage which goes from planorbiform (no. 3 and 5) to trochiform (10, 11, 12, 16) and back to planorbiform (no. 18) shells. In 1866 he discovered that no. 18 is followed by slightly trochiform morphs at the top of the section. (The bifurcation of the main lineage is revoked in later publications). There is one major branch lineage consisting of planorbiform shells (nos. 7, 13, 14, 15, 19) and one minor branch lineage also consisting only of planorbiform shells (no. 6 and 8). The main lineage and the major branch lineage end with the end of sedimentation. The minor branch lineagege fused with the main lineage (solution of 1863) or went extinct before the end of sedimentation (solution of the later publications). HILGENDORF (1863) revised the taxonomy of V. KLEIN in the discussion of his own phylogenetic tree and in later publications he added more morphs and side branches but no new lineages to his tree. In 1879 he emphasized that the differences between individual morphs were of species-level, sometimes even of subgeneric-level, significance!

Hilgendorf's phylogenetic tree was confirmed by the bed by bed sampling of the two major students of the Steinheim snails in the first decades of this century, W. WENZ and F. GOTTSCHICK. F. GOTTSCHICK's (1919/1920) tree has three lineages (main lineage, major branch lineage and minor branch lineage) with only one side branch (on the major-branch lineage). WENZ (1922) found three lineages with 5 side branches on the main lineage and on the major branch lineage (Fig. 6). The Department of Geology and Paleontology, Tübingen, houses a phylogenetic tree (cardboard with attached snails) which is labelled "after GOTTSCHICK 1920". This tree was probably constructed on the basis of personal contributions by Gottschick. It is also found as "after GOTTSCHICK 1920" in KRANZ et al. (1924). Since I could not find it in the copy of GOTTSCHICK's 1919/1920 paper available to me, it will be referred to as "Gottschick unpublished". In this tree 24 morphs are distinguished. The main lineage has three side branches and the major branch lineage has 5 side branches (Fig. 8).

Wenz and Gottschick applied more or less the same methods as Hilgendorf had. Despite the fact that they confirmed by and large Hilgendorf's tree (they were probably strongly influenced by his work), there is one important difference, Hilgendorf assumed that the planorbids evolved in the basin, whereas Wenz and Gottschick regarded all the morphs as ecophenotypic morphs produced directly by the assumed hot volcanic springs in the basin (see below). This difference in opinion may be the reason for an important difference in the trees of Hilgendorf on the one hand and of Gottschick and Wenz on the other. Wenz and Gottschick regarded the stem species of the three lineages as belonging to a single species: *steinheimensis* is the stem species of the main lineage; "*steinheimensis* (mostly small shells)" is the stem species of the major branch lineage and "*steinheimensis* involutus" is the stem species of the minor branch lineage. HILGENDORF (1867 and 1879), in contrast, had clearly argued that the founders of the three lineages were not conspecific.

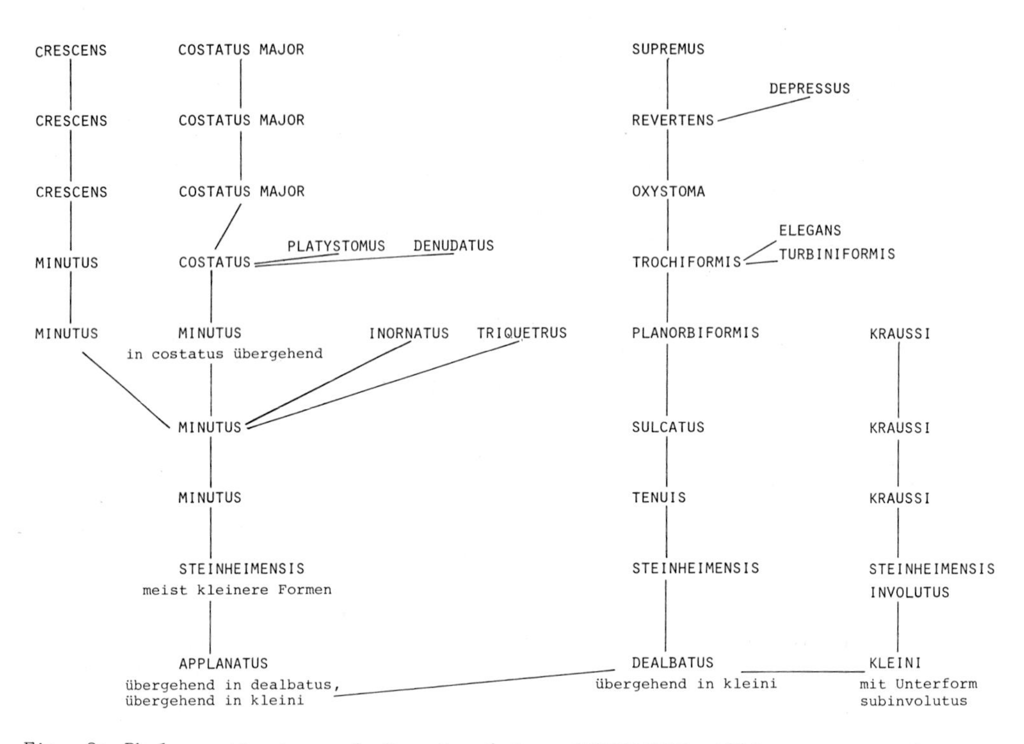

Fig. 8: Phylogenetic tree of *Gyraulus* (after GOTTSCHICK, 1920, see text for explanation; from KRANZ et al., 1924).

The polymorphism of *G. kleini* discerned by Gottschick (1919/1920) formed the basis for his trichotomy of the phylogenetic tree. "*Dealbatus* with transitions to *kleini*" gives rise to the main lineage; "*applanatus* with transitions to *dealbatus* and *kleini*" gives rise to the major branch lineage and "*kleini*, with the submorph *subinvolutus*" gives rise to the minor branch lineage (GOTTSCHICK 1919/1920 and unpublished). This is a very interesting model, but it is doubtful whether biometrical studies would support it. (Unfortunately there are no exposures today to collect the necessary samples).

Only two authors have advocated the possibility of a fusion of lineages (WIGAND, 1874-77; LUBOSCH, 1920), but they were never supported by the specialists.

Two authors based their contradictions to the Hilgendorf-Gottschick-Wenz-phylogeny on extensive collections. HYATT (1880) assumed that four "varieties" of one species (identified as *P. laevis)* populated the lake and gave rise to four independent lineages, one of them with side branches. All subsequent authors agree that Hyatt's stratigraphical information was insufficient and, taking this fact into account, Hyatt's tree can easily be transformed into Hilgendorf's tree (see KRANZ et al., 1924, for a critical literature review). KLÄHN (1922, 1923) provided extensive sedimentological and mineralogical data, which also formed the basis for his interpretation of the *Gyraulus* lineages. He assumed that the lake dried up after *kleini*-times and was repopulated by three independent, not closely related species: *steinheimensis*, *kraussi* and *minutus*, which formed three independent lineages. After the extinction of the main lineage (with *trochiformis*) and the major branch lineage (with *kraussi*) the lake was populated by two more new species,

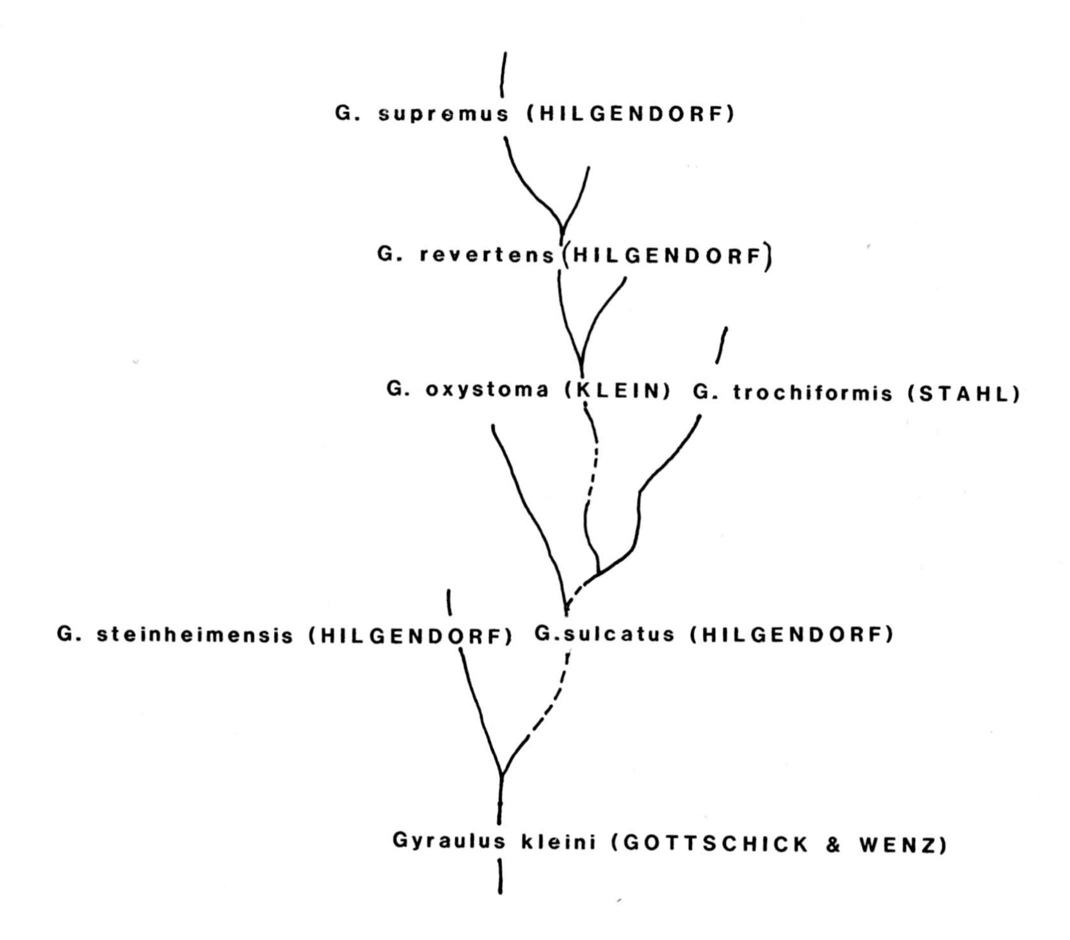

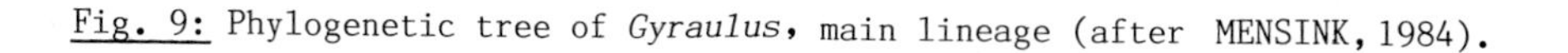
Fig. 9: Phylogenetic tree of *Gyraulus*, main lineage (after MENSINK, 1984).

each of which established a short lineage until the final drying up of the lake. Later authors did not confirm these discontinuities in the sedimentary record or in the lineages as described by Klähn. By a process of elimination, we come to the conclusion that the Hilgendorf-Gottschick-Wenz-phylogeny is the best available. However, one has to keep in mind that this phylogeny is based only on strictly phenetic and typological methods and data and has no biometrical-statistical basis.

The first biometric test of the phylogenetic tree was attempted by MENSINK (1967, unpublished; 1984). In a composite but virtually complete section (18 m thickness) he collected 50 samples and took measurements of the specimens from the main lineage. The section was combined from four different parts of the basin. The branch lineages were neglected because specimens were too rare in the samples. For each sample Mensink drew univariate diagrams for shell width and height (after second whorl), height/width, width of crest (after second and after third whorl), height of crest (after second and after third whorl), crest-height/crest–width (after second and after third whorl), but a crest occurs only on some morphs. Mensink's histograms for all parameters are unimodal or polymodal but the parameters were not adequate to separate individual morphs. The multivariate analyses of LINDENBERG & MENSINK (1979) did not solve these problems and hence will not be considered here. MENSINK (1984) identified individual maxima of his histograms with individual morphs of the classical nomenclature by taking into account additional morphological information which is not reflected in the histograms. In several cases he united morphs which had been kept separate in the classical nomenclature.

Mensink's data confirm by and large the sequence of morphs of the main lineage as reconstructed by Hilgendorf, Gottschick and Wenz. Contrary to the results of the older authors, however, the high planorbiform *trochiformis* is not ancestral to *oxystoma* but represente a side branch going extinct. The ancestor of *oxystoma* is sulcatus.

Nevertheless, the pattern within the main lineage as found by Mensink differs considerably from that of the earlier authors. GOTTSCHICK (1920, unpublished) found in the lineage 9 morphs successively replacing each other and additionally 3 morphs which are clearly side branches. Mensink on the other hand found 7 morphs which change in size and proportions through time, but can be traced as independent elements of the lineage. According to his phylogenetic diagram (my fig. 9) the morphs arise by splitting and each morph forms an independent side branch. In the text, however, Mensink spoke of transformations of the morphs: *sulcatus* shells are transformed into *trochiformis* shells (p. 43); *revertens* is "reshaped" into *supremus* (p. 47); *kleini* is transformed into *steinheimensis* and *sulcatus* (p. 47). On other pages Mensink consistently talks of splitting off of the new morphs. In the univariate diagrams the splitting is very gradual and starts with an increase in variation of a morph (p. 47). "Short times of rapid transformation alternate with longer periods of stability, in which there is a nearly constant rate of evolution" (MENSINK, 1984, p. 47).

I conclude that the different interpretations provided by Mensink are not consistent with each other. The text is not quite clear whether the new morphs arise by transformation or by splitting. The parameters of the univariate diagrams do not sharply discriminate the morphs. The phylogenetic diagram (fig. 9) does not follow unequivocally from the univariate diagrams but represents Mensink's conclusion after taking all available information into account. Though it may be highly subjective, I will use this tree as the basis for further discussion.

The interpretation is further complicated by what Mensink called "Rückschläge" (setbacks). The splitting of *kleini* (samples no. 1 to 3, fig. 4) is interrupted in sample no. 4 and then reoccurs. Mensink gave the following interpretation for this (p. 41 ff; p. 47): The trichotomous splitting in the samples no. 1 to 3 leads to *steinheimensis*, *sulcatus* and *tenuis* (a submorph of no taxonomic rank). In sample no. 4 *kleini* continues and is the only morph present. It is gradually transformed into *steinheimensis* (sample no. 5 to 8). After sample no. 8 *sulcatus* reappears again. Typical *kleini* shells can be traced up into the lower *sulcatus* beds (sample no. 1O). They disappear at the same .time as *steinheimensis* (MENSINK, 1984, Fig. 1O). The relationship between *kleini* and *steinheimensis* is thus far from clear. The only solution seems to be to regard both *steinheimensis* and *sulcatus* as side branches of *kleini* which arose before sample no. 1. Their disappearance in sample no. 4 is not a "set back" of any kind. (Mensink did not explain this term, but simply stated that they are "einschneidende, nahezu reversible Abläufe", drastic, almost reversible modes). It can simply be explained by a local extinction in the area of the lake which was studied. The geographic control of Mensink's samples is not sufficient to exclude the possible survival of *steinheimensis* and *sulcatus* in other parts of the lake.

Other setbacks are shown by *trochiformis* and *oxystoma*. *Trochiformis* appears (sample 12 to 16), then disappears (sample 17 to 2O), and the separation of *sulcatus* and *trochiformis* is only firmly established after sample 23 (p. 43). *Oxystoma* occurs in samples no. 14 to 2O, then disappears and is then present again between samples 28 to 36. Shortly after the splitting of *supremus* from *revertens* both morphs lose their crests (only in sample 43), but crests reoccur in sample 43. In none of these cases is it necessary to interpret these patterns as evolutionary patterns. It suffices to invoke local extinctions and migrations of morphs.

5. SEDIMENTARY RECORD

Sedimentation was virtually continuous during the lake's history. While water level changed considerably through time (MENSINK, 1984, Fig. 14, 28), the lake never dried up completely. The characteristic mass accumulations of the planorbid snails were caused by winnowing and transport in the marginal facies of the lake, although water turbulence was never very strong (MENSINK, 1967, p. 19). Only two levels of strong reworking have been found, both easily identified by sorting and strong wear of the shells. In addition, three sedimentation gaps were discovered by abrupt changes in the biometry of the lineages. These gaps could be verified and filled by more complete sections in neighbouring sand-quarries (MENSINK, 1967, p. 22; 1984).

In summary, the following history of the lake emerges (BAJOR, 1965; WOLFF & FÜCHTBAUER, 1976; MENSINK, 1984):

1) *Kleini*-beds: aragonitic Characeae may indicate that evaporation started already in the lowermost beds.

2) *Steinheimensis*- and *sulcatus* -beds: The water level of the lake rose. Mg/Ca-ratio was at least 2-3, which indicates strong evaporation.

3) *Trochiformis*-beds: Evaporation probably reached a peak and the water level of the lake dropped considerably.

4) *Oxystoma*-beds: at the end of the *oxystoma* times a strong freshwater influx reduced the salinity. This is indicated by a decrease of delta ^{18}O and delta ^{13}C values, the lack of dolomite, a reduction of aragonite and a decrease of Sr-content of the *Gyraulus* shells. The water level rose.

5) *Revertens* and *supremus* beds: The sediments are free of dolomite; aragonite cement is rare, which probably indicates freshwater conditions. The water level remained constant.

The reconstructed Mg/Ca-ratio, and the values of delta ^{18}O and of Sr (in the gastropod shells) are used as the main indicators of salinity. However, all three values fluctuate strongly during time and it is difficult to discover clear correlations between the three curves (BAJOR, 1965; MENSINK, 1984, fig. 28).

As was pointed out above, several authors assumed that there is parallelism between salinity and shell shape and that the evolution of shell shape was controlled by the chemistry of the lake water: "Evaporation of the lake water was probably the factor causing change in the shell shape of *Gyraulus*. Hence this is the first time where connections between chemistry of the rocks, environment and shell morphology of fossil organisms have been demonstrated" (BAJOR, 1965, p. 356). WOLFF & FÜCHTBAUER (1976) were more careful: "In the Steinheim Basin, the organisms (especially *Gyraulus*) display evolutionary changes that can be tentatively correlated with the mineralogical facies and, consequently, with the inferred chemical changes of the lake" (p. 4). "The influence of the change in the chemical milieu on the biota is indicated first in the smallness, then in the extinction of the lymnaeids during *oxystoma*-times and also possibly in the extinction of *Gyraulus trochiformis* which was discovered by MENSINK (1967) and the *Gyraulus* forms in general. Only *Gyraulus oxystoma*, which originated much earlier, survives this period. Yes, and it is even stimulated to a forceful further evolution" (WOLFF & FÜCHTBAUER, 1976, p. 36). *Lymnaea (=Radix)* occurs throughout the section and does not go extinct before the end of sedimentation (MENSINK, 1984, p. 51).

6. ENDEMIC EVOLUTION OF *GYRAULUS*?

During the first half of this century (since MILLER, 1900) the predominant hypothese has been that the morphs of *Gyraulus* (*Planorbis* of earlier authors) in Steinheim are ecophenotypic morphs. In the preceding chapters is was, however, assumed that *Gyraulus kleini* evolved within the Steinheim basin. The whole issue will be discussed more explicitly here.

MENSINK (1984, p. 51-56; see also REIF, 1983b) advanced the following arguments in favour of the evolution hypothesis (and he gave these morphs which form independent branches in his phylogenetic tree the status of species, Fig. 9):

1) There is no clear correlation between environmental changes in the lake and changes in shell form. Rapid environmental changes are not correlated with significant changes in the gastropods.
2) The "set-backs" were never complete (hence cannot be interpreted as a release of of environmental pressure).
3) Contrary to the observation of other authors, the main lineage does not, in the upper parts of the section, return to the shell forms of the lower part. Rather, *oxystoma* and *revertens* can be discriminated from *kleini* and *steinheimensis*. In other words, the comparable medium salinities of the lower and of the upper beds did not produce the same shell forms.
4) New morphs arise spontaneously in small numbers within one sample (one population) and then increase continuously in relative frequency.

Several more arguments can be given (see below). However, it is first necessary to clarify some concepts and terms, because several misunderstandings hinder a fair evaluation of the literature. MENSINK (1984, p. 52) assumed from the statement of BAJOR (1965) and WOLFF & FÜCHTBAUER (1976) quoted above, that these authors had a modificatory (=ecophenotypic) change in mind. My reading of their sentences is that these authors regard the water chemistry as a Darwinian selection factor that produced predictable shell shapes. (However, I do not find any genetic law which would make such predictions possible). Numerous paleontologists today still accept the ecophenotypy hypothesis (e.g. ZIEGLER, 1972).

There are three possible (but not necessarily mutually exclusive) explanations for the "Tree of the Steinheim Morphs" (which has been reconstructed from HILGENDORF, 1863 to MENSINK, 1984):

1) Ecophenotypy
2) Genetic polymorphism
3) Interspecific variation

It is necessary to discuss the concept of ecophenotypy of *Gyraulus* to some extent. Ecophenotypy is defined as a phenotypic variation that is created solely by the environment of the individual organism. ("Ecophenotype: A nongenetic modification of the phenotype in response to an environmental condition". MAYR, 1963, p. 665). Thus, it is not genetic differences between individuals that contribute to ecophenotypic variation, but rather environmental factors that provide stimuli for certain developmental pathways. These "aquired characters", however, are not heritable (according to Weismann's fundamental theorem of neo-Darwinism).

MILLER (1900) was the first to advance the ecophenotypy hypothesis. Dr. Konrad Miller, a theologian and natural historian, was a professor at a Gymnasium in Stuttgart. He rejected Hilgendorf's phylogenetic tree and split the Steinheim morphs into two genera, three subgenera, seven species and numerous varieties. He rejected the idea that the Steinheim snails could in any way support the theory of descent. He admitted that the variability of planorbid species is exceptionally high, but emphasized that this variability had to be regarded as a characteristic of all freshwater snails in Steinheim. Without being very specific, he pointed to several environmental factors that might have caused the variation: hot springs, water turbulence, high CO_2 content due to volcanism and plant-life. Hence the ecophenotypy hypothesis was first proposed by an anti-evolutionist to account for variation within a narrowly defined species.

GOTTSCHICK (1911), in an important contribution to the stratigraphy and paleontology of the basin, demonstrated with numerous arguments that the lowest beds (*kleini*-beds sensu MENSINK, 1984) were formed at normal water temperatures and all the upper beds in warm water heated by hot springs. He fully accepted Hilgendorf's tree of planorbid morphs, not discussing Miller's rejection of Hilgendorf's result nor Miller's ecophenotypy hypothesis. It is obvious that Gottschick was not an anti-evolutionist. "There can be no doubt that the hot springs were the cause of the change of the aquatic gastropods" (GOTTSCHICK, 1911, p. 524). From this and other statements in the same paper it is not clear, whether Gottschick had ecophenotypy, neo-Lamarckian mechanisms ("soft inheritance", of acquired characters) or Darwinian selection in mind.

In his later monograph (1919/1920), in which he provided an extensive documentation of the tree of the planorbid morphs, GOTTSCHICK referred to numerous examples of Recent gastropods living in warm water. After an extensive discussion he came to the conclusion (p. 214): "By and large I am inclined, to refer the amazing transformations of the Steinheim snails predominantly to the physiological effects of the thermal water conditions which changed several times". Again it is not clear whether this implies only ecophenotypy or inheritance of acquired characters. It is only clear from his discussion that Gottschick did not accept Darwinian selection for most of the morphological changes, because he was unable to attribute any adaptive significances to them. He grouped all the shells into one species with several subspecies.

In a comment to Gottschick's paper, PLATE (1919/1920), a very important neo-Lamarckian author, emphasized that the planorbid tree is an especially complete example of an "orthogenetische Reihe" (p. 218). (The terminological confusion increases! See REIF, (in press b), for a historical analysis of the orthogenesis concept). According to Plate, selection played virtually no role in the morphological transformations, because one could not demonstrate the gradual replacement of an old morph by a new (better adapted) one; rather, new characters occurred convergently in very many animals. (Compare the opposing results of MENSINK, 1984). In all three lineages a "zig-zag-orthogenesis" can be observed. The branch lineages are characterized by "Kümmerformen" ("scanty forms"). These Kümmerformen suffered from the warm water and one lineage even went extinct. The main lineage on the other hand was stimulated by the warm water to form bigger shells. (If members of the three lineages reacted differently to the same environmental conditions there must have been heritable differences between them! This is not made explicit by Plate). A trend from planorbiform to trochiform shells parallels this size increase. This is supposedly because of a functional (=physiological) correlation in the morphogenetic apparatus of these gastropods. The secondary return to planorbiform shells is according to Plate a clear exception to Dollo's irreversibility theorem (p. 222). The changes in shell shape are not caused by Darwinian mechanisms of mutation and selection. Plate argued that on the one hand directed mutations are highly unlikely and that on the other hand only continuous orthoselection could lead to long term trends. What remained for Plate are "Somationen": over many generations the organisms reacted more and more to a constant stimulus (water temperature). The new characters are not heritable. However, as soon as the environment changes back to the original conditions they are only gradually, rather than immediately lost. In other words "Somationen" are a neo-Lamarckian mechanism to explain evolutionary changes in the seeming absence of environmental ones.

The confusion about ecophenotypy and Lamarckian mechanisms is inherent in the writing of many authors of that time. W. WENZ (1922), a professional mollusc specialist, went a step further. He based his data largely on Hilgendorf and accepted Plate's idea of Kümmerformen in the branch lineages. He referred to similar Recent gastropods living in waters of high temperature and/or high salinity, but provided no experimental data that these populations are simply ecophenotypic morphs of populations living under normal freshwater conditions. He emphasized that in the upper part of the section the snails of the main lineage return almost to their starting point. Nevertheless the shapes are clearly "Reaktionsformen" (modifications) caused by abnormal environmental conditions. According to Wenz, the return to original shells did not contradict Dollo's Law, because this law applies in essence only to rudimentary and lost organs. In order to under stand this seeming confusion one has to take into consideration that Wenz accepted soft inheritance, i.e. the gradual genetic fixing of acquired characters. This fixing would take a long time and the duration of the lake was too short to fix the characters securely ("die erworbenen Eigenschaften zu befestigen", p. 24) except in the *costatus*-lineage. Once it had acquired its characteristic shape, this lineage underwent no changes.

From this review it is obvious that the ecophenotypy hypothesis was never formulated unequivocally. Rather there is a considerable ambiguity in the writings of Plate, Gottschick and Wenz. This ambiguity is largely due to the fact that neo-Lamarckism had a strong influence in the first decades of this century and that this theory, which has soft inheritance as its basis, does not distinguish sharply between ecophenotypy and genetic polymorphism. The rejection of an evolutionary pathway controlled by spontaneous mutations and selection of optimal shell shapes did not cause any problems for the authors mentioned above, because evolution by soft inheritance was a sufficient explanation.

7. ECOPHENOTYPY AND GENETIC VARIATION IN RECENT GASTROPODS

The current discussion of Neogene freshwater gastropod evolution (Steinheim Basin, Turkana Basin etc.) is hindered by the lack of a critical review of the population genetics of Recent gastropods: Even if it is not possible to compare Recent and fossil cases directly (often we deal with different species and the fossil environments and population structures cannot be reconstructed with sufficient details), the data from the Recent are the only significant basis for the interpretation of fossil cases.

It is almost text-book knowledge that gastropods have a high degree of phenotypic variability. Many authors equate this phenotypic variability with ecophenotypy. In the case of the Plio-Pleistocene fresh-water gastropods (and bivalves) from the Turkana Basin (Kenya) several authors proposed ecophenotypy as an alternative to WILLIAMSON's (1981a) evolutionary (punctuated equilibria) model. MAYR (1982) and BOUCOT (1982) clearly favoured ecophenotypy, but provided no data from the Recent to support their interpretation of the Turkana gastropods. KAT & DAVIS (1983) emphasized that shell shape is a very unreliable taxonomic character in freshwater molluscs compared to soft tissue characters, radula, enzyme chemistry, chromosome structure and shell microstructure. They argued that a taxonomy based on these last-mentioned characters demonstrates the degree of environmentally modified variability in shell shape characters. This is, however, not a valid argument, because it simply shows that shell shape characters change more often in gastropod evolution. These changes may very well be genetically controlled. One would simply find that in gastropod development shell shape characters are less canalized than other characters. KAT & DAVIS (1983) then point to several publications (WETHERBY, 1882; BAKER, 1928; PILSBRY & BEQUAERT, 1927; HAAS, 1936) which, as they claim, have shown strong ecophenotypic effects that could even result in non-overlapping phenotypes.

However, the publications referred to by KAT & DAVIS (also the interesting contributions by DI CESNOLA, 1906 and WELDON, 1901, which are quoted by Williamson's critics CHARLESWORTH & LANDE, 1982) only document, often quantitatively, phenotypic variation of shell shape. Even the demonstrations that shell shape changes following an artificial environmental change (artificial lake; BAKER, 1928) do not suffice to establish ecophenotypy without genetical control. To prove ecophenotypy beyond doubt it is necessary to breed the animals with known genetic constitution under different conditions. However, if it can be shown that phenotypic variation and allelic variation are correlated, this is a good case for genetically controlled variation (or polymorphism).

CLARKE et al. (1978) reviewed the literature on genetics of pulmonates up until 1974 (see also RAVEN, 1966; RUSSELL-HUNTER, 1978 and FRYER et al., 1983). They distinguished between continuous variation and discontinuous variation (=polymorphism). There is no sharp boundary between these two phenomena. Theoretically both can be either gene-controlled or environment-controlled. Or, in the case of incomplete heritability, they can, to a varying degree, be gene- and environment-controlled. There are very few cases of land-snails where the relative contribution of genotype and environment is known to a certain degree. The status of the literature on freshwater pulmonates is much poorer.

The classical studies on continuous variation in different species of *Lymnaea* (WADDINGTON, 1975, p. 92; CLARKE et al., 1978, p. 222) show that in addition to ecophenotypy, genetic control plays a significant role in the variation of shell characters. This is also true for the variation of shell size in the land-snails *Arianta*, *Cepaea* and *Partula* (p. 224), in which the (incomplete) heritability is even known quantitatively.

Discontinuous variation (polymorphism) has been found with respect to colour and banding patterns of the shell and body, shell-coiling, electrophoretic mobility of proteins and the numbers of chromosomes. A priori there is little doubt as to the high heritability of electrophoretic mobility of proteins and the chromosome numbers. CLARKE et al. (1978, p. 227) list more than 50 species in which colour polymorphism was described, but (usually scarce) data on the genetics are available for no more than 12 species. In all the better studied cases it is clear that colour polymorphism in pulmonates has a very strong genetic control; ecophenotypy seems to be rather unimportant. The same is true for chirality (direction of shell coiling).

BROWN (1980) provided some information on continuous variation in African freshwater prosobranchs and pulmonates; quantitative data on shell characters, however, are scarce.

A classical example of ecophenotypic polymorphism of shell colour pattern which is often quoted in the modern literature (e.g. FRYER et al., 1983; WILLMANN, this volume) is the freshwater prosobranch *Theodoxus fluviatilis*. The original experimental studies on the influence of salinity on the pattern were carried out by NEUMANN (1959a, b). Neumann's account, however, clearly shows that there is a strong genetic component involved. The distributions of salinity (in rivers and estuaries of northern Germany) and of colour patterns do not strictly correlate. Some morphs are lacking in some populations, which suggests the fact that some genes are missing in these populations.

CLARKE (1978) based his extensive study of the adaptive significance of polymorphism in marine molluscs on the assumption of a strong genetic control of this polymorphism. Experimental data, however, were not available to him.

In recent years careful population studies have been carried out on several land-snails and coastal snails. All of them adduce direct or indirect evidence to show that shell character variation and polymorphism are largely gene-controlled.

Nucella lapillus (Prosobranchia, habitat: rocky shores) forms populations of small areal extent with a conspicuous intrapopulational and interpopulational shell shape variation. Few experimental data are available. Lack of correlation between the distribution of shell shapes and environmental factors indicate a strong genetic component in this variability (CROTHERS, 1982; 1983a; 1983b).

Enzyme studies showed that the land-snail *Triodopsis albolabris* consists of at least two species, both of which have a high genetic variability. Shell shape, however, is comparatively uniform throughout the area of distribution of the species (MCCRACKEN & BRUSSARD, 198O).

Littorina rudis and *Lirroeina arcana* (Prosobranchia) occur sympatrically in the British isles, they have both a polymorphism of shell colour. Experiments show that the colour is inherited. The distribution of the colour patterns of both species is strongly correlated in natural populations. This indicates that the colour polymorphism is maintained by natural selection (ATKINSON & WARWICK, 1983).

Biochemical and morphological analyses of a complex of five species of the land-snail *Sphincterochila* in Israel (12 sampled populations) have revealed a high degree of morphological and allozymic variation within and between the populations. Allozymic and morphological variation are partly correlated (NEVO et al., 1983), which indicates that morphological variation cannot be completely ecophenotypical. Breeding experiments were not carried out.

GRÜNEBERG (198O 1981) studied the ontogeny of colour patterns and the distribution of the patterns in several populations of *Umbonium* spp., *Clithon oualaniensis*, and *Nerita polita* (marine and brackish prosobranchs). In all three genera sharply defined colour morphs occur. For one of the morphs of *Clithon oualaniensis* the genetic basis has been confirmed by breeding experiments in the laboratory. However, the distribution and morphogeny of the morphs indicate that all of them have a simple genetic basis. In addition to the specimens with these genetically controlled colour patterns, other specimens with less distinct and definable colourations and patterns occur, which grade into each other and which even change during ontogeny. This suggests that "there is no simple correspondence between genotype and phenotype: a single genotype may be moulded into several phenotypes by influences of the environment and presumably by interaction with a polygenic genetic background, a situation here designated as pseudo-polymorphism" (GRÜNEBERG, 1980, p. 533).

To conclude this review, ecophenotypy seems to play a much smaller role in gastropods than is usually assumed in the discussion of fossil lineages. However, data are scarce and they are most badly lacking in freshwater pulmonates.

8. ECOPHENOTYPY, POLYMORPHISM AND SPECIATION IN *Gyraulus*.

It is commonly the case in natural history that there are no answers which are clear-cut and mutually exclusive. Rather, often the only answer to be given is a rough estimate of the relative influences of various factors which contribute to the phenomenon under study. This is also true for *Gyraulus* in the Steinheim Basin. The three possible factors that may have contributed to the "tree of morphs" (ecophenotypy, genetic polymorphism and intra-lacustrine speciation) are by no means mutually exclusive. Rather it is necessary to estimate the relative importance of these factors.

From the discussion of the preceding chapter it follows that I do not consider the differences in shell shape of synchronous morphs to be environmentally controlled. The role of ecophenotypy was probably restricted to a contribution to the continuous variation within morphs. This within-morph variation was studied by MENSINK (1984), but only histograms and no numerical data are available. It is impossible to estimate the relative importance of ecophenotypy and genetic control for this within-morph variation..

Thus one can assume that there were significant genetic differences between the morphs. The next question is, what factors created these genetic differences? For all genetic differences in organisms there are two possible factors which again are not mutually exclusive, but may act together to varying degrees:

(1) selection and

(2) chance effects like drift and founder effects (see e.g. ROUGHGARDEN, 1979; LEWONTIN, 1974; GOULD & LEWONTIN, 1979; NEVO, 1983).

It is one of the most basic questions of evolutionary theory how important chance effects and how important selection have been. Land snails have played an important role in the discussion of this question.

In *Cepaea nemoralis*, for example, ecology, population structure, behaviour and genetics are so well known that selection was studied under both natural and experimental conditions (for data and reviews see JONES, 1973; FORD, 1975; JONES et al., 1977; CLARKE et al., 1978, and TILLING, 1983). In general, founder effects and drift play a significant role in small populations, in populations with drastic size changes, in the colonization of new populations and in gene exchange between panmictic local populations.

It is impossible to determine population sizes in the *Gyraulus* morphs. The great abundance of the shells in most beds point to large numbers of snails living at the same time. However, there are no data as to how large panmictic populations were and how the populations were distributed over the lake. Because of the limited mobility of the snails panmictic populations were much smaller than the total extent of the lake (see below). Hence the lake (which was very shallow) must have contained a complex mosaic of populations of various morphs. As far as is known, the major morphs occurred all over the lake, whereas it is very likely that there were numerous minor morphs (more than have been discovered so far in the lateral branches of the tree) which were restricted to certain areas.

Even if we grant chance effects a significant role in the evolution of *Gyraulus*, selection was by far the most important factor. However this does not mean that we are able to identify the selective factors. Water chemistry was only one factor. The ecology of Recent planorbids is so poorly known that it is impossible to give a list of possible selection factors. In the recent discussion of non-adaptive factors in evolution (see e.g. GOULD & LEWONTIN, 1979; NEVO, 1983 and NEVO et al., 1983 for references) one often finds a disturbing confusion: non-adaptive evolutionary pathways (by random walk) are not kept clearly separate from non-adaptive differences between organs of related taxa. Even if we grant selection the major role in the evolution of *Gyraulus*, it is not necessary to assume that there were significant differences in function between all of the various shell morphs. It may well have been that two different shell shapes evolved in parallel and served the same function (multiple evolutionary pathways, sensu BOCK, 1959). Also changes in shell shape could have had neutral selective value and yet the overall change of the organism was directed by selection. To conclude this, we have to emphasize that our failure to find adaptive differences between the shell morphs does not prove that there were none, and hence this does not (contra PLATE, 1919/1920, see above) refute selection as the operating process.

Endogeneous factors that direct evolution have never been demonstrated experimentally. MENSINK (1984, p. 48) regarded the high variability of all measured characters of *supremus* as an indicator "of a phylogenetic ageing at the end of evolution". This implies that if the lake had not fallen dry, the *Gyraulus* main lineage would eventually have died from "old age". This is clearly an orthogenetic explanation and not in accordance with the Synthetic Theory of Evolution.

In land snails, panmictic populations have a very small extent due to the low mobility of the animals. At a distance of a few meters to a few hundreds of meters one passes gradually from one panmictic population to the next one (CLARKE et al., 1978). Nevertheless these same species have a wide distribution (e.g. *Cepaea nemoralis* and *Cepaea hortensis* occurring all over central and western Europe). There are two possible explanations for this:

(1) despite high intrapopulational variability (CLARKE et al., 1978) there is little tendency to evolve isolating mechanisms;

(2) the divergent evolution of interpopulational differences is continuously counterbalanced by the dispersal of the snails.

Two possible dispersal mechanisms have been discussed:

(a) active migration and
(b) passive dispersal by birds, either as eggs which stick to the birds' feet or as adult snails which had fallen prey to birds but were then lost.

To the best of my knowledge dispersal rates have never been measured in *Cepaea*, the best studied pulmonate, nor in any other gastropod.

There is no doubt that passive dispersal mechanisms are effective over a much wider distance than active ones. If there are no passive dispersal mechanisms sufficient to balance divergent evolution and the trend to evolve mechanisms for reproductive isolation, then one would expect a pattern of species in which each has a small areal extent; the regional distributions of the species should overlap only to a certain degree. Such a case is the genus *Partula* which occurs on several Pacific islands. It was extensively described by CRAMPTON (1932 and earlier monographs). The species complex occurring on the small island of Moorea (near Tahiti; maximum diameter 10 miles) is now well studied with respect to distribution ecology, genetics, biochemistry, and population genetics (CLARKE & MURRAY, 1969; JOHNSON et al., 1977; LIPTON & MURRAY, 1979; MURRAY & CLARKE, 1966, 1968, 1976a, b; MURRAY et al., 1981).

Among gastropods, the information on *Partula* is second in quality and quantity only to the classic case of *Cepaea*. However, despite the long and intensive interest in *Partula* from Moorea I have found no information on possible passive dispersal mechanisms. There are now 9 species of *Partula* on Moorea, most of which are restricted to small parts of the island. Hence there is a considerable degree of allopatry (see maps in CLARKE & MURRAY, 1969; LIPTON & MURRAY, 1979). Up to 4 species can occur sympatrically with clearly partitioned niches (MURRAY et al., 1981). The mobility of the animals is so low that there are genetic differences between natural populations only short distances (20 m or less) apart (CLARKE & MURRAY, 1969). It seems that this high degree of intra-island endemism (= high degree of allopatry) is largely due to the lack of passive transporting mechanisms (birds that feed on gastropods; REIF, in preparation).

It is doubtful whether the very low dispersal of *Partula* is a model for *Gyraulus* in Steinheim. There are no data in the literature on active and passive dispersal of pulmonates in a lake of the size of the Steinheim Basin. Possible predators of the small but thick-shelled snails could have been birds, reptiles and fishes. Floating leaves and water turbulence created by storms are other possible dispersal agents. Despite these dispersal mechanisms there is no reason to assume that *Gyraulus* formed large homogeneous, amphimictic populations extending over the whole lake. Rather one can assume that there was a high degree of intra-lake endemism. Except for Mensink's "setbacks" there is no direct evidence for allopatric populations. This may be partly due to post-mortem transport of the shells. Basically, however, Mensink did not provide geographic data. His standard section was combined from four different parts of the lake; the individual sections hardly overlap and there are two gaps of unknown extent in the section. In the more than 1100 short boreholes (0,5 to 3,5 m) and the additional exposures Mensink did not find any regional differentiation. However, one has to take into consideration that endemic populations may have been short-lived and would not be found, and, additionally, that Mensink had no independent stratigraphic control. The only guide-fossils he could use are the *Gyraulus* morphs themselves.

If we accept a pattern of synchronous populations of various morphs which have different areal extents and which partly overlap in area, we must pose the question whether there were reproductive barriers between the morphs or not. In other words, how many *Gyraulus* species existed at the same .time in the lake? Again *Partula* in Moorea can serve as a model.

Genetical and population genetical complexity is found in *Partula* on several different levels of integration. Local populations are highly polymorphic with respect to allozymes and shell colour pattern (JOHNSON et al., 1977; CLARKE & MURRAY, 1969; MURRAY & CLARKE, 1966). In some populations dextral and sinistral shells co-occur. There is little allozymic variation among species. There are two species groups (*P. suturalis* and *P. taeniata*). Within each group there is a set of species. Several of the within-group species pairs behave like good species (no hybridization) in some places, while in others there is introgression (hybridization). One species pair is connected by a continuous series of intergrading populations. In another case a species pair coexists without apparent exchange of genes, but a third sympatric species forms a link between them. There may be even some gene exchange between members belonging to the different species groups. In the laboratory, some species of the two groups hybridize. Hence it is very difficult to decide which of the nine species are good species.

The dynamics of this pattern are not well understood. On the one hand behavioural differences (LIPTON & MURRAY, 1979) and niche partitioning (MURRAY et al., 1981) will contribute to a strenghtening of the genetic barriers. On the other hand introgression may be caused by a secondary sympatry after the two populations had been separated for some time and had developed incomplete reproductive barriers (MURRAY & CLARKE, 1968). It may eventually lead to a complete fusion of the two "species"-

Speciation patterns in *Partula* can serve as a model for *Gyraulus* from Steinheim. Self-fertilization is rare in *Partula* (see literature quoted above). Recent Planorbidae reproduce by cross-fertilization. Some species self-fertilize when bred in isolation. Several species of the family fail to self-fertilize (DUNCAN, 1975).

From data on *Partula* we expect in Steinheim a complex pattern of local temporary endemism, migration of clines, hybridization by introgression, incomplete reproductive isolation controlled by behavioural differences, changing population sizes and selection-controlled genetic polymorphism which is only partly expressed in the shell shapes. Even if the sections were more complete and if the exposures were significantly better, it is doubtful whether very careful biometrical analyses (using appropriate shell parameters which optimally distinguish the morphs) would ever document the complex evolution in the three lineages. At the moment all we can do is to regard each morph which can be biometrically defined and which behaves as a stable entity through time as a species. Undoubtedly there were plenty of opportunities for allopatric speciation and for parapatric speciation (extinction of intermediate populations in a chain of races; clinal speciation; area-effect speciation; WHITE, 1978). Mensink's tree (Fig. 9) is thus an acceptable preliminary solution. Taking the main lineage and the branch lineages together, 6 to 10 species may have occurred at any time. However, as there seems to be no objective way to define evolutionary species (see literature review and discussion in REIF, in press a) there is also no way to determine how many species *Gyraulus kleini* gave rise to.

9. THE PROBLEM OF *LYMNAEA DILATATA* AND *PSEUDAMNICOLA PSEUDOGLOBULUS*

Both species occur throughout the section (MENSINK, 1984). *L. dilatata* is rare and no biometrical study was carried out. Mensink simply documented the wide phenotypic variability with photographs (MENSINK, 1984, pl. 10). *P. pseudoglobulus* is rare below sample no. 10; above it is very common and often dominant. Univariate diagrams (for height/width, after second, third and fourth whorl) do not show any significant changes through time. Mensink's observation that in the *supremus*-beds the variability of *P. pseudoglobulus* is higher than in the lower beds may not be statistically significant.

Hence the shell characters of the two species are not sufficient to document evolutionary changes during the time of the lake, intra-lake endemism or reproductive isolation between the populations in Steinheim and populations of lakes in the vicinity.

10. PLANORBID EVOLUTION IN "ANCIENT LAKES"

"Ancient lakes" have long caused interest because of their high degree of endemicity in fresh-water gastropods. The best-known ancient lakes are Lake Baikal (USSR), Lake Tanganyika (Africa), Lake Ohrid (Jugoslawia), Lake Titicaca (Peru), Lake Biwa (Japan) and Lake Inlé (Burma). The history of all these lakes dates back to the Pleistocene, to the Miocene or even to the Oligocene. The classical explanation of the high degree of endemism is that in these old lakes a fauna survives which is extinct everywhere else (WESENBERG-LUND , 1939, p. 719). However, this model does not explain why these "old species" have not recently dispersed into other lakes.

A recent review (BOSS, 1978) of the ancient lakes provides a completely different picture. All the ancient lakes are very deep and have two different faunas:

(1) a shallow-water fauna, consisting only of pulmonates and
(2) a deper-water fauna dominated by prosobranchs but also with pulmonates.

The shallow-water pulmonates are easily dispersed from lake to lake and are hence not endemic. The deep-water prosobranchs are stenotopic and philopatric. Additionally, the prosobranchs seem to speciate more readily than the pulmonates. The deep-water snails are not easily transported to other lakes; hence their high degree of endemicity.

Among the pulmonates, the Planorbidae have many more endemic species than any other family (in Lake Baikal, Lake Ohrid, Lake Titicaca and Lake Biwa). These species prefer deeper water and, according to BOSS (1978), are stenotopic. Perhaps the successful radiation of the planorbids in ancient lakes has, in part, been facilitated by the presence of hemoglobin, which is lacking in other pulmonates, allowing the planorbids to exploit deeper habitats, where conditions of reduced oxygen supply or even near anaerobic situations might obtain. The planorbids also have a secondary respiratory faculty in the development of respiratory folds and projections within the pulmonary cavity (BOSS, 1978, p. 418). Planorbids have comparatively thick egg-capsules which may be an important factor of tolerance to environmental conditions.

Boss' data thus lead to the conclusion that the degree of endemicity of these "ancient lakes" is not so much controlled by their age but rather by their depth. Hence, I suggest one should no longer call them "ancient lakes" (however fascinating their age is) but "deep lakes".

This model allows the prediction that deep lakes have a high degree of endemicity almost regardless of their age. It is doubtful, however, whether enough deep lakes of varying ages are known to test this prediction.

Gyraulus (Carinogyraulus) trapezoides in Lake Ohrid has a significant variability in shell shape ranging from planorbiform to trochiform shells (BOSS, 1978). However, there is no information available on whether this variability is ecophenotypic or not.

The "ancient lakes" are no model for the Steinheim Basin despite the fact that they have a radiation of endemic planorbids in common. The Steinheim Basin was neither old nor deep. However, the environmental tolerance (hemoglobin, thick egg-capsules and other physiological aspects which are less well studied) of the Planorbidae may not only have contributed to the radiation in the old lakes but also in the Steinheim basin. Of 11 or more pulmonate species, in the Steinheim basin, only 2 survived when the salinity increased, *Gyraulus kleini* and *Lymnaea dilatata*. Of the 5 prosobranchs *Pseudamnicola pseudoglobulus* survived (KRANZ et al., 1924). From what is known about the "ancient lakes", it is perhaps not surprising that *Gyraulus* is among the survivors. However, the numbers do not reflect Boss' finding in the ancient lakes that the pulmonates are more eurytopic than the prosobranchs.

11. CONCLUSIONS

The discussion has shown that despite rather favourable conditions in the Steinheim Basin the study of gastropod microevolution is hampered by several factors:

(1) Incomplete sections;
(2) post-mortem transport of the snails;
(3) insufficient information on the polymorphism, evolution and biogeographic distribution of the founding species *G. kleini*;
(4) lack of geographic control within the Steinheim Lake;
(5) problems of defining evolutionary species;

(6) an inability to distinguish between primitive and advanced characters in the Steinheim snails;

(7) limited sample sizes;

(8) insufficient information about the genetics and population genetics of the living representatives of the genus;

(9) insufficient biometric methods to describe objectively the changes of the morphs.

The literature reviews demonstrated how much the perception of the authors was influenced by the various evolutionary theories to which they adhered. Darwinism (in its original version, in the case of Hilgendorf), neo-Lamarckism and orthogenesis have played the most important roles in the literature on the Steinheim snails.

Acknowledgements: I thank Karl Flessa for critically reading the manuscript. John Maynard Smith drew my attention to the population genetics of *Partula* on Moorea.

REFERENCES

Adam, K.D. 1980: Das Steinheimer Becken - eine Fundstelle von Weltgeltung.- Jh. Ges. Natkd. Württemb. 135: 32-144.

Atkinson, W.D. & Warwick, T. 1983: The role of selection in the colour polymorphism of Littorina rudis Maton and Littorina arcana Hannaford-Ellis (Prosobranchia: Littorinidae).- Biol. J. Linn. Soc. 20: 137-151.

Bajor, M. 1965: Zur Geochemie der tertiären Süßwasserablagerungen des Steinheimer Becken, Steinheim am Albuch (Württemberg).- Jh. geol. Landesamt Bad.-Württemb. 7: 355-386.

Baker, F.C. 1928: Influence of a changed environment in the formation of new species and varieties.- Ecology 9: 271-283.

Bock, W.J. 1959: Preadaptation and multiple evolutionary pathways.- Evolution 13: 194-211.

Boss, K.J. 1978: On the evolution of gastropods in ancient lakes.- In: Fretter, V. & Peake, J: Pulmonates, pp. 385-428, London (Academic Press).

Boucot, A.J. 1982: Ecophenotypic or genotypic?- Nature 296: 609-610.

Brown, D.S. 1980: Freshwater Snails of Africa and their Medical Importance.- 487 pp., London (Taylor & Francis).

Charlesworth, B. & Lande, R. 1982: Morphological stasis and developmental constraint: no problem for Neo-Darwinism.- Nature 296: 610.

Cohen, A.S. & Schwartz, H.C. 1983: Speciation in molluscs from Turkana Basin: Discussion.- Nature 304: 659-660.

Clarke, A.H. 1978: Polymorphism in marine mollusks and biome development.- Smithson. Contrib. Zool. 274, 14 pp., Washington.

Clarke, B., Arthur, W., Horsley, D.T. & Parkin, D.T. 1978: Genetic variation and natural selection in pulmonate molluscs.- In: Fretter, V. and Peake, J.: Pulmonates, vol. 2A, pp. 219-270, London (Academic Press).

Clarke, B. & Murray, J. 1969: Ecological genetics and speciation in land snails of the genus Partula.- Biol. J. Linn. Soc., London 1: 31-42.

Crampton, H.E. 1932: Studies on the variation, distribution and evolution of the genus Partula. The species inhabiting Moorea.- Carnegie Inst. Wash. Publ. 410: 1-335.

Crothers, J.H. 1982: Shell shape variation in dog-whelks, Nucella lapillus (L.) from the West Coast of Scotland.- Biol. J. Linn. Soc. 17 (4): 319-342.

Crothers, J.H. 1983a: Some observations on shell-shape variation in North American populations of Nucella lapillus (L.).- Biol. J. Linn. Soc. 19 (3): 237-274.

Crothers, J.H. 1983b: Variation in dog-whelk shells in relation to wave action and crab predation.- Biol. J. Linn. Soc. 20(1): 85-102.

Dehm, R., Gall, H., Höfling, R., Jung, W. & Malz, H. 1976: Die Tier- und Pflanzenreste aus den obermiozänen Riessee-Ablagerungen in der Forschungsbohrung Nördlingen 1973.- Geologica Bavarica 75: 91-110.

Di Cesnola, A.P. 1906: A first study of natural selection in "Helix arbustorum" (Helicogena).- Biometrica 5: 387-399.

Dingus, L. & Sadler, P.M. 1982: The effects of stratigraphic completeness on estimates of evolutionary rates.- Syst. Zool. 31: 400-412.

Duncan, C.J. 1975: Reproduction.- In: Fretter, V. & Peake, J.: Pulmonates, vol. 1, 309-366, London (Academic Press).

Ford, E.B. 1975: Ecological Genetics.- 4. edition, 447 pp., London (Chapman and Hall).

Fryer, G., Greenwood, P.H. & Peake, J.F. 1983: Punctuated equilibria, morphological stasis and the paleontological documentation of speciation: a biological appraisal of a case history in an African lake.- Biol. J. Linn. Soc. 20: 195-205.

Gentner, W., Lippolt, H.J. & Schaeffler, O.A. 1963: Argonbestimmungen an Kaliummineralien. XI: Die Kalium-Argon-Alter der Gläser des Nördlinger Rieses und der böhmisch-mährischen Tektite.- Geochim. Cosmochim. Acta 27: 191-200.

Ginzburg, L.R. & Rost, J.D. 1982: Are "punctuations" artefacts of time-scales?- Nature 296: 610-611.

Gottschick, F. 1911: Aus dem Tertiärbecken von Steinheim a.A.- Jh. Ver. vaterl. Natkd. Württemb. 67: 496-534.

Gottschick, F. 1919/1920: Die Umbildung der Süßwasserschnecken des Tertiärbeckens von Steinheim a.A. unter dem Einflusse heißer Quellen.- Jenaische Z. Naturwiss. 56: (N.F. 49): 155-216.

Gottschick, F. 1925: Noch einmal: Die Umbildung der Süßwasserschnecken des Tertiärbeckens von Steinheim a.A. unter dem Einfluß heißer Quellen.- Cbl. Mineral. Geol. Paläontol. 1925: (Abt. B): 8-16, 43-51.

Gottschick, F. & Wenz, W. 1916: Die Sylvana-Schichten von Hohenmemmingen und ihre Fauna.- Nachrbl.dt.malakozool. Ges. 48: 17-31; 55-74; 97-113.

Gould, St.J. 1980: Darwinism and the expansion of evolutionary theory.- Science 216: 380-387.

Gould, St.J. & Lewontin, R.C. 1979: The spandrels of San Marco and the Panglossian paradigm: A critique of the adaptationist programme.- Proc. R. Soc. Lond. B 205: 581-598.

Groschopf, P. & Reiff, W. 1966: Neue Untersuchungen im Steinheimer Becken.- Fortschr. Mineral. 44: 141-142.

Groschopf, P. & Reiff, W. 1979: The geological trail in the Steinheim basin.- In: Reiff, W. (ed.): Guidebook to the Steinheim Basin Impact Crater, Stuttgart, pp. 19-23.

Grüneberg, H. 1980: On pseudo-polymorphism.- Proc. R. Soc. Lond. B 210: 533-548.

Grüneberg, H. 1981: Pseudo-polymorphism in Nerita polita (Neritacea, Archaeogastropoda).- Proc. R. Soc. Lond. B 212; 53-63.

Haas, F. 1936: Binnen-Mollusken aus Inner-Afrika.- Abh. Senckenb. naturf. Ges. 431: 156 pp., Frankfurt/M.

Hilgendorf, F. 1863: Beiträge zur Kenntnis des Süßwasserkalkes von Steinheim.- 42 pp., unpubl. Ph.-D. thesis, Philosophical Fac., Univ. of Tübingen.

Hilgendorf, F. 1867: Über Planorbis multiformis im Steinheimer Süßwasserkalk.- Monatsber.kgl. preuss. Akad. Wiss. Berlin, for 1866, pp. 474-504(communicated by H.E. Beyrich), Berlin.

Hilgendorf, F. 1879: Zur Streitfrage des Planorbis multiformis.- Kosmos 5: 10-22, 90-99, Leipzig.

Hyatt, A. 1880: Tertiary species of Planorbis at Steinheim.- Ann. Mem. Boston Soc. Natur. Hist. 1980, 114 pp.

Jankowski, B. 1981: Die Geschichte der Sedimentation im Nördlinger Ries und Randecker Maar.- Bochumer geol. und geotechn. Arb. 6, 115 pp., Bochum.

Johnson, M.S., Clarke, B. & Murray, J. 1977: Genetic variation and reproductive isolation in Partula.- Evolution 31: 116-126.

Jones, J.S. 1973: Ecological genetics and natural selection in molluscs.- Science 182: 546-552.

Jones, J.S. 1981: An uncensored page of fossil history.- Nature 293: 427-428.

Jones, J.S., Leith, B.H. & Rawlings, B. 1977: Polymorphism in Cepaea: A problem with too many solutions?- Ann. Rev. Ecol. Systematics 8: 109-143.

Kat, P.W. & Davis, G.M. 1983: Speciation in molluscs from Turkana Basin: Discussion.- Nature 304: 660-661.

Klähn, H. 1922: Das Steinheimer Becken.- Z. Deutsch. Geol. Ges. 74: (Abh. No. 1): 26-161.

Klähn H. 1923: Paläontologische Methoden und ihre Anwendung auf die paläobiologischen Verhältnisse des Steinheimer Beckens.- 127 pp., Berlin (Gebrüder Borntraeger).

v. Klein, A. 1846: Conchylien der Süßwasserkalkformationen Württembergs.- Jh. Ver. vaterl. Naturkd. Württemb. 2: 60-116, Stuttgart.

Kranz, W. 1926: Wanderungen, Probleme und weitere Forschungen im Becken von Steinheim am Albuch.- Blätter Schwäb. Albverein 38: 68-80.

Kranz, W. 1936: Nachtrag zu den Begleitworten zur Geognostischen Spezialkarte von Württemberg. Atlasblatt Heidenheim, 2. Auflage, IV. Abschnitt: Das Steinheimer Becken, 17 pp., Stuttgart (Ernst Klett) (Württ. Statistisches Landesamt).

Kranz, W., Berz, K.C. & Berckhemer, F. 1924: Begleitworte zur geognostischen Spezialkarte von Württemberg. Atlasblatt Heidenheim. 2. Auflage.- 138 pp., (Württ. Statistisches Landesamt) Stuttgart (W. Kohlhammer).

Lewontin, R.C. 1974: The Genetic Basis of Evolutionary Change.- 346 pp., New York and London (Columbia University Press).

Lindenberg, H.G. & Mensink, H. 1979: Multivariate Gruppierungsmethode in phylogenetisch orientierter Paläontologie (am Beispiel der Gastropoden aus dem Steinheimer Becken).- Berliner Geowiss. Abh. A 15: 30-51.

Lindsay, D.W. 1982: Punctuated equilibria and punctuated environments.- Nature 296: 611.

Lipton, C.S. & Murray, J. 1979: Courtship of land snails of the genus Partula.- Malacologia 19 (1): 129-146.

Lubosch, W. 1920: Das Problem der tierischen Genealogie nebst einer Erörterung des genealogischen Zusammenhangs der Steinheimer Schnecken.- Arch. mikr. Anat., Festschrift Hertwig, pp. 459-499.

Mayr, E. 1963: Animal Species and Evolution.- 797 pp., Cambridge, Mass. (The Belknap Press of Harvard University Press).

Mayr, E. 1982: Questions concerning speciation.- Nature 296: 609.

McCracken, G.F. & Brussard, P.F. 1980: The population biology of the white-lipped land snail, Triodopsis albolabris: genetic variability.- Evolution 34: 92-104.

Mensink, H. 1967: Zur Entwicklungsgeschichte der tertiären Panorben aus dem Steinheimer Becken in Süddeutschland.- Bochum (Unpubl. Habilitationsschrift, 66 pp.).

Mensink, H. 1984: Die Entwicklung der Gastropoden im miozänen See des Steinheimer Beckens (Süddeutschland).- Palaeontographica, Abt. A, 183: 1-63.

Miller, K. 1900: Die Schneckenfauna des Steinheimer Obermiozäns.- Jahresh. Ver. vaterl. Naturk. Württemberg 56: 365-406.

Murray, J. & Clarke, B. 1966: The inheritance of polymorphic shell characters in Partula (Gastropoda).- Genetica 54: 1261-1277.

Murray, J. & Clarke, B. 1968: Partial reproductive isolation in the genus Partula (Gastropoda) in Moorea.- Evolution 22: 684-698.

Murray, J. & Clarke, B. 1976a: Supergenes in polymorphic land snails. I. Partula taeniata.- Heredity 37: 253-269.

Murray, J. & Clarke, B. 1976b: Supergenes in polymorphic land snails. II. Partula suturalis.- Heredity 37: 271-282.

Murray, J., Johnson, M.S. & Clarke, B. 1981: How do Partula spp. partition their habitat?- Virginia Journal of Science 32 (3): 96.

Neumann, D. 1959a: Experimentelle Untersuchungen des Farbmusters der Schale von Theoxus fluviatilis L.- Verh. deutsch. zool. Ges. 53: 152-156.

Neumann, D. 1959b: Morphologische und experimentelle Untersuchungen über die Variabilität der Farbmuster auf der Schale von Theodoxus fluviatilis L.- Z. Morph. Ökol. Tiere 48: 349-411.

Nevo, E., Bar-El, C. & Bar, Z. 1983: Genetic diversity, climatic selection and speciation of Sphincterochila landsnails in Israel.- Biol. J. Linn. Soc. 19: 339-373.

Nevo, E. 1983: Population genetics and ecology: The interface.- In: Bendall, D.S. (ed.): Evolution from Molecules to Men, Cambridge (Cambridge Univ. Press), pp. 287-322.

Pilsbry, H. & Bequaert, J. 1927: The aquatic mollusks of the Belgian Congo, with a geographical and ecological account of Congo malacology.- Bull. Amer. Mus. nat. Hist. 53: 69-602.

Plate, L. 1919/20: Bemerkungen über die deszendenztheoretische Bewertung der Umwandlungen von Planorbis multiformis.- Jenaische Z. Naturwiss. 56 (N.V. 49): 218-224.

Prothero, D.R. & Lazarus, D.B. 1980: Planktonic microfossils and the recognition of ancestors.- Syst. Zool. 29: 119-129.

Raven, C.P. 1966: Morphogenesis: The Analysis of Molluscan Development.- 2nd ed., London (Pergamon).

Reif, W.-E. 1975: Die Erforschung des Steinheimer Beckens. Ein Beitrag zur Geschichte der Erdwissenschaften in Süddeutschland.- In: Ackermann, M. (ed.): 75 Jahre Heimat- und Altertumsverein Heidenheim 1901-1976, pp. 66-85, Heidenheim.

Reif, W.-E. 1983a: Hilgendorf's (1863) dissertation on the Steinheim planorbids (Gastropoda; Miocene): The development of a phylogenetic research program for Paleontology.- Paläont. Z. 57: 7-20.

Reif, W.-E. 1983b: The Steinheim snails (Miocene; Schwäbische Alb) from a Neo-Darwinian point of view: A discussion.- Paläont. Z. 57: 21-26.

Reif, W.-E. (in press a): Art-Abgrenzung und das Konzept der evolutionären Art in der Paläontologie.- Z. zool. Systematik u. Evolutionsforsch.

Reif, W.-E. (in press b): The search for a macroevolutionary theory in German biology.- J. Hist. Biol.

Reif, W.-E. (in preparation): Why are there so many snails in Hawaii?

Reiff, W. 1974: Einschlagkrater kosmischer Körper auf der Schwäbischen und Fränkischen Alb.- Aufschluß 25: 368-380, Heidelberg.

Roughgarden, J. 1979: Theory of Population Genetics and Evolutionary Ecology: An Introduction.- 634 pp., London (Collier Macmillan).

Russel-Hunter, W.D. 1978: Ecology of freshwater pulmonates.- In: Fretter, V. & Peake, J.F.: Pulmonates, vol. 2A, pp. 335-384, (Academic Press) London.

Schindel, D.E. 1982: Resolution analysis: A new approach to the gaps in the fossil record.- Paleobiology 8: 340-353.

Seemann, R. 1941: Geologische und paläofaunistische Untersuchungen am Goldberg im Ries.- Festschr. württ. Naturaliensamml.: 49-62, Stuttgart.

Schwabe, G. & Murray, J. 1970: Electrophoresis of proteins in natural populations of Partula (Gastropoda).- Evolution 24: 424-430.

Stanley, S.M. 1979: Macroevolution. Pattern and Process.- 332 pp., San Francisco (W.H. Freeman).

Storzer, D. & Gentner, W. 1970: Spaltspuren-Alter von Riesgläsern, Moldaviden und Bentoniten.- Jahresber. Mitt. Oberrh. Geol. Ver. N.F. 52: 97-111.

Sudhaus, W. 1978: Der "Kryptospezies"-Begriff zur Kennzeichnung genetisch isolierter Populationen einer Morpho- und Ökospezies am Beispiel von Rhabditis spiculigera (Nematoda).- Z. zool. Syst. Evolut.-forsch. 16: 102-107.

Tilling, S.M. 1983: An experimental investigation of the behaviour and mortality of arteficial and natural morphs of Cepaea nemoralis (L.).- Biol. J. Linn. Soc. 19; 35-50.

Van Valen, L.M. 1982: Integration of species: Stasis and biogeography.- Evol. Theory 6: 99-112.

Waddington, C.H. 1975: The Evolution of an Evolutionist.- 328 pp., Ithaca, New York (Cornell University Press); pp. 92-96: "Genetics assimilation of Limnaea".

Weldon, W.F.R. 1901: A first study of natural selection in Clausilia laminata (Montagu).- Biometrica 1: 109-124.

Wenz, W. 1922: Die Entwicklungsgeschichte der Steinheimer Planorben und ihre Bedeutung für die Deszendenzlehre.- Berichte senckenberg. naturf. Ges. Frankfurt a.M. 52: 135-158.

Wenz, W. 1923: Gastropoda extramarina tertiara II: Pulmonata.- Foss. Catalog. Animalia 22, Berlin (W. Junk).

Wesenberg-Lund, C. 1939: Biologie der Süßwassertiere. Wirbellose Tiere.- 817 pp., Wien (Julius Springer).

Wetherby, A.G. 1882: On the geographical distribution of certain freshwater mollusks of North America and the probable causes of their variation.- Amer. J. Sci. 23: 203-212.

White, M.J.D. 1978: Modes of Speciation.- 445 pp., San Francisco (W.H. Freeman).

Wigand, A. 1874-1877: Der Darwinismus und die Naturforschung Newtons und Cuviers.- vols. 1-3, Braunschweig (references to Steinheim: I: 288 f., 427 to 434; III: 132, 134.)

Williamson, P.G. 1981a: Paleontological documentation of speciation in Cenozoic molluscs from Turkana basin.- Nature 293: 437-443.

Williamson, P.G. 1981b: Morphological stasis and developmental constraint: Real problems for neo-Darwinism.- Nature 294: 214-215.

Williamson, P.G. 1982: Williamson replies.- Nature 296: 611-612.

Williamson, P.G. 1983: Speciation in molluscs from Turkana Basin: Reply.- Nature 304: 661-663.

Willmann, R. 1981: Evolution, Systematik und stratigraphische Bedeutung der neogenen Süßwassergastropoden von Rhodos und Kos/Ägäis.- Palaeontographica Abt. A. 174: 10-235.

Wolff, M. & Füchtbauer, H. 1976: Die karbonatische Randfazies der tertiären Süßwasserseen des Nördlinger Ries und des Steinheimer Beckens.- Geol. Jahrb., Reihe D 14: 1-53.

Ziegler, B. 1972: Allgemeine Paläontologie. 245 pp., Stuttgart (Schweizerbart).

RESPONSES OF THE PLIO-PLEISTOCENE FRESHWATER GASTROPODS OF KOS (GREECE, AEGEAN SEA) TO ENVIRONMENTAL CHANGES

Rainer Willmann

Kiel

Abstract: In the Plio-Pleistocene freshwater gastropods of Kos, three different kinds of faunal responses to the changing environment can be referred to:
1) Varying species numbers as responses of the fauna as a whole,
2) evolutionary changes in shell morphology, and
3) non-hereditary modifications in shell colour as a reaction to varying salinity.

Evolutionary changes in shell sculpture must be explained as an expression of adaption to certain environmental factors, which, however, are still unknown. Nevertheless, some extrinsic forces important for gastropod evolution can be determined. Separating mechanisms within the basin caused splitting of populations, and the populations separated from each other had different evolutionary trends (microgeographical differentiation, e.g. *Mikrogoniochilus minutus*). Micro-allopatry can also be observed in *Rhodopyrgula rhodiensis* from the Pliocene of Rhodes. Some more wide spread populations were split by the separation of the eastern Kos lake from inland waters in central Kos (*Melanopsis gorceixi*, *Theodoxus doricus*), and in the latter species they became reconnected, when there was subsequent contact between these waters. A similar development seems to have occurred in the Rhodian *Viviparus rhodensis*.

On the other hand, lack of separating mechanisms within the basin led to gradual evolution along non-splitting lineages (e.g. *Prososthenia sturanyi* and large portions of the *Viviparus* and *Melanopsis* lineages). Evolutionary rates increased when the populations became relatively small (bottleneck effect). In the case of *Melanopsis*, reduction of population size was caused by marine transgressions, in the case of *Viviparus brevis* by the shrinkage of lake size.

INTRODUCTION

Neogene freshwater gastropods are among the classic examples of evolutionary lineages. Lineages from a small lake on Kos island in Greece, from the Steinheim basin in Germany and from Slavonia have become especially famous (TOURNOUER, 1876;

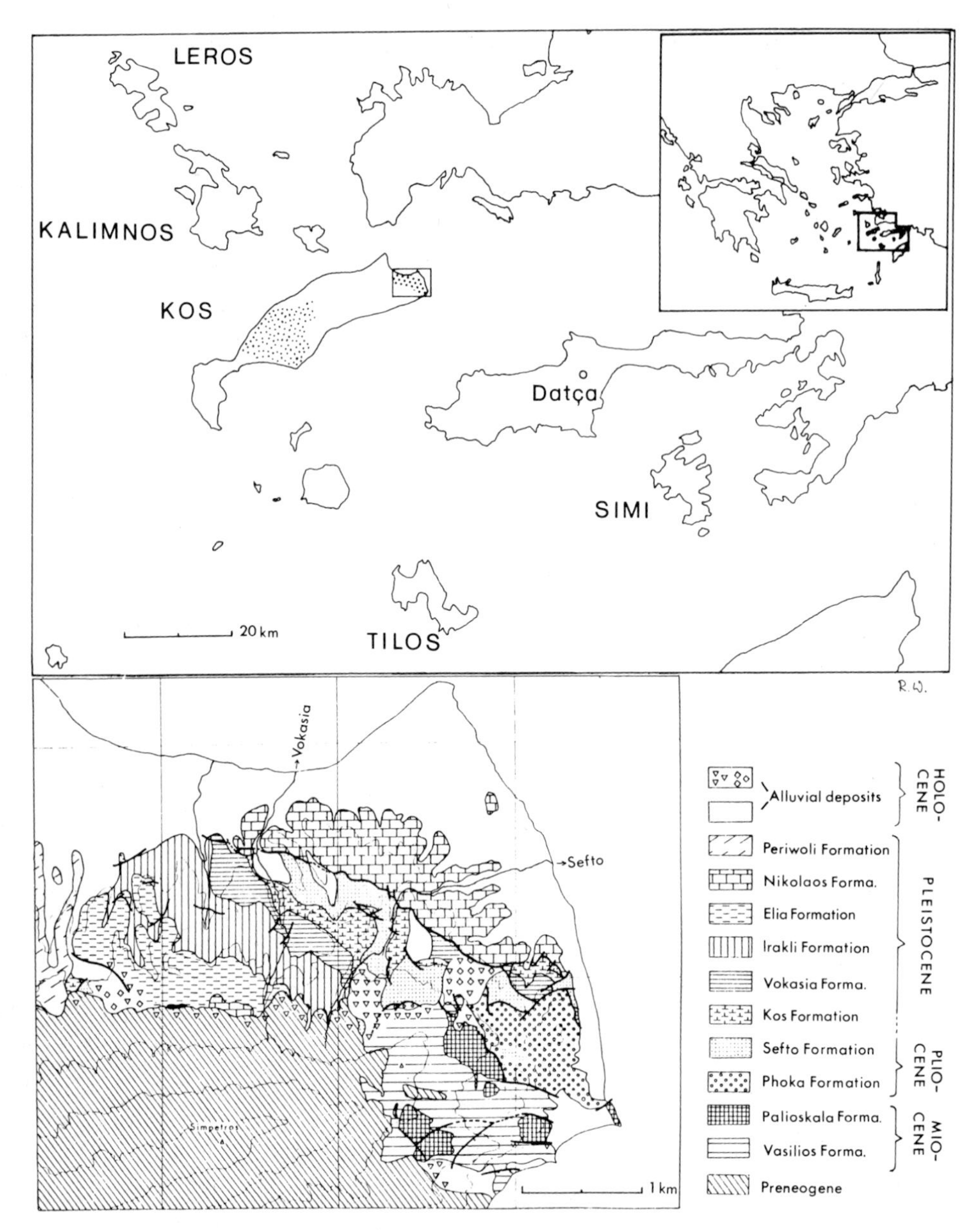

Fig. 1: Maps showing location of Kos and the lacustrine Plio-Pleistocene sedimentary basins (stippled). b. Geological map of eastern Kos. Modified after BÖGER et al. (1974).

NEUMAYR, 188O, 1889; HILGENDORF, 1866, 1867, 1879 etc.; NEUMAYR & PAUL, 1875) from studies in the 19th century. Many more endemic Neogene gastropod faunas -- showing gradual transformations -- were described from southern Europe later on (e.g. BUKOWSKI, 1893, 1895; GORJANOVIC-KRAMBERGER, 1923; OLUJIC, 1936; JEKELIUS,

1932; GIROTTI, 1966). Of these examples, the faunal history of snails from the lake in the eastern part of Kos (Fig. 1) seems to be best documented.

Lake basins function as very effective sedimentary traps. This favors a complete geological record, and especially in lacustrine sediments may biological responses to ecological changes be preserved in detail. Some other factors make lacustrine fossils useful tools for evolutionary palaeontology and palaeoecology:

(1) Lacustrine species evolve in a restricted area. This simplifies procedures in reconstructing their development. The problem of whether we are dealing with evolution or with replacement of geographic subspecies in time is not as important as in marine sediments.

(2) In small lakes especially there is very little redeposition of sediments.

(3) Population size is often only small. This favors rapid evolutionary transformations, and for that reason morphological changes may be visible in relatively short sedimentary sequences. In the Plio-Pleistocene of Kos all of these suppositions are largely fulfilled.

1. NEOGENE HISTORY OF KOS

Kos was part of the Aegean landmass from the Upper Miocene to the Lowermost Pleistocene, which once linked Anatolia and the Greek mainland. In the beginning of the Upper Pliocene fluviatile and terrestrial sediments were deposited - the Gurniati Formation of central Kos and the contemporaneous Phoka Formation of east Kos (Fig. 2). Repeatedly small, stagnant, and ephemeral waters were formed.

The biological history of these waters, was completely integrated into that of the remaining landmass. Endemic species did not yet occur - apart from one exception: *Melanopsis inexspectata* from parts of the Gurniati Formation. Other gastropods were widespread: *Melanopsis sporadum*, *M. delessei*, *Melanoides tuberculata*, *Theodoxus hellenicus*, *Valvata hellenica*, *Lymnaea megarensis*, *Gyraulus aegaeus*, *Planorbis planorbis*, *P. carinatus* etc. (Fig. 3).

As many of these species have been recorded from other localities under different species names, their exact geographic and stratigraphic range is still difficult to estimate. On the other hand, many species have been recorded from other places erroneously (see below, section 3). Revision of Neogene freshwater gastropods from the Balkans and Anatolia is urgently needed.

During the Pliocene the inland waters of Kos gained a more marginal position on the landmass. The Aegean landmass did not exist any longer in its Miocene extension, due to partial subsidence and transgression of the sea from the south. During the Upper Pliocene the sea reached the environs of Kos, and Kos became situated at the edge of

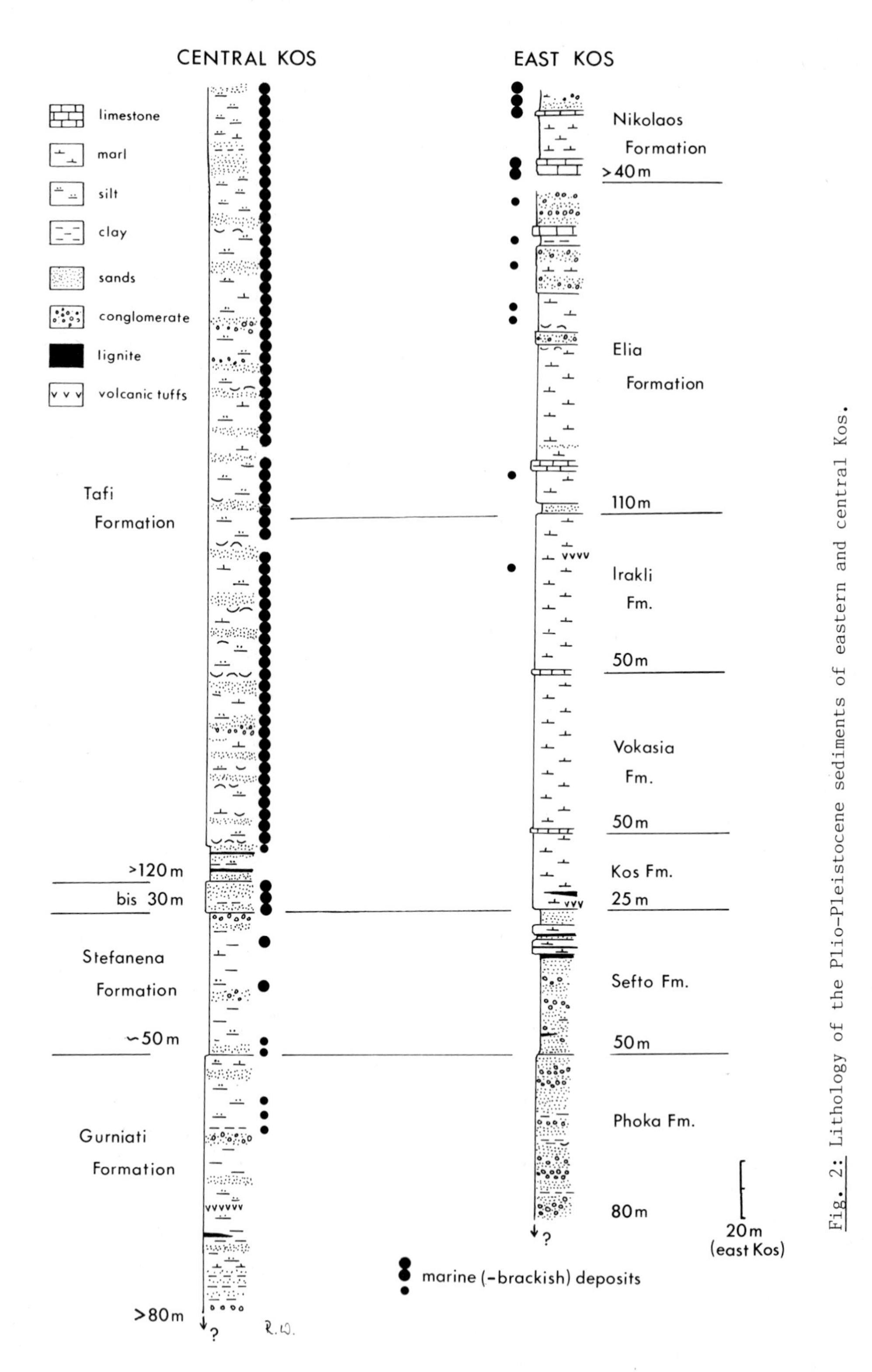

Fig. 2: Lithology of the Plio-Pleistocene sediments of eastern and central Kos.

Anatolia. Furthermore, the limnic systems of Kos seem to have been cut off to a greater extent from those of the Anatolian hinterland when uplift of several mountain chains began - e.g. the Dikeos mountains, which today form the "back-bone" of Kos island.

During this time, the fluviatile-terrestrial Stefanena Formation was deposited in the central region of Kos. It, however, contains several marine intercalations. Simultaneously, in eastern Kos the Sefto Formation was formed (Fig. 2). It does not show any marine influence. In the Upper Sefto Formation lignites and diatomitic calcareous marls evidence the existence of temporary pools. Later, one of these pools becomes transformed into the lake of east Kos. Its size increased rapidly up to 3 km in diameter. Calcareous marls, lignites, and four layers of volcanic tuffites were deposited (Lower Kos Formation). During the Middle Kos Formation the diameter of the lake decreased to about 600 m, not taking into account the marginal zone of vegetation, which is preserved as lignites (Fig. 9B, C). Afterwards, the size of the lake again increased, and over 200 m of calcareous marls, clay, coarser clastics and volcanics accumulated (Upper Kos, Vokasia, Irakli, and Elia Formations) (Fig. 2). Deposition of such a thick sedimentary column was apparently made possible by the continuous subsidence of the area of east Kos.

15 km further west, marine sediments in the contemporaneous Tafi Formation predominate. Repeated marine influences can also be traced in the lake of eastern Kos - by means of the diatom flora and several layers containing *Cardium edule*, *Rotalia beccarii* and others (Fig. 11). Marine intercalations become more numerous in the younger sediments of eastern Kos (Elia Formation). The freshwater fauna, however, surrvived many marine ingressions. It was not completely destroyed until the sea finally invaded east Kos (Nikolaos Formation). After deposition of the Nikolaos Formation, the Plio-Pleistocene sediments of Kos were displaced and elevated above sea-level. At the same time, Kos became separated from Anatolia and was transformed into an island. Erosion of the Dikeos-Simpetros mountain chain led to deposition of Pleistocene slope debris (Periwoli Formation; named herein Periwoli, German transcription of Greek το Περιβολι: garden; named for the gardens surrounding the town of Kos; "ältere Alluvionen" of BÖGER et al., 1974). The Periwoli Formation is restricted to the area south of Kos town, where it forms wide fans. These include the eastern extensions of a Pleistocene volcanic tuff which covers central Kos with a layer of up to 30 m thickness ("Plateau Tuff"). For details concerning the Neogene stratigraphy and history of Kos see BESENECKER & OTTE, 1978, 1979; BÖGER, 1978, WILLMANN 1983.

2. SPECIES NUMBERS IN RELATION TO THE ENVIRONMENT

The Plio-Pleistocene freshwater fauna of Kos has become especially famous due to the evolutionary changes visible in the gastropod fauna (see section 3). Evolution, as a result of adaptation to the environment, appears not to be the only response; there appear also non-hereditary (phenotypic) modifications in reaction to environmental stimuli (section 4), and also changes in species numbers.

Interaction between species numbers and environmental changes can be detected (Fig. 4). In the predominantly fluviatile Phoka Formation, the oldest sequence of Pliocene sediments in eastern Kos, about 8 gastropod species have been found. These are *Theodoxus hellenicus*, *Melanopsis sporadum*, *M. delessei* , *Prososthenia sturanyi* , *Iraklimelania levis*, *Bithynia* cf. *leachi*, *Valvata hellenica*, and *Planorbis* (?) sp. (WILLMANN, 1981, 1983) (many of these species are shown in Fig. 3). Species number increased when

---

Fig. 3: Freshwater gastropods from the Plio-Pleistocene of Kos. d = diameter, h = height

a *Viviparus calverti* (NEUMAYR), Stefanena Formation, h 26 mm

b *V. brevis brevis* TOURNOUER, Lower Kos Formation, h 22 mm

c *V. brevis trochlearis* TOURNOUER, Elia Formation, h 32 mm

d *Theodoxus hellenicus* (BUKOWSKI), Phoka Formation, h 7 mm

e *Th. doricus cous* (NEUMAYR), Lower Kos Formation, h 8 mm

f *Melanopsis sporadum* TOURNOUER, Gurniati Formation, h 30 mm

g *M. gorceixi gorceixi* TOURNOUER, Sefto Formation, h 24 mm

h *M. gorceixi heldreichi* NEUMAYR, Elia Formation, h 22 mm

i *M. delessei* TOURNOUER, Stefanena Formation, h 20 mm

k *Melanoides tuberculata* ssp.?, Upper Kos Formation, h 12,5 mm

l *M. tuberculata dadiana* (OPPENHEIM), Elia Formation, h 13 mm

m *Valvata hellenica* TOURNOUER, Tafi Formation, h 2,8 mm

n *Bithynia* cf. *leachi* (SHEPPARD), Lower Kos Formation, h 4,6 mm

o *Xestopyrguloides neumayri* WILLMANN, Irakli Formation, h 3,1 mm

p *Marticia cosensis* (MAGROGRASSI), Irakli Formation, h 10 mm

q *M. brusinai* TOURNOUER (ssp. formosa WILLMANN), Tafi Formation, h 9,9 mm

r *Ferrissia* cf. *illyrica* (NEUMAYR), Sefto Formation, length 1,6 mm

s *Prososthenia sturanyi communis* WILLMANN, Lower Kos Formation, h 3,6 mm

t *P. st. coa* WILLMANN, Elia Formation, h 3,1 mm

u *Lymnaea megarensis* GAUDRY & FISCHER, Gurniati Formation, h 11,5 mm

v *Acroloxus lacustris* (L.), Tafi Formation, length 3,2 mm

w *Gyraulus aegaeus* WILLMANN, Gurniati Formation, d 3,3 mm

x *Planorbis planorbis* (L.), Tafi Formation, d 4 mm

y *Planorbis carinatus* MULLER, Gurniati Formation, d 6 mm

a
b
c
d
e
f
g
h
i
k
l
m
n
o
p
q
r
s
t
u
v
w
x
y

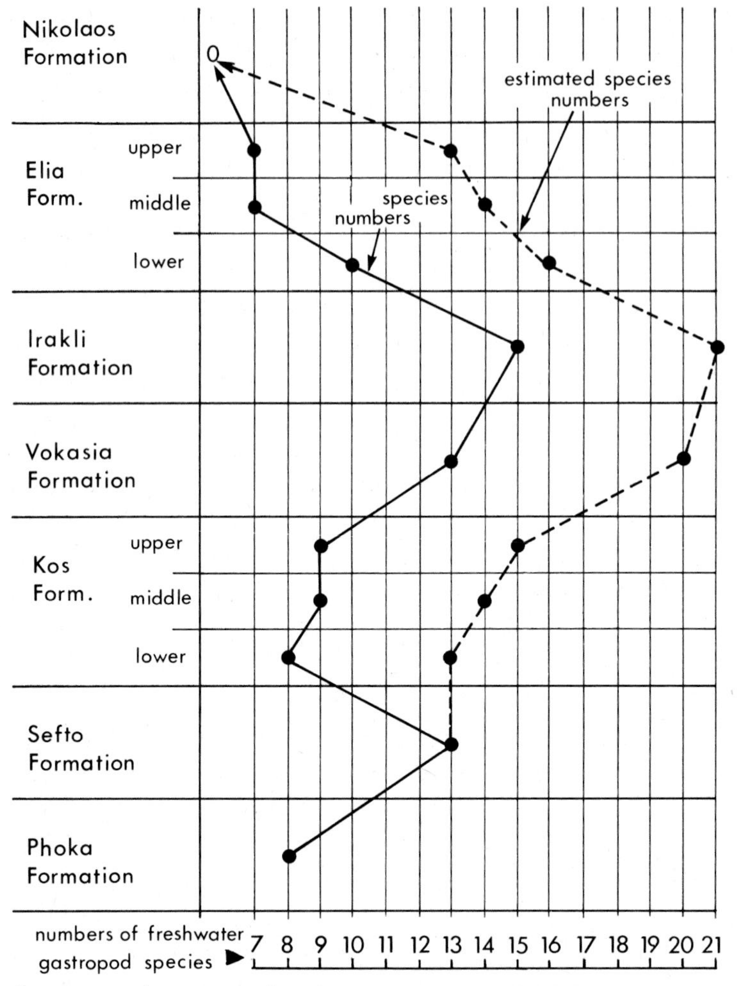

Fig. 4: Species numbers as related to environmental changes. In the predominantly fluviatile Phoka Formation 8 species of freshwater gastropods have been found. Species number increased up to 13 in the Sefto Formation, and even more species inhabited the lake of eastern Kos (Kos up to the Elia Formation). In the younger deposits, species numbers are reduced because of repeated marine ingressions. Finally, fully marine sediments of the Nikolaos Formation accumulated, and the Koan freshwater fauna died out.

ponds became more numerous (Sefto Formation). At that time *Planorbarius corneus*, *Acroloxus lacustris*, *Ferrissia* cf. *illyrica*, *Gyraulus* cf. *aegaeus*, *Lymnaea* sp., and *Pseudamnicola dodecanesiaca* appear in east Kos. Many of these species doubtlessly lived during the accumulation of the younger formations in eastern Kos also, although no sediments of their habitat have been preserved: This is due to the fact that they live preferably in the vegetational rims of the waters - and these are not documented except for the Middle Kos Formation.

Deposition of the lowermost Kos Formation marks the beginning of the existence of the lake of eastern Kos. At that time, *Viviparus brevis* appears in the basin, immediately becoming one of the dominant elements in the gastropod fauna. In the Upper Kos Formation, the lake size increased considerably, and another species arrived: *Melanoides tuberculata*. (A new bivalve - *Dreissena* - arrived shortly afterwards, being unknown from the older sediments of Kos). During deposition of the Vokasia and the Irakli Formations the lake was of considerable size and depth. The Vokasia Formation contains five new species: *Mikrogoniochilus minutus*, *Marticia cosensis*, *M. brusinai*, *Xestopyrguloides neumayri*, and *Valvata heidemariae*, giving a total of 13 species. This number increased in the Irakli Formation, as 15 species are recorded from its sediments. The actual species numbers must have been about 19 and 21, respectively (Fig. 4).

At this time the first massive ingression of the sea took place. A marine - brackish fauna invaded the basin. *Cardium edule* was especially abundant. The marine intercalation does not contain freshwater gastropods, except for reworked specimens. The freshwater fauna, however, was not destroyed. In the eastern portion of the basin, a marine ingression is not indicated by fossils apart from changes in the diatom flora (GERSONDE in prep.). In this region freshwater species survived. They then reappeared in the western part of the basin above the marine intercalation.

Similar conditions are visible in the numerous marine ingressions in the Elia Formation. Now, however, former species numbers were not maintained any longer in the limnic sequences. The freshwater gastropods species disappeared successively, and the endemics became extinct (see chapter 5 for details).

3. EVOLUTIONARY TRANSFORMATIONS IN RELATION TO ENVIRONMENTAL CHANGES

In the former lake of east Kos, several gastropod lineages are preserved without any important gaps, and sometimes evolutionary changes can be traced in the field by means of hundreds of thousands individuals. Among them are the famous evolutionary lineages of *Viviparus ("Paludina") brevis*, *Melanopsis gorceixi* and *Theodoxus ("Neritina") doricus*, the first true fossil lineages ever mentioned: They were described by FORBES & SPRATT as early as 1846 and 1847 and later by Melchior Neumayr in his celebrated monograph of 188O (see NEWTON, 1911; WILLMANN, 1978). Further good examples of lineages are those of *Prososthenia sturanyi* and *Mikrogoniochilus minutus*. They were either endemic to the lake of Kos or gave rise to endemic subspecies. In addition, the evolution of two further gastropods will be mentioned, namely *Rhodopyrgula rhodiensis* (BUKOWSKI) and *Viviparus rhodensis* BUKOWSKI, both from the Pliocene of Rhodes.

21 gastropod species were found in the lakes' sediments. About one third of them were endemic (WILLMANN, 1983: 852). Some species had several microgeographically separated subspecies and numerous chrono-subspecies. (Some more freshwater species are known from sediments which were deposited in the surroundings of the lake). A number of environmental changes influenced the evolution of these gastropods. The most important and initial event was the subsidence of parts of eastern Kos at the end of the Pliocene. This led to the formation of a sedimentary basin wherein pools and, later on, the lake of eastern Kos originated. The development of a gastropod fauna with a high degree of endemicity was the result of the separation of the inhabitants of this basin from surrounding populations.

After the formation of this basin further events influenced the development of the snails: The rise and fall of water level changed both the morphology of the lake as well as transforming the lake into a marginal marine basin. The deposition of volcanic ashes and repeated increases in salinity do not seem to have affected gastropod evolution. Last, the invasion of the sea finally destroyed the freshwater fauna.

3.1 Phylogenetic Patterns

Several different modes of phylogenetic differentation, connected with environmental changes, can be observed. These are transformation of species along single lineages without branching, branching of evolutionary lineages, and the fusion of such branches (WILLMANN, 1981). The splitting of one population into two and their subsequent fusion can only be explained by the establishment and disappearance of external separating mechanisms. In the Irakli and Elia Formations, a complex lake morphology is documented by marine intercalations which appear as lenses in the limnic sequence: *Cardium* layers cannot be traced all over the basin, and the diatom flora indicates local differences of salinity during one and the same time within the east Kos basin (GERSONDE, in prep.). This factor - varying lake morphology - may have favoured intraspecific differentiation within the basin. Another possible factor is zones of vegetation, which either formed island-like habitats within the lake or which separated populations of species not inhabitating these zones. In no case, however, can the nature of the extrinsic forces favouring i n t r a l a c u s t r i n e splitting of evolutionary lineages be established with any degree of certainty. On the other hand, it is certain that separation of the lake from neighbouring inland-waters initiated splitting of previously wide-spread populations. Subsequent contact between these waters facilitated the fusion of formerly separated lineages.

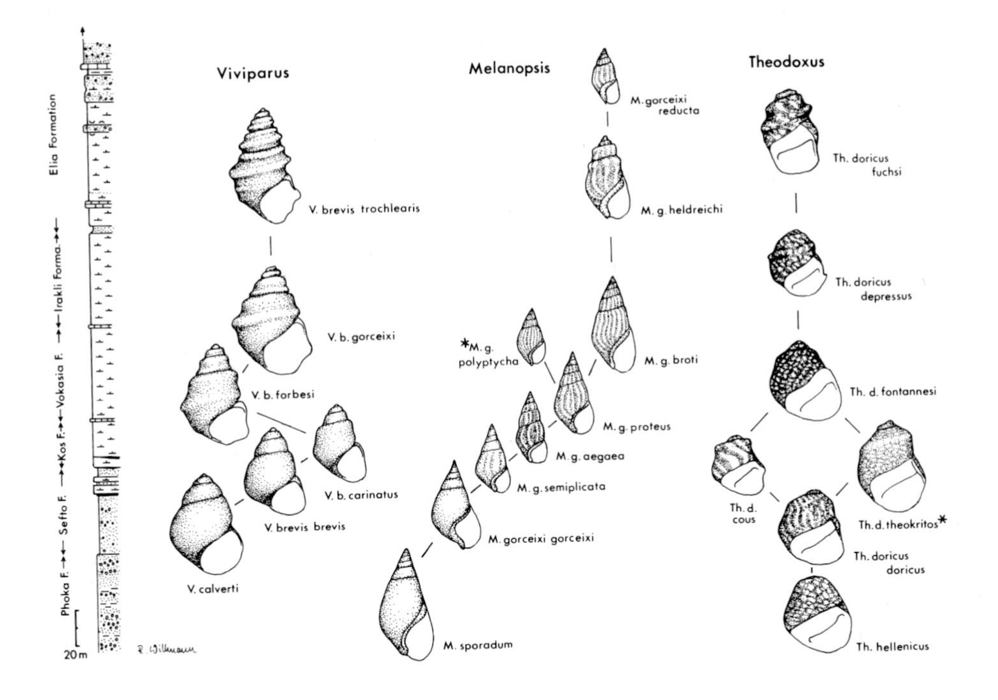

Fig. 5: The evolutionary lineages of *Viviparus brevis*, *Melanopsis gorceixi*, and *Theodoxus doricus*. The forms marked by an asterisk are restricted to the basin in central Kos. For exact stratigraphic ranges see text.

3.1.1 Gradual transformation and splitting of lineages

Viviparus brevis (see also section 3.2):

NEUMAYR (1889) considered the lineage of *Viviparus* from Kos to be the best-documented fossil example of gradual morphological transformation of them all. The ancestral species is *Viviparus calverti*, from the Stefanena Formation of central Kos (Figs. 3a, 5). *V. calverti* gave rise to *V. brevis brevis*, a form with flattened flanks to the whorls (Figs. 3b, 5). This is the oldest evolutionary step of the *Viviparus* lineage, which is endemic to the basin in eastern Kos. Its origination is clearly related to the formation of the eastern Kos lake.

Within a short period of time, *V. brevis brevis* TOURNOUER changed gradually into *V. b. carinatus* TOURNOUER (Middle Kos Formation) and *V. b. carinatus* into *V. b. forbesi* TOURNOUER (Upper Kos and Lower Vokasia Formations) (Figs. 5, 10). In *V. b. forbesi*, there are two well developed spiral keels to be found. In the next chrono-subspecies, *V. b. gorceixi* TOURNOUER (Fig. 5), the second carina is very strong and slightly shifted towards the apex. Finally, in *V. b. trochlearis* TOURNOUER (Irakli, Elia and Nikolaos Formations) there are three prominent keels.

The lineage of *Viviparus brevis* underwent intralacustrine splitting within the lake basin. In the Upper Kos Formation *V. b. forbesi* consisted of two microgeographically separated populations, characterised by slightly different shell forms: one form shows two keels equal in size whereas in most specimens of the other the basal keels are larger than the upper ones. This subdivision vanished in upper parts of the Upper Kos Formation (WILLMANN, 1981).

Melanopsis gorceixi

Melanopsis gorceixi (Figs. 3, 5) presents the second-best documented lineage in the Aegean Neogene. This species originated from the slender, smooth-shelled *M. sporadum* (Fig. 3f), which is abundant in the Phoka and Gurniati Formations. In the first recognisable evolutionary step the last whorl became inflated (*M. gorceixi gorceixi* TOURN., see section 3.2). In the immediately following *M. g. semiplicata* NEUMAYR (Middle Kos Formation) this whorl shows axial ribs. While most of the other subspecies of this lineage are widely distributed on Kos, *M. g. semiplicata* is restricted to the basin of eastern Kos. In the Upper Kos Formation the earlier whorls became costate and the number of the ribs increased to about 10 per whorl (*M. g. aegaea* TOURN.). Possibly, origination of the costate forms must be seen in connection with a drastic reduction of the population size during the Middle Kos Formation (bottlenecking; section 3.2). In the uppermost Kos and the Lower Vokasia Formations (eastern Kos) and in the Tafi Formation (central Kos) there are about 15 (*M. g. proteus* TOURN.) and later on up to 23 costae to be found (*M. g. broti* NEUMAYR; Vokasia, Irakli and Tafi Formations). At the same time the flanks of the whorls became flattened.

Later in the lineage the number of ribs is again reduced. In *M. g. heldreichi* NEUM. (Elia, Nikolaos and Upper Tafi Formations; Neogene of Datca) there are at times only eight costae per whorl. A further morphological innovation in this subspecies is a node which is situated at the adapical end of each rib. In the youngest chrono-subspecies, *M. g. reducta* WILLMANN, the number of ribs is slightly increased again.

In the southern area of central Kos another subspecies occurs, namely *M. g. polyptycha* NEUMAYR, a descendant of *M. g. proteus*. Like *M. g. broti* this form has numerous axial costae but the sides of the whorls are not flattened. *M. g. polyptycha* and *M. g. broti* were two synchronous geographic subspecies. Hence it follows that the

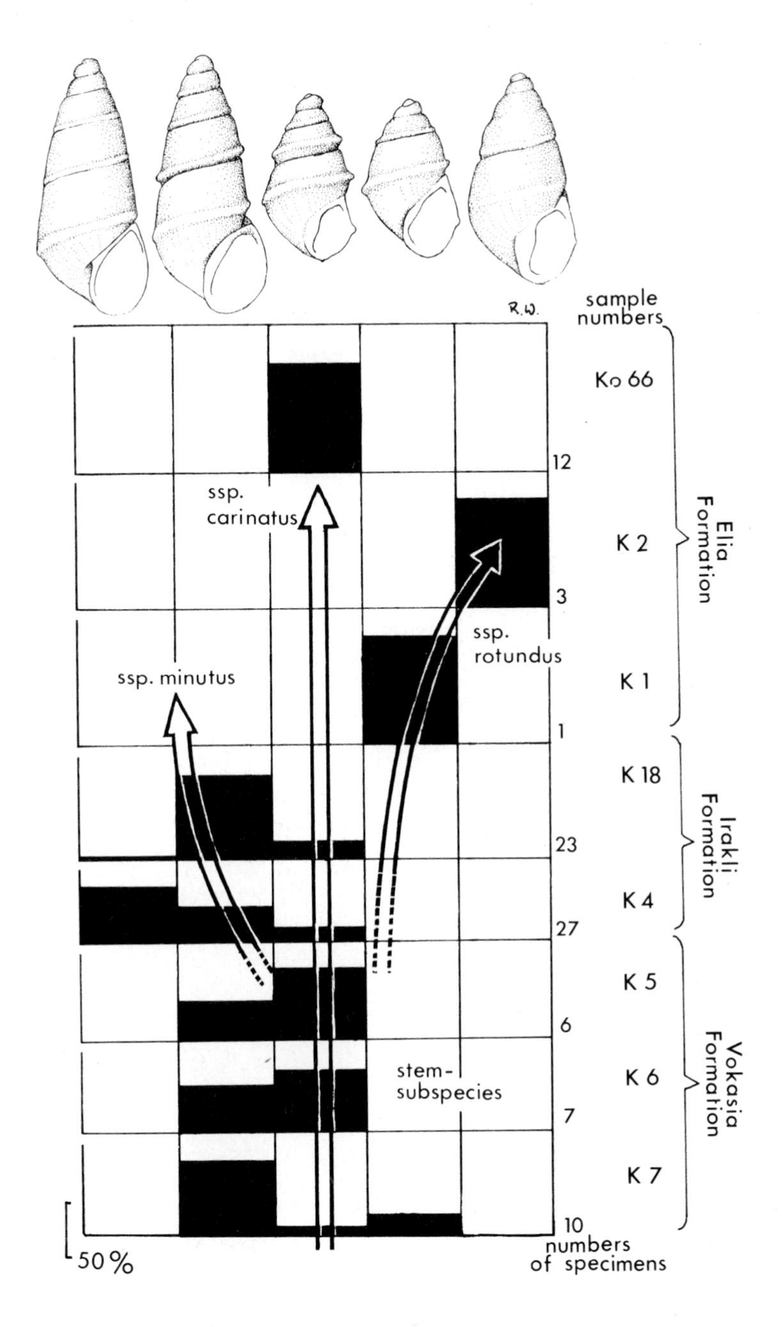

Fig. 6: Evolution of *Mikrogoniochilus minutus* in the Plio-Pleistocene of east Kos. The sequence of samples is interpreted in terms of intralacustrine splitting of a stem population (samples K 5 – K 7) into the three subspecies *minutus*, *carinatus*, and *rotundus*. Differentiation of these populations must have been initiated by external separating mechanisms, e. g. establishment of morphological barriers or insular habitats. Height of the specimen figured at the left margin 2.5 mm.

lineage of *M. gorceixi* has branched. This must have been initiated by the separation of the southern Kos inland waters from those of the remaining area.

Mikrogoniochilus minutus

In the small *Mikrogoniochilus minutus* (Fig. 6) as many as three coeval geographical subspecies, all within the former lake of eastern Kos, can be distinguished. They all have their origins in populations of the Vokasia and Lower Irakli Formations, which show a wide range of morphological variety (sample K 5 - K 7, see Fig. 6). Younger populations have either retained the ancestral morphology *(M. minutus carinatus)* or modified it in different directions. In *M. m. rotundus* the flanks of each whorl became slightly roundish. In *M. m. minutus* the suture was nearly closed by the keel and the flanks of each whorl were flattened.- Splitting must have been caused by fragmentation of the lake, which was followed by separation of the once united populations.

Prososthenia sturanyi

In the Plio-Pleistocene sediments of eastern Kos, this species shows gradual transformation of shell form along a single lineage. Originally, in *P. s. communis* (Fig. 3s), the whorls were flattened laterally. This form is widely distributed on Koso (Phoka, Sefto, Gurniati, Stefanena and Kos Formations). The populations in central Ko later died out, when the sea covered this area (Tafi Formation). A comparatively small population survived in the eastern Kos basin and developed into *P. st. coa* (Fig. 3t, Vokasia up to the Elia Formation). This chrono-subspecies is endemic to east Kos.

3.1.2 Branching and fusion of lineages

Theodoxus doricus

Theodoxus is very abundant in Greek Neogene strata. The roundish *Th. hellenicus* (BUKOWSKI) developed into *Th. doricus doricus* on Kos, which is characterised by a weak sulcus on the sides of each whorl (Fig. 5). *Th. d. doricus* (NEUMAYR) gave rise to two separate evolutionary lines: In east Kos, *Th. doricus cous* (NEUMAYR) (Sefto and Kos Formation) developed a prominent spiral keel. During that time, *Th. d. theokritos* WILLMANN in central Kos (Tafi Formation) retained the morphological features of their common ancestor. Thus, *Th. d. cous* and *Th. d. theokritos* were two geographical subspecies which occurred at a distance of only about 15 km. Shortly afterwards, fusion of the two subspecies led to weakening of the keel in eastern Kos populations.

In central Kos, where smooth forms had once lived, the hybridization caused carination. A wide range of variation in this character can be observed in the populations of this hybrid-form, namely the chrono-subspecies *Theodoxus doricus fontannesi* (NEUMAYR). Later, *Th. d. fontannesi* gave rise to *Th. doricus depressus* (MAGROGRASSI), which had a well developed keel. *Th. d. fuchsi* (NEUMAYR) from the Elia and Upper Tafi Formation, the latest representative of this lineage, had a second keel.

3.1.3 Evolution of some freshwater gastropods from the Pliocene of Rhodes

Viviparus rhodensis

A similar development as in the early *Viviparus brevis* can be observed in *V. rhodensis* BUKOWSKI from the Pliocene Apolakkia/Monolithos Formation. Originally *V. rhodensis* had no shell sculpture, whereas the younger *V. rh. sulcatus* MAGROGRASSI is characterised by a deep sulcus in the middle of the whorls. The uppermost sediments contain a form resembling the earliest representatives of this lineage. This is perhaps due to regressive evolutionary steps, but it is more likely that hybridization with contemporaneous populations, these having retained the primitive features, took place (WILLMANN, 1981: 18, 96). Although such populations have not been found, their existence is probable. There were many small inland waters in the Pliocene of Rhodes, and the evolution of their faunas can have easily had its own tendencies.

Rhodopyrgula rhodiensis

The lower part of a particular section at the Pliocene Salakos Formation near Kamiros (Rhodes) contains smooth or weakly carinated individuals of the small *Rhodopyrgula rhodiensis* BUKOWSKI (Fig. 7, sample R 47). Higher in the succession the pattern of variation changes. An increasing number of strongly sulcated forms occur (samples R 45 - R 21; morphological classes III - VI, Fig. 7). Samples R 46 to R 40 also contain the subspecies *Rh. rh. solida* Willmann, a form with extremely prominent carinae. The biological relationship between these two forms could not be interpreted with any degree of certainty. If *Rh. rh. solida* had arrived from somewhere else and lived sympatrically with *Rh. rh. rhodiensis* the two forms must be regarded as two biospecies. The author, however, is not inclined to this interpretation. Perhaps two microgeographically separated subspecies were brought together postmortally. These microgeographical subspecies may have occurred at a distance of only a few hundred meters or less in different parts of the same waters: in the case of a contemporaneous section of the Salakos Formation, only 100 m away from the section figured in Fig. 7, neither a shifting of the morphology of *Rh. rh. rhodiensis* could be traced nor did *Rh. rh. solida* occur (WILLMANN, 1981: 19, 32). This shows how small the habitats of popula-

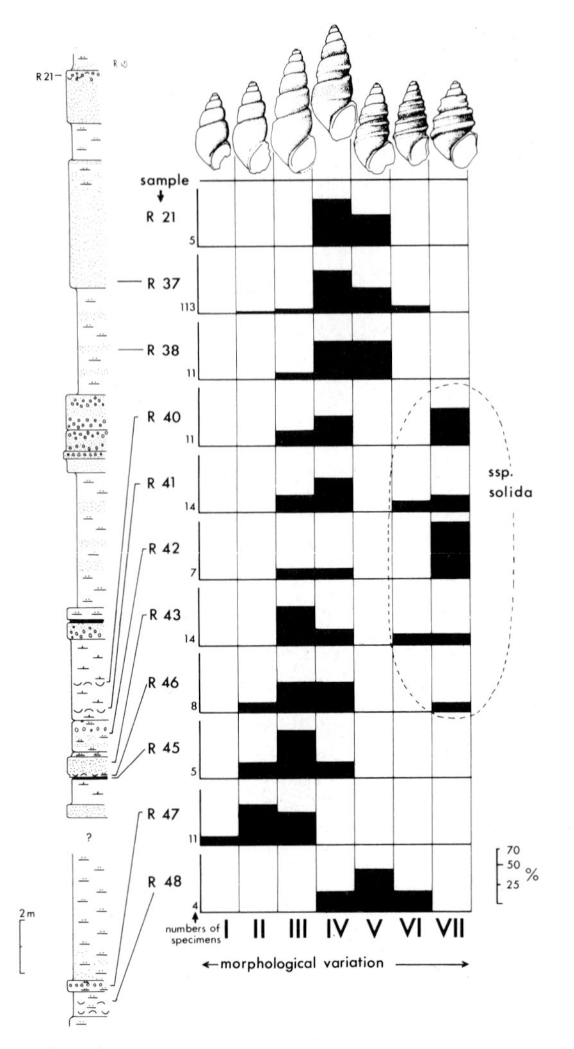

Figs. 7: Changes in the morphological variation of *Rhodopyrgula rhodiensis* in a particular section of the Salakos Formation (Pliocene, Rhodes).

tions of this species must have been. Interaction between these populations may easily have caused complex evolutionary patterns. It is also possible that *Rh. rh. rhodiensis* and *Rh. rh. solida* were only ecologically separated and lived sympatrically - though ecological separation also can be interpreted as a micro-allopatry.

It is uncertain whether an earlier population from the same section (sample R 48, Fig. 7) represents the common ancestral stock of *Rhodopyrgula rhodiensis rhodiensis* and *Rh. rh. solida* or not. This population covers the morphological classes IV to VI and is thereupon intermediate to the early *Rh. rh. rhodiensis* and *Rh. rh. solida*.

3.2 THE BOTTLENECK-EFFECT

One of the most interesting results of the investigation of freshwater gastropods from Kos is that evolutionary rates increased several times whenever populations became relatively small. For the first time the so-called bottleneck-effect (MAYR, 1949; STANLEY, 1979) can thereupon clearly be demonstrated in the fossil record (WILLMANN, 1978: 237, 1981, 1983). A direct evolutionary response to the changing environment thus becomes visible. This bottleneck-effect can be traced in both *Melanopsis* and Viviparus..

Population size was influenced by rise and fall of lakewater level in the first example (*Viviparus brevis*). Shrinkage of the population led to rapid morphological changes. Marine transgressions had the same effect, in that they restricted the formerly widespread populations of *Melanopsis sporadum* to a small area in east Kos.

Viviparus brevis

Viviparus brevis evolved especially rapidly in the Middle Kos Formation, when its area within the basin of eastern Kos was reduced by at least 80% (Figs. 8-9). During this period the transformation of *V. brevis brevis* into *V. brevis carinatus* and thence into *V. b. forbesi* took place (Fig. 5). No significant changes can be observed previously or subsequently over periods of time at least equivalent to the duration of this event. E. g., in the Lower Kos Formation, no morphological transformation is visible over a sequence of 6.5 m of calcareous marls. Then, in a sequence of 5 m at the most (center of the basin, Middle Kos Formation) *V. brevis* developed as mentioned above (WILLMANN 1981: 38-48). Condensation of these sediments in relation to those of the Lower Kos Formation does not seem to have taken place in the center of the lake or at least did not play a major role. (Sediments of the Middle Kos Formation are much more condensed near the margins of the basin; compare Fig. 8: thickness of the sediments between b and c in section 13 and in section 8, respectively).

BÜTTNER (1982) gives a good impression of the increase of the evolutionary rate in *Viviparus brevis* (see Fig. 10), although Büttner himself was not aware of the dramatic acceleration of evolutionary change during the Middle Kos Formation.

Reduction of population size was caused by a fall of the waterlevel. This led to movement of the peripheral vegetation towards the center of the lake basin (Figs. 8, 9). As *Viviparus* did not inhabit this vegetation, its population shrank.

Lake size in the various phases can be estimated. The Lower Kos Formation, which contains the oldest form of *Viviparus brevis (V. b. brevis)* is widespread in east Kos. It was deposited in an area of at least 1,800 x 400 m (720 000 m^2) in the area of the Sefto and Vokasia valleys (Figs. 1, 9A) and of at least 900 x 200 m (180 000

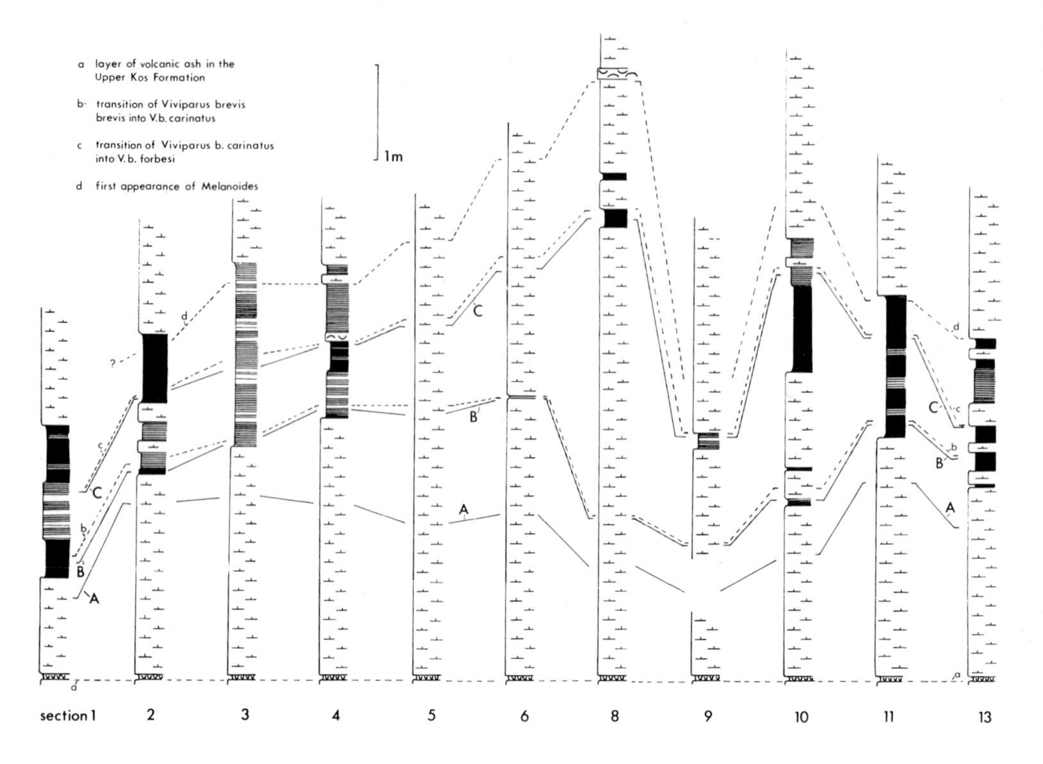

Fig. 8: Sections of the Middle Kos Formation (locations see Fig. 9A). Lignites and lignitic marls indicate former zones of vegetation surrounding the center of the eastern Kos basin. Repeatedly, this zone was extended into the lake, reducing the area of deposition of calcareous marls. Only section 5 in the center of the lake (cf. Fig. 9 A) does not include lignitic layers. a-d: Geological and biological events used for stratigraphic coordination; A-C refer to the stages in the development of the lake as illustrated in Fig. 9.

m^2) in the easternmost portion of the island, making a total of 900 000 m^2. In the Middle Kos Formation, however, the lake was only about 600 m long and between 100 and 200 m wide (about 90 000 m^2, Fig. 9c), so that lake size was reduced by 90%. As abundance of *Viviparus* does not seem to have increased per volume of sediment, a comparable reduction of the population size must have taken place.

In undisturbed sequences the number of *Viviparus* is about 180 individuals per m^2, a sedimentary column of 1.5 cm taken into account (1.5 cm equals the width of *Viviparus* shells). If we assume that 1.5 cm of marls was deposited in about 100 years and take into consideration a duration of life of 4 years then $\frac{180 \times 4}{100}$ (about 7) individuals had lived on each square meter at any time. Hence the population size was reduced from about 900 000 x 7 (6 300 000) to about 90 000 x 7 (630 000) individuals. These numbers may be higher as 1.5 cm of sediment may have accumulated in less than 100 years. On the other hand, in some parts of the lake density of *Viviparus* shells may have been lower than 180 specimens per m^2.

FRÖMMING (1956: 255) reported a maximum age of 2 1/2 years for reared individuals of the European *Viviparus viviparus*. The males of the North American *Viviparus malleatus* usually die after their third or right after their fourth year of existence, while females reach an age of up to five years (STANCYKOWSKA et al. 1971:1433).

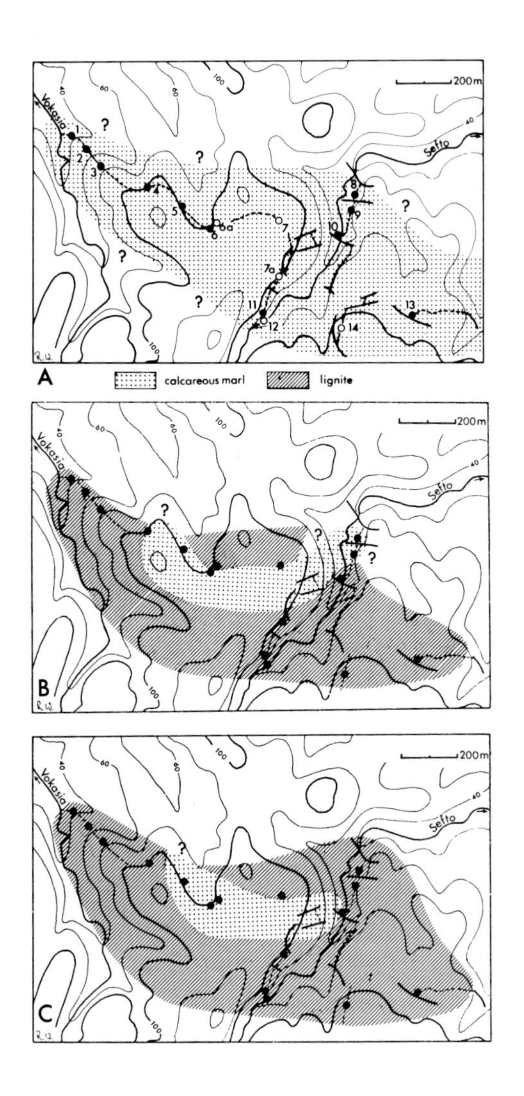

Fig. 9: Three stages of the palaeogeographic development of the eastern Kos lake during the Middle Kos Formation. For location, compare Fig. 1b. A - C refer to the time sections as indicated in Fig. 8. Area of marls corresponds to the distribution area of *Viviparus brevis*. During phase C, this area had its minimal extension.

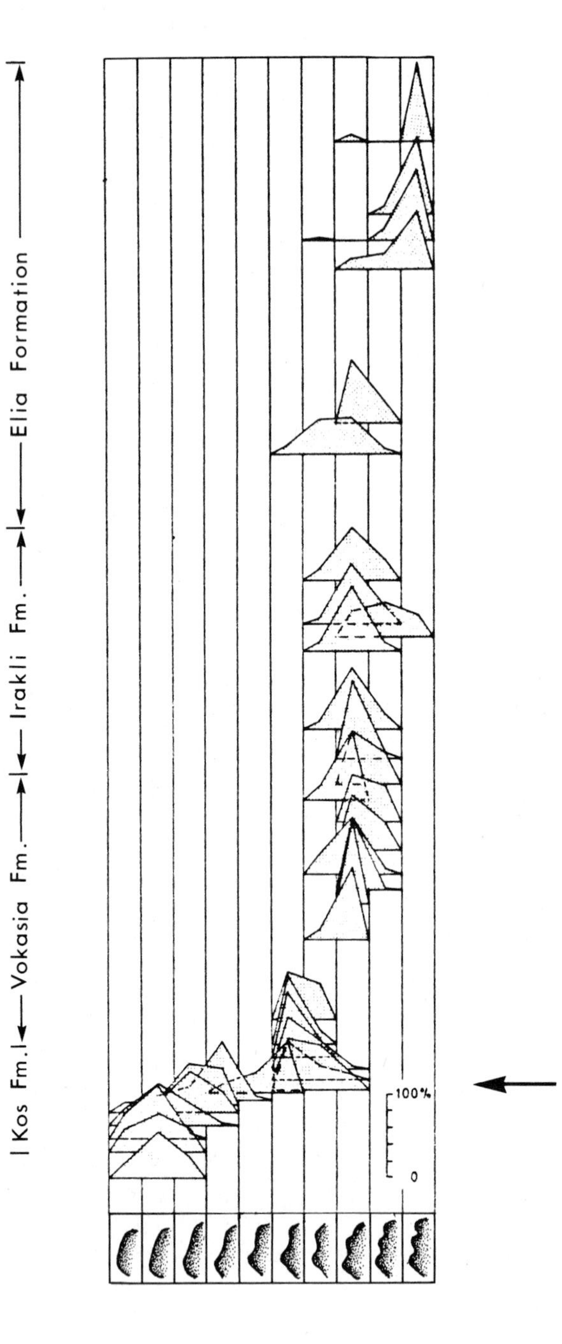

Fig. 10: Changes in shell sculpture of *Viviparus brevis*. Ten morphological classes are distinguished (below). In the Middle Kos Formation, transformation was especially rapid (arrow). Modified after BÜTTNER (1982).

Melanopsis

A second example of the bottleneck effect is shown by the transformation of *Melanopsis sporadum* into *M. gorceixi*. Originally, *Melanopsis* was spread all over the island (Gurniati and Phoka Formations). In central Kos, transition of the Gurniati into the Stefanena Formation is marked by a marine intercalation (WILLMANN, 1983). As a result of this transgression the inland waters of Kos were restricted in area, and the population size of *Melanopsis* diminished considerably. The rest of the population probably survived in eastern Kos and developed a slightly broader shell. As the sea regressed from middle Kos, this early form of *Melanopsis gorceixi gorceixi* expanded its range over the whole area of Kos. Shortly afterwards the sea returned. The population size of *Melanopsis* was reduced again, and in the remaining individuals the morphology of the typical *M. gorceixi gorceixi* was developed. While *M. sporadum* did not change over a sedimentary sequence of at least 70 m, transformation into the typical *M. gorceixi* can be traced within a sequence of 3 m only (WILLMANN, 1983). These sediments are condensed in relation to those containing *M. sporadum*. Condensation, however, was not of an amount for which the assumption of an increase in evolutionary rate would have to be ruled out.

Some other cases of rapid change seem to be caused by bottlenecking, e. g. the change of *Melanopsis gorceixi gorceixi* into *M. g. semiplicata* and *M. g. aegaea* in the lake of eastern Kos (see above, section 3.1.1). This was a parallel development, compared to that of *Viviparus brevis brevis* into *V. b. forbesi*.

4. ENVIRONMENTAL CHANGES AND PHENOTYPIC MODIFICATIONS

The opinion sometimes expressed that the lineages of Neogene freshwater snails are not due to true phyletic evolution, but are rather direct modifications of the phenotype by environmental stimuli, cannot be here supported (WILLMANN, 1978: 237, 1981; BÖGER et al., 1979: 156; BÜTTNER, 1982). As mentioned above, in no case can such a relationship be established between morphological transformation and changes in the environment (e. g. variations of salinity).

Only in *Theodoxus doricus* from Kos can a direct influence of the environment be detected. This response, however, does not involve shell morphology, but coloration: in lacustrine environments *Theodoxus doricus* had a pattern of brown and white. Whenever salinity was slightly increased, the pattern became grey and white, and considerable salinity (of 7 ‰ or more) sometimes caused black patterns (Fig. 11). Influence of

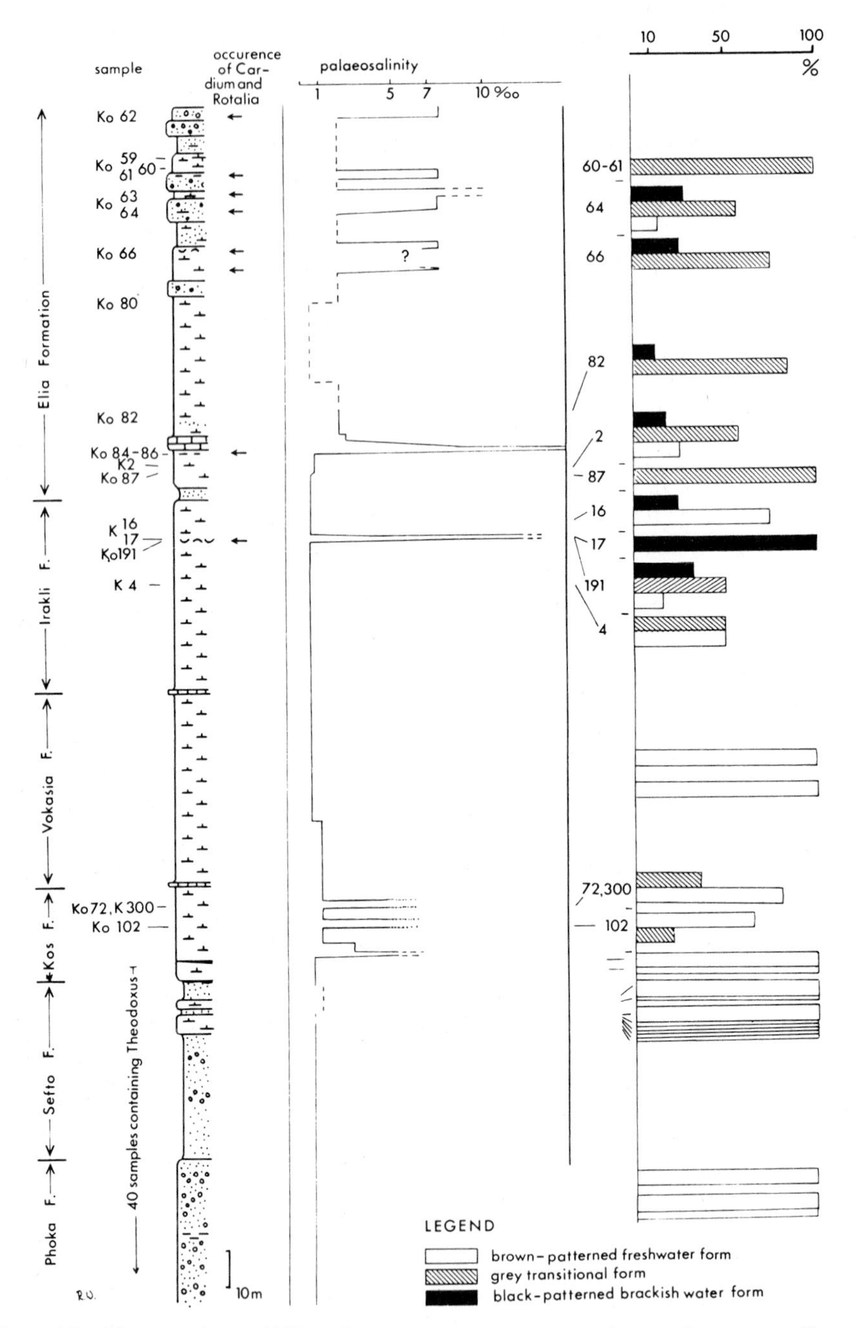

Fig. 11: Phenotypic modifications as responses to the environment. Changes of palaeosalinity in the Plio-Pleistocene lake of east Kos and occurrence of differently coloured *Theodoxus* specimens. Colouration of *Theodoxus* is clearly related to salinity. Palaeosalinity as indicated by fishes, marine-brackish mollusks, foraminifera, characeans and preliminary diatom analyses (BÖGER et al., 1979; WILLMANN, 1981). In two cases the increase of salinity is concluded solely from the occurrence of brackish-water specimens of *Theodoxus* (niveaus of samples Ko 102, Ko 72, K 300).

varying salinity on coloration in Recent *Theodoxus* was reported by NEUMANN (1959).

5. THE FINAL RESPONSE OF THE KOAN FRESHWATER GASTROPODS TO CHANGES IN THE ENVIRONMENT: EXTINCTION

The endemic fauna of Kos was finally destroyed when the sea invaded the basin in the Lower Pleistocene. As mentioned above, the sea had covered central Kos repeatedly in former times, but conditions in the eastern Kos basin had not been significantly influenced up to this time.

Marine influence increased especially during deposition of the Upper Irakli and the Elia Formations. The freshwater gastropods, however, did not disappear suddenly but rather step by step. Among the first species to become extinct were *Xestopyrguloides neumayri* and *Valvata heidemariae*. In the Lower and Middle Elia Formation four more species disappeared: *Melanopsis delessei* , *Bithynia* cf. *leachi* , *Iraklimelania levis* , and *Prososthenia sturanyi*. The Upper Elia Formation includes several layers containing *Cardium edule*, *Rotalia beccarii* etc., which indicate an increase of salinity up to 7°/₀₀ at the very least (Fig. 11). The lake thus became a lagoon.

The last freshwater gastropods in the lake of eastern Kos - or portions of it - were *Theodoxus doricus*, *Melanopsis gorceixi* , *Viviparus brevis*, *Melanoides tuberculata*, *Mikrogoniochilus minutus* , *Marticia cosensis* , *M. brusinai*, and *Valvata hellenica*. They became extinct during deposition of the uppermost Elia and the Nikolaos Formation. The Nikolaos Formation includes fully marine sediments (BÖGER et al., 1974; BESENECKER & OTTE, 1978, 1979: 456; WILLMANN, 1983: 854-855).

In other areas of the Aegean, extinction of freshwater gastropods was caused by different factors. On Rhodes, the fauna of the Pliocene Salakos-Formation disappeared when the lake dried out. This extinction also included many endemic species; e. g. *Viviparus clathratus* (DESHAYES), *Micromelania orientalis* WILLMANN, *Valvata kamirensis* WILLMANN, *Rhodopyrgula rhodiensis* (BUKOWSKI), and *Melanopsis vandeveldi* BUKOWSKI (WILLMANN, 1981: 110-129).

6. CONVERGENCE IN SHELL MORPHOLOGY

As mentioned above, the endemic fauna of eastern Kos was not by far the only one in the Neogene of the Balkans and of Anatolia. Interestingly, in many basins the same morphological types occur, e. g. Viviparids with two keels, Melanopsines with axial ribs, and species of *Theodoxus* with sulcated flanks. Though these sculptures often developed independently, the resulting forms cannot often be told apart. In fact, numerous earlier authors considered many of these forms to be taxonomically identical.

According to KÜHN (1951, 1963) for example, "*Melanopsis* of Bosnian type" occurs near Athens. This idea led to further extensive palaeogeographical conclusions. KÜHN (1951: 190) wrote:

> Die beschriebene Fauna (from Athens and district) zeigt zum ersten Male ein Weiterreichen der miozänen dalmatinisch-herzegowinischen Süßwasserfauna über diesen Bereich hinaus ... bis nach Mittelgriechenland."

The specimens Kühn believed to belong to the Yugoslavian *M. astrapaea* BRUSINA and *M. bicoronata* BRUSINA, however, are a different species: *M. oroposi* PAPP 1979, which is endemic to a restricted area in Attika.

> Many more examples of misidentifications due to convergences in shell morphology can be listed. The two Koan forms *M. gorceixi proteus* and *M. g. polyptycha* were erroneously recorded from Korinthos (GILLET, 1963) and from the Neogene of Patras (FUCHS, 1900). The Slavonian *M. harpula* was mentioned from Rhodes (DESIO, 1931: 255). In reality the Rhodian species is *M. vandeveldi*. *M. vandeveldi* was more than once believed to occur near Datca in Anatolia and on Kos (CHAPUT, 1955: 44; BECKER-PLATEN, 1970: 220; MAGROGRASSI, 1928: 252) - this, however, is *M. gorceixi heldreichi*. The Koan *Theodoxus doricus fontannesi* was recorded many times from the Rhodian Salakos Formation (where only *Th. hellenicus pseudomicans* occurs), and *Viviparus brevis forbesi* from Kos was erroneously mentioned from the southern Pliocene basin of Rhodes (e. g. BUKOWSKI, 1893; in reality *V. rhodensis sulcatus*). The endemic *V. clathratus* from the Rhodian Salakos Formation was recorded from Joannina in northern Greece (DOLLFUS, 1922; GILLET, 1962) - this, however, is still an undescribed species - and so on.

Unfortunately, a number of stratigraphic correlations were based on erroneous taxonomic identifications.

Similar sculptures were possibly caused by similar selection pressures. However, nothing is known about the function of the sculptures observed. Therefore, many hypotheses about their adaptational value have been published, none of them, however, being satisfactory. Among the most popular explanations are

(1) adaptation to water current—but many times sculptures developed in waters with minimal movements,

(2) adaptation to mineralized waters, or

(3) as a shelter against birds of prey (recent Viviparids are reported to be eaten by the heron, *Ardea cinerea*; FRÖMMING, 1956: 258; WILLMANN, 1981: 21-22).

At the moment, it can only be said for sure that in each of the many Neogene lakes representatives of the same genera occurred. The genetic background did not allow any major change in shell morphology: Sculptures usually consist of few elements, namely axial or spiral ribs, nodes, and spines, or combinations of these. Because of these limitations the same patterns must have developed repeatedly.

REFERENCES

Becker-Platen, J.P. 1939: Lithostratigraphische Untersuchungen im Känozoikum Südwest-Anatoliens (Türkei).- Beih. Geol. Jb. 97: 244 S., Hannover.

Besenecker, H. & Otte, O. 1978: Late Cenozoic Development of Kos, Aegean Sea.- In: Closs, H. et al. (eds.): Alps, Apennines,Hellenides: 506-509, Stuttgart.

--- 1979: Late Cenozoic sedimentary history and paleogeography of Kos, Aegean Sea.- Proc. VI. colloquium Geol. Aegean Region, Athens 1977: 451-457, Athen.

Böger, H. 1978: Sedimentary History and Tectonic Movements During the Late Neogene.- In: Closs, H. et al. (eds.): Alps, Apennines, Hellenides: 510-512, Stuttgart.

Böger, H., Gersonde, R. & Willmann, R. 1974: Das Neogen im Osten der Insel Kos (Ägäis, Dodekanes) - Stratigraphie und Tektonik.- N. Jb. Geol. Paläont. Abh. 145: 129-152, Stuttgart.

--- 1979: Paläoökologie der Diatomeen im Neogen von Kos (Dodekanes, Griechenland) und die Evolution limnischer Gastropoden.- Ann. Géol. Pays Hellén., Tome hors série 1: 149-157.

Bukowski, G. 1893: Die levantinische Molluskenfauna der Insel Rhodus 1.- Denkschr. Kaiserl. Akad. Wiss., Math.-Nat.Cl., 60: 265-303.

--- 1895: Die levantinische Molluskenfauna der Insel Rhodus 2.- Denkschr. Kaiserl. Akad. Wiss., Math.-Nat. Cl., 63: 1-70.

Büttner, D. 1982: Biometrie und Evolution der Viviparus-Arten (Mollusca, Gastropoda) aus der Plio-Pleistozän-Abfolge von Ost-Kos (Dodekanes, Griechenland).- Berliner geowiss. Abh. A 42: 1-79, Berlin.

Chaput, G. 1955: Contribution à l'étude de la faune pliocène de la Péninsule de Cnide (Turquie).- Bull. Sci. Bourgogne 15: 39-51.

Desio, A. 1931: Le isole Italiane dell' Egeo.- Mem. descr. carta geol. Ital. 24: 547 S., Roma.

Dollfus, G.F. 1922: Faune malacologique du miocène supérieur de Janina en Epire.- Bull. Soc. géol. France: 101-123, Paris.

Forbes, E. & Spratt, T.A.B. 1846: On a Remarkable Phaenomenon Presented by the Fossils in the Freshwater Tertiary of the Island of Cos.- 15th Ann. Rep. Assoc. Adv.Sci. London: 59, London.

Frömming, E. 1956: Biologie der mitteleuropäischen Süßwasserschnecken.- 313 S., Berlin.

Fuchs, Th. 1900: Über einige von Custos O. Reiser in Griechenland gesammelte Tertiärfossilien.- Ann. k.k. naturhist. Hofmus. 15: 1-4.

Gillet, S. 1962: Remarques sur des gastropodes de quelques gisements du Pliocène et du Quaternaire d'Epire.- Prakt. Akad. Athenon 37: 260-273.

--- 1963: Nouvelles données sur le gisement villafranchien de Néa - Corinthos.- Prakt. Akad. Athenon 38: 400-419.

Girotti, O. 1966: Un Viviparus pleistocenico dell' Italia centrale. Confronto con le species fossili e viventi europee.- Arch. Moll 95: 255-268.

Gorjanovic-Kramberger, K. 1923: Über die Bedeutung der Valenciennesiiden in stratigraphischer und genetischer Hinsicht.- Paläont. Z. 5: 339-344.

Hilgendorf, F. 1866: Planorbis multiformis im Steinheimer Süßwasserkalk. Ein Beispiel von Gestaltveränderung im Laufe der Zeit.- 36 S., Berlin.

--- 1867: Über Planorbis multiformis im Steinheimer Süßwasserkalk.- Monatsber. Preuß. Akad. Wiss. 1866: 474-504.

--- 1879: Zur Streitfrage des Planorbis multiformis.- Kosmos 5: 10-22, 90-99.

Jekelius, E. 1932: Die Molluskenfauna der dazischen Stufe des Beckens von Brasov.- Memoriile Inst. Geol. Romaniei 2: 5-37.

Kühn, O. 1951: Süßwassermiozän von bosnischem Typus in Griechenland.- The Geology of Greece 5: 185-192.

--- 1963: Das Süßwassermiozän von Attika.- Prakt. Akad. Athenon 38: 370-400.

Magrograssi, A. 1928: La fauna levantina di Coo e di Rodi.- Atti. Soc. Ital. Sc. Nat. 67: 249-264, Milano.

Mayr, E. 1949: Speciation and Systematics.- In: Jepsen, G.L., Mayr, E. & G.G. Simpson (eds.): Genetics, Paleontology and Evolution. Princeton Univ. Press.: 281-298.

Neumann, D. 1959: Experimentelle Untersuchungen des Farbmusters der Schale von Theodoxus fluviatilis L.- Zool. Anz. 23. Suppl.: 152-156.

Neumayr, M. 1880: Über den geologischen Bau der Insel Kos und über die Gliederung der Jungtertiären Binnenablagerungen des Archipels.- Denkschr. k. Akad. Wiss. 40: 213-314, Wien.

--- 1889: Die Stämme des Thierreiches 1, Wien, Prag.

Neumayr, M. & Paul, C. 1875: Die Congerien- und Paludinenschichten Slavoniens und deren Faunen.- Abh. k.k. geol. Reichsanst. 7: 1-111, Wien.

Newton, R.B. 1911: On the Modifications in Form of the Upper Tertiary Lacustrine Shells of the Island of Cos, as First Observed by Edward Forbes and T.A.B. Spratt.- Proc. Mal. Soc. London 9: 363-368.

Olujić,J.P.1936: Über die geschlossenen, progressiven Entwicklungsreihen der Schalen der pontischen Prososthenien (Vorläufige Mitteilung).- Arch. Moll. 68: 118-120.

Spratt, T.A.B. & Forbes, E. 1847: Travels in Lycia, Milyas and the Cibyratis 2, London.

Stanczykowska, A. Magnin, E. & Dumouchel, A. 1971: Étude de trois populations de Viviparus malleatus (Reeve) (Gastropoda, Prosobranchia) de la région de Montréal I. Croissance, fécondité, biomasse et production annuelle.- Can. J. Zool. 49: 1431-1441.

Stanley, S.M. 1979: Macroevolution - Pattern and Process.- 332 S., San Francisco.

Tournouer, R. 1876: Étude sur les fossiles tertiaires de l'île de Cos.- Ann. scient. école norm.sup. sér. 2,5: 445-475, Paris.

Willmann, R. 1978: Die Formenreihen der pliozänen Süßwassergastropoden von Kos (Ägäis) und ihre Erforschungsgeschichte.- Natur u. Museum 108: 230-237, Frankfurt.

Willmann, R. 1981: Evolution, Systematik und stratigraphische Bedeutung der neogenen Süßwassergastropoden von Rhodos und Kos/ Ägäis.- Palaeontographica A 174: 10-235, Stuttgart.

--- 1983: Neogen und jungtertiäre Entwicklung der Insel Kos (Ägäis, Griechenland).- Geol. Rdsch. 72: 815-860.

THE STEINHEIM BASIN AS A PALEO - ANCIENT LAKE

A. Gorthner & C. Meier-Brook

Tübingen

Abstract: The convergent evolution of "thalassoid" planorbid gastropod shells suggests that the Miocene impact-crater lake of Steinheim (S. Germany) is a fossil example of an Ancient Lake (i.e., a "Paleo-ancient Lake"), comparable to the modern Lake Ochrid (Yugoslavia). In these and other cases it was probably the long-term stability of such environments that allowed an unusual degree of evolutionary diversification and enniche-ment of resident species. Ecological close-packing, rather than a hydrochemical barrier, is suggested to eventually stop the faunal exchange of such ancient lakes with adjacent, less stable fresh water biotopes.

INTRODUCTION

After its origin by meteorite impact in the Upper Miocene the crater of Steinheim in south Germany formed a lake, whose gastropod inhabitants underwent a well documented evolutionary transformation (for details, see REIF, this volume; MENSINK, 1984; ADAM, 1980). Recent discussion has focussed on three points:

1. Why did Steinheim snails remain endemic once they had evolved? Correspondingly, why did widely distributed gastropods fail to enter, from the surroundings, into the biocenosis of newly evolved gastropods?
2. To what extent did the transformation have a merely modificatory character (the question of heritabily of shell characters)?
3. Of what nature was the selective pressure, if any, exerted on the colonizing species?

Answering these questions poses problems since, as in any ecosystem, all of them are in a complex of interdependance. Nevertheless they deserve special attention in view of the recent studies by WILLIAMSON (1981) and WILLMANN (1981) (see REIF, this volume) on similar fossil cases. All of these study objects have in common that they cannot be solved with paleontological methods alone, particularly with respect to the question of heritability. It was the actualistic principle that led us to look for a Recent model of the Steinheim ecosystem. A lake showing analogous phenomena will be offered here.

THALASSOID GASTROPODS AND ANCIENT LAKES

The Steinheim planorbid snails of the genus *Gyraulus* are descendants of the stem species, *Gyraulus kleini* (GOTTSCHICK & WENZ, 1916), which is widely encountered in the South German Freshwater Molasse and resembles extant ubiquitous *Gyraulus* species (Fig. 1). In the Steinheim section the small, thin-shelled, rounded-planispiral snails of *G. kleini* gave rise to a multitude of larger, thick-shelled, angled or carinate, temporarily trochoid endemic forms. Freshwater snails with such characters, called thalassoid because of their similarity with marine groups (HUBENDICK, 1952), are found in only a few Recent lakes (fig. 1; compare BOSS, 1978). Evolution of these shell characters has presumably taken place essentially within these lakes. Anatomical evidence for convergent transformation into thalassoid *Gyraulus* shells in Lakes Ochrid, Prespa, and the Japanese L. Biwa (also in L. Baical?) is given by MEIER-BROOK (1979, 1983). All extant water bodies harbouring thalassoid gastropods are characterized by their high age and therefore classified as "Ancient Lakes" (Fig. 1), the vast majority of them having persisted in the order of ten thousands of years before they are silted up or end as swamps. These lakes present unusually high numbers of endemic species in molluscs as well as other phyla (as for gastropods, see Table 1). In the Steinheim lake, e.g., most ostracods were endemic (SIEBER, 1905).

Ancient lakes can form in areas of continuous geotectonic depression or where sedimentation - and thus silting - is drastically reduced. The latter may have been the case with the meteorite crater of Steinheim, which lies in a pronounced karst region and did not develop a major inflow for a long time. The life span of Steinheim lake is not known precisely. For the meteorite crater lake of Nördlinger Ries which is close to Steinheim and owes its origin to the same impact event, JANKOWSKI (1981) assumes a span of 0.3 to 2 Myr. For the nearby Randecker Maar, which was even smaller than the Steinheim Lake, but due to volcanic activity at about the same time, he estimates no less than 0.1 - 0.5 Myrs. A few hundreds of thousand years render a lake tenfold older than "normal" ones. In this sense Steinheim Lake certainly qualifies as an ancient lake.

Among extant ancient lakes we also encounter world's deepest lakes. This automatically means a prolonged span of existence particularly if depression continues to deepen by tectonic processes. This coincidence of age and depth lead REIF (this volume) to regarding depth to be the only reason for ancient Lakes' high degrees of endemism

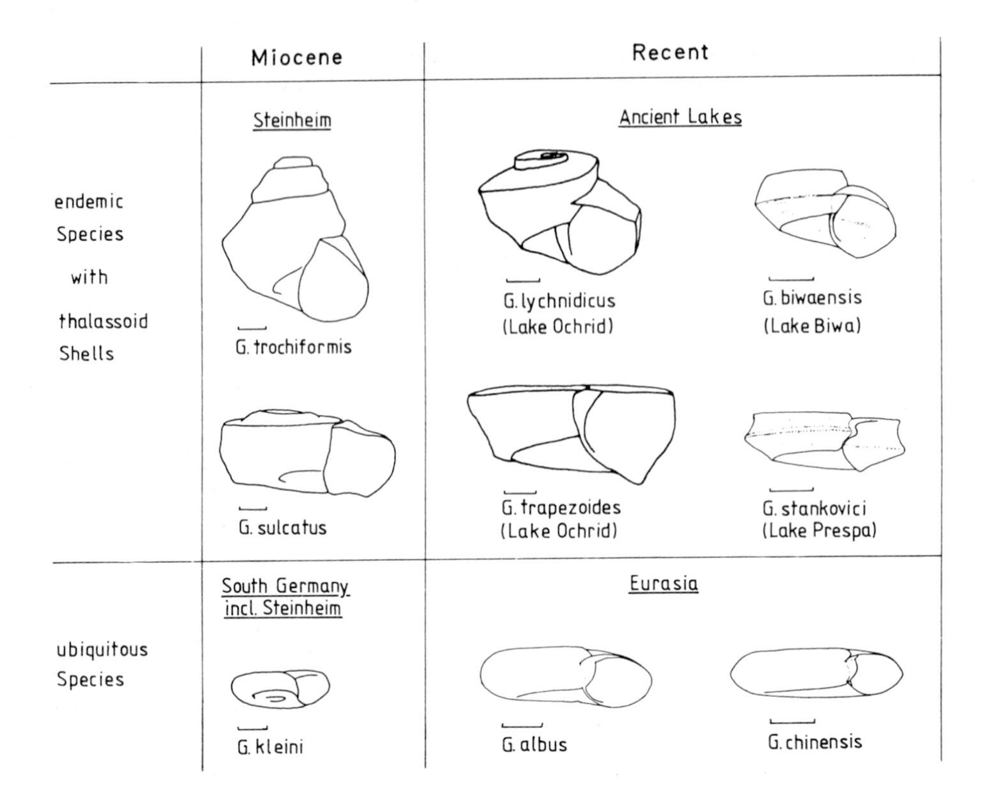

Fig. 1: Shells of the gastropod genus *Gyraulus*. Unlike ubiquitous species, the endemic species in ancient lakes (Recent) as well as in Steinheim Basin (Miocene) mostly have thalassoid shells (large, carinate, sometimes trochoid). Evolution from normal and ubiquitous to thalassoid and endemic snails can be followed in the Steinheim Basin (from MEIER-BROOK, 1983, and GORTHNER, 1984).

Table 1: Characteristics of ancient lakes (modified after BOSS, 1978).

Ancient Lakes	Ochrid	Baikal	Biwa	Tanganyika	Titicaca	Inlē
Location	Balkans	Siberia	Japan	Eastafrica	Peru/Bolivia	Burma
Area (km²)	340	31500	675	34000	7600	20
Depth (m)	286	1741	104	1470	281	7
Age	Pliocene	Mio-Pliocene	prePliocene	Pliocene	post Miocene	Pliocene?
Gastropoda species	75	72	24	55	26	25
Endemics (%)	76 *)	76	37	56	76	40

*) 96% of 46 spp. in the lake proper

in their gastropod faunas. This view, based on a questionable hypothesis of BOSS (1978), is easily rejected. (1) Species numbers enormously decrease towards deeper zones. (2) Lake Inlê in Burma, which is an ancient lake with no less than 40% endemic species, does not exceed the depth of seven meters. (3) The huge number of postglacial "young" lakes comprise many deep lakes. Many a relic from the early post-Pleistocene has survived for more than 10.000 years, and that in the profundal zone (though in Europe not in gastropods) - the littoral environment being a barrier to dispersal. Still, no endemic species have evolved in these cases.

In contrast with REIF (this volume) we claim that it was the relatively long span that classifies the Steinheim lake as an "ancient lake" in terms of intralacustrine evolution. Besides, the Steinheim crater, with a maximum depth of up to 150m, was, at least temporarily, also a very deep lake, as the densely laminated stratification and the chemical characteristics of sediments in the basin indicate (MENSINK, 1984).

"Ancient lakes" are a well known phenomenon in Recent biology and limnology, which paleontologists have not considered so far. The fact that paleontology exclusively deals with "ancient" subjects, may explain this lack. In this case, a term like "long lasting lake" would do; but since "ancient lake" is an established term, we propose, as an adequate compromise, to use the term "Paleo-Ancient Lake". Besides the Steinheim Basin, we expect a series of other extinct water bodies will be recognized as paleo-ancient lakes in the future. The huge sediment masses accumulated in such water bodies might favour their being discovered, recognized, and studied. Suitable and important candidates are the former lakes of Kos (WILLMANN, 1981) and the Turkana Basin (WILLIAMSON, 1981).

LAKE OCHRID AS A MODEL FOR THE STEINHEIM BASIN

Among modern ancient lakes, the Makedonian Lake Ochrid is best suited for a comparison with the Steinheim Basin (Fig. 2), as indicated by the following features. (1) Its geographical proximity in the same faunal province. (2) Its topography in a karst limestone area with probably similar water regime, mineral input and probably similar living conditions for molluscs. (3) A *Chara*-belt plays a dominant role in the vegetation of both lakes, so that habitat patterns may resemble each other. (4) In both gastropod faunas are rich in individuals, even as associations are concerned: The genera *Gyraulus* and *Radix* (Lymnaeidae) as well as the family of Hydrobiidae are represented in close association in both lakes. As the only difference, Lake Ochrid contains a variety of other gastropods coexisting. In particular, it is the existence of thallasoid "*Gyraulus*" species that both lakes have in common, with details of shell

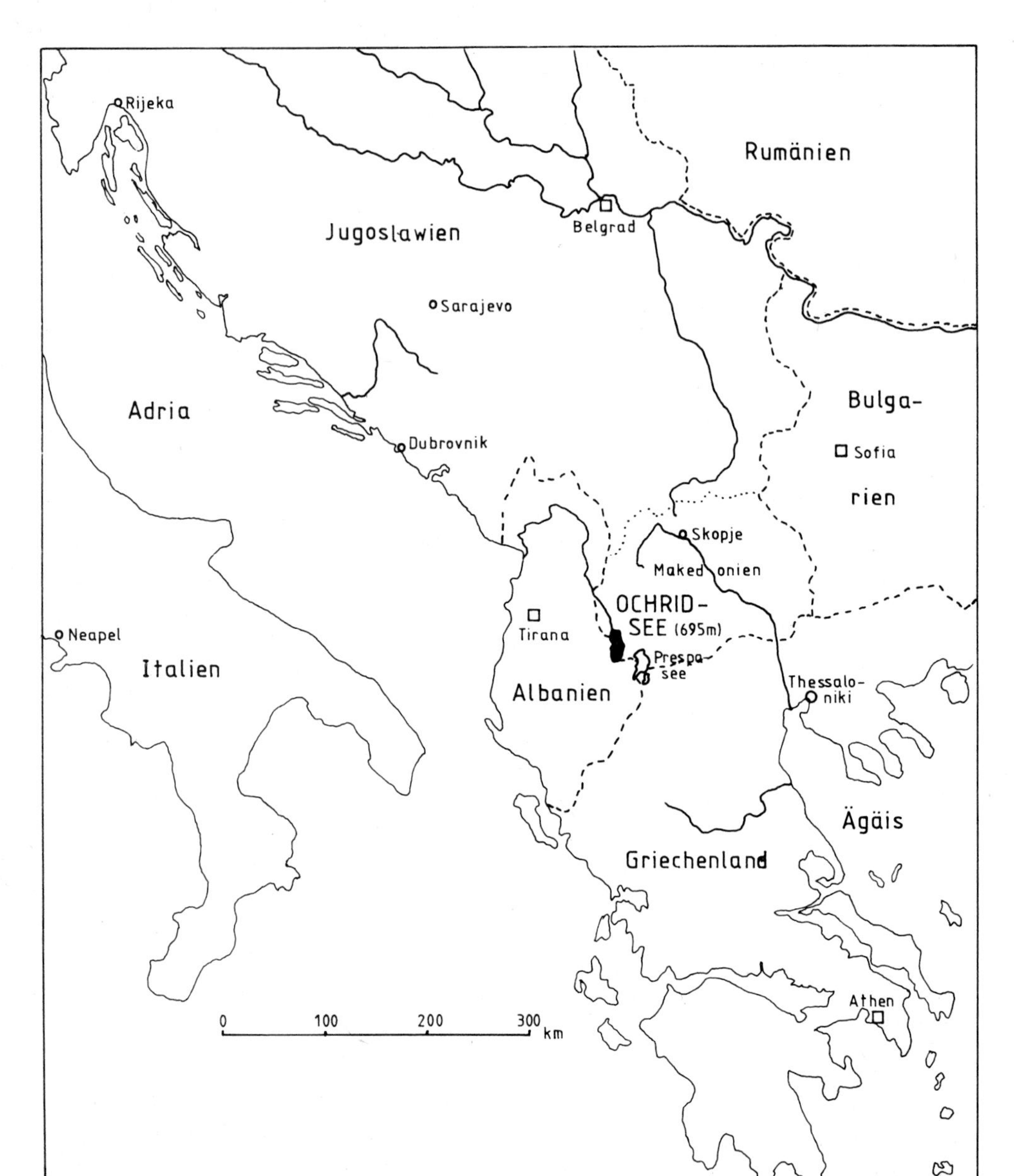

Fig. 2: The position of Lake Ochrid in Makedonia, Southeast-Europe.

shapes (a rectangular spire section and turricular forms (Fig. 1, for example) which makes comparison of Lake Ochrid with Steinheim suitable. We must, however, stress once more that shell similarity is very likely due to convergent evolution, as evidenced by studies of embryonic shell structures (GORTHNER, 1984). Nevertheless, the external resemblance of *Gyraulus* snails implies similar selection pressures, i.e. environmental

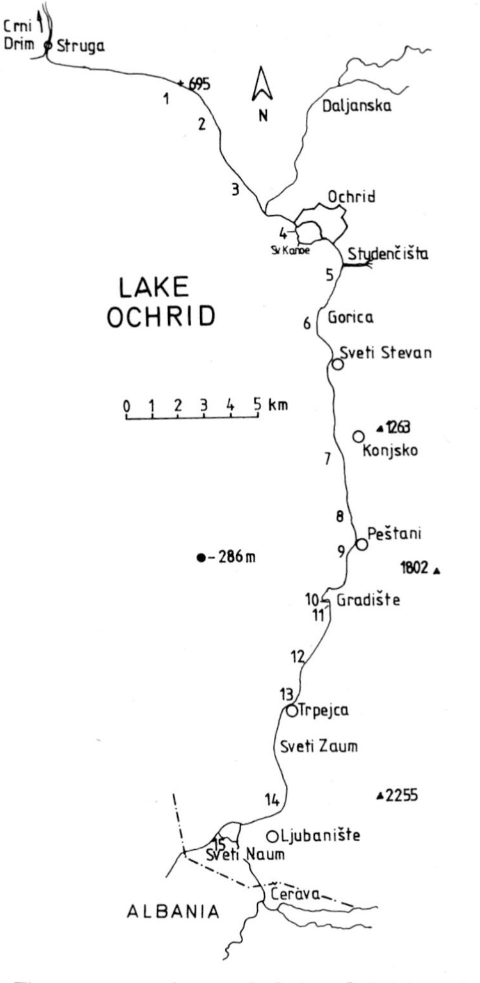

Fig. 3: The eastern shore of Lake Ochrid with sampling stations (1-15), at which *Gyraulus* was collected at various depths.

parameters that independently resulted in the evolution of thalassoid shell geometries. Since the biology of Ochridan species was far from being known, one of us (A.G.) performed ecological investigations in situ as well as in the laboratory.

GYRAULUS IN LAKE OCHRID - AN EXAMPLE OF INTRALACUSTRINE SPECIATION

Five *Gyraulus* species have at present been described from the Ochrid Basin (HUBENDICK & RADOMAN, 1959), four of which are thalassoid (subgenus *Carinogyraulus*); the systematic position of the fifth (*G. albidus* Radoman) is doubtful. Only two of the described *Carinogyraulus* species live in the lake proper, in which they are widely distributed and abundant. *Gyraulus lychnidicus* HESSE lives near the shore, *G. trapezoides* (POLINSKI) in deeper zones. The studies were aimed to characterize the ecological ennichement which implied a morphological delimitation of forms, to the description of variation, and to distribution. Only some aspects can be dealt with here.

Snails were sampled - mostly quantitative - at various littoral stations and at various depths (Fig. 3). Of the four shell parameters, two are presented here (Fig. 4) that allow easy discrimination of typical populations of the two lacustrine *Carino-*

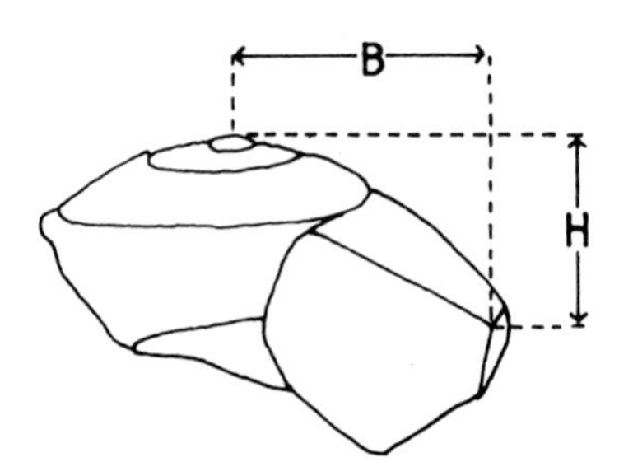

Fig. 4: Measuring shells with the parameters spire height (H) and distance between the axis and the outer apertural wall (B).

gyraulus species (Fig. 5). Intraspecific variation shows significant morphological difference, even between populations not far distant from each other (Fig. 6). The height/ width ratio (H/B), e.g., increases from sampling station no. 11 to no. 5. This means there is a North-South-cline along the axis of the lake, in which spatial proximity is correlated with morphological proximity. Such a cline must be the result of intralacustrine divergent evolution. Population differences occurring within a few kilometers in the same lake imply restricted gastropod dispersal, whether active or passive, in this water body. At any rate, even a relatively small lake such as L. Ochrid (its longest extension being some 30 km), appears to offer inhabitant species the separation of local populations required for speciation.

Gyraulus lychnidicus and *G. trapezoides* of Lake Ochrid can be distinguished as "good" species not only by shell form (Fig. 5) and anatomy (HUBENDICK & RADOMAN, 1959). They also occupy, in spatial separation, special niches. The first species lives on littoral limestones, usually no deeper than 1 m, while the latter inhabits soft bottoms, between 6 and 15 m deep, covered by tall *Chara*-meadows. These two habitats are nowhere in contact to each other. At station 4, however (cp. Fig. 3 and 7), a soft bottom zone 3 to 5 m deep is devoid of *Chara*. Here empty shells of a *Gyraulus* population were found that cannot be definitely assigned to either of the two species (Fig. 7), because shell shape and habitat are positioned "between" the two species. This population approaches population no. 5 of *G. trapezoides* and can be arranged

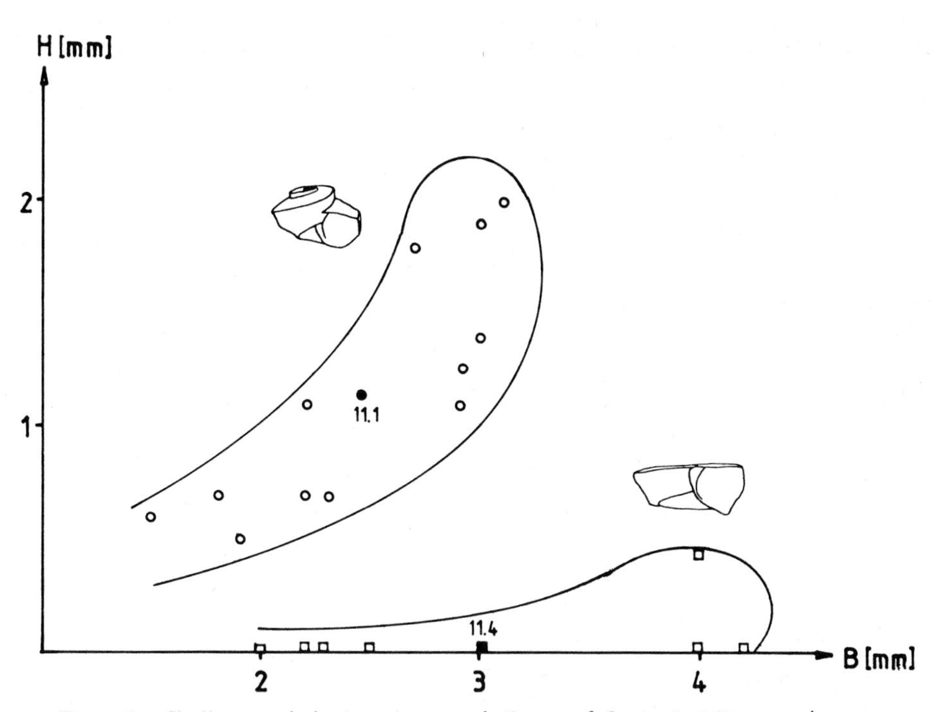

Fig. 5: Shell morphologies in populations of *G. lychnidicus* and *G. trapezoides* at sampling station no. 11. Individuals in areas of 900 cm^2 each, near the shore (11.1) and in 8 m depth (11.4). The species live allopatrically without habitat overlapping.

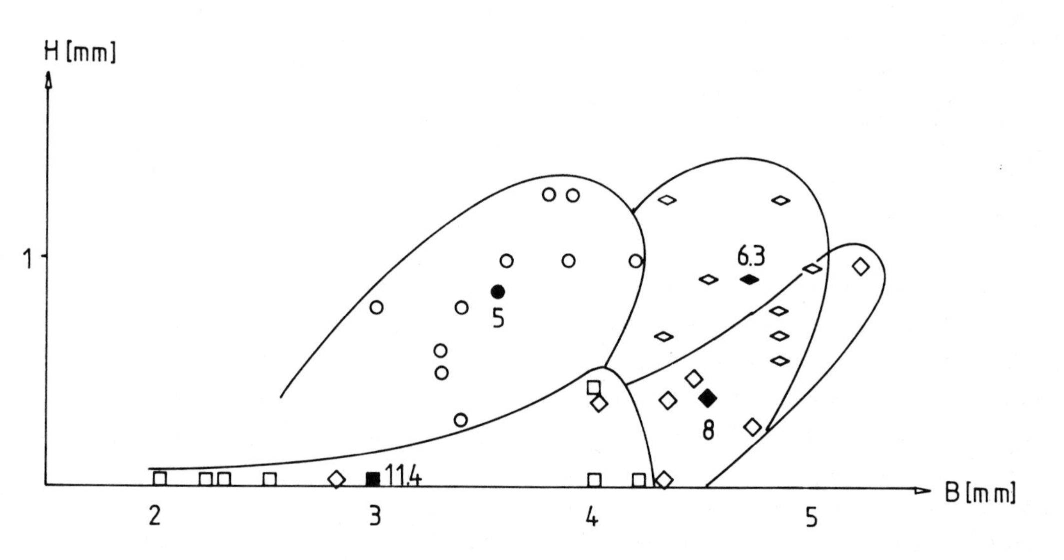

Fig. 6: Populations of *Gyraulus trapezoides*. Geographical proximity corresponds to morphological similarity in mean shell parameters.

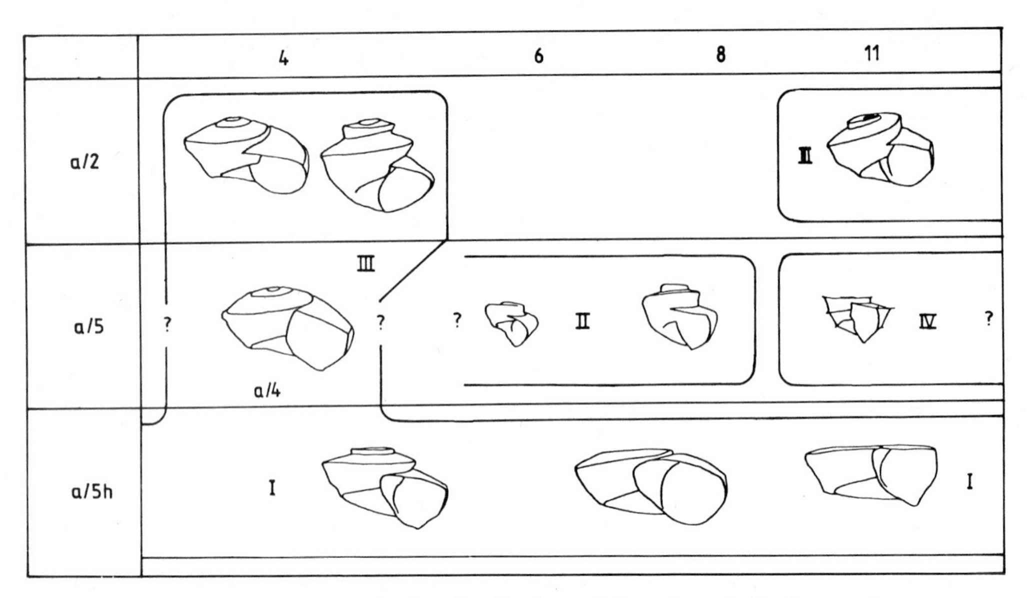

Fig. 7: An overview of the distribution of *Gyraulus* shell shapes at sampling stations 4 to 11 (horizontal axis) in different habitats and depths (vertical axis). Depicted are average shell shapes of the respective populations. Various populations united to species. Nos. II and IV were previously regarded to be modifications of I and III (cp. HUBENDICK & RADOMAN, 1959). They are, however, strictly separated, by geography, ecology and morphology, from their "stem species", and will have to be regarded as good species (not yet described).
At station 4 a transitional form between I and III occurs; the littoral population shows high variation in spire height.

Habitats and depths:			
a/2	-	stony shore	0 - 1m
a/4	-	fine mud	varying
a/5	-	low *Chara*	3 - 6 m
a/5h	-	tall *Chara*	6 - 15m

Gyraulus species:
I - *G. trapezoides*
II - *G.* spec. undescribed 1
III - *G. lychnidicus*
IV - *G.* spec. undescribed 2

within the morphological cline shown above (Fig. 6 and 7). On the other hand there is, in close neighbourhood to the (dead) population in question, a littoral occurrence of *G. lychnidicus* (station no. 4), where *G. lychnidicus* lives in its typical habitat, i.e. on limestones, but shows broad variation in spire height (which is the most reliable species character!). Also the intermediate population is highly variable; the shell, however, is more depressed, and, thus, more resembling *G. trapezoides*. The highly variable littoral population (no. 4) of *G. lychnidicus* appears to be much restricted, since it is bound to an isolated occurrence of limestone. Its nearest other population lives as far as 10 km away (SE-shore) where *G. lychnidicus* shows the typical shape and low variation and lives in a larger population. Here no problems arise as to species distinction (Fig. 5 and 7). Explanation as a hybrid population between *G. lychnidicus* and *G. trapezoides* is not excluded a priori, but shell shapes are not intermediate between the two species, when their normal shapes, as in sampling station no. 11, are regarded. The "in-between"-population, on the contrary, has a higher spire than normal *G. lychnidicus*. These morphological-biogeographical analyses, done for the first time in L. Ochrid, open a promising field for evolutionary theory. On the other hand, they cast some light on possible evolutionary processes that might have taken place in the Steinheim Basin with respect to variability, population size and interpopulational isolation. The heritability of shell characters has been tested by breeding experiments. Investigation of the autecology of *Carinogyraulus* also allows conclusions about the mode of life of the Steinheim snails. A detailed account with a discussion of observations will be presented in a separate paper (GORTHNER, in prep.).

EVOLUTION IN ANCIENT LAKES

Once the Steinheim Basin is recognized as a paleo-ancient lake, theories of evolution in ancient lakes may be applied to this example. The "stability-time hypothesis" (SANDERS, 1968) has been based on diversity patterns in marine benthos biota, but it is claimed to apply not only to abyssal and tropical marine shallow water basins, but also to tropical rain forests and to ancient lakes, i.e. for habitats with a minimal temporal fluctuation of abiotic environmental factors. It also accounts for isolation and diversification in the faunas of such habitats. This hypothesis, which is little known even among biologists, will here be adapted to the ancient lake situation as follows. Limnic organisms normally live in ephemeric habitats that are subject to frequent environmental changes. Thus, the physiological tolerance of such organisms is usually high (euryoecous; generalists), and their distribution is rather wide (eurytopic; ubiquists). Ancient lakes, on the other hand, are conservative environments, in which physical and chemical factors remain constant over long periods of time, so that organisms have a chance to adapt more and more completely and become stenoecous specialists with stenotopic (i.e. endemic) distribution. Species specializing in narrow niches avoid interspecific competitions, a situation favorable for any species. In fact, environmental stability is the sine-qua-non for such a degree of specialization. Thus, in an ancient lake specialists, diversified by intralacustrine adaptive radiation, are able to coexist. How long the "inoculum"-species needs to become reproductively isolated from surrounding populations and how this can be accomplished will require further studies.

The high diversity of endemic species in ancient lakes also poses another problem. Evolutionary adaptation is related not only to the physical and chemical environment, but particularly to the co-evolving species in a process of mutual feed-back; i.e. a unique ecosystem of its own develops, in which potential niches are tightly occupied and in which biotic selective pressure predominates. In normal, ephemeric water bodies, in contrast, biocenoses are mainly abiotically selected and generalists occupy "broader" niches that overlap each other to a greater extent. Consequently they are poorer in species. This evolutionary strategy is complementary to that in ancient lakes, where there is almost no exchange between biocenoses: an endemic specialist that has lost its tolerance to fluctuating environmental conditions has become unable to compete with ubiquitous generalists. On the other hand, no ubiquist "specialized" to tolerate a wide range of conditions can effectively immigrate into an ancient lake, because with even narrow niches being tightly occupied it finds no ecological "license" for euryoecous animals. Dispersal of individuals from and into ancient lakes certainly does take place again and again, but it fails to result in population founding and species remain where they are.

This may explain why Lake Ochrid -- though surrounded by various other fresh-water bodies -- has developed and preserved a gastropod fauna with 96% of its species being endemic. There is no geographical isolation of the lake at all; sufficient marginal waters, such as swamps or springs and the lake outflow do exist. Yet, ubiquitous species faill to successfully enter the lake, although their water bodies do communicate and are not fundamentally different in physiography (STANKOVIC & RADOMAN, 1955). It is likely that living conditions in an ancient lake are sufficient to account for the diversification and uniqueness of the Steinheim *Gyraulus*. Fluctuations of salt concentration that have been evoked as a promotor of evolutionary processes or as a barrier in the Steinheim Lake possibly loose their significance if the paradigm of Lake Ochrid is applied. Could they perhaps even have disturbed the continuous ennichement of gastropods temporarliy? A possibility is that the process of mineral concentration caused extinction of 13 out of the 16 gastropod species existing earlier in the Steinheim history, but it proceeded slowly enough to enable the remaining three species, particularly *Gyraulus* , to adapt to the changing hydro-chemical conditions. On the other hand, fluctuations of salt content and water level in the contemporary lake of the Nördlinger Ries were probably so fast and extensive (JANKOVSKI, 1981) that its fauna could not switch over to the evolutionary strategy of an ancient lake.

In summary, studies in the Recent Lake Ochrid, with its spatial population patterns, and in the fossil Steinheim Basin, with its temporal dimension, appear to open a promising field of cooperation between neontologists and paleontologists.

Acknowledgements: We are indebted to Prof. A. Seilacher and Dr. U. Bayer, Institut für Geologie und Paläontologie, Tübingen, for invitation of this paper and to Prof. Seilacher for critically reading the ms.

REFERENCES

Adam, K.D., 1980: Das Steinheimer Becken - eine Fundstätte von Weltgeltung.- Jh.Ges. Natkd.Württemb., 135: 32-144, Stuttgart.

Bajor, M., 1965: Zur Geochemie der tertiären Süßwasserablagerungen des Steinheimer Beckens, Steinheim am Albuch (Württemberg).- Jh.geol.Landesamt Baden-Württ., 7: 355-386.

Boss, K.J., 1978: On the evolution of gastropods in ancient lakes.- In: Fretter, V. & Peake, J. (Ed.): Pulmonates, Vol. 2A: 385-428, Academic Press, London.

Brooks, J.L., 1950: Speciation in ancient lakes.- Quart.Rev.Biol., 25: 30-60, 131-176.

Gorthner, A., 1984: *Gyraulus* (Gastropoda; Planorbidae) im Steinheimer Becken und Ochridsee - ein Vergleich als Beitrag zur Kenntnis der Ökologie und Evolutionsbedingungen.- Dipl.Arb. (Biologie) Tübingen; 1-138.

Hilgendorf, F., 1867: Über *Planorbis multiformis* im Steinheimer Süßwasserkalk.- Mber.kgl.preuss.Akad.Wiss. Berlin 1866: 474-504.

Hubendick, B., 1952: On the evolution of the so-called thalassoid molluscs of Lake Tanganyika.- Ark.Zool.Stockholm (Ser. 2) 3: 319-323.

Hubendick, B., 1960: The Ancylidae of Lake Ochrid and their bearing on intralacustrine speciation.- Proc.zool.soc.Lond. 133(4): 497-529.

Hubendick, B. & Radoman, P., 1959: Studies on the *Gyraulus* species of Lake Ochrid. Morphology.- Arkiv för Zoologi (Ser. 2) 12: 223-243.

Jankowski, B., 1981: Die Geschichte der Sedimentation im Nördlinger Ries und Randecker Maar.- Bochumer geol.u.geotechn.Arb. 6: 315 pp, Bochum.

Lemcke, K., 1981: Das Nördlinger Ries: Spur einer kosmischen Katastrophe.- Spektrum der Wissenschaft, 1: 111-121, Weinheim.

Meier-Brook, C., 1979: The planorbid genus *Gyraulus* in Eurasia.- Malacologia 18: 67-72.

Meier-Brook, C., 1983: Taxonomic studies on *Gyraulus* (Gastropoda: Planorbidae).- Malacologia, 24(1/2): 1-113.

Mensink, H., 1984: Die Entwicklung der Gastropoden im miozänen See des Steinheimer Beckens (Süddeutschland).- Palaeontographica, A, 183: 1-63, Stuttgart

Miller, K., 1900: Die Schneckenfauna des Steinheimer Obermiocäns.- Jh.Ver.vaterl. Naturk.Württ., 56: 385-406, Stuttgart.

Polinski, W., 1932: Die reliktäre Gastropodenfauna des Ochrida-Sees.- Zool.Jb.Abt. Syst., 62: 611-666.

Radoman, P., 1960: Two sibling species of *Pseudamnicola* in Ohrid Lake.- Basteria, 24, 1/2: 1-28.

Radoman, P., 1978: Beispiele der mikrogeographischen Speciation im Ohrid-See und die neue Gattung *Adrioinsula*.- Arch.Moll.,109(1/3): 45-50.

Raup, D.M., 1966: Geometric analysis of shell coiling: General problems.- J.of Paleont., 40,5: 1178-1190.

Reif, W.-E., 1983a: Hilgendorf's (1863) dissertation on the Steinheim planorbids (Gastropoda; Miocene): The development of a phylogenetic research program for Paleontology.- Paläont.Z., 57(1/2): 7-20.

Reif, W.-E., 1983b): The Steinheim snails (Miocene, Schwäbische Alb) from a Neo-darwinian point of view: A discussion.- Paläont.Z., 57(1/2): 21-26.

Reif, W.-E., (this volume): Endemic evolution of *Gyraulus kleini* in the Steinheim Basin (Planorbid Snails, Miocene, Southern Germany).

Sanders, H.L., 1968: Marine Benthic diversity: A comparative study.- The American Naturalist, 102, 925: 243-282.

Sieber, G., 1905: Fossile Süßwasser-Ostrakoden aus Württemberg.- Jh.d.Ver. f. vaterl.Naturk., 61: 321-346.

Stankovic, S. & Radoman, P., 1955: Le peuplement des eaux littorales adjacentes du bassin d'Ohrid.- Arch.sci.biol, Beograd, 7: 1-20.

Stankovic, S., 1960: The Balkan Lake Ohrid and its Living World.- 357 pp., Junk, Den Haag.

Williamson, P.G., 1981: Palaeontological documentation of speciation in Cenozoic molluscs from Turkana Basin.- Nature 293: 437-443.

Willmann, R., 1981: Evolution, Systematik und stratigraphische Bedeutung der neogenen Süßwassergastropoden von Rhodos und Kos/Ägäis.- Paläontographica Abt. A: 174: 10-235, Stuttgart.

PART 5

THE LOWER HIERARCHY OF CYCLES

SPATIAL AND TEMPORAL SUBSTRATE GRADIENTS

Cyclic sediments commonly exhibit a hierarchy of cyclic and periodic patterns, possibly reflecting a similar hierarchy of sea-level changes. The distribution and evolution of substrate conditions is the main object of this part:

(1) The sedimentological record of cycles, their recognizable 'amplitude', depends on their position relative to major trends, either temporally or spatially within the basin, and each hierarchical level provides another scale in terms of basin analysis and allows to gather information about the average hydrodynamic regime.

(2) Alternating deposition and erosion within cyclic sequences provide the base for substrate gradients within the basin. Early diagenetic processes provide nodules which reworked provide patched hardgrounds allowing specific benthic communities to settle. Shell production may also initiate further substrate evolution.

(3)Epibenthic communities, particularly bivalves, crinoids, brachiopods, etc. provide a biological source for coarse sediment material, allowing further substrate changes or even biological-sedimentological feedback loops.

The processes considered here are still above the level of single events, they reflect the 'short term' stationary state of substrates; however, a level is reached which allows to apply the actualistic principle.

** *By an actualistic example A. Seilacher reminds us that our usually stationary concept of substrate may have to be replaced by more dynamic thinking -- a shell bed may evolve under repeated mud covering and winnowing.*

** *The temporal hierarchy of cycles in terms of a conceptual classification is discussed by T. Aigner;*

** *Bayer, Altheimer & Deutschle discuss substrate gradients in relation to sedimentary cycles and basin gradients;*

** *Aigner & Kidwell discuss the dynamics of shell accumulation and the ecological response of living organisms in terms of taphonomic feedback.*

The contributions of this part tend to view the sedimentological content of basins as expression of temporally and spatially changing substrate conditions: Certain beds are not necessarily the expression of a certain hydrodynamic regime, but may be the product of processes involving primary sedimentation, early diagenesis, and biological response. Thus, the situation is comparable to an evolutionary process which also depends on the changing boundary conditions.

THE JERAM MODEL: EVENT CONDENSATION IN A MODERN INTERTIDAL ENVIRONMENT

A. Seilacher
Tübingen

Abstract: Observations through several years show that an intertidal shell bed established by a catastrophic storm event about 60 years ago has set the stage for sedimentary dynamics ever since. During more quiet periods of several years the area became gradually mud.covered; bus subsequent storms repeatedly stripped off the mud, projected its mainly infaunal shell content onto the previous shell bed and allowed a mainly epifaunal community to develop on it before the mud took over again. The model suggests that seemingly uniform fossil shell beds from other environments may reflect similarly compley burial/unburial histories, but at considerably larger time scales.

Event condensations as we observe them in geological sections usually happen at time scales inaccessible to actualistic observation. This is particularly true for temporal changes in the sedimentary record, to which an evolutionary response can be expected. If we accept, however, the notion that there is a general similarity betweeen stratigraphic cycles of different magnitude and that associated faunal changes are the integral of minor ecologic responses, it makes sense to compare them to short-term cyclicities in which the interaction between sedimentary and ecologic processes can be directly observed.

The described example provides such an opportunity. I owe it to the hospitality of the Geologic Department of the University of Malaya in Kuala Lumpur (Malaysia) during a sabbatical semester in 1979 and during subsequent visits. In particular, it was through the efforts of Dr. Peter STAUFFER that the studies could be pursued in spite of logistic and other difficulties

A. The modern example

Jeram is a small village at the west coast of the Malayan Peninsula, about 14 kilometers south of the mouth of Selngor River. Due to its sedimentological nature and unpredictability, the local beach (Pantai Jeram) never developed into a major resort, but it allowed generations of local geology students to observe modern sedimentary environments and their faunal content.

Fig. 1: During the muddy periods (upper level) sessile epibenthos is restricted to barnacles that settle on living substrates such as burrowing clams, brachyuran crabs and gastropod shells inhabited by hermit crabs. In the post-event periods (lower level) shell lags of reworked mud-burrowing bivalves allow other sessile forms (solitary corals, bryozoa, sabellariid worms, small oysters) to become established. Also the barnacles grow larger and sabellariids may develop into larger reeflets. Over the years, repeated storm reworking produces one amalgamated shell bed, in which the two communities are mixed and only the last, mud-smothered generation of shell-bottom encrusters is preserved in life position. Only the larger sabellariid reeflets (not shown in the picture) may resist reworking over several periods, so that they can continue to grow from new spatfall after every exhumation.

When I first went to the place in June 1979, it was in the hope to find living *Lingula*, which had been common there and easily accessible in previous years. To our disappointment however, *Lingula* had disappeared, and the half mile stretch of the tidal zone was completely covered by soft, sticky mud, too deep to use rubber boots. But barefoot wading was also strenuous because of the cutting edges of a shell layer buried underneath.

Under-mud mapping showed that this shell layer consisted mainly of reworked valves of *Anadara* and other mud-dwelling bivalves, whose convex-up surfaces were heavily encrusted by large barnacles, small oysters, corals, bryozoans and sabellariid worms tubes (Fig. 1) and increasingly so away from the beach. In a zone about 10 m wide the agglutinated *Sabellaria* tubes grew up to form head-like reeflets about 20cm in height.

The buried shelly bottom fauna contrasted sharply with the one that by then inhabited the muddy surface, with epifaunal browsers (cerithiid gastropods, crabs and the fish *Periophthalmus*) predominating in addition to infaunal bivalves and worms. In general, the mud-fauna was less diverse, particularly with respect to the encrusters. These were represented only by acorn barnacles of smaller size than on the buried shell surface, and only by ones encrusting living hosts such as gastropod shells inhabited by hermit crabs, the rear ends of burrowing bivalves and crabs (Fig. 1).

Considering the contrast between shelly and muddy substrates, this difference was to be expected, but there were some details that did not readily fit the model of a simple event deposit, such as the heavy encrustation of the shell bed surface that should have been covered by the terminal mud fallout had it been an individual autochthonous storm coquina.

But a storm deposit it was, as interviews with local fishermen revealed: the large blocks and a concrete tank, now in the middle of the tidal flat, are the remains of the old village school that stood there before a heavy storm around 1920 claimed the whole area back to the sea and forced the inhabitants to protect the rest of the village by an artificial dam. What is actually going on, came out during subsequent visits two years later.

In July 1981 the mud-blanket was again stripped away except for parallel ridges of remnant mud perpendicular to the shore (Fig. 2) indicating that this erosion was largely the work of wave-swept shell debris similar to the mechanism suggested for gutter casts in subtidal environments (AIGNER & FUTTERER 1978). Further out the mud was completely gone and the re-exposed shell bed had become again colonized by encrusters. Even the *Sabellaria* reeflets had come to life again , evidently by the selective settlement of larvae on the dead, exhumed structures made by earlier generations of the same species.

In February 1982, the situation was essentially the same. The mud remnants in the nearshore zone had gone; but from the seaward edge, newly deposited soupy mud began covering the shell bed again. In December 1983, eventually, things had returned to what they had been 4 1/2 years before: The whole tidal flat was covered again by about 30cm of mud.

Since observations were spotty, the cycle may have actually been shorter than

Fig. 2: Parallel ridges of remnant mud at right angle to the shore line show that mud erosion is not completed during one storm and that it works by shell-grinding similar to the erosion of gutter casts (AIGNER & FUTTERER 1978). In the following year (1982) these mud ridges had disappeared so that the shell lag, which in the picture is seen only between the ridges, was exposed over the whole surface.

four years and it may also be very irregular. Nevertheless it is clear that shelly and muddy bottom periods and their respective faunas did alternate many times during the past 60 years. It is also probable that faunal spectra were not exactly the same during corresponding periods, as expressed by the absence of *Lingula* during the last shelly phase. But their shell residues have all been winnowed and projected down on the reference level of the 1920 storm to form a single shell bed of ever increasing resistivity.

B. Application to fossil examples

It is quite obvious that the Jeram model can not be applied to fossil examples on a one to one basis. Firstly, because intertidal deposits as a whole have a very low fossilization potential. Secondly, overall sedimentation rates would not allow the record of such a short time span to be preserved, because still larger events would wipe them out in the long run. Nevertheless the model can tell us what we should expect to happen over much longer time spans in deeper environments, where sedimentation rates as well as event frequencies are considerably lower.

Following the deeply entrenched habit to read the stratigraphic record in terms of positive rather than negative sedimentation, we would tend to relate individual shell beds to single events. Our model, however, tells us that event-condensation on top of an initial erosion-resistant reference horizon is a very common case. In fact, a more careful analysis may reveal within such beds slight stratinomic or faunistic changes that demonstrate their amalgamated nature (MUNDLOS 1978, KIDWELL & AIGNER, this volume).

The same is true for biohermal buildups. The stromatolite-like reeflets of the small encrusting bivalve *Placunopsis ostracina*, for instance, have been used by WAGNER (1936) to measure sedimentation rates in the German Muschelkalk sea. Based on the correct assumption (BACHMANN 1979) that the previous layer of encrusters had to be dead and stripped of the free upper valves before the next generation could settle on top, he came to an order of 10 cm/1000 yrs. But this estimate is probably too high. Careful studies (HAGDORN 1982) have revealed in these reeflets levels with many microborings that mark interruptions in reef growth. The fillings between the reeflets also show signs of interrupted sedimentation. Thus we probably deal with a situation like in the Jeram *Sabellaria* reeflets: repeated burial and selective recolonization after the structures had been exhumed again by erosive events.

Another example is a condensed horizon in the Bajocian (Mid-Jurassic) of the Normandy coast. In its upper, oolithic part, it contains a mixed assemblage of ammonites belonging to 2-3 zones (SEILACHER 1982, Fig. 2). Their perfect preservation, however, is in conflict with the notion that these shells had been lying at or near the sea floor for millions of years during a time of non-sedimentation. A detailed study (FÜRSICH 1971) has in fact shown that the heterochronous assemblage is not only mixed, but also size-grained. Since reworking could have happened in one or only a few events during a million years of burial time, with exposure only for a few hours, shell preservation would have been affected very little and only in a mechanical sense. Particularly interesting in this case is the fact that other expressions of event condensation preceed the oolite deposition, and that all are better understood if repeated burial/unburial cycles are added to the one-way picture. But the example also shows that event-condensation is not an isolated feature but characterizes, in various expressions, larger cycles of subsidence or sealevel-fluctuations.

C. Possible evolutionary effects

As has been pointed out earlier, it is only through such larger cycles, combined with the effect of taphonomic feedback (KIDWELL & JABLONSKI 1983, KIDWELL & AIGNER, this volume), that we may expect an influence of relatively small and local events (such as storms) on evolution. The change from muddy to shelly bottom conditions over longer periods will primarily affect the benthic biota. But through trophic chains the substrate change may also influence organisms that are normally attributed to the nekton. Whether the biotic response is only ecologic in nature or whether it also reaches the genetic level, will depend largely on the size of the populations and their degree of isolation. It is one of the aims of this book to examine cases in which such mechanisms could have worked.

References

Aigner, T. & Futterer, E., 1978: Kalk-Töpfe und -Rinnen (pot and gutter casts) im Muschelkalk - Anzeiger für Wattenmeer?.- N.Jb.Geol.Pal., Abh., 156: 285-304.

Bachmann, G.H., 1979: Bioherme der Muschel *Placunopsis ostracina* v. SCHLOTHEIM und ihre Diagenese.- N.Jb.GeolPaläont.Abh., 158: 381-407.

Fürsich, F., 1971: Hartgründe und Kondensation im Dogger von Calvados.- N.Jb.Geol. Paläont.Abh., 138: 313-342.

Hagdorn, H., 1982: The "Bank der kleinen Terebrateln" (Upper Muschelkalk, Triassic) Near Schwäbisch Hall (SW Germany) - a tempestite condensation horizon.- In: EINSELE & SEILACHER (eds.): Cyclic and event stratification, Springer-Verlag: 263-285.

Kidwell, S. & Jablonski, D., 1983: Taphonomic Feedback. Geological Consequences of Shell Accumulation. In: TEVESZ, M.J.S. & MCCALL, P.L (eds.): Biotic Interactions in recent and fossil benthic communities.- Plenum Publ.Corp. 195-248.

Mundlos, R., 1978: Terebratulid shell beds. In: AIGNER, T., HAGDORN, H. & MUNDLOS, R.: Biohermal, biostromal, and stem-generated coquinas in the Upper Muschelkalk.- N.Jb.Geol.Pal.Abh., 157: 45-47.

Seilacher, A., 1982: General remarks about event deposits. In: EINSELE & SEILACHER (eds.): Cyclic and event stratification, Springer Verlag: 161-173.

Wagner, G., 1936: Riffbildung als Maßstab geologischer Zeiträume.- Aus der Heimat, 49: 157-160.

DYNAMIC STRATIGRAPHY OF THE UPPER MUSCHELKALK, SOUTH-GERMAN BASIN

SUMMARY

T. Aigner
Tübingen

Although the descriptive stratigraphy of many epicontinental sequences is well established, dynamic processes in such marginal seas is far from being fully understood. This study presents an approach towards a "dynamic stratigraphy" (MATTHEWS, 1984) of the Upper Muschelkalk in SW-Germany that may be applicable to other storm-dominated basins. The lithostratigraphic subdivision of the Upper Muschelkalk (Middle Triassic) is well known from earlier workers (for summary see GEYER & GWINNER, 1968). Dynamic processes in the shallow-marine Upper Muschelkalk sequence are reconstructed in a three-level stratigraphic analysis (Fig. 1).

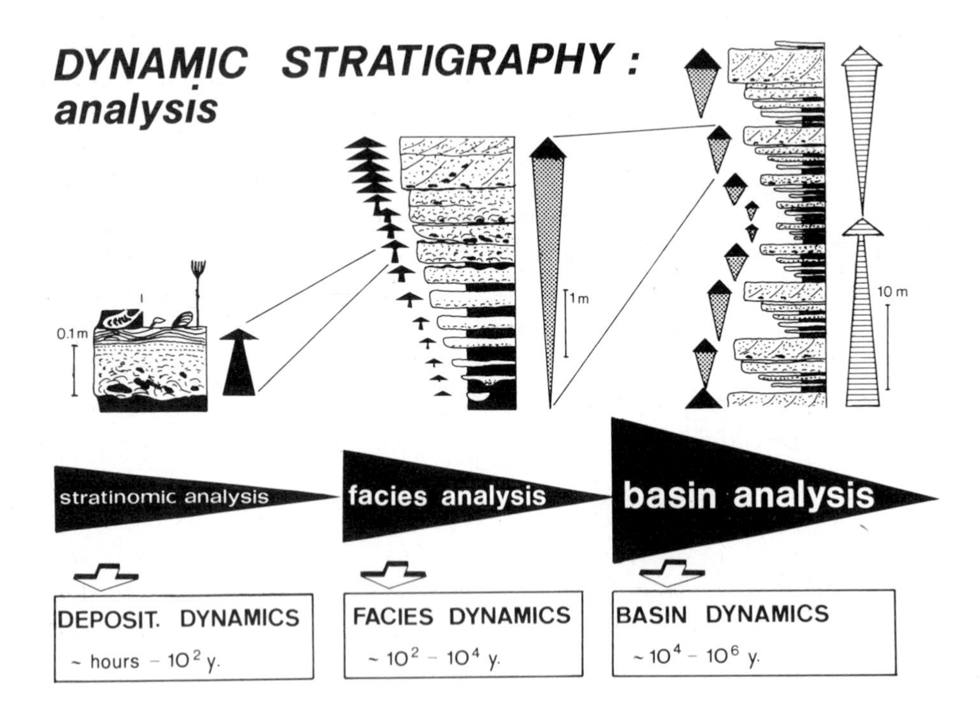

Fig. 1. Three-level hierarchical approach towards a dynamic stratigraphy of storm-dominated basins.

1. Stratinomic Analysis

Most stratification types in the Upper Muschelkalk show evidence for episodic, storm-related processes. In the open-marine facies, bedding is dominated by 1-30 cm limestone sheets with sharp, erosional bases marked by bipolar tool marks and overlain by fining-up sequences as well as parallel, low-angle, hummocky and wave-ripple lamination. Bed tops are commonly bioturbated. Many beds are clearly composite and include the amalgamation of several events. The particular association of sedimentary structures indicates deposition from storm-induced combined oscillatory/unidirectional flows under waning energy (tempestites, see AIGNER, 1982). Hydraulic parameters deduces from these tempestites are compatible with those known from modern continental shelf storm sedimentation. Lateral proximality trends and facies transitions suggest a gently inclined carbonate ramp depositional system.

2. Facies Analysis

Commonly, storm beds in the Upper Muschelkalk are cyclically arranged into 1-7 m thick coarsening- and thickening-upwards sequences, that record an upward transition from distal to proximal tempestites, i.e. progressive shallowing. Several types of shallowing-upward sequences (e.g. oolite grainstone cycles etc) are commonly characterized by systematic changes in the molluscan and trace fossil associations in response to changing substrate conditions. Widespread changes of soft into firm and shelly substrate associated with shallowing allowed in several instances for virtually instantaneous and geographically widespread colonisation of cycle tops by specific brachiopods and crinoids. The amaglamated and condensed tops of these "ecologically fingerprinted" shallowing-up sequences (e.g. Spiriferina-Bank; Holocrinus-Bank, see HAGDORN, this volume) have been used as marker beds, although their cyclic context had so far not been recognized. Similarly, prominent marlstone horizons have long been used in lithostratigraphic correlation and named as "Tonhorizont α, β, γ ...". The recognition of cyclicity now identifies these marlstone horizons as the basal units in shallowing-upwards sequences.

3. Basin Analysis

Minor shallowing-upward sequences can be correlated over many ten's of km, many of them over the entire basin. Towards the basin margin, however, some cycles are only locally developed, others show "amalgamation" or "condensation" of a higher order in areas of paleohighs.

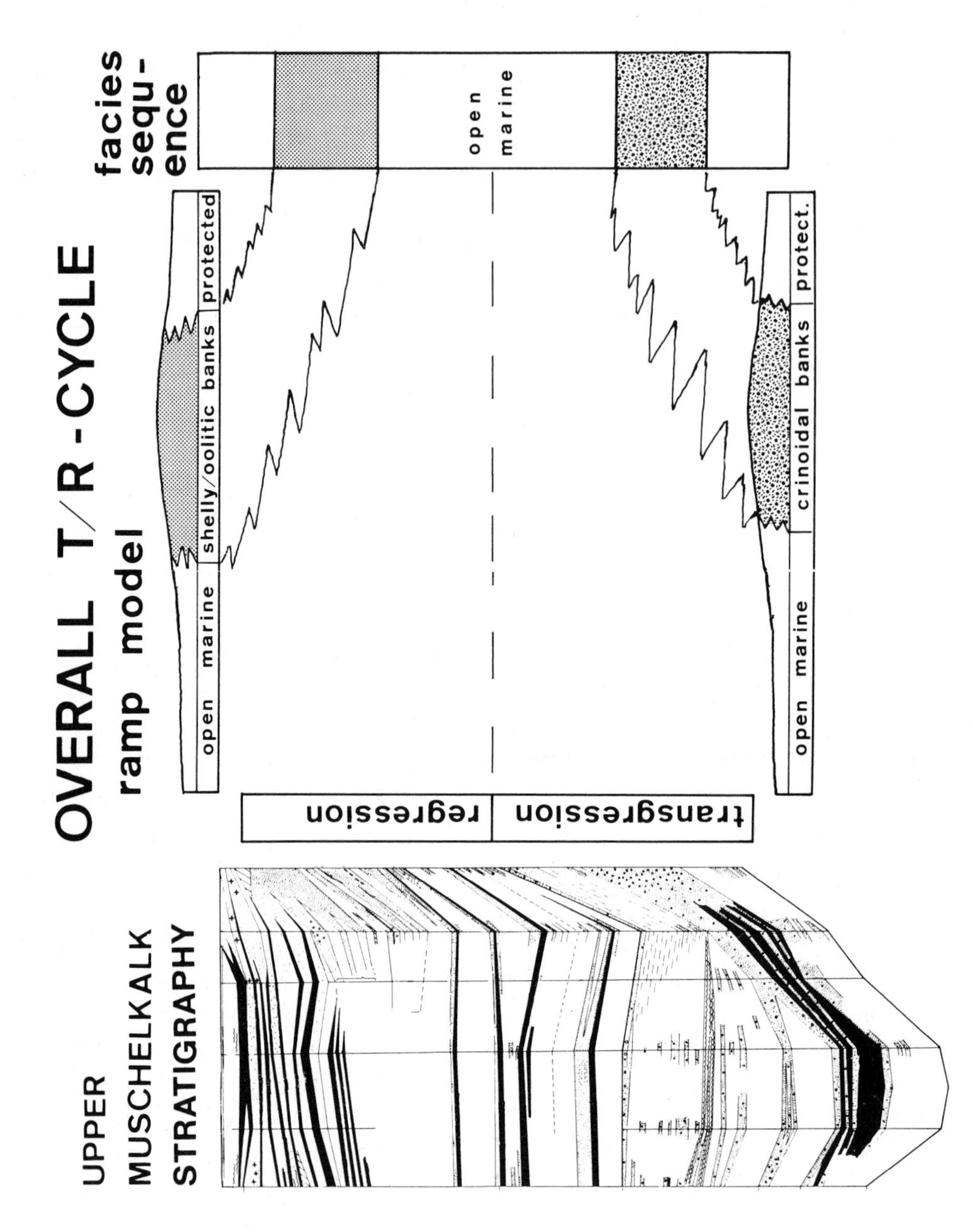

Fig. 2. In the established Upper Muschelkalk lithostratigraphy of the South-German Basin (left, after GEYER & GWINNER, 1968), marlstone horizons (black, most prominent in the basin center) can now be regarded as transgressive bases, and massive skeletal/oolitic units (stippled, most prominent at the basin margin) as regressive tops of minor transgressive/regressive sequences. These asymmetrical cycles punctuate the overall, nearly symmetrical Upper Muschelkalk trans/regressive cycle (right, strongly schematic).

Vertically, large parts, if not all, of the Upper Muschelkalk is composed of successive asymmetrical shallowing-up cycles of various types, similar to "Punctuated Aggradational Cycles" (GOODWIN & ANDERSON, 1980). Successive minor cycles, however, constitute in turn an overall, transgressive/regressive cycle forming the entire Upper Muschelkalk (Fig. 2). In contrast to the minor cycles, this overall cycle is nearly symmetrical and comparable in thickness and duration to the "third-order cycles" of VAIL et al. (1977). Since a large-scale late Anisian/Ladinian transgression is recognized around the entire Mediterranean, it is possible that the overall Upper Muschelkalk cycle is eustatically controlled (BRANDNER, 1984).

Conclusions
(Fig. 3)

1. Within the carbonate ramp setting of the Upper Muschelkalk, most stratification types show evidence of episodic storm-related processes, comparable to actualistic models.
2. Minor asymmetrical shallowing-upward cycles record repeated small-scale transgressive/regressive shifts of the carbonate ramp system.
3. The overall symmetrical Upper Muschelkalk T/R-cycle is punctuated by these minor asymmetrical cycles. The recognition of this cyclicity allows to understand the dynamic causes of the purely descriptive lithostratigraphic subdivision currently used in the Upper Muschelkalk of the South-German Basin.

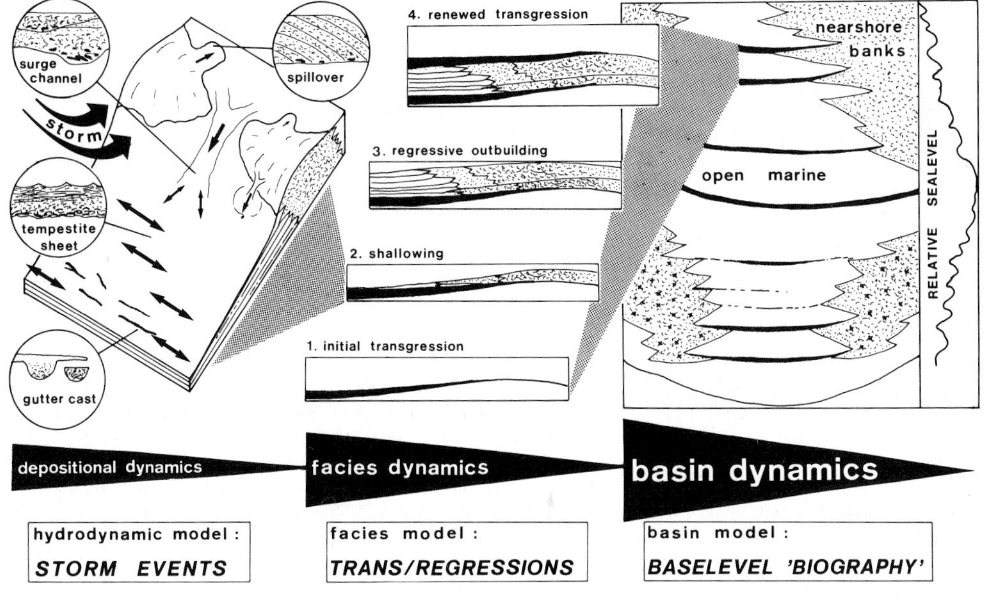

Fig. 3. Interpretation of dynamic processes on three levels of stratigraphic sequences.

REFERENCES

Aigner, T. 1982: Calcareous tempestites: storm-dominated stratification in Upper Muschelkalk limestones (Middle Triassic, SW-Germany), in G. Einsele & A. Seilacher (Eds.), Cyclic and Event Stratification, p. 180-198. Springer.

Brandner, R. 1984: Meeresspiegelschwankungen und Tektonik in der Trias der NW-Tethys: Jb. Geol. B.-A., 126: 435-475.

Geyer, O.F. & Gwinner, M.P. 1968: Einführung in die Geologie von Baden-Württemberg, Schweizerbart.

Goodwin, P.W. & Anderson, E.J. 1980: Punctuated aggradational cycles: a general hypothesis of stratigraphic accumulation, Geol. Scoc. Am., Abstr. with Progr., 12: 436.

Matthews, R.K. 1984: Dynamic stratigraphy, Prentice-Hall (2nd Ed.).

Vail, P.R., Mitchum, R.M. & Thompson, S. III. 1977: Seismic stratigraphy and global changes of sea level, Part 4: Global cycles of relative changes of sea level, Am.Assoc. Petrol. Geol., Mem. 26: 83-97.

ENVIRONMENTAL EVOLUTION IN SHALLOW EPICONTINENTAL SEAS:

SEDIMENTARY CYCLES AND BED FORMATION

Ulf Bayer, Ewald Altheimer, Walter Deutschle

Tübingen

Abstract: Asymmetric cycles or coarsening-upward sequences are reconsidered as sedimentological evidence for changing physical environments. Cyclicity by itself, however, is not a sufficient control parameter because a cycle usually consists of a hierarchical system of superimposed cyclic and rhythmic patterns in space and time, down to the individual depositional, erosional and ecological events. This hierarchical pattern needs to be carefully analysed until a chart of regional cyclicity can be established. Temporally and spatially, cycles may become reduced to single condensed beds or split into thickening-upward sequences of "event beds" towards the distal ends.

The spatio-temporal pattern of facies associations projects only partially onto proximity gradients. Short term erosional and ecological events occasionally drive the sedimentary system out of its equilibrium and induce biological and diagenetic reactions. Depending on the magnitude of the event, their frequency and the general state of the system, we expect either a rapid return to the original sedimentary equilibrium or the establishment of a new sedimentary system. The typical 'roof beds' of cycles (which are usually oolitic) are considered to represent such a new sedimentary system, from which ferruginized oolites may be dispersed over the basin to form ironstones.

Reworking by repeated erosional events and reactivation of earlier diagenetic and biological responses to events are considered major processes in changing the sedimentary environment. The formation of ferruginous ooids and of ironstones is discussed as a product of fluctuating sedimentation, erosion and winnowing.

INTRODUCTION

The concept of cycles provides a useful tool in both sedimentology and paleontology. Repetitive patterns of facies and faunas provide repeated natural experiments under similar constraints, from which likely generalizations can be drawn. The concept of cycles has, of course, been criticized for several reasons as READING (1978) pointed out: The establishment of cycles is too often subjective, and the cycle commonly becomes more important than the sedimentological (or faunistic) patterns; and the discussion of cycles leads directly to a discussion of their causes, which likely terminates in speculations about the ancient dynamic sedimentological regime. Finally, the concept of cycles is *a priori* a gradualistic concept (READING, 1978, p. 5):

> *"Since the use of cycles is based on the idea that there is a regularity to sedimentary sequences and that sedimentation is a normal steady process apparently random events are commonly neglected, although, in some environments, they may dominate sedimentation".*

The cycles discussed here are classical asymmetric 'Klüpfel-cycles'-- coarsening --upward sequences -- which are commonly paralleled by increasing carbonate content and ammonite faunal changes (KLÜPFEL, 1917; HALLAM, 1961; BAYER & McGHEE, 1984; this volume).

Cycles, indeed, have usually been viewed as gradualistic sequences, and so have the associated phenomena been viewed. Explanations of the cyclic pattern have usually been given on the regional level (e.g. KLÜPFEL, 1917; ALDINGER, 1957, 1965). Much less attention has been drawn to the local discontinuous patterns within the sequences and to the interaction between deposition, erosion, diagenesis and benthic faunas.

A cycle, or coarsening upward sequence, indicates at least a local temporal change in the sedimentary conditions and, therefore, a change of the local bottom conditions. The question, indeed, is how far such a cycle reflects a gradual environmental trend or alternatively a discontinuous accumulation of short time events which disturbed the usual bottom conditions only locally. The sediment is the only system -- besides the fauna -- from which we can gather environmental information.

Environmental conditions affect organisms, and vice versa organisms affect the bottom conditions, the uppermost layer of the sediment, its homogeneity, its chemistry -- various organisms produce shells and other hard parts which then for example provide a source for coarse material. One should expect that locally some feedback mechanism exists -- interactions between sediment, depositional processes, fauna, and diagenesis -- and that some external disturbances will have a remarkable impact on this system.

To illustrate some of these complex relationships is our aim although this is only a preliminary paper. More detailed descriptions will be given elsewhere; the examples from the Lower Muschelkalk discussed herein are from E. ALTHEIMER (thesis in preparation).

1. CYCLIC PATTERN

The relation between asymmetric cycles and basin configuration will be illustrated using the example of the South German basin during Upper Aalenian/lowermost Bajocian times. The simplified paleogeography and facies distribution is given in Fig. 1 (for details see BAYER & McGHEE , 1984, this volume). A cross-section (Fig. 1) along the present

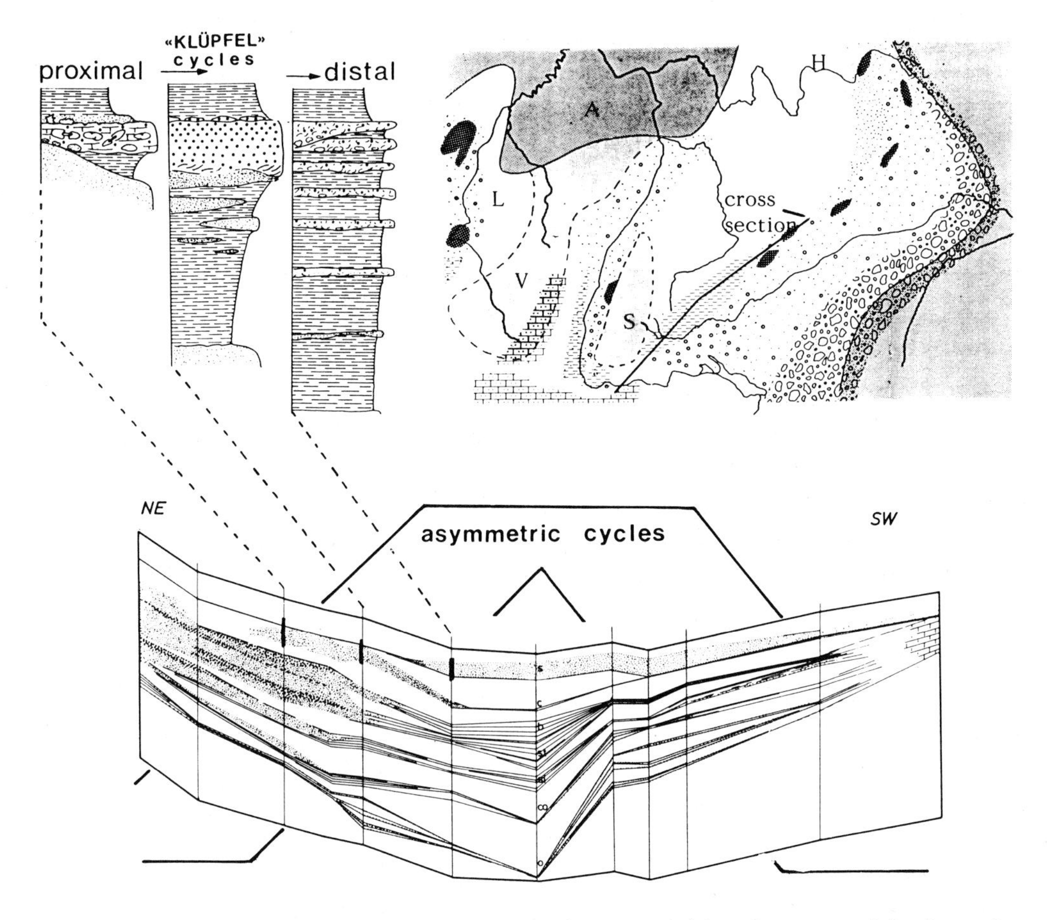

Fig. 1: Generalized facies distribution within the upper Aalenian of South Germany and parts of Lorraine (L: Lorraine, V: Vogesian swell, S: Black Forest swell, A: Ardennen island, H: Hessian strait). The dominating patterns are iron-oolitic offshore bars which parallel the coastline. Finer details of lithofacies patterns are illustrated in a cross--section of upper Aalenian -- lowermost Bajocian strata along the present outcrop line of the Swabian Alb. The stratigraphic and geographic distribution of major sand bodies (stippled regions), muddy carbonate beds (black lines), shale horizons (white regions) and ironstones (heavy stippled) are given. The offlapping of sandstone bodies indicates a major regression trend which is subdivided into minor asymmetric cycles or coarsening-upward sequences (upper left corner) which occur within a well defined facies association. Modified after BAYER & McGHEE (1984, this volume).

outcrop line at the Swabian Alb provides the essential lithostratigraphic information. A dominant pattern in the facies are iron-oolite bars and sand bars which parallel the coast line. These offshore bars occur usually as "roof-beds" at the top of asymmetric sedimentary cycles (WERNER, 1959; SAUER, 1956; BAYER, 1970). The iron ores of the 'Minette' (Lorraine, western margin of the paleogeographic map of Fig. 1) show the same cyclic pattern with the iron ores in the position of "roof beds" (LUCIUS, 1945, 1948; THEIN, 1975). The cycles are minor structures, as indicated in Fig. 1, which may repeat several times within a single profile. On the other hand, the entire sequence represents one major regressive cycle (WEBER, 1967; BAYER & McGHEE, 1984, this volume). Thus, cyclic patterns occur at various scales.

1.1 Minor asymmetric cycles

The cross-section along the Swabian Alb (Fig. 1) illustrates that nearly gradual coarsening-upward sequences or asymmetric cycles are restricted to a certain facies belt, which is likely related to water depth. The minor cycles occur symmetrically in the northern sand-facies and the southern clay-facies. Through time (Upper-Aalenian/ lowermost Bajocian in the cross-section) the zone of cycles shifts towards the center of the basin. This causes a typical offlap of sandstones in the northeastern part of the basin (WEBER, 1967; BAYER & McGHEE, 1984), reflecting a general regressive trend throughout the time interval. The repeated minor asymmetric cycles, therefore, represent a specific sedimentary pattern within a spatio-temporal facies association.

Proximally, the minor cycles reduce to simple condensed beds with common signs of reworking and concentration of prefossilized ammonites. This pattern caused ÖCHSLE (1958) to assume a catastrophic ammonite extinction during the "*sowerbyi*-Zone" (= *laeviuscula*-Zone, lower Bajocian). A sedimentological analysis of the bed, however, shows that the condensed ammonites were prefossilized and overgrown by serpulids, oysters, bryozoans and sessile foraminifera. Serpulids on the lower side of the prefossilized ammonites indicate repeated reworking and cast formation. The degree of condensation in "roof beds" increases as beds are traced into proximal areas, and also increases within vertical sections with the regression. The "*sowerbyi*-bed" marks the extreme regression state in a proximal position.

Distally, the cycles split up into a sequence of single beds -- rhythms -- which ideally are thickening-upward sequences (Fig. 1). Redeposited clasts from shallower areas (Fig. 2, rounded sandstone pebbles, angular carbonate clasts with ooides for which no source rock is locally available), occasionally graded beds and 'rolling-up' structures of sheet sands indicate an "event-deposition" within the deeper parts of the basin. The thickening-upward sequences of the event beds and the usually much better developed thinning-upward sequences of the interbedded clays indicate an increasing event intensity and an increasing event frequency within the distal record of the minor asymmetric cycles.

This spatio-temporal pattern of minor asymmetric cycles portrays a proximity gradient (SEILACHER, 1982). However, the interpretation of locally observed coarsening-upward

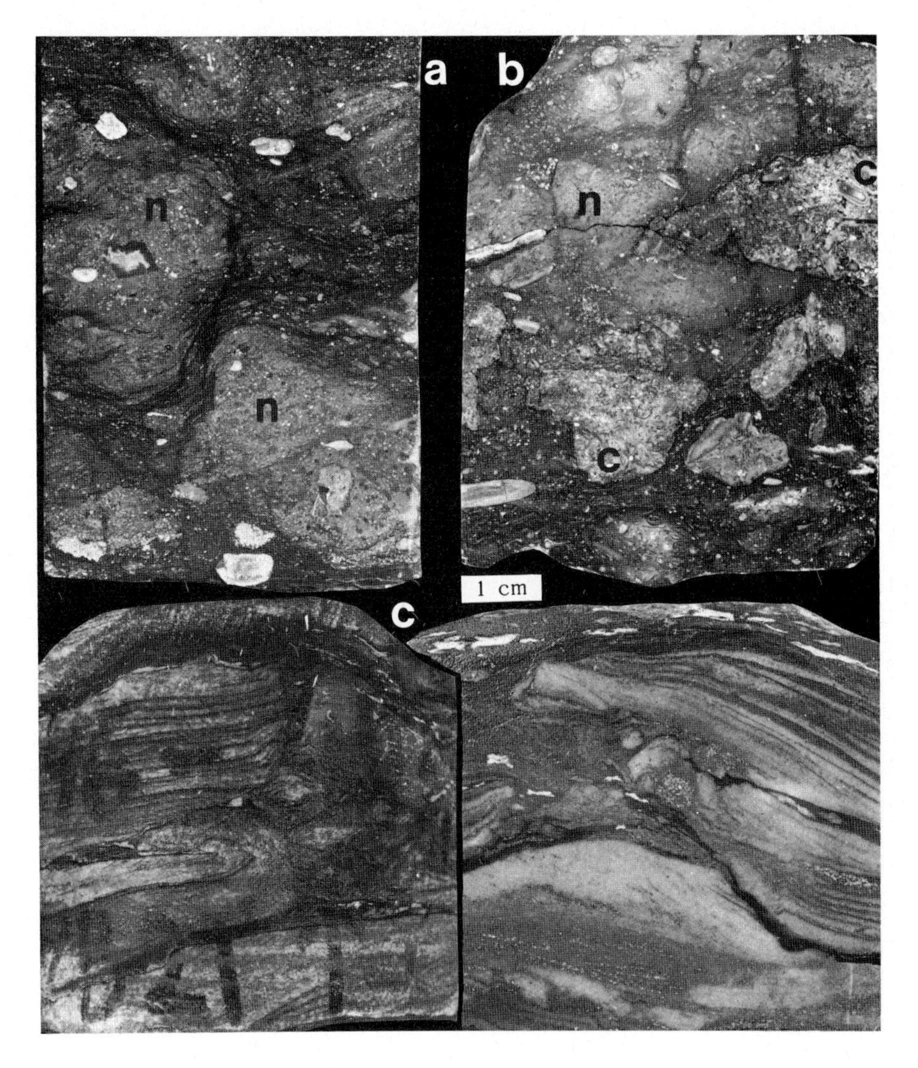

Fig. 2: Distal event beds of the argillaceous facies. a,b: Redeposited clasts (c - larger clasts), shells, and ooids which have been early cemented by formation of sideritic nodules (n). c: 'Roll up' structures of distal sheet sands.

cycles as proximity trends should be handled with care. Two profiles from the upper Aalenian and the lower Muschelkalk of South Germany are given in Fig. 3 which are nearly identical in terms of cyclicity. For both profiles an overall regressive trend is well established: The Aalenian profile reaches from the *comptum*-Zone to the *concava*--zone; the lower Muschelkalk profile terminates at the transition to the salinar facies of the Middle Muschelkalk. An increasing condensation of the smaller cycles should be expected because of the general regressive trends. In these profiles, however, every minor cycle has a long initial phase of clay sedimentation, and the overall trend becomes only visible in the "roof beds". The thickness of these beds increases from the first

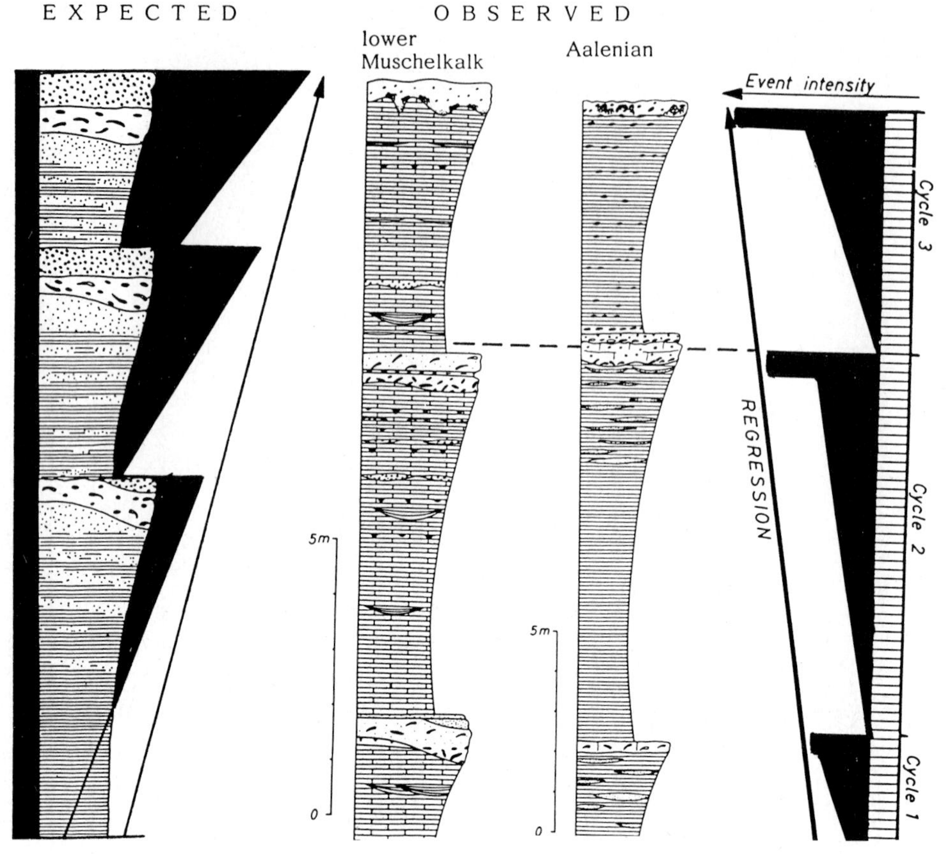

Fig. 3: Asymmetric cyclic patterns in the lower Muschelkalk and the upper Aalenian. Although an overall regression trend is well established for both profiles, the expected 'condensation' of minor cycles during the course of the regression does not occur. In contrast, the erosional base of the 'roof beds' appears as a sharp break in event intensity (near bottom flow velocity) as indicated.

to the second cycle. While the erosional phase at the base of the first two cycles is equally developed, oolites become dominant in the second phase. The third cycle, then, is terminated by stromatolites in the Muschelkalk example and by algal mats and small serpulid-foraminiferal bioherms in the Aalenian example. In both cases, the third bed has a lagoonal characteristic with oncoids and algal mats. Thus, the three cycles terminate with quite different sediments which, in sequence, support the otherwise recognized overall regressive trend, but the sediments, which precede these beds within each cycle do not reflect this trend, the sequences do not reflect a proximity gradient as would be expected. The minor cycles appear truncated. After an initial phase within a normal, gradual coarsening-upward sequence, a discontinuity in deposition terminates the cycle abruptly -- the "roof bed" follows over an erosional base and within the "roof bed" a fining-upward sequence is usually developed: After a more or less sharp erosional junction follows a zone of reworked pebbles and coarse shell material which subsequently is replaced by oolites. This pattern could be interpreted as the initial phase of the transgression (EINSELE, this volume). However, the "roof bed" is usually terminated by a final erosional surface with lithofied pebbles from the oolitic bed -- a feature which points to the highest erosional energy at the top of the oolites which themselves are commoly crossbedded bars which point to high energy levels (HALLAM & BRADSHAW, 1979). In the two discussed cases, the cycles reduce nearly to rhythms with rather abrupt facies replacements. The erosional base of the "roof beds" marks environmental changes which precede oolite formation on one hand and faunal replacements (BAYER & McGHEE, this volume) on the other.

1.2 Major Klüpfel-cycles

Within the Upper Aalenian/lower Bajocian profile the minor asymmetric cycles (Fig. 1) form a succession which itself appears as a major thickening-upward sequence. In the course of the overall regressive trend, the "roof-beds" increase in thickness and contain an increasing number of erosional and reworked horizons. Local sections -- single profiles -- of the entire sequence appear, therefore, as a major asymmetric cycle. Such cyclothems have been described by KLÜPFEL (1917) from Lorraine. Klüpfel's examples are redrawn in Fig. 4. The dominant pattern within his examples is a gradual lithological change, gradual in the sense of a thickening-upward sequence of the limy beds. The typical Klüpfel-cycle (HALLAM, 1961), therefore, is a rhythmic sequence -- oscillations of

Fig. 4: Major Klüpfel cycles modified from KLÜPFEL (1917). The cycles are composed of clay-carbonate rhythms. From left to right the clays are increasingly replaced by carbonates as they are during the course of individual cycles.

argillaceous rock and carbonates -- whereby the quantities of the two alternating lithologies change through time. The thickness of clay beds decreases as the thickness of limy beds increases. If sands are present, the rhythms develop usually into minor cycles (Fig. 1) whereby the initial argillaceous phase is suppressed during the course of the major cycle while the final "roof-bed" phase becomes more pronounced.

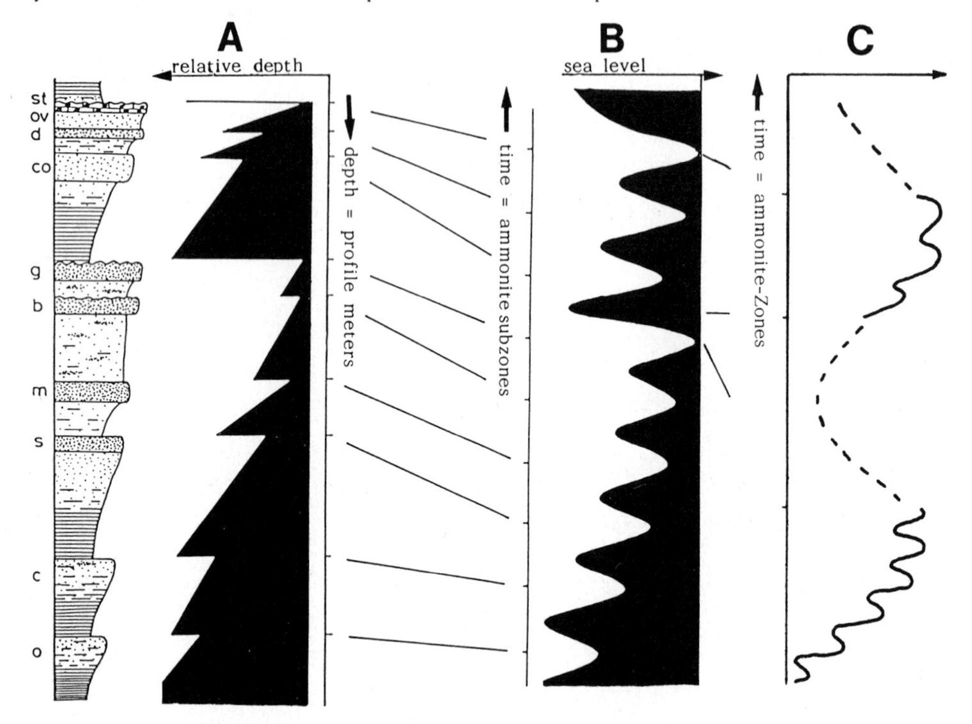

Fig. 5: Cyclic pattern of sedimentation during the upper Aalenian and lowermost Bajocian (after BAYER & McGHEE, 1984). On the stratigraphic level (A), the early transgressions appear as sharp events due to non-deposition. On the time scale (B,C) symmetric cycles are likely whereby condensation and non-deposition phases cause gaps on the level of ammonite zones (see text). Ammonite zones indicated by letters: o, *opalinum* Zone; c, *comptum* Zone; s, *sinon* subzone; m, *staufensis* subzone; b, *bradfordensis* subzone; g, *gigantea* subzone; co, *concava* s.s. subzone; d, *discites* Zone; ov, *ovalis* subzone; st, *stephani* subzone.

A hierarchical cyclic pattern may be found if the thickening-upward sequence consists of minor cycles. Such a 'hierarchical structure' of Klüpfel cycles is especially well observable in the Aalenian example (Fig. 5). The major regressive cycle breaks down into a series of smaller cycles -- repeated thickening-upward sequences -- which again consist of repeated minor cycles or coarsening-upward sequences. As the generalized section (Fig. 5) illustrates, at least three magnitudes of cycles can be recognized:

** the large scale regression trend = stage-level,

** an intermediate cyclic pattern = ammonite-Zones,

** the small scale cycles = ammonite-subzones.

Two (locally three) of the intermediate cycles are recognizable. They terminate in the *bradfordensis-gigantea* subzones (in the *concava*-Zone) and in the *laeviuscula*-Zone. The terminal "omission" of these cycles is well developed and regionally widespread. For example the *bradfordensis-gigantea* discontinuity terminates the major phase of iron-ore formation throughout South Germany and Lorraine.

Thus, we have not only coarsening-upward sequences which are typical for a certain spatial facies zone and not -- as Klüpfel pointed out -- a single major cycle, but a hierarchical system of cyclic patterns of different magnitudes which are superposed one upon the other.

The usual interpretations of asymmetric cycles are changing water depth or varying strength of near bottom flow (Fig. 5). If relative water depth is drawn along the stratigraphic column (Fig. 5A), the transgressions appear as sharp events because of the asymmetry of the lithological sequence where clays overlay the terminal hardground of the cycles. If the sequence is mapped onto a relative time scale, however, a rather symmetric, gradual oscillation appears. Thus, on the level of ammonite subzones (BAYER & McGHEE, 1984), a sinusoidal pattern of near-bottom flow energy appears likely for the minor asymmetric cycles while the intermediate cycles are still asymmetric.

When ammonite zones are applied as time scale, the medium scale cycles transform also into a sinusoidal pattern (Fig. 5c). Irregularity in the pattern results from the different numbers of minor cycles within each medium cycle, and a clear condensation trend throughout the course of the medium cycles becomes visible (Fig. 5C). However, the different number of minor cycles and ammonite-subzones can simply be related to the general condensation trend. Most of the ammonite-subzones are, of course, local features -- e.g. subzones are defined by the genus *Staufenia* , an endemic evolutionary lineage (BAYER & McGHEE, this volume). Such a regular pattern of superposed, inferred sinusoidal cycles suggests regular underlying oscillations which could have their causes in astronomical periods (SCHWARZACHER & FISCHER, 1982; FISCHER, 1981, 1982; EINSELE, this volume). However, the previously discussed proximity gradient of cyclic patterns and the progressive condensation of minor cycles in the course of larger cycles show that a locally observed cycle is only a local pattern, the expression of a certain sequence of sedimentological states at that particular point.

1.3 Causes of cyclicity

The fractional pattern of cycles, the dimension or scale one uses to "measure" cycles, provides one root for confusion. Klüpfel originally explained the cyclic patterns by local epeirogenetic movements. Local, synsedimentary tectonics is well known from his study area Lorraine (LUCIUS, 1940; MULLER et al. 1976), and this local tectonics controlled to some extent the spatial and temporal facies distribution. Klüpfel's cycles, however, are of the magnitude of stages while other cyclic patterns have much less duration times (ammonite zones or subzones). The epeirogenetic explanation was mainly criticized by those who studied these minor cycles (e.g. ALDINGER, 1957, 1965). On the level of minor cycles rather rapid epeirogenetic movements would be required. Other causes, therefore, have been considered, including:

changes in current strength,
sea level,
climate,
rate of subsidence etc.

Most of these possible causes have been already controversially discussed by KLÜPFEL (1917). Fig. 6 gives a list of several possible causes which are grouped into three major processes which can be used to explain single asymmetric cycles, coarsening-upward sequences or thickening-upward sequences:

A: falling sea level,
B: relatively rising sea floor,
C: increasing depth of storm wave base.

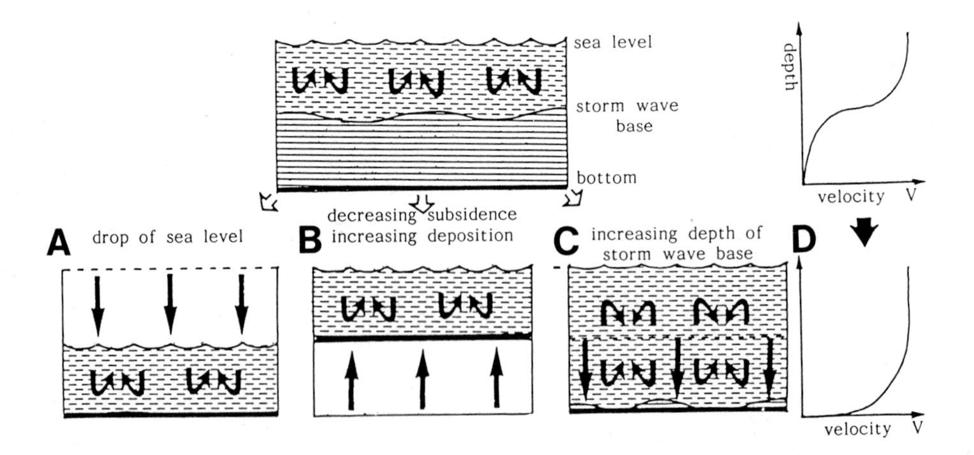

Fig. 6: Three possible major causes for cyclic sedimentary patterns -- they all can be subsumed under the relevant parameter 'near bottom flow velocity'.

But these different causes can be subsumed under the single relevant parameter "near bottom flow velocity" (Fig. 6D) which, of course, is affected by all cited causes and which is lastly responsible for the spatial and temporal distribution of grain sizes:

A locally observed coarsening-upward sequence indicates an increase in the relative parameter "local flow intensity".

Another widely discussed feature is the asymmetric nature of the cycles. Although the change in flow intensity is likely a smooth process, the sediment reflects usually only the increase phase of flow intensity change. Again various explanations of the asymmetric sediment supply are possible which, of course, are correlated to the general explanation of cyclicity. Such explanations are:

** retainment of sediments near the coast during a transgression;

** interruption of sediment supply during the high regression state if e.g. a strait closes (e.g. the Hessian strait or the Eifel depression in South Germany and Lorraine);

** redeposition of muds and sands into deeper parts of the basin and, therefore, reduced availability during the early phase of the transgression (hysteresis in sediment supply);

** change of weathering conditions correlated with the depth of storm wave base in the course of climatic changes.

All these factors can be summarized in terms of 'sediment availability' or in terms of the relationship between local sediment supply and erosion/redeposition. On the level of a kinematic description, a sedimentary cycle reflects the local evolution of two relevant parameters:

1) the "mean maximal flow power", the strength of near bottom currents which was summed over long time periods, i.e. the stratigraphic sequence may well have resulted from purely episodic or random depositional and erosional events.

2) The local equilibrium between sediment supply and erosion/redeposition.

These parameters are independent of the magnitude of the cycle while the term "locally" depends in its meaning on the magnitude of a cycle -- e.g. the sedimentary basin in the case of Klüpfel cycles or a very restricted area within the basin in the case of minor cycles. At least minor cycles, therefore, can be described in terms of a local kinematic model -- by some suitable equilibrium conditions -- which in detail are discussed by EINSELE (this volume). On the other hand, a local facies sequence or cycle reflects local temporal environmental changes which usually are related to 'erosional energy' (HALLAM & BRADSHAW, 1979). An important aspect is the hierarchy of cycles -- the superposition of oscillations of different phases -- or, alternatively, the accumulation of events of different magnitudes -- event beds, minor cycles etc. -- what results in a discontinuous non-gradualistic evolution of sedimentary environments. Such discontinuous patterns are commonly mature "marker-beds" and oolitic "roof beds" which will be discussed in detail in the next section.

2. SEDIMENTARY CYCLES AND BED FORMATION

A cyclic sedimentary pattern in the stratigraphic record, whatever its magnitude or composition, documents a local change in depositional conditions. These changes in substrate conditions affect the benthic community (e.g. BAYER & JOHNSON, this volume). On the other hand, benthic activities and erosion affect the physical and chemical properties of the upper substrate layer, its homogeneity and stability, porosity and permeability, and the magnitude and direction of local diffusion gradients. The local physico-chemical system and early diagenesis have again important effects on the erodibility of the sediment, the structure of the post-event substrate and, therefore, on the subsequent benthic community. A single event or a sequence of depositional events, therefore, may likely start a process which develops an internal dynamics, a feedback mechanism which results in a specific sedimentary pattern within otherwise monotonous series -- bioherms, of course, are the extreme case of a 'sedimentary' system which is driven by its own internal feedback mechanism between organisms and sediment.

Erosional and depositional events may disturb the surface layer within a short time interval and , therefore, are capable of pushing the physico-chemical system out of its equilibrium. However, the general facies type becomes another important factor.

2.1 Sedimentary patterns of minor cycles

Within a minor cycle (Fig. 7) a sequence of four facies units can ideally be recognized:

** In the early phase (A) muddy sediments dominate which occasionally are interrupted by layers of carbonate (siderite) concretions which, in some cases, can be related to bioturbation events.

** In the second phase (B) storm sand sheets are interbedded with clays. Excellent preserved crinoids and graded shell beds (Fig. 8) indicate event deposition.

** The third phase or the "roof bed" (C) follows ususally over a discontinuous, erosional base (Fig. 8). Commonly the sediment then turns into carbonates, mainly oolites.

** The final phase (D) is, usually, erosional -- an omission or "emersion" horizon which is well documented by hardgrounds or by a layer of concretions which

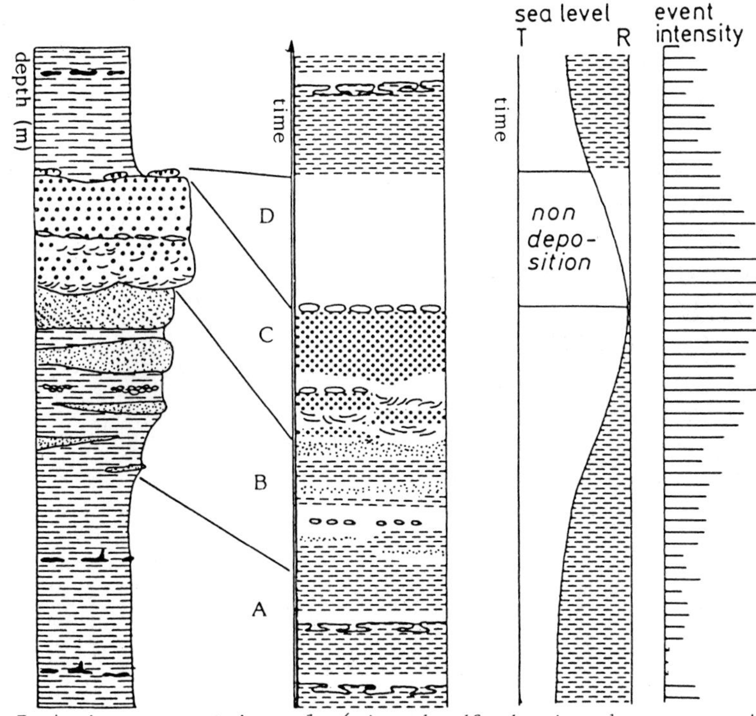

Fig. 7: A minor asymmetric cycle (*sinon-bradfordensis* subzones, Achdorf near Donaueschingen), its stratigraphic appearance, a likely transformation into the temporal sequence and a depositional model -- deposition and erosion events superposed onto a regressive-transgressive cycle. A to D: the four major facies types which form the sequence (A: claystones; B: claystones and sheet sands; C: oolitic "roof bed", D: terminal omission horizon and non-deposition phase of the early transgression).

were reworked from the "roof bed" (ALDINGER, 1957; WERNER, 1959; BAYER, 1970).

Depending on one's point of view, the aim of the study, and the magnitude of the patterns analysed, such sedimentary cycles appear either

** as a regular sedimentary sequence with gradual transitions between facies units,

** or as a sequence of distinct facies units replacing one another,

** or as an accumulation of more or less random events.

The succession of events in minor cycles is discussed here as a main source of environmental changes. In any case, some gradual trend is necessary to produce cyclic patterns. In the case of event sedimentation, this trend has to change either the magnitude or the frequency of events. The two factors are illustrated in Fig. 7: A gradual change of (relative) sea-level alters locally the intensity of events (which are assumed to occur rather regularly, e.g. "seasonally") -- whereas the intensities of events may change randomly. However, the gradual trend could alternatively be viewed as a gradual climatic change and the event intensities of Fig. 7 as 'average number of events per unit time'. The two possible interpretations are just a smooth transformation of an amplitude modulated system into a phase modulated one (BAYER, 1983). In either system neither the temporal picture of the stratigraphic column nor the distribution of erosional and non-depositional phases would change much.

A) Small scale erosion and bioturbation

The first phase (A in Fig. 7) of asymmetric cycles is usually dominated by muddy 'background' sedimentation. In the argillaceous facies the monotonous clays are occasionally interrupted by layers of carbonate (siderite) concretions. In some cases, a biological origin of these nodules is indicated by their horizontal constancy, their nearly circular cross-section and strong elongation, and by common horizontal and vertical branching of the nodules (Fig. 9a). These patterns are most simply explained if the nodules are related to burrows -- perhaps *Thalassinoides* in the Jurassic. The nodules were commonly eroded and accumulated. They then formed patched hardgrounds which allowed oysters, sepulids etc. to settle (Fig. 9b,c). Features, which would indicate deep erosion, have not been observed, and a rather shallow formation of the nodules, therefore, is likely; although the process of concretionary growth continued usually down to rather high overburdens (SEIBOLD,1962; FÜCHTBAUER & MÜLLER, 1977). It was also found by OERTEL & CURTIS that some carbonate concretions started to grow within the pore space of

---

Fig. 8: Shell beds and sheet sands. a: Cross-bedded shell bed at the base of an iron oolite ('roof bed') of the Minette (Lorraine, Thionville). The cross-bedded sequence consists out of *Liostrea* shells. b: Graded *Liostrea*--shell bed (lower Bajocian) with a sequence: larger left valve -- smaller right valves -- crinoid fragments. Cross-section and lower side of bed. c: Cross-bedded sheet sand with layers of crinoid fragments which in part are rather well preserved (arrow).

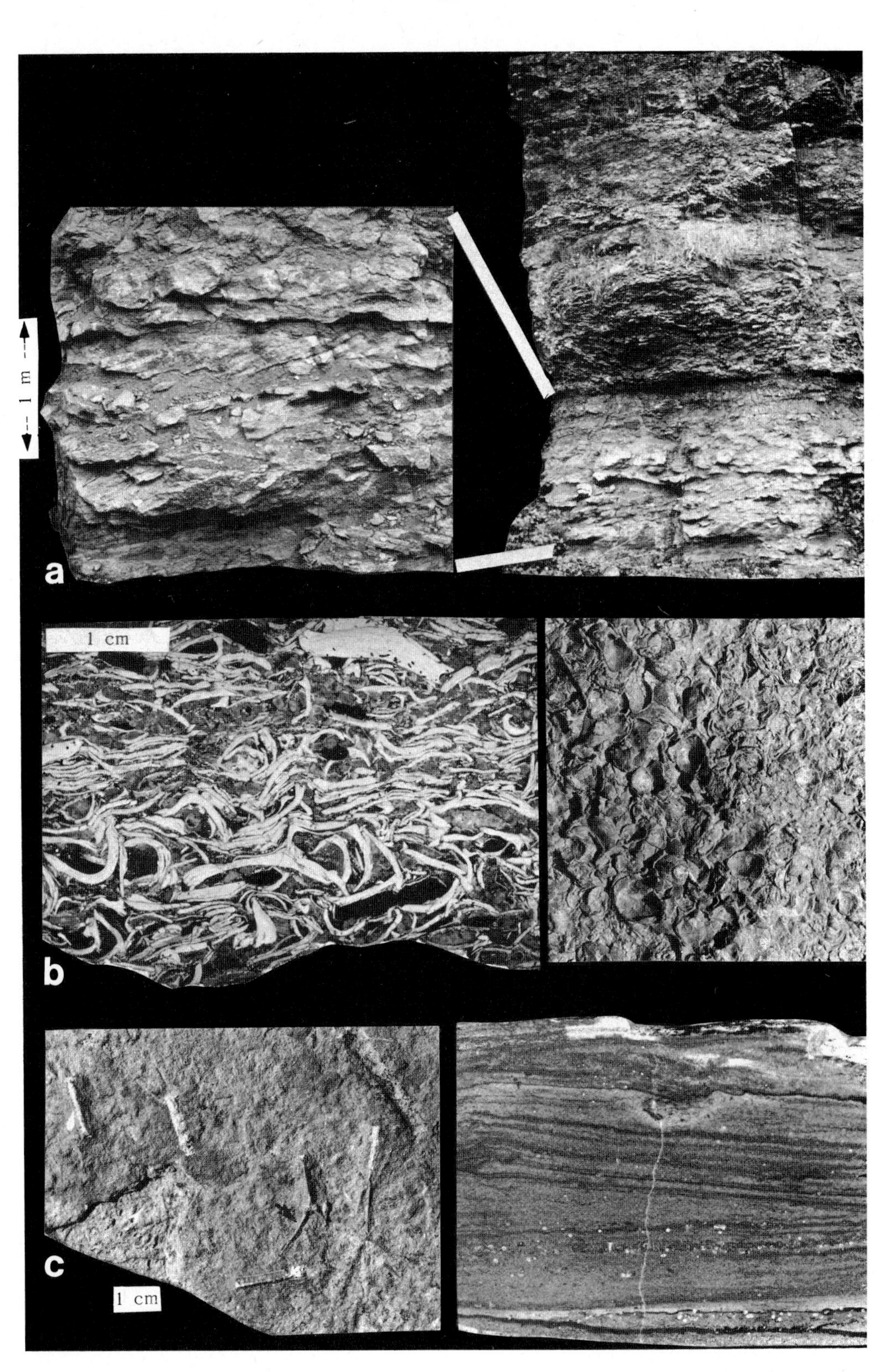
1 m
a
1 cm
b
c
1 cm

Fig. 9: Nodules, 'snuff boxes' and pebble accumulations (upper Aalenian). a: siderite nodules related to *Thalassionoides* burrows. b: reworked pebbles overgrown by oysters and serpulids and boreholes of bivalves. c: 'Snuff box ' -- a sideritic concretion overgrown by serpulids and sessile Foraminifera. d: Sideritic nodules with *Nerites*. e: Pebble agglomerate (rounded sandstones) overgrown by serpulids and oysters. Overgrowth by serpulids at the lower side indicates cast formation and early internal cementation of the pebble accumulation.

a flocculated clay sediment soon after deposition. The early induction of cementation within burrows perhaps can be related to local changes in permeability and to an associated shifting of the sediment/seawater interface into the sediment. The impact of benthic activity on sediments was well recognized by SCHÄFER (1956) and summarized by WEBB et al. (1976). SCHINK & GUINASSO (1977) showed e.g. how silica diagenesis depends on changes in bioturbation rates and how effects of bioturbation on sediment/-seawater interaction influence dissolution. The local horizontal consistency of the concretions and the otherwise missing bioturbation traces indicate a short time physical event which may have caused an ecological event either by altered water or bottom conditions (i.e. by changes of the O_2-content of near bottom water) or by formation of a stiff to firm ground (e.g. by packing under wave or current action). The original extremely high water content of the muds is illustrated by pseudonodules and gravity induced roll--up structures of sheet sands in the argillaceous facies, and by convolute bedding in the muddy carbonates (Fig. 2). Subsequent to a physical event, an ecological "infaunal event"

Fig. 10: Channels and "gutter casts" in the Lower Muschelkalk. a,b: Interfingering channel system without accumulation of relict sediments. c: Bifurcating "gutter cast" filled with shells. d: "Gutter cast" with a central filling tube.

can disturb locally the sediment chemistry and caused localized cementation due to a higher chemical diffusion gradient (Fig. 11) of carbonates and iron from the undisturbed clays (low pH) to the burrows (higher pH). Such ecological events with their diagenetic consequences thus provided the base for patched hardground formation whenever an erosional event reached the bioturbated and prefossilized horizon (Fig. 9 b,c,d).

The equivalent sedimentary patterns within the muddy carbonate facies are again bioturbated horizons (Fig. 10) and local shell accumulations in strongly elongated "gutter casts" (AIGNER & FUTTERER, 1978; EINSELE & SEILACHER, eds., 1982). Not uncommonly meandering "gutter casts" are found, and others show Y-like bifurcations (Fig. 10 d). Because there exists no evidence for an intertidal formation of these structures, it is a hydrodynamic problem to explain the formation of these minor channels under permanent water coverage (LEEDER, 1982) unless one assumes some already existent depression which acted as a trap for shells. Again, the usual association of these structures with burrows and the above mentioned morphology relates them to infaunal activity. They take a position similar to the eroded trace fossils at the base of turbidites. However, the erosional mechanism is different: It is more linear under wave and current actions than in a gravity flow -- consequently under linear flow conditions the burrows parallel to the flow are better and more commonly preserved than branches oblique to the flow. This view is supported by the occasional occurrence of "gutter casts" with a central filling tube which contains micritic sediment (Fig. 10 d). This pattern can clearly be related to the filling of a hollow tube inside the sediment (SEILACHER, 1967).

In contrast to the argillaceous facies, diagenetic cementation in muddy carbonates always begins within the sediment: Burrows, therefore, appear larger because of a cementation halo (which also surrounds the "gutter casts"). No data are yet available which would allow us to speculate about the induction time of cementation although bioturbated horizons were cemented to some degree before erosion, as reworked pebbles at the base of the "roof beds" show.

The type of early diagenetic cementation turns out to be the major difference between the argillaceous and the muddy carbonate facies. While in clays cementation occurs inside areas of increased porosity, the diffusion gradient is reversed in micritic carbonates. This can be sufficiently explained by inverse pH (EDER, 1982) and perhaps Eh gradients, whereby the latter is mainly responsible for iron migration. As a consequence a common initial structure becomes likely for the different sedimentary structures in the two major facies types. A sediment-internal tubular structure turns either into a positive or a negative image, dependent on the direction of early cementation. In any case, local burial activity changes the porosity and dislocates the sediment-seawater boundary into the substrate (Fig. 11), a short term process which locally disturbs the diffusion gradient and the chemical equilibrium which then has to be restored. However, the effect of bioturbation depends on its intensity:

A single bioturbation event increases the inhomogeneity of the sediment while intensive bioturbation may have the contrary effect, e.g. a stratified sediment becomes homogenized.

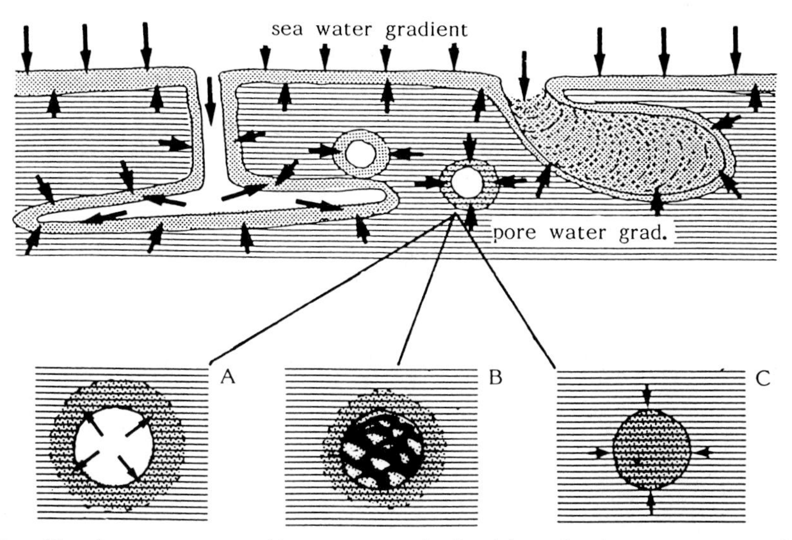

Fig. 11: Pore water gradients near single bioturbation structures in the uppermost sediment layer. The gradient depends on the overall chemical system of the sediment: A, diffusion from the burrow into the sediment (carbonates); B, cementation of the disturbed area and formation of a diagenetic halo; c, formation of concretions within the burrow (claystones).

B) Repeated erosional events -- shell and pebble accumulations

The second phase (B in Fig. 7) of a cycle in the argillaceous facies is characterized by increasing erosional intensity and by the intercalation of storm sands. The finely laminated and partially cross-bedded sands (Fig. 2 c) commonly contain rather well preserved parts of crinoids which are not disarticulated and, therefore, prove a short term deposition event. Convolutions -- as they commonly occur in the muddy carbonates of the Lower Muschelkalk -- point to the same direction (ARKETELL et al., 1970). Increasing erosional energy is documented by well rounded sandstone pebbles with sizes of about 1 cm (Fig. 9 e). Overgrowth and borings again indicate early diagenetic, perhaps concretionary cementation of the sands. Phosphatized pebbles point to restricted sedimentation and repeated reworking, what locally resulted in accumulations of such pebbles together with oyster shells (*Gryphaea*) of comparable size. Such accumulations in casts formed again the base for patched hardgrounds. Overgrowth by serpulids on the lower side of such pebble accumulations shows that they had been worked out by flute cast formation, and that they had already been internally cemented at this time. Once again, deposition, erosion, diagenesis and biological response interact in the formation of sedimentary patterns. But now, however, the increasing interaction between deposition and erosion at the same locality initiates a feedback mechanism which starts to alter the sediment by its local intrinsic properties. The new factor is the increasing importance of the biological response, the mass-production of coarse skeletal material by incrusting and free living oysters. These shells provide a new sedimentary source which can be redeposited in often graded shell beds with the sequence:

larger left valve - smaller right valve -- crinoid fragments

(Fig. 8 b). This new sedimentary source becomes of increasing importance, as the top part of the cycle is approached. At the base of the "roof bed" such shells may form thick cross-bedded beds, channel fillings, or built-up ridges (Fig. 2 a).

In the Lower Muschelkalk the dominant pattern in this second phase are channels which reach sizes of one to a few meters (Fig. 10 a,b), filled with finely laminated micrite. Usually intersections of such channels show again repeated phases of deposition and erosion. Bed formation, as discussed above, does not occur in these channel systems because no coarse material was available, i.e. no relict sediments were formed. The fluctuation of deposition and erosion turns out to be the main pattern of this second phase, and early diagenesis provides the base for substrate changes and increased shell production.

The new sediment material "shells", however, alters the substrate conditions and initiates early diagenetic processes again -- the formation of nodules occurs whenever coarse material is locally accumulated (Fig. 2 a,b), i.e. if the sediment porosity is locally increased. This relationship occurs throughout the sequence including the "roof bed". However, the diagenetic gradient is reversely oriented in the two major facies types: In the silicoclastic sediments nodules are formed; in the muddy carbonates a cementation halo surrounds areas of increased porosity. Three factors are likely of importance:

1) The local pH - Eh gradient, as discussed above -- which is reversely oriented in the two major facies types - controls the direction of cementation.

2) The local increase in porosity provides the physical framework for precipitation or dissolution as controlled by the diffusion gradient.

3) Local accumulations of shells and perhaps of organic material change the microchemical system -- whereby the Ca/CO_2 system acts at least as a pH buffer.

C) The final facies transition -- "roof beds"

Many "roof beds" of cycles represent a sedimentation phase which is well separated from the monotonous muddy 'background' sedimentation, and if studied in detail, the complex history of a condensed section becomes visible which consists of a number of burial, encrustation and reworking events (e.g. SCHLOZ, 1972; FÜRSICH, 1982; BRANDT, this volume). Within local sections the temporal sequence of facies types seems to reflect environmental changes along which the faunal response and facies dependences of fossils can be analysed. However, the question is how far the sequence reflects a gradual spatio-temporal process of facies replacements.

The gradual facies transition within cycles are commonly disturbed at the base of the "roof bed", and such erosional discontinuities are usually overlain by a fining-upward sequence within the "roof bed"; i.e. the "roof bed" starts with accumulations of shells, reworked concretions and pebbles which are followed by oolites. Three points are of interest:

** commonly "roof beds" have an erosional contact;

** the fining-upward sequence within the "roof bed" indicates decreasing flow intensity from the viewpoint of grain-size distribution;

** the transition to oolites marks a major facies change and quite different sedimentary regimes.

At the erosional base of the "roof beds" channel structures occur commonly (Fig. 12). The formation of a channel with relict sediments causes locally a rather dramatic substrate change -- the muddy sediments are locally replaced by firmgrounds, patched hardgrounds, shell and pebble beds. Consequently to the new bottom conditions new benthic communities appear -- incrusting and byssate bivalves with usually thick shells but also

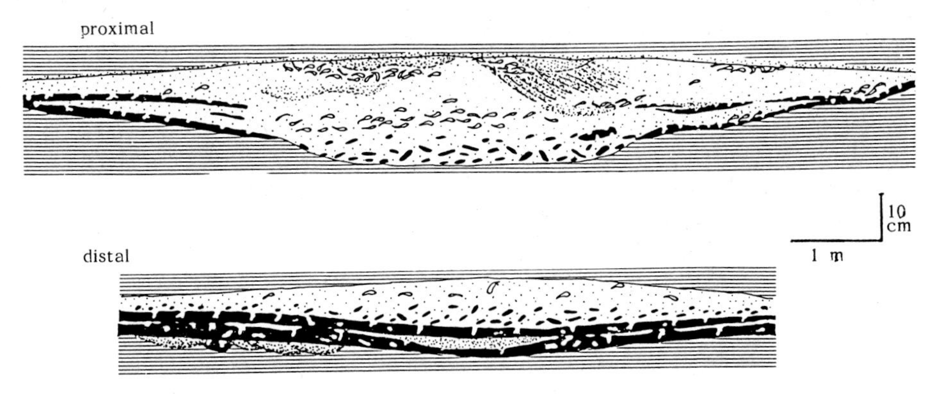

Fig. 12: Channel patterns in the lower Muschelkalk which accumulate spatially and temporally into a 'roof bed'. Such channel complexes are likely related to storm surge flow (JAMES, 1980).

semi-infaunal forms like Trigoniids. As expected, the faunal composition indicates a patchy arrangement of firmgrounds, shell and pebble beds.

The consequences are twofold: the locally altered benthic communities produce increasing amounts of coarse clastic carbonates (Fig. 8 a,b), and the local accumulation of shells further changes the substrate conditions. Repeated localized erosional events (channels) are likely stopped and horizontally dislocated whenever previously accumulated relict sediments are reactivated. Thus, in the course of time one can expect an increasing accumulation of coarse material (shells) and a lateral dislocation of channels which are filled with relict sediments. At the same time, the increasing patches of shell beds permit epibenthic communities to spread, which increase further the production of skeletal material. The role of coarse relict sediments is visible in a lower Muschelkalk example (Fig. 12). Channels with laminated or cross-bedded micrite filling (Fig. 10) are rather

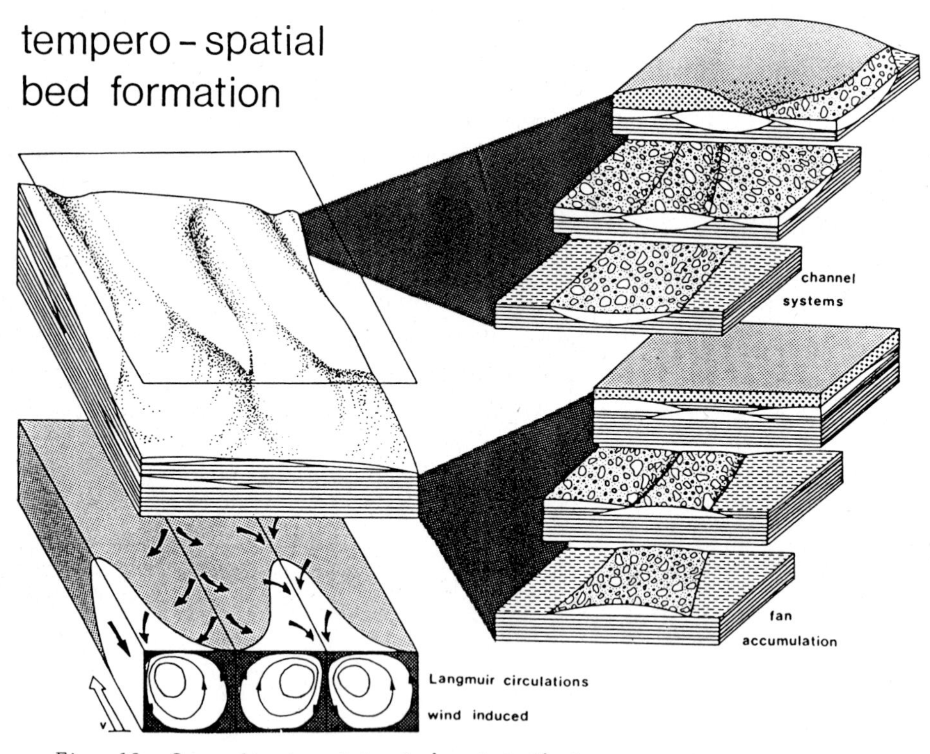

Fig. 13: Generalized model of 'roof bed' formation by the accumulation of channels and fans. A near bottom bifurcation of storm surge back-flow -- similar to a Langmuir circulation (POLLARD, 1977) of surface waters -- is assumed to cause channel erosion, i.e. a secondary helical flow pattern with alternating zones of convergence and divergence (FLOOD, 1981).

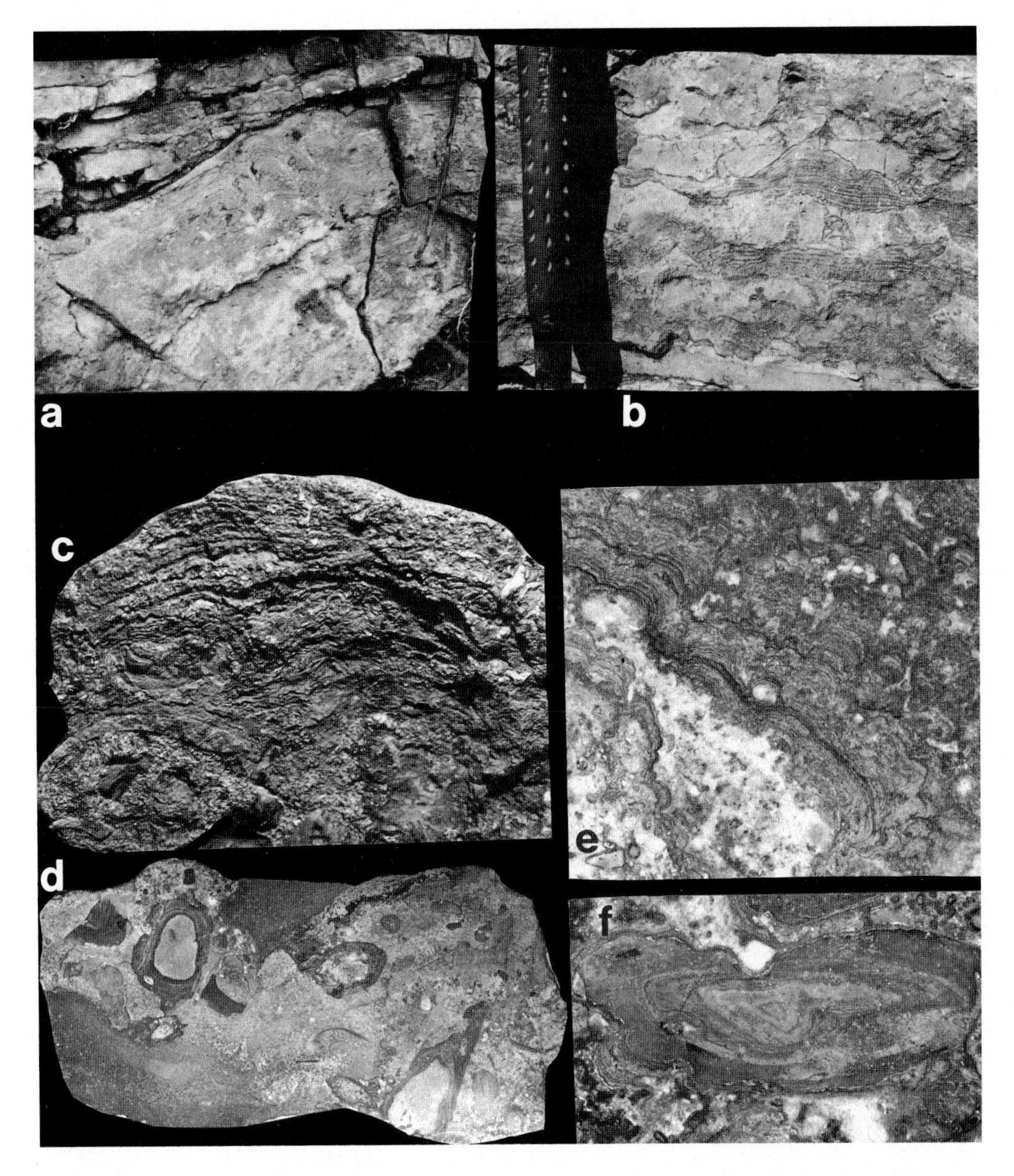

Fig. 14: Biogenetic build-ups. a,b: Stromatolites in the uppermost lower Muschelkalk. c-f: serpulid-foraminiferal (? algal) overgrowth and 'snuff boxes' (usually ferruginized by chamosite).

irregularly distributed within the profile. They intersect one another without respect to the previous structure and thus generate a random channel system. In contrast, channels with relict sediments (Fig. 12) tend to align in one horizon and to approach single beds. As local erosion produces sedimentation in adjacent neighborhoods, channels produce fans with a proximity gradient in grain size distribution (Fig. 12; RICKEN, this volume). A single erosional event, therefore, should produce a variety of -- gradually changing -- environmental conditions in its neighborhood.

Fig. 13 illustrates this dynamic view of bed formation. The local channels and their associated fans accumulate through time and space until they condense into a 'marker bed'. The larger linear erosional patterns are likely storm surge channels which formed along convolutions in the near bottom boundary back-flow. Helical flow patterns with alternating zones of convergence and divergence (i.e. a periodical pattern of flow intensities) are well known as boundary flows. The Langmuir circulation (cf. POLLARD, 1977 for a

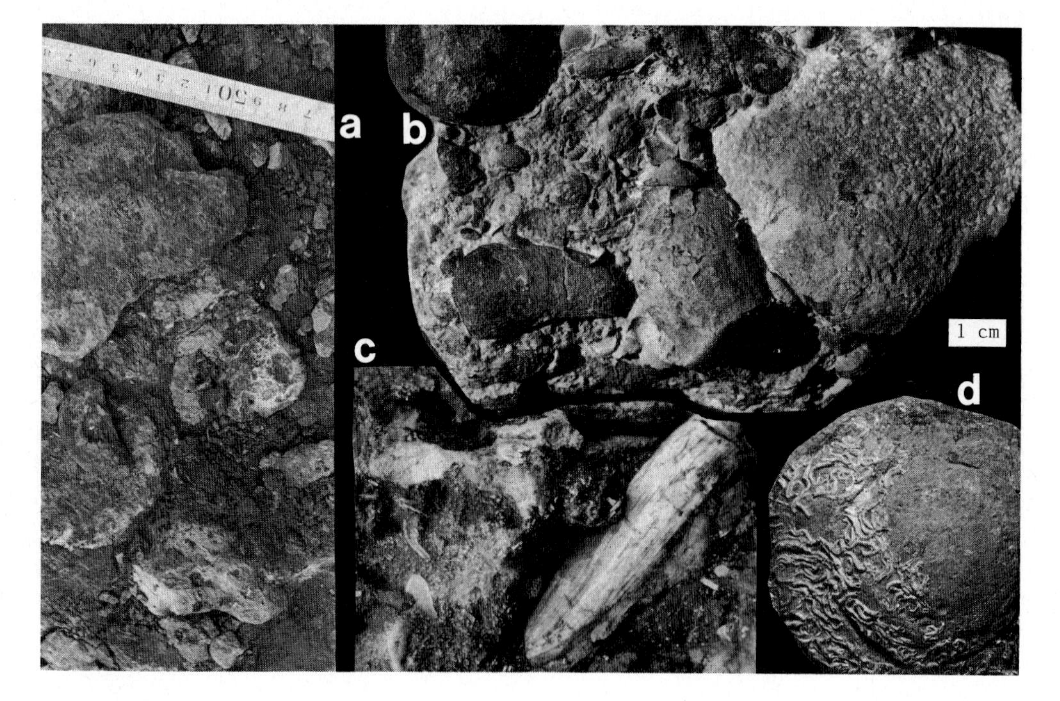

Fig. 15: Omission surfaces of 'roof beds'. a: flat concretions and prefossilized ammonites from an oolitic bed. b: flat to spherical reworked nodules of a sandstone. c: hardground with reworked phosphatic fossils. d: serpulid overgrowth of a reworked concretion (sandstone).

discussion) of wind driven surface waters is such an example which has been applied to near bottom flows by FLOOD (1981). A different channel system is described by RICKEN (this volume) from the Upper Jurassic carbonates. These channels are spatially stationary during long time intervals and, therefore, may have been formed under topographical control like rip channels.

A fining-upward sequence within 'roof beds' must be carefully interpreted. As the sedimentary regime usually changes within such beds, grain size may not be a proper measurement of flow intensity. The reworked concretions at the base of such beds are likely worked out by channel and cast formation (SCHLOZ, 1972; BAYER & McGHEE, this volume; RICKEN, this volume; BRANDT, this volume), a process which, perhaps, needs less energy than the repeated reworking of oolite bars and the accumulation of shell beds. Therefore, the erosional energy may still increase within the "roof bed" although grain size decreases. Again early cementation within "roof beds" stabilizes the oolite flats as indicated by erosional surfaces within and at the top of "roof beds" which contain reworked pebbles from the bed (Figs. 14 , 15). Additionally, cementation by algae and incrusting organisms occurs (Fig. 14).

D) The terminal omission -- spreading of oolitic beds

Oolite formation is also restricted to a certain 'energy band' -- only above a certain energy level reworking does occur frequently enough to allow oolite formation. And as the upper boundary is reached, the oolites are redeposited into deeper parts of the basin, and locally the production of new material is replaced by erosion. This state is reached at the terminal erosive surface of cycles, the "omission" or "emersion" horizon at the top of "roof beds".

Local erosion of oolite beds and bars requires their redeposition eventually in deeper areas, a situation which is illustrated in Fig. 16. A locally well developed cycle turns into a rhythmic sequence of clays and interbedded shell beds only a few kilometers away. This more distal sequence is terminated by an oolitic bed which, lithologically, is identical with the "roof bed" of the proximal cycle. However, the ammonite faunas belong to two different subzones, indicating that the distal bed corresponds with the erosional or non-depositional phase at the top of the cycle. This relationship can be verified by comparing different sections throughout the basin. The transition between the two subzones can be put together like a puzzle, as sometimes relicts of the later subzone are present in the terminal omission horizon, and sometimes relicts of the earlier subzone are found at the base of distal oolite fans or showers.

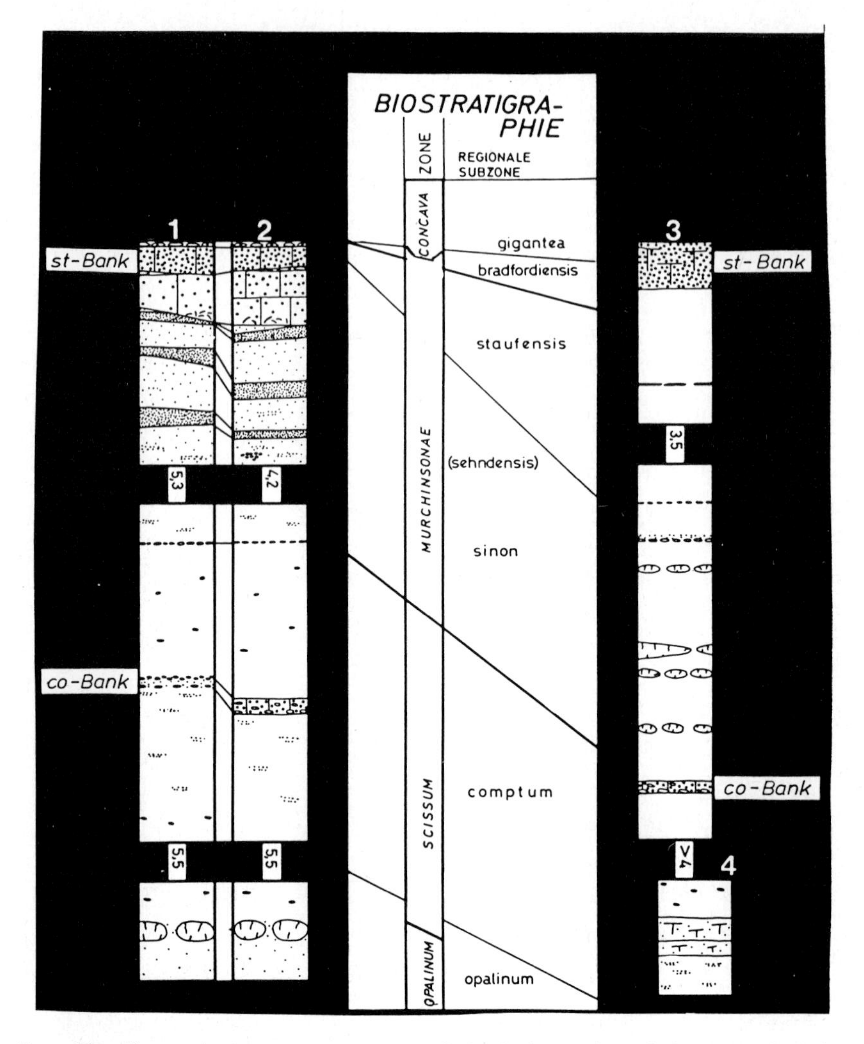

Fig. 16: Short distance variation of lithological and biostratigraphical sequences (1,2: Achdorf; 3: Geisingen; distance 15 km). The 'roof beds' are lithologically equivalent, but stratigraphically the distal one corresponds with the proximal non-deposition phase (see text for discussion).

While the previous discussion focused on an increase in environmental diversity during a sedimentary cycle, this final phase reduces the diversity of substrates. Oolite beds and bars have little production of organic skeletal material, and hardgrounds prevent any of the previously discussed mechanisms which enhance biotic activity. Sedimentologically and biologically a cycle of increasing and then again decreasing diversity is passed during a time interval of hundredthousands to millons of years, a process which is likely to influence biological evolution (BAYER & McGHEE, this volume). The central point, however, is the slowly changing environment, the long-time accumulation of short-time events which finally induces facies changes. The formation of typical 'marker beds' -- mature oolitic carbonates as they commonly occur in the middle European Jurassic -- can be related to such processes, either to 'condensed cycles' in proximal areas or to redeposition of material which was reworked at the 'omission' horizons.

E) Summary of sedimentary patterns

In a detailed analysis of minor Klüpfel cycles, the erosional events turn out to be a major property of the entire sequence. Channel and cast formation occur at different

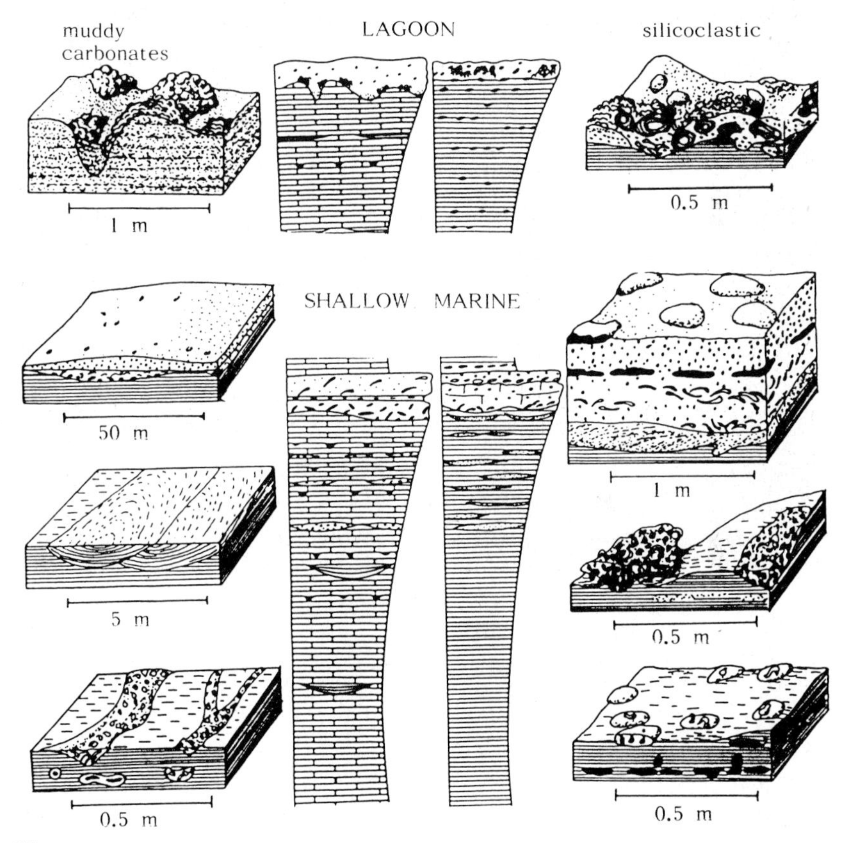

Fig. 17: Erosional patterns of the two major facies types. In the muddy carbonates (lower Muschelkalk) a sequence is found from meandering and bifurcating "gutter casts", small scale channel systems without relict sediments, and large scale channels which accumulate into beds and are followed by oolite deposition. In the argillaceous facies (Aalenian) the sequence starts with nodule layers which occasionally are reworked; multiple reworking and accumulation of pebbles follow , and the sequence terminates in the 'roof bed' with commonly erosional base and top. During the extreme regression, algal mats and stromatoliths occur in both cases, indicating a lagoonal situation.

energy levels with different scales (Fig. 17). In the early phase "gutter casts" are common in muddy carbonates while reworked carbonate concretions take a similar position in claystones. As the erosional energy increases, the dimensions of the linear erosive structures increase: channels occur, and pebbles are repeatedly reworked. In any case, independent of the major facies and of erosional energy, these erosional events change local substrate conditions and induce reactions in the local benthic community, and, as previously mentioned, local substrate changes will depend on the relationship between biological 'sediment production', mud supply and reworking rate. In the case that mud supply dominates the other processes, a disturbed system will rapidly return to the initial conditions. This situation usually occurs in the early phase of asymmetric cycles where reworked concretions form sporadically patched hardgrounds, or "gutter casts" develop into shell lenses (Fig. 17). If the biological production of skeletal material dominates mud supply then sediment conditions change more markedly. Algal mats appear, the activity of microboring organisms and repeated reworking provide a wide spectrum of carbonate grains, and organic and inorganic carbonate precipitation induces the formation of ooliths which so often overlay the initial erosive phase. Finally, erosion and reworking may be enhanced during the course of larger cycles and the different factors -- especially diagenetic processes -- depend on the major facies type (Fig. 11).

2.2 Oolitic ironstones

The sedimentological patterns of the "roof bed" were discussed in some detail in the last chapter. In the micritic facies of the lower Muschelkalk the "roof beds" consist mainly of calcite (aragonite) oolites which clearly point to an extreme regression state, i.e. the highest energy level. In the Jurassic claystones, however, oolitic ironstones are usually found in this position and, therefore, were likely to have been formed under similar hydrodynamic conditions (HALLAM & BRADSHAW, 1979) although there is little doubt that the mechanisms of ooid formation are quite different for iron- and aragonite-oolites. KIMBERLEY's (1979) resurrection of Sorby's and Cayeux's old idea that the ironoolites were formed by diagenetic replacement of originally calcareous ooids (HALLAM & BRADSHAW, 1979; KIMBERLEY, 1979), therefore, is ad hoc an attractive concept. However, "to ferruginize an oolitic bed ..., subaerial weathering and erosion ... of overlying marine mud" (KIMBERLEY, 1979) is necessary -- which thus returns to KÜPFEL's (1917) "emersion" cycles. There is no evidence for exposure and weathering in the middle European 'Minette'-type iron ores. The erosional surfaces, which usually terminate the oolitic beds, are much too continuous, too planar -- they can only be "explained by marine abrasion" as LUCIUS (1945, 1948) pointed out. Further, why should the calcareous ooids -- which were accumulated in a bar -- be perfectly replaced by iron-minerals before pore lining and closure starts.

However, there exists a common consensus that the ferruginous oolites were formed in very shallow water and that they are closely related to argillaceous sediments. This relation can be well elucidated by the facies evolution of the South German and the Lorraine basin through (Toarcien) Aalenian -- Bajocian times. In the Lorraine basin the terrigenous input declines throughout the Bajocian, the sediment changes to skeletal carbonates, and locally coral reefs appear (e.g. at the Ardennen island). At the same time, the ferruginous oolites are replaced by calcareous ones. Within the South German basin the same trend occurs on its western boundary (in the depression of the Oberrhein Graben) while the terrigenous supply of clays continues in its eastern part until the Oxfordian. Within this silicoclastic belt only ferruginous oolites are deposited whereby the sedimentary structures in any case "point to the ironstone having been deposited at the highest energy level" (HALLAM & BRADSHAW, 1979; BRADSHAW et al., 1979). However, an important sedimentary structure of these beds is the common repeated reworking and reactivating of the ferruginous oolites bearing beds (WERNER, 1959; ALDINGER, 1957, 1965; FÜRSICH, 1971).

As was already pointed out by WILD (1951) and WERNER (1959), the South German 'Minette'-type iron ores are clearly local bars which were accumulated by repeated winnowing, i.e. the place of deposition is likely not the place of origination and, of course, the present petrography does not reflect the sediment conditions under which the oolites formed. Indeed, grain size is extremely homogeneous (WERNER, 1959), and the hydrologic equivalent diameter is slightly above that of sand grains (ALDINGER, 1957, 1965). In the last chapter we tried to point out how the usual sheet-like occurrence of iron-oolitic beds is related to regressive cycles, i.e. to a final dispersion over the basin during the extreme regression state.

The oolites are usually less well sorted whenever they occur within a muddy matrix (Fig. 18). In these cases additional features occur: the oolites reach larger diameters than in the ores, the grain size is less homogenous, composite and irregular oolites are common and the mineralogy is less homogeneous, limonitic and chamositic layers occur commonly within the same ooid. As the size of the ooids increases, overgrowth by sessile foraminifera in distinct layers is a common feature and clearly indicates repeated reworking. Shell fragments as nuclei are heavily bored by algae. Along these microborings dissolution of the aragonite and replacement by limonite usually takes place (Fig. 18c), a feature which points to diagenetic ferruginization (for a discussion of diagenetic models see e.g. MAYNARD, 1983). However, the most significant pattern may be shrinkage of the ferruginous oolites (Fig. 18 a,b) which causes radial fracturing and fractures parallel to the layers. The pattern itself is likely related to the dewatering of gel-like iron-hydroxides, i.e. true limonite; the important point, however, is that pieces of such broken oolites occur isolated in the muddy matrix and as the nuclei of other oolites -- all stages from superficial ooids to true ooids and small foraminiferal-(algal) onkoids occur. There is little doubt

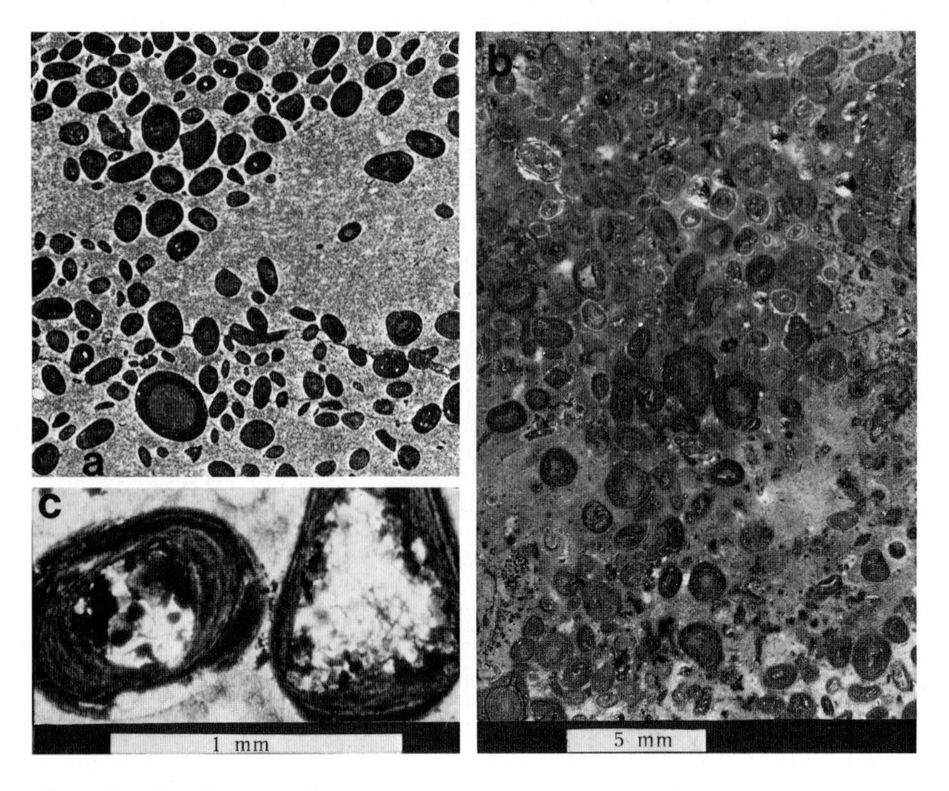

Fig. 18: Shrinkage and fracturing patterns of iron ooids (Aalenian -- Bajocian). a,b: Shrinkage and fracturing of ooids which together with isolated fragments swim in a muddy matrix. The same fragments occur as nuclei (Bajocian, same scale for both figs.). c: Microborings in the shells which form the nuclei. Along the microborings the carbonate is dissolved and replaced by iron oxides.

that these oolites which now 'swim' isolated in a muddy matrix were repeatedly reworked; as indicated by shrinkage, fractioning and overgrowth. It is, in addition, likely that they were already ferruginized and consisted in part of chamosite. Within the typical minor cycles oolites of this type are usually restricted to the uppermost part of the "roof bed", i.e. to the final omission surface (WERNER, 1959) and, therefore, to the 'highest erosional energy'. However, it is in just this part where early diagenetic iron migration and precipitation becomes obvious in two ways: The usually reworked ferruginous concretions at the top of the roof bed show that iron migration and percipitaion took place before the erosional event, and limonitic and chamositic incrustations of the lower side of the concretions show that this process was still going -- especially as serpulid and foraminiferal overgrowths are ferruginized.

Therefore, we suggest that early diagenesis may play an important role for iron-oolite formation, perhaps by replacement of aragonitic particles. We have found that there is much evidence that the iron-oolitic beds were formed under fluctuating deposition and erosion, under reworking and winnowing. Relicts of clay layers are common even within the iron-ores (WERNER, 1959), and repeated covering by muds is very likely for most iron-oolitic beds. Identical features occur in the calcareous oolites of the lower Muschelkalk, where oolite bars and channel fillings laterally interfinger with micritic carbonate layers (ALTHEIMER, thesis in preparation). Whether calcareous or ferruginous oolites occur seems mainly a function of the general facies type or the overall chemistry of the sediment. Of course, it is no problem to transport iron in connection with clays and to reactivate the iron even in the uppermost few centimeters of the sediment where biological activity influences the relevant chemical control parameters. Fluctuating deposition and erosion together with winnowing provide the framework which allows for the formation of chamosite (which is sometimes assumed to be the primary iron-mineral of the oolites -- KIMBERLEY (1979), HALLAM & BRADSHAW (1979)) and limonite, of organic and anorganic processes hand in hand. As was pointed out earlier in the discussion of cycles, time is not a problem: The iron-stones at the top of minor cycles are, in any case, condensed, and as BERNER (1980) points out such early diagenetic processes may need only thousands of years -- a short time interval with respect to the cycles with duration times of at least hundred thousands of years. Still less time may be necessary if the diagenetic model is not attributed to the iron-oolitic bed but to the individual ooid which -- swimming in a muddy matrix -- forms in a locally disturberd microchemical system. Chamositic oolites are commonly associated with chamositic clays (HALLAM & BRADSHAW, 1979).

With respect to the formation of siderite concretions we stressed the importance of the local diffusion gradient which -- dependent on the general facies type -- may point either to local disturbance or to the opposite direction. The same question arises in the context of early diagenetic formation of ferruginous oolites. An aragonitic (calcareous) particle can either be replaced by iron minerals, as indicated by dissolution and limonitic lining of microborings, or it may affect the alteration of nearby clay minerals -- the early autogenetic formation of chamosite from clay-minerals. Petrographic, chemical and theoretical analysis of suitable systems -- likely primary deposits of ooids outside the ores -- will be necessary to test the model of early diagenetic formation and to elucidate the possible processes in more detail.

CONCLUSIONS

The concept of sedimentary cycles is reconsidered in a classical area (South Germany and Lorraine) and in a classical time interval (Aalenian and Bajocian). In comparison with another major facies -- the muddy carbonates of the Lower Muschelkalk -- general patterns of minor Klüpfel cycles (or "coarsening-upward sequences") and the facies-dependent differences in sedimentary patterns can be determined. The direction and mode of early diagenesis, controlled by general parameters like subsurface, pH and Eh gradients, is likely a major factor in facies-controlled pattern formation of minor erosional structures.

The concept of cycles has been criticized, as the definition of a cycle is a highly subjective process -- the superposition of cycles of various phases may provide an even more complicated picture which -- because of the various condensation trends -- may hardly be analysed by 'objective' methods especially because a local vertical sequence may not be representative for the evolution of the basin.

Spatial and temporal patterns exhibit some common features like increasing condensation. However, to interpret temporal cycles as a proximity gradient -- as the vertical arrangement of a spatial facies succession -- may be erroneous even if a regression trend can be proved by basin wide features such as offlapping patterns etc.

The time-dimension allows us to consider additional features: A single disturbance drives the system locally and temporally out of its equilibrium -- its sedimentological, biological and chemical equilibrium. Resulting structures such as bioturbation, increased shell production, and the formation of nodules become part or may become part of the sedimentary process as well as they may influence all other factors involved. Dependent on the intensity of the disturbance and the general state of the sedimentary system, a single event can cause a short term response which either returns the system to its original state, or a new equilibrium is achieved which then persist for some time. The accumulation of such events, the reactivation of previous events and their condensation may be importsnt factors along the time-axis of sedimentary processes in terms of sedimentary pattern formation. There seems to exist no simple relationship between spatial and temporal facies associations.

REFERENCES

Anketell, J.M., Ceyla, J., Dzulyuski, S. 1970: On the deformational structures in systems with reversed density gradients. Rocz. Polsk. Tow. Geol. Ann. Soc. Geol. Pologne, XL, 1, 3-30.

Aigner, Th. & Futterer, E. 1978: Kolk-Töpfe und -Rinnen (pot and gutter casts) im Muschelkalk - Anzeiger für Wattenmeer? - N. Jb. Geol. Paläont. Abh., 156(3), 285-304.

Aigner, Th. & Reineck, H.-E. 1982: Proximality trends in modern storm sands from the Helgoland Bight (North Sea) and their implications for basin analysis.- Senckenbergiana marit., 14(5/6), 183-215.

Aldinger, H. 1957: Eisenoolithbildung und rhythmische Schichtung im süddeutschen Jura.- Geol. Jb., 74, 87-96.

--- 1965: Über den Einfluß von Meeresspiegelschwankungen auf Flachwassersedimente im Schwäbischen Jura.- Tschermarks min. u. petrogr. Mitt., 10, 61-68.

Bayer, U. 1970: Das Profil des Erz-Tagebaus Ringsheim (Ober-Aalenium/ Unter-Bajocium).- Neues Jahrb. Geol. Paläont. Abh., 155: 162-215.

--- 1983: The influence of sediment composition on physical properties interrelationship. In: Ludwig, W.J., Krasheninikov, V.A., et al., Initial Reports of the Deep Sea Drilling Project, LXXI, 1111-1132.

--- 1983b: Pattern Recognition Problems in Geology and Paleontology.- Habil. Schr. Univ. Tübingen.

Bayer, U., McGhee, G.R., Jr. 1984: Iterative evolution of Middle Jurassic ammonite faunas.- Lethaia, 17: 1-16.

Berner, R.A. 1980: Early diagenesis. A theoretical approach. (Princeton University Press) Princeton, 241 pp.

Bradshaw, M.J. et al. 1980: Origin of oolitic ironstones - Discussion.- Journ. Sed. Petrology, 50: 295-304.

Eder, W. 1982: Diagenetic redistribution of carbonate, a process in forming limestone-marl alternations (Devonian and Carboniferous, Rheinische Schiefergebirge, W. Germany). 98-112. In: Einsele, G. & Seilacher, A. (eds.): Cyclic and event stratification. (Springer) Berlin, 536 pp.

Einsele,G. & Seilacher, A. 1982: Cyclic and event stratification. (Springer) Berlin, 536 pp.

Fischer, A.G. 1981: Climatic Oscillations in the Biosphere. 103-131. In: Biotic crises in ecological and evolutionary time. (Acad. press).

--- 1982: Long-Term Climatic Oscillations recorded in stratigraphy. 97-104. In: Studies in Geophysics, Climate in Earth History. (Nat. Aca. Press).

Flood, R.D. 1981: Distribution, morphology, and origin of sedimentary furrows in cohesive sediments, Southampton Water. Sedimentology, 28: 511-529.

Füchtbauer, F. & Müller, G. 1977: Sedimente und Sedimentgesteine, Teil II (Sediment-Petrologie). (E. Schweizerbart) Stuttgart.

Fürsich, F. 1971: Hartgründe und Kondensation im Dogger von Calvados.- N. Jb. Geol. Paläont. Abh., 138, 313-342.

Hallam, A. 1961: Cyclothems, Transgressions and Faunal Change in the Lias of North-West Europe.- Transact. Edinburgh Geol. Soc., 18 (2): 124-174.

Hallam, A. & Bradshaw, M.J. 1979: Bituminous shales and oolitic ironstones as indicators of transgressions and regressions.- Journ. Geol. Soc. London, 136(2): 157-164.

James, W.C. 1980: Limestone channel storm complex (Lower Cretaceous) Elkhorn Mountains, Montana.- J. Sed. Petrol. 50: 447-456.

Kimberley, M.M. 1979: Origin of oolitic iron formations.- Journ. Sed. Petrology, 49: 111-132.

Klüpfel, W. 1917: Über die Sedimente der Flachsee im Lothringer Jura.- Geol. Rdsch., 7, 98-109.

Leeder, M.R. 1982: Sedimentology - Process and Product. (G. Allen & Unwin) London, 344 pp.

Lucius,M. 1940: Der Luxemburger mesozoische Sedimentationsraum und seine Beziehung zu den herzynischen Bauelementen.- Veröff. Lux. geol. Landesaufn., II, 41-102.

--- 1945: Die Luxemburger Minetteformation und jüngere Eisenerzbildungen unseres Landes.- Serv. Carte Géol. Luxembourg.

--- 1948: Das Gutland.- Serv. Géol. Luxembourg, V.

Maynard, J.B. 1983: Geochemistry of sedimentary ore Deposits. New York.

Müller, A. et al. 1976: Tektonische Richtungen und Faziesverteilung im Mesozoikum von Luxemburg-Lothringen.- Jber. u. Mitt. Oberrhein. geol. Ver. NF., 58, 153-181.

Oechsle, E. 1958: Stratigraphie und Ammoniten-Fauna der Sonninien-Schichten des Filsgebietes unter besonderer Berücksichtigung der sowerbyi-Zone.- Paleontographica 111(A): 47-129.

Oertel, G. & Curtis, Ch. D. 1972: Clay-Ironstone concretion Preserving Fabrics Due to progressive Compaction.- Geol. Soc. Am. Bull. 83: 2597-2606.

Pollard, R.T. 1977: Observations and Models of the Structure of the Upper Ocean. In: Kraus, E.B., Ed.: Modelling and prediction of the upper layers of the ocean. (Pergamon Press) Oxford, 102-117.

Reading, H.G. 1978: Sedimentary environments and facies.- (Blackwell Scientific Publications) Oxford, 569 pp.

Sauer, K. 1956: In: Führer zu den Exkursionen anläßlich der Frühjahrstagung der Deutschen Geolog. Ges. Mitt. Arb. Geol.-Pal. Inst. TH Stuttgart, NF 40.

Schäfer, W. 1956: Wirkungen der Benthos-Organismen auf den jungen Schichtverband.- Senckenbergiana lethaea, 37(3/4), 183-263.

Schink, D.R. & Guinasso Jr., N.L. 1976: Effects of bioturbation on sediment-seawater interaction. Mar. Geol., 23: 133-154.

Schloz, W. 1972: Zur Bildungsgeschichte der Oolithenbank (Hettangium) in Baden-Württemberg.-Arb. Inst. Geol. Paläont. Univ. Stuttgart NF 67: 101-212.

Schwarzacher, W. & Fischer, A.G. 1982: Limestone-shale bedding and perturbations of the Earth's orbit. 72-95. In: Einsele, G. & Seilacher, A. (eds.): Cyclic and event stratification. (Springer) Berlin, 536 pp.

Seibold, E. 1962: Kalk-Konkretionen und karbonatisch gebundenes Magnesium.- Geochimica et Cosmochimica Acta, 26: 899-909.

Seilacher, A. 1967: Sedimentationsprozesse in Ammonitengehäusen.- Abh. math.-naturwiss. Klasse, 9, 191-203, Tafel I.

--- 1982: General remarks about event deposits. 161-174. In: Einsele, G. & Seilacher, A. (eds.): Cyclic and event stratification (Springer) Berlin, 536 pp.

Thein, J. 1975: Sedimentologisch-stratigraphische Untersuchungen in der Minette des Differdinger Beckens.- Pub. Serv. Géol. Luxembourg, XXIV, 60 pp.

Webb, J.E. et al. 1976: Organism sediment relationships. In: McCave, I.N.: Benthic boundary layer. (Plenum Press) New York.

Weber, H.-S. 1967: Zur Westgrenze der ostschwäbisch-fränkischen Fazies des Braunjura (Dogger) beta in der Schwäbischen Alb (Württemberg).- Jber. u. Mitt. Oberrhein. geol. Ver. NF., 49, 47-54.

Werner, F. 1959: Zur Kenntnis der Eisenoolithfazies des Braunjura beta von Ostwürttemberg.- Arb. Geol. Pal. Inst. TH Stuttgart, NF., 23, 169 pp.

Wild, H. 1951: Zur Bildungsgeschichte der Braunjura-beta-Flöze und ihrer Begleitgesteine in NO-Württemberg.- Geol. Jb., 65, 271-298.

SEDIMENTARY DYNAMICS OF COMPLEX SHELL BEDS: IMPLICATIONS FOR ECOLOGIC AND EVOLUTIONARY PATTERNS

Susan M. Kidwell & Thomas Aigner
University of Arizona and Universität Tübingen

Abstract: The complex dynamics of shell accumulation by sediment aggradation, erosion, and omission (condensation) can proceed by three basic pathways, each having different implications for post-mortem bias and the ecologic response of living organisms to the accumulation of dead hardparts (taphonomic feedback). In evolutionary time scales, regimes of condensation can maintain shell gravel habitats for sufficiently prolonged periods to record microevolutionary changes in species, but morphometric trends through shell bed sequences can also reflect ecophenotypy in response to changing environments, e.g. during transgressive-regressive cycles. These implications are illustrated by complex shell beds from the Miocene of Maryland (U.S.A.) and the Eocene of Egypt.

INTRODUCTION

Internal complexity is a common feature of skeletal accumulations and is expressed (a) in the admixture of hardparts in different stages of degradation, (b) in the subdivision of single beds by discontinuity surfaces, and (c) in the admixture of hardparts from different environments or ages. However, aside from biostratigraphically complex condensed sequences and hardgrounds (HEIM, 1924; JENKYNS, 1971; FÜRSICH, 1971; WENDT, 1970), complex shell beds have received little attention from the perspective of their physical, sedimentary dynamics and their implications for ecological and evolutionary analysis.

Independent studies of complex shell beds from Miocene strata of Maryland (U.S.A.; KIDWELL, 1982, 1984) and from Eocene strata of Egypt (AIGNER, 1982) led us to similar conclusions regarding the dynamics of skeletal accumulation at two scales:

1. In terms of the small-scale, bed-by-bed processes of sediment aggradation and erosion, which can influence patterns of biotic colonization and faunal changes over short, ecological time scales. This is the significance of the dynamics of shell concentration for paleoecological patterns.

2. In terms of larger scale sedimentary regimes of stratigraphic condensation during transgressive-regressive cycles. Such regimes can maintain shell gravel habitats over

evolutionarily significant periods of time, but can also confound the record of genetically controlled evolution with ecophenotypic changes in response to environmental changes. This is the significance of sedimentary dynamics for evolutionary patterns.

Both the Miocene and Eocene strata contain internally complex shell beds of wide distribution that rest on disconformities and serve as marker beds for regional correlation (KIDWELL, 1984; AIGNER, 1982; STROUGO, 1977). These skeletal accumulations lie at the base of disconformity-bounded depositional sequences of similar scale and are interpreted as the stratigraphically condensed records of minor transgressions. The six depositional sequences of the Miocene Calvert (Plum Point Member) and Choptank formations (SHATTUCK, 1904) together record ca 2.5 m.y. and 5 diatom zones in the late Burdigalian to Serravallian stages, they are each 5-10m thick (KIDWELL, 1984). The 6-7 complex shell beds in the Eocene Quasr-es-Sagha and Maadi formations subdivide about 1.5 m.y. in the Priabonian Stage (STROUGO, 1977); each transgressive-regressive cycle being 3-8 m thick.

The basal disconformities of these complex shell beds are marked by firmgrounds characterized by *Thalassinoides* burrows. The four Miocene shell beds can each be traced 2500 to 7600 km^2 along the basin margin. They vary from 0.5 to 10 m in thickness, depending on pre-existing relief on the basal disconformity, with shell beds thinnest over paleotopographic highs. Evidence for their stratigraphic condensation and primarily sedimentologic origin includes: (1) the lateral tracing of discrete shell horizons into amalgamated sections over paleohighs; (2) the winnowing of the find sand matrix; (3) the reorientation and close-packing of parautochthonous infauna; and (4) the preservation of pods of original unreworked silty sand. Eocene complex shell beds are thinner, ranging from 0.5 to 4 m, but exhibit similar patterns of amalgamation over a paleohigh, winnowed matrix, and evidence for hydraulic reworking and concentration. The fossil assemblages of both settings are dominated by molluscan species.

1. ECOLOGICAL-SCALE DYNAMICS

1.1. Pathways of Condensation and Ecologic Response

Dead hardparts can influence the structure of living benthic communities by providing substrata for attachment and by changing the mass properties of sedimentary substrata. Dead hardparts thus facilitate colonization by firm-bottom and epifaunal species while they inhibite the success of infaunal species through the reduction of suitable habitat space (Fig. 1). The entire spectrum of live/dead interactions in benthic communities has been termed "taphonomic feedback" (KIDWELL & JABLONSKI, 1983), since the living benthos not only contribute eventually to the dead assemblage but

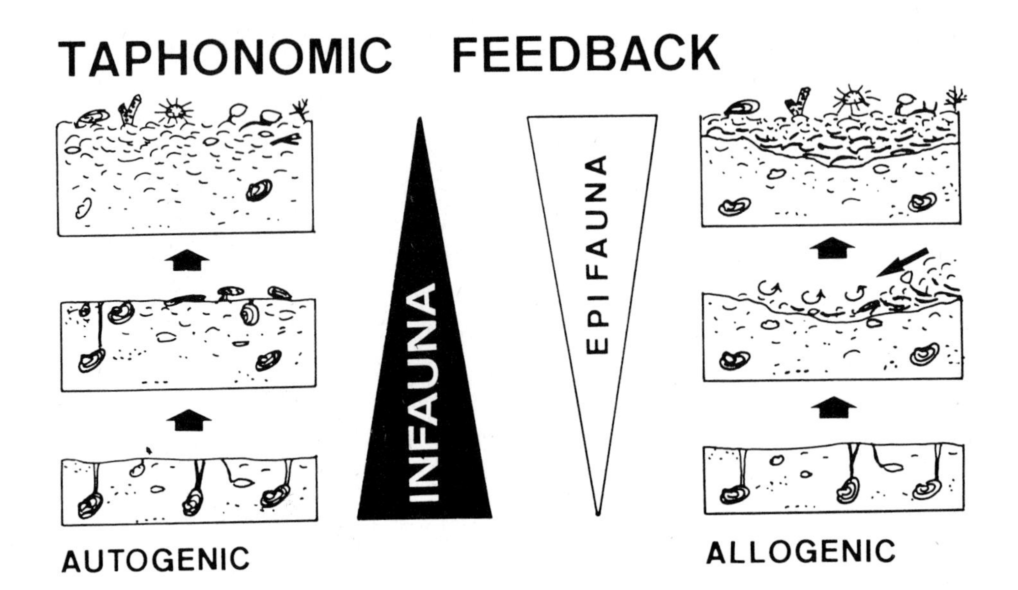

Fig. 1. Schematic diagram illustrating two principal modes of "taphonomic feedback" (KIDWELL & JABLONSKI, 1983) and resulting trends in the composition of benthic assemblages. In the autogenic mode (left), hardparts from the initial soft-bottom community transform the substrate into a progressively coarser, shell-rich substratum, thus facilitating firm-bottom and epifaunal species while inhibiting infaunal colonization. In the allogenic mode (right), changes in the physical environment (winnowing, shell introduction) produce shell concentrations that facilitate epifaunal colonisation.

are also influenced by it. Since sedimentary processes are to a large degree responsible for the availability of hardparts on and shallowly buried within the seafloor, sedimentary dynamics have direct significance for the ecology of benthic communities.

Stratigraphic condensation -- the process of accumulating a relatively thin stratigraphic record under conditions of reduced net sedimentation -- can proceed by several dynamic pathways of sediment aggradation, erosion, and omission. These are reduced here to three basic patterns (Figure 2). Each has different consequences for the taxonomic composition of the shell bed owing to differences in selective post-mortem destruction of skeletal elements and in pathways of taphonomic feedback.

Pathway 1 is the very simple situation of continuous accumulation of a shell gravel because sedimentation fails to keep up with hardpart accumulation. This can result (a) from negligible sediment supply (starvation) or bypassing of an appreciable supply, or (b) from high rates of biological production of autochthonous hardparts or (c) from an abundant supply of allochthonous hardparts from outside. Relatively infrequent and thin depositional increments (indicated by notches on the otherwise smooth

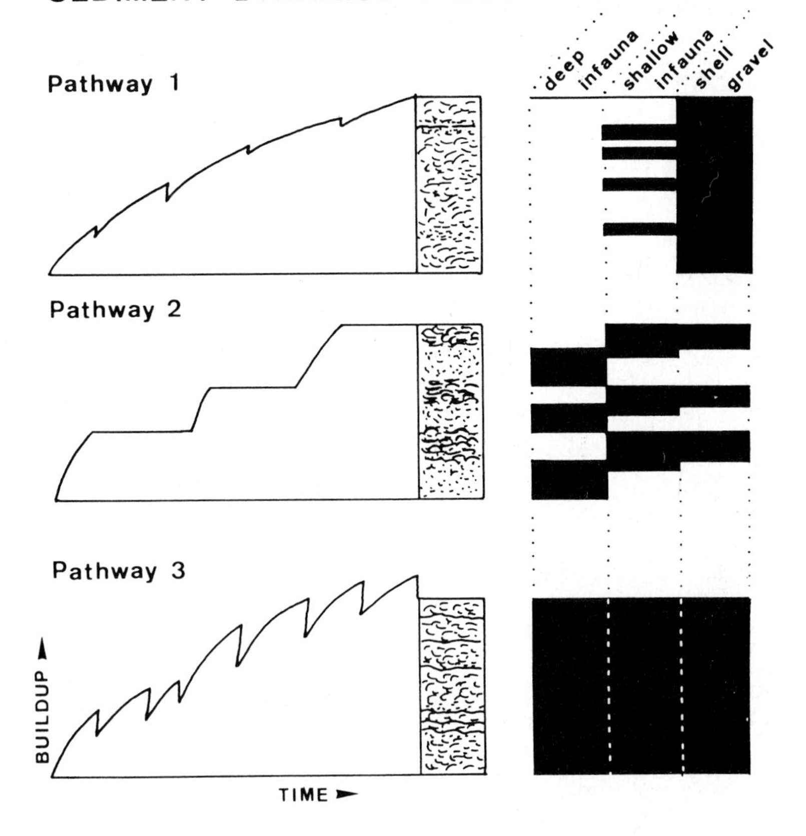

Fig. 2. Three basic pathways in the dynamics of complex skeletal accumulations and expected ecological responses. For further explanation see text.

curve, Fig. 2) permit the episodic colonization by shallow-burrowing infauna, but the community and final fossil assemblage will be dominated by the continued successful colonization of shell gravel species. The complex shell bed that results from pathway 1 will contain few discrete discontinuity surfaces and will have a winnowed, well-sorted sedimentary matrix.

Pathway 2 is characterized by a background condition of zero or low sedimentation relative to hardpart accumulation which is interrupted intermittently by the rapid accumulation of relatively thick depositional increments of muddy sediment. These increments provide an opportunity for colonization by deep- as well as shallow-burrowing infauna. After each "mud event", shell gravel conditions are gradually established as the hardparts from the initial soft-bottom community transform the substrate into a shell-reich and thus progressively coarser and firmer substratum (taphonomic feedback), in which infauna is inhibited by the gradual exclusion of a suitable habitat. In the idealized situation illustrated in Fig. 2, each depositional increment records this temporal change in community composition in its sequence of assemblages, which contain an increasing diversity and abundance of shell gravel species. A complex shell bed built up by a series of such depositional events will contain a series of omission surfaces marked by shell gravel communities and will consist of the original sediment matrix, except where it was winnowed along the omission surfaces.

Pathway 3 is through alternating episodes of aggradation and erosional reworking: soft-bottom colonizers of depositional increments are reworked in-situ into a shell gravel lag, which provides opportunities for colonization by epifaunal taxa. This physically-driven mode of taphonomic feedback produces ecologically mixed assemblages of soft-bottom and shell gravel species which are both homogenized and then amalgamated onto older assemblages by reworking events. In addition to the mixed ecologic character, the final complex shell bed should be characterized by a series of minor internal erosional surfaces and imperfectly winnowed sedimentary matrix.

1.2 Examples

1.2.1. Miocene

The different complex shell beds of the Miocene Calvert and Choptank formations exhibit similar vertical sequences: a basal fragmental shell hash, grading into the main body of the shell bed consisting of both whole and broken closely packed shells, and an upper interval of closely spaced but discrete shell horizons, by which the shell bed grades into less fossiliferous overlying strata (KIDWELL, 1984; KIDWELL & JABLONSKI, 1983). In detail, however, the four complex shell beds had very different histories of sediment accumulation and biotic response. For example, the Camp Roosevelt shell bed is dominated by molluscan taxa preferring shell gravel conditions as judged from species morphology, facies occurrence, and the ecology of modern congeners (KIDWELL & JABLONSKI, 1983). These include free-living bivalves (*Glycymeris*, *Chesapecten*) byssate nestlers (*Anadara*, *Carditamera*), muricid gastropods, and a rich assortment of encrusting bryozoa, hydractinids, barnacles, boring polychaetes, and clionid sponges. Discrete discontinuity surfaces within the shell bed are rare, the sedimentary matrix is extremely well-sorted, and infaunal species of the assemblages are almost all small-bodied, shallow burrowers (venerids, *Astarte*, *Bicorbula*, *Turritella*) that could inhibit the interstices of a shell gravel or could colonize thin, temporary sedimentary veneers and tolerate episodic exhumation. The low incidence of epizoans (ca 5-6 % of all shells; high compared to other shell beds, but low in absolute terms) may itself be another indication for the existence of a shifting bottom such as might be produced by migrating ripple fields. These features indicate condensation under conditions of low total and net sedimentation (pathway 1 in Figure 2).

By contrast, the Drumcliff shell bed contains a diverse assemblage of soft-bottom, deeply burrowing infauna throughout most of its thickness, excluding only the basal 1 m thick shell hash, which is dominated by epifauna (*Crucibulum*, *Balanus*). Shell gravel species and shallow-burrowing species occur abundantly in those parts of the shell bed that are most densely packed (KIDWELL & JABLONSKI, in prep.) suggesting that the ecologically mixed fossil assemblages of the Drumcliff do record repeated conversion of soft-bottom sedimentary substrata into shell gravel habitats. An alternation of aggradation and seafloor reworking probably best describes the shell bed history (pathway 3 in Figure 2), accounting for the numerous scoured and burrowed discontinuity surfaces within the shell bed, the pods of silty sand representing the otherwise winnowed matrix, the thorough admixing of soft-bottom and shell gravel assemblages, and the lack of internal "successions" within the complex accumulation.

Individual reworking events were less effective in winnowing shells from sedimentary increments in the upper part of the shell bed, probably related to the nearing attainment of maximum transgression and maximum water depths for the depositional sequence. Thus, whereas the Camp Roosevelt shell bed records pathway 1 with shell gravel conditions maintained by taphonomic feedback during a period of sedimentary omission, the Drumcliff shell bed records only a short, initial interval of persistent shell gravel conditions, followed by an interval of condensation through pathway 3, with repeated reworking of depositional increments into shell gravel habitats. This presented a far more variable habitat to benthos on ecological time scales than did the pathway 1 dynamics of the Camp Roosevelt shell bed.

1.2.2. Eocene

The most striking feature of the Eocene complex shell beds is the frequently repeated change in faunal composition within the beds. Commonly, the basal firmground is overlain by a zone of mostly articulated, epifaunal anomid bivalves (*Carolia*), which are replaced upwards by *Plicatula* or by ostreid oysters that are partly cemented onto the large *Carolia* shells. In the *Ostrea* bed, the oysters are in turn encrusted by corals, leading to small coral banks. The larger coral colonies commonly show episodic growth restrictions with zones of borings, indicating an alternation of coral growth, anastrophic burial or growth slowdown due to turbidity, and recovery. These "systematic" vertical sequences in faunal composition are restricted to only the lower parts of some shell beds, and record taphonomic facilitation of benthic colonization during an initial period of low total sedimentation (pathway 1 in Figure 2). Most of each shell bed, however, is characterized by unsystematic shifts in assemblage composition. Assemblages include thick layers of (frequently glauconitic) shell hash with *Ophiomorpha*, and intercalations of muddy sand containing infaunal soft-bottom organisms such as *Turritella* and burrowing echinoids. Subsequent reworking of these infaunal assemblages is indicated by post-mortal encrustation, for example of *Turritella* by bryozoa. Although these intercalations suggest episodic aggradation and omission (pathway 2 of Figure 2), erosional reworking of assemblages was a major factor in the condensation as evidenced by the internal discontinuities and erosion surfaces found within the bed sequences. These surfaces are commonly overlain by *Carolia* in a colonization pattern similar to that found on the basal firmground of the complex shell bed.

2. EVOLUTIONARY-SCALE DYNAMICS

2.1 Expected Patterns

Complex shell beds generated by stratigraphic condensation can record shell gravel habitats that persisted over evolutionarily significant periods of time. For example, each of the Miocene complex shell beds are estimated to have accumulated over thousands to tens of thousands of years (KIDWELL, 1982). Successive assemblages within complex shell beds may thus record changes in species morphometry produced by true genetic microevolution (e.g. BAYER & MCGHEE, this volume). However, morphometric trends may also reflect ecophenotypic variation in response to environmental changes, brought about both by the short-term dynamics of sedimentation during condensation events and by longer term cyclicity in condensation with transgression-regression.

2.1.1. Condensation and Information Loss

Samples collected from condensed shell beds will contain more kinds and greater degrees of post-mortem bias than those collected from non-condensed beds, owing to repeated events of hardpart reworking and to prolonged time-averaging (KIDWELL, 1982). For example, the dominance of robust forms (greater shell thickness, more compact shapes) in assemblages condensed through episodic erosion (pathway 3 in Fig. 2) may be a taphonomic artefact of the selective post-mortem destruction of less robust variants from a single genetic population. Prolonged time-averaging can also obscure original morphologic trends or compositions by mixing specimens from successive populations. Layer by layer dissection of complex shell beds might yield samples roughly equivalent in degree of time-averaging to those from a series of discrete, non-amalgamated concentrations. However, the admixture of assemblages by repeated events of reworking, omission, and colonization during the period of condensation will introduce to the samples an additional level of taphonomic bias. Thus it must diminish confidence in the direct comparison of samples between condensed and non-condensed beds as well as samples taken within a single condensed bed.

2.1.2. Condensation and Ecophenotypy

Many benthic species vary morphometrically with habitat parameters such as water depth, water turbulence, and substratum characteristics (EKMAN, 1953; ALEXANDER, 1974; CISNE et al., 1982). Consequently, a major concern in the evolutionary analysis of faunas from lithologically variable sequences is the problem of distinguishing true evolutionary change from ecophenotypic response to environmental change.

In sequences of alternating condensed and non-condensed strata, the condensed shell beds can look sufficiently similar to one another to be categorized as a single lithofacies type. However, treating complex shell beds as the expression of identical habitats in the hope of avoiding ecophenotypic variation among samples can be misleading on several scales.

1. On the scale of samples collected layer by layer through a single complex shell bed, ecophenotypic variation can be confounded with genetic morphologic change because of the alternation of soft-bottom, shell gravel, and ecologically mixed assemblages (Fig. 2).

2. A series of samples from a complex shell bed can record ecophenotypic variation among different bathymetric environments, since the shell bed may be the condensed record of transgressive shoreline migration.

3. Ecophenotypic variation among samples from successive complex shell beds in a stratigraphic sequence can be expected since the shell beds can record different pathways of condensation and condense a different spectrum of environments.

In addition to these problems of scale, several different patterns of morphometric variation can be expected from ecophenotypic responses to, for example, transgressive-regressive cycles (Fig. 3):

A. Species exhibit an excursion in some aspect of morphometry through a transgressive-regressive cycle, tracking water depth or distance from shore.

B. Species exhibit a directional shift in average morphometry through a transgressive-regressive cycle. The trend bears no relationship to water depth, but instead tracks substratum change as reflected in the lithofacies sequence.

C. Morphometric variance of a species decreases upward through a transgressive-regressive cycle from a maximum value within the basal condensed shell bed. High morphometric variability early in the complex shell bed would reflect prolonged time-averaging and mixing of specimens from a series of bathymetric

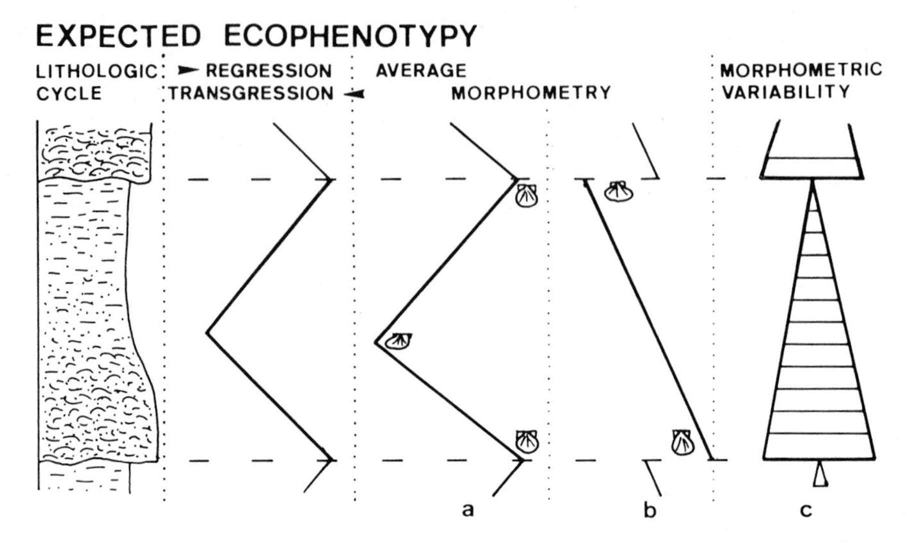

Fig. 3. Several patterns of morphometric variation can be expected from ecophenotypicpic responses to transgressive/regressive cycles that form asymmetric lithologic sequences. Further explanation see text.

environments through stratigraphic condensation. Morphometric variance decreases higher in the cycle because of the lesser time-averaging of fossil assemblages and the greater probability that specimens from a single sample represent the same environment.

2.2 Examples

The morphometric patterns expected from ecophenotypy are in fact apparent within complex shell beds of the Miocene Calvert and Choptank formations, as indicated by KELLEY's (1979; 1983) study of eight mollusk genera. *Lucina anodonta*, for example (Fig. 4), exhibits monotonic decreases in five out of eight measured features over a series of samples from the Camp Roosevelt shell bed (Zone 10 of SHATTUCK, 1904), which records deepening open marine waters (KIDWELL, 1984). This pattern is repeated through the Drumcliff shell bed (Zone 17) and Boston Cliffs shell bed (Zone 19), both also through transgressive deepening-up records. Although KELLEY (1983) interprets these patterns, which all range within 10 to 20% of the mean for the species, as microevolutionary in origin, they may as likely be ecophenotypic. If they are, the Miocene mollusks provide evidence of even greater morphometric stasis than previously thought.

Carolia placunoides, the characteristic bivalve of the Eocene shell beds, also varies in morphology, but on the scale of shell bed to shell bed through the regressive Qasr-es-Sagha Formation. This variation could similarly be interpreted as either a microevolutionary or an ecophenotypic phenomenon. The ecophenotypic explanation is supported by the occurrence of two basic colonization strategies. In most instances, *Carolia* forms dense in-situ pavements colonizing discontinuity surfaces at the base of, or within, complex shell beds. But in portions of the shell beds that are dominated by infaunal molluscs such as *Turritella*, *Carolia* commonly forms stacks of several individuals byssally attached on top of one another. Here, colonization by the stacking strategy seems to be the ecophenotypic response of the bivalve to a scarcity of other firm substrata in a soft-bottom habitat.

Unfortunately, information on morphometric variance is not available for either the Miocene or Eocene mollusks in these sections. However, the trend expected from ecophenotypy (Fig. 3c) is of evolutionary interest, because it mimics a pattern inferred by some workers (e.g. CARSON, 1975; WILLIAMSON, 1981; SYLVESTER-BRADLEY, 1977) for speciation -- an initial burst of variation followed by a more narrowly-defined range of variation. The expected ecophenotypic trend reflects both ecophenotypic and time-averaging effects during stratigraphic condensation, and so should apply to sequences of shell beds condensed over any time scale.

2.3 Other Cyclic Patterns in Condensation

In the Miocene and Eocene sequences, fining-up depositional sequences, bounded by disconformities generate the basic cyclic pattern. Condensed intervals, marked by complex shell beds, lie at the base of each sequence and record transgressive events; they grade upward into shell-poor (i.e. carbonate-poor) muddy prograding facies. This pattern is opposite to that frequently observed in carbonate records, which are characterized by coarsening-up depositional sequences with condensed intervals at the top of the sequence marking the end of regression. Because discontinuity surfaces are used to define the depositional sequences in both situations, the different patterns in grain size and in the position of condensed intervals within cycles are not simply a problem of semantics or conceptual bias.

The timing of condensation during early transgression in the Miocene and Eocene situations can be explained in terms of the models of SWIFT (1968) and RYER (1977): during a rapid relative rise in sealevel, terrigenous sediment supplied by rivers is trapped in coastal estuaries owing to the rise in baselevel, thus starving the shelf. Seventy

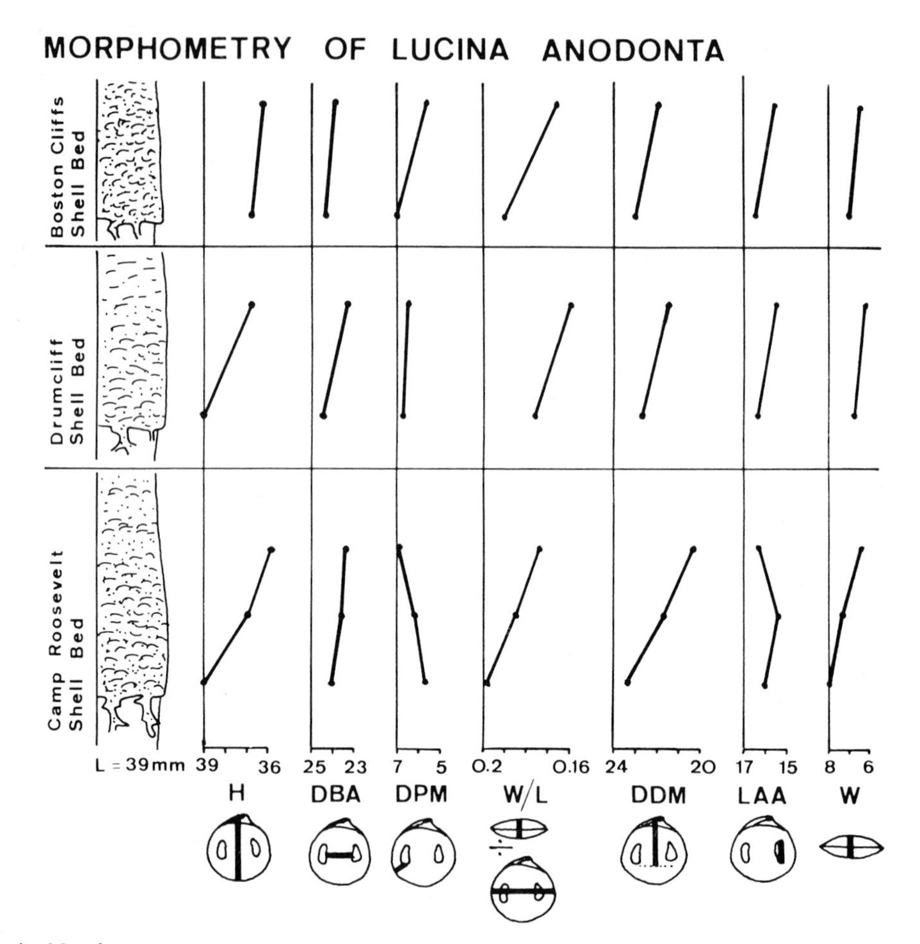

Fig. 4. Morphometric patterns of Lucina anodonta within complex shells beds of the Miocene Calvert and Choptank formations (from KELLEY, 1979, 1983). These patterns may be true genetic microevolutionary changes but they may just as well represent the ecophenotypic response to environmental changes.

percent of modern continental shelf areas have such starved regimes today as a consequence of the Holocene transgression (EMERY, 1968). The relatively thick regressive phase of the cycle records the resumption of terrigenous sediment supply to the shelf once estuaries and other coastal sediment sinks fill to grade, plus the effects of falling baselevel which permits erosional reworking and redistribution of older sediments.

By contrast, carbonate systems are largely fed by autochthonous sediment production, and can thus rapidly aggrade during initial baselevel rise. Hence, they generate a shallowing-up, regressive sequence (e.g. JAMES, 1980). Phases of baselevel still-stand and fall would be characterized by condensation of the upper part of the depositional sequence through winnowing. The lower (BAYER et al., this volume) and upper Muschelkalk (AIGNER, in press) of Germany provide excellent examples of subtidal shallowing-up cycles capped by complex, amalgamated, and condensed shell beds.

Early cementation of carbonate sediments on the seafloor impedes or prevents erosional destruction of the condensed record during latest regression and lowest base-level positions. In this way, not only is the end-regressive condensed phase preserved, but it can be utilized as a substratum by faunas of the following early transgressive phase, for instance as widespread hardground surfaces.

Thus, the opposite patterns in condensation through transgressive-regressive cycles -- base-of cycle shell beds in terrigenous systems and end-cycle shell beds in carbonate systems -- probably derive from qualitative differences in sediment supply between the largely allochthonous terrigenous and largely autochthonous carbonate systems.

CONCLUSIONS

Ironically, rich skeletal accumulations -- long perceived as bonanzas by paleontologists -- are among the most difficult subjects for ecological and evolutionary studies, owing to their comples biological and taphonomic histories.

In ecological analysis, the primary complexity involves the significance of ecologically mixed assemblages. Do they record:

(a) the admixing of allochthonous hardparts;

(b) condensation of successive, unrelated communities; or

(c) in-situ change in community composition due to biotically or physically driven taphonomic feedback?

This ambiguity can be resolved by the analysis of sedimentologic and stratigraphic features (e.g. matrix type, presence of discontinuity surfaces, proximity of appropriate source facies for allochthonous hardparts) as well as paleoecologic and taphonomic features (e.g. KIDWELLL & JABLONSKI, 1983, in prep.). Understanding the ecology of ancient shell gravel habitats, however, is further complicated by the longer-term condensation of assemblages from migrating environments, as in the Miocene and Eocene examples. These patterns can usually be resolved by examining the larger stratigraphic and paleoenvironmental context of the complex shell bed.

Evolutionary analysis is also complicated by the several scales of complexity in shell bed accumulation. This ranges from the ecophenotypic variation among populations occupying successive soft-bottom and shell gravel substrata, to the complication of each shell bed comprising populations from a bathymetric spectrum of environments. Even seemingly directional trends can be generated by an ecophenotypic response to progressive change in substratum type within an asymmetric cycle (Fig. 3). Factoring out such confounding effects depends on careful sampling along bathymetric gradients (e.g. CISNE et al., 1982) or, as GOULD & ELDREDGE (1977), emphasized, sampling among basins within a species geographic range but exhibiting different environmental histories (e.g. JOHNSON, 1982).

ACKNOWLEDGEMENTS

This paper was written while SK was a temporary research fellow with the SFB 53, Tübingen. We thank A. Seilacher and D. Jablonski for useful discussions. This is publication no. 60 within the project Fossil-Lagerstätten of the SFB 53.

REFERENCES

Aigner, T. 1982: Event-stratification in nummulite accumulations and in shell beds from the Eocene of Egypt. In: Cyclic and Event Stratification (G. Einsele & A. Seilacher, Eds.), pp. 248-262, Springer.

Aigner, T. in press: Dynamic stratigraphy of epicontinental carbonates, Upper Muschelkalk (M. Triassic), South-German Basin.- N. Jb. Geol. Paläont., Abh.

Alexander, R.R. 1974: Morphologic adaptations of the bivalve Anadara from the Pliocene of the Kettleman Hills, California.- J. Paleont., 48: 633-651.

Carson, H.L. 1975: The genetics of speciation at the diploid level.- Am. Nal., 109: 83-92.

Cisne, J.L., Chandlee, G.O., Rabe, B.D. & Cohen, J.A. 1982: Clinal variation, episodic evolution, and possible parapatric speciation the trilobite Flexicalymene senaria along an Ordovician depth gradient.- Lethaia 15: 325-341.

Ekman, S. 1953: Zoogeography of the sea.- Sidgwick and Jackson, London, 417 pp.

Emery, K.O. 1968: Relict sediments on continental shelves of world.- Am. Assoc. Petrol. Geol. Bull., 52: 445-464.

Fürsich, F. 1971: Hartgründe und Kondensation im Dogger von Calvados.- N. Jb. Geol. Paläont., Abh., 138: 313-342.

Gould, S.J. & Eldredge, H. 1977: Punctuated equilibria: The tempo and mode of evolution reconsidered.- Paleobiology, 3: 115-151.

Heim, A. 1924: Über submarine Denudation und chemische Sedimente.- Geol. Rdsch. 15: 1-47.

James, N.P. 1980: Shallowing-upward sequences in carbonates. Geoscience Canada, Reprint Ser. 1, 109-119.

Jenkyns, H.C. 1971: The genesis of condensed sequences in the Tethyan Jurassic.- Lethaia 4: 327-352.

Johnson, J.G. 1982: Occurrence of phyletic gradualism and punctuated equilibria through geologic time.- J. Paleont. 56, 1329-1331.

Kelley, P.H. 1979: Mollusc lineages of the Chesapeake Group (Miocene).- PhD-Thesis, Harvard University, 220 pp.

Kelley, P.H. 1983: The role of within-species differentiation in macro-evolution of Chesapeake Group bivalves.- Paleobiology, 9: 261-268.

Kidwell, S.M. 1982: Time scales of fossil accumulations: Patterns from Miocene benthic assemblages.- Proc. 3rd North Am.Paleo. Conv. I: 295-300.

Kidwell, S.M. 1984: Basin margin unconformities in the lower Chesapeake Group (Middle Miocene), Atlantic Coastal Plain.- Am. Assoc. Petrol. Geol. Mem. (in press).

Ryer, T.A., 1977: Patterns of Cretaceous shallow-marine sedimentation, Coalville and Rockport areas, Utah.- Geol. Soc. Am. Bull.88: 177-188.

Strougo, A. 1977: Le "Biarritzien" et le Priabonien en Egypte et leurs Faunes de Bivalves.- Trav. Lab. Paléont. Univ. Paris, Fac. des Sc. d'Orsay.

Swift, D.J.P. 1968: Coastal erosion and transgressive stratigraphy.- J. Geol. 76: 444-450.

Sylvester-Bradley, P.C. 1977: Biostratigraphical est of evolutionary theory.- In: Kauffman, E.G. & Hazel, J.E.; Concepts and Methods of Biostratigraphy.- Dawden, Hutchinson & Ross, 41-63.

Wendt, J. 1970: Stratigraphische Kondensation in triadischen und jurassischen Cephalopodenkalken der Tethys.- N. Jb. Geol. Paläont. Mh. 1970: 433-448.

Williamson, P.G. 1981: Morphological stasis and developmental constraint: Real problems for neo-Darwinism.- Nature 294: 214-215.

PART 6

ECOLOGICAL AND MORPHOLOGICAL GRADIENTS

At the lower end of the 'cyclic resolution' -- usually still in the order of one million years -- replacement of faunas and faunal associations appears to predominate over the evolutionary response. At this level, 'morphologically plastic' groups provide the problem to distinguish evolutionary changes from frequency fluctuations of morphotypes in polymorphic species and from simple ecophenotypic reaction.

** *The relation between faunal associations and lithofacies is analyzed by P. Ward by means of multivariate cluster strategies. While the benthic organisms are closely related to lithofacies, as would be expected, his Cretaceous ammonite data provide a counter example to the previously discussed evolutionary trends; a result of special interest because his study area, a fore arc basin, probably was less isolated than the otherwise discussed epicontinental seas.*

** *Oysters are a particularly 'plastic' group. Seilacher, Matyja & Wierzbowski contribute examples of morphological successions within single beds in terms of temporal and spatial morphological gradients. The application of constructional morphology allows to reconstruct the complex environmental conditions of condensed and repeatedly winnowed beds. On the other hand, they show that the paleontological contribution to evolutionary theory is limited by temporal and 'genetic' resolution.*

** *Gryphaea is a classical object for evolutionary studies. Bayer, Johnson & Brannan reconsider the Middle Jurassic 'Gryphaea story' and recognize substrate related morphological gradients which are of the same magnitude as previously observed phylogenetic trends. Besides this, the record of European Gryphaea is discontinuous with rather long gaps separating the occurrences in time. This aspect returns to the discussion of migrational events in the first part.*

In summary, we find evidence that the 'punctuation' of the paleontological record can, at least in parts, be related to faunal substitutions triggered by external physical changes. Morphological gradients may well result from selection of certain morphotypes of polymorphic species -- spatially and temporally. Such changes can be considered as 'ecophenotypic' reactions because the gene pool remains stable in the sense that it is not altered by the interaction of mutation and selection.

UPPER CRETACEOUS (SANTONIAN-MAASTRICHTIAN) MOLLUSCAN FAUNAL ASSOCIATIONS BRITISH COLUMBIA

Peter D. Ward

Davis

Abstract: Multivariate analyses of Upper Cretaceous (Santonian-Campanian) molluscan assembladges from the Nanaimo Group of British Columbia, Canada, and northwestern Washington, USA, differentiates substrate controlled associations of species. Ammonites and inoceramid bivalves dominate offshore muds and silts, while infaunal bivalves and gastropods are most common in more nearshore facies. Associations of ammonites viewed in successive zones show striking functional inequality even within similar lithofacies, indicating that ammonites evolution is occuring more on a province wide than basin wide level.

INTRODUCTION

The Cretaceous North Pacific Biotic Province was first defined by JELETZKY (1965) as the area of distribution of late Berriasian to mid-Valanginian marine faunas of the Pacific slope north of Mexico. This definition was later expanded to include this region for the entire Cretaceous (JELETZKY, 1971). KAUFFMAN (1973) considered this same area a subprovince of a larger North Pacific Province, which included Japan, the Soviet Far East, and the western coast of North America. As has been pointed out by both of these authors, the Late Cretaceous macrofauna of these regions is quite distinct from Tethyan and boreal faunas. Among ammonites, phylloceratids, desmoceratids, and

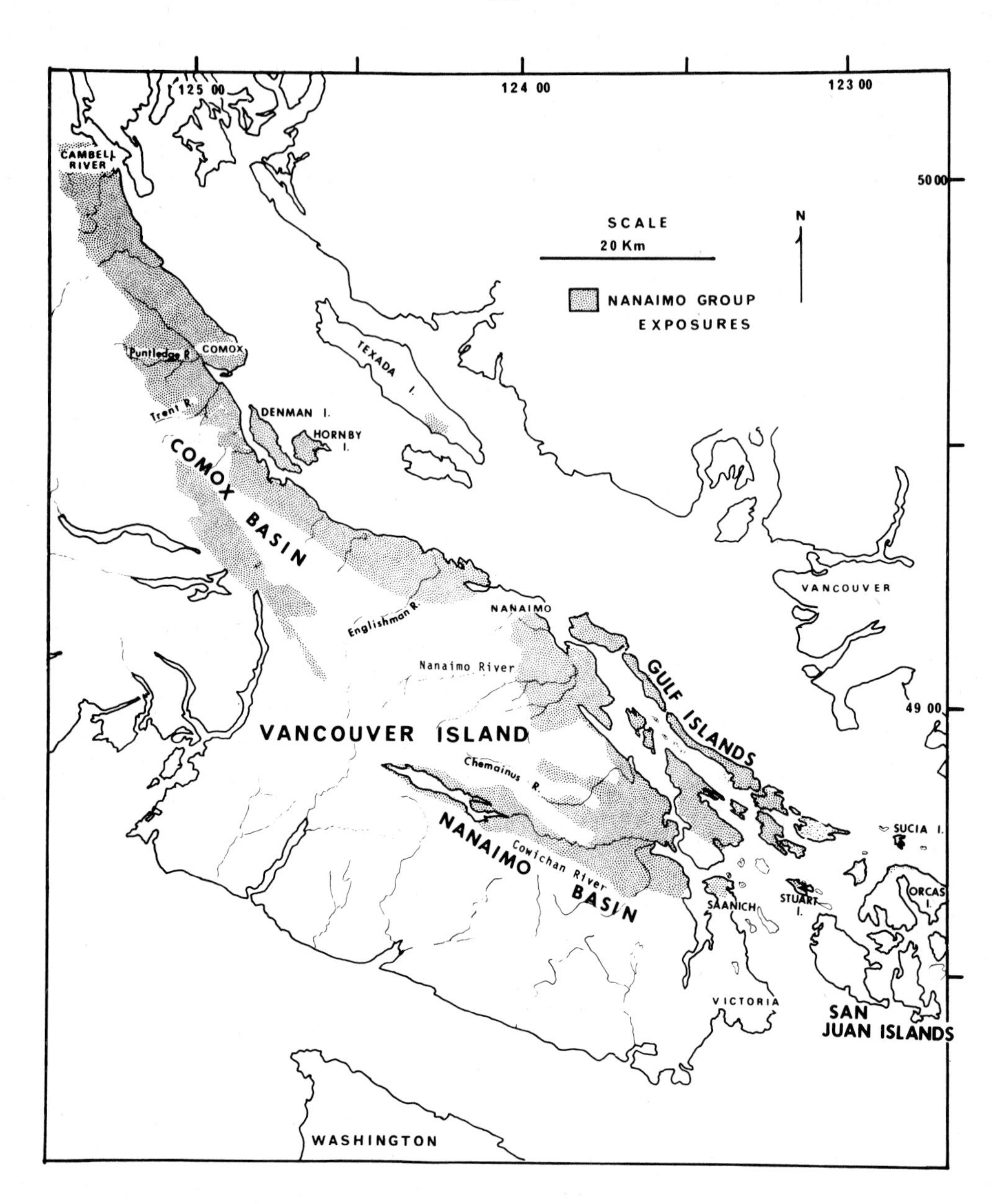

Fig. 1: Outcrop of the Nanaimo Group in the working area.

planispiral lytoceratids are dominant in the North Pacific Province, with pachydiscids and baculitids especially numerous. Scaphitid ammonites, neoceratites, and belemnites are rare or absent. Trigoniids and inocerami are common but not diverse. Colonial corals, rudistids, echinoids, and crinoids are rare.

Detailed work on the systematics and biostratigraphy of ammonites and bivalves of the North Pacific Province during the last two decades, mainly by Jeletzky, Jones, and Matsumoto, has resulted in a large number of papers dealing with regional and inter-regional correlation. Much less attention has centered on paleoecologic characteristics of the Cretaceous North Pacific Province macrofaunas. SAUL (1960) discussed faunal assemblages of the Chico Creek sections of California. MATSUMO (1960) made comparisons of Japanese and Californian Upper Cretaceous ammonite assemblages, and the facies in which they occur.

In this paper macrofaunal assemblages from the Upper Cretaceous (Santanian-Maastrichtian) Nanaimo Group of southeastern British Columbia (Fig. 1) are discussed from a paleoecologic, rather than biostratigraphic standpoint. The Nanaimo Group offers a number of advantages for this type of study, including a refined macrofaunal biostratigraphy (JELETZKY, in MULLER & JELETZKY, 1970), and the fact that it was deposited in a relatively small depositional basin. The Nanaimo Group lies intermediate between the eastern (California) and western (Japan) portions of the North Pacific Province, and as such contains many faunal elements (and possibly faunal associations) common to both California and Japan. The major disadvantage to the Nanaimo Group as a subject for study is the antiquated taxonomy available for most of its bivalves and gastropods. Inoceramids from the Nanaimo Group are especially in need of taxonomic revision, as KAUFMANN (1977) has pointed out. In spite of this obstacle, faunal assemblages can be differentiated, and the observations noted here constitute a first step toward recognition of North Pacific biotic associations near the end of the Cretaceous.

STRATIGRAPHIC SETTING

The Nanaimo Group has been divided into eleven formations (MULLER, in MULLER & JELETZKY, 1970; WARD, 1978, Table 1). The oldest Nanaimo strata (Comox Formation) are non-marine sandstone and conglomerate unconformably resting on either Permian and Triassic volcanics, or on Jurassic intrusives. The age of the lowest Comox Formation is uncertain; the oldest diagnostic marine fossils, from near the top of the formation, are late Santonian (JELETZKY, in MULLER & JELETZKY, 1970). The youngest strata of the Nanaimo Group (Gabriola Formation) are coarse, cross-bedded sandstones and conglomerates that rest on lower Maastrichtian mudstones.

Macrofossil biostratigraphy for the Nanaimo Group was first established by USHER (1952) on the basis of ammonite ranges. Revisions were later made by JELETZKY (in MULLER & JELETZKY, 1970) and WARD (1978). Five macrofossil zones are recognized here (Table 1).

Table 1: Stratigraphy of the Nanaimo Group in the working area.

MULLER AND JELETZKY, (1970)

European Stages	FORMATION	ZONE and SUBZONE	
Maestrichtian	GABRIOLA		
—?——?—			
Upper Campanian	SPRAY	*P. suciaensis* Zone	*N. hornbyense* Subzone
Upper Campanian	GEOFFREY	*P. suciaensis* Zone	*M. pacificum* Subzone
Upper Campanian	NORTHUMBERLAND	*P. suciaensis* Zone	*M. pacificum* Subzone
Upper Campanian	DE COURCY	*H. vancouverense* Zone	
Upper Campanian	CEDAR DISTRICT	*H. vancouverense* Zone	
Upper Campanian / Lower Campanian	EXTENSION PROTECTION	*H. vancouverense* Zone / *I. schmidti*	
Lower Campanian / Santonian	HASLAM	*B. elongatum* Zone	*E. haradai* Subzone
Santonian	COMOX	*B. elongatum* Zone	*I. naumanni* Subzone

HERE

FORMATION	ZONE, SUBZONE and ZONULE	
GABRIOLA	? ? ?	
SPRAY	*P. suciaensis* Zone	*N. HORNBYENSE* ZONULE
GEOFFREY	*P. suciaensis* Zone	
NORTH-UMBERLAND	*P. suciaensis* Zone	
DE COURCY	*PACIFICUM-SUCIAENSIS* BARREN INTERZONE	
CEDAR DISTRICT	*B. REX* ZONULE / *M. pacificum* Zone / *H. vancouverense* Zone	
PROTECTION	*H. vancouverense* Zone	
PENDER	*B. chicoensis* Zone	
EXTENSION	*I. schmidti* Zone	
HASLAM	*B. elongatum* Zone	*E. haradai* Subzone / *P. VANCOUVERENSE* ZONULE
COMOX	*B. elongatum* Zone	*P. VANCOUVERENSE* ZONULE / *I. naumanni* Subzone

The Upper Cretaceous Nanaimo Group was deposited in a fore-arc basin on the exotic Insular Belt (Wrangellia terrane) and represents the oldest recognized sedimentary rocks in the Insular Belt of southern Vancouver Island, the Gulf Islands, British Columbia and San Juan Islands, Washington that contain debris derived from the mainland (EISBACHER, 1974; MULLER, 1977; MULLER & JELETZKY, 1970). The Wrangellia terrane is believed to have formed in the late Paleozoic - early Mesozoic Pacific Ocean and was probably accreted on to the western margin of North America during the late Mesozoic (DANNER, 1977; JONES et al., 1977; MULLER, 1977). DICKINSON (1976) has suggested that the intra-oceanic elements of the Insular Belt were tectonically welded on to the continental margin by middle to late Jurassic time, but that the deformation associated with this crustal collision and suturing may have continued into the early Cretaceous when fore-arc basins to the east were deformed and the magmatic arc jumped westward. After the westward jump of the magmatic arc, rocks from older subduction-arc complexes, such as those now exposed in the western Cascades and San Juan Island, became an upland source of sediment for a newly formed late Cretaceous fore-arc basin on the Insular Belt. Deposits of this fore-arc basin are preserved in the Comox and Nanaimo basins of Vancouver Island, the Gulf Islands, and San Juan Islands (Fig. 1). The basal formations of the Nanaimo Group in the Nanaimo Basin, the Comox and the Haslam Formations, are the oldest rocks that provide information about the unroofing of the pre-Upper Cretaceous subduction-arc complexes at the southern margin of the Insular Belt, the initial supply of detritus from the magmatic arc in the Coast Plutonic Belt, and the initial phase of sedimentation in the fore-arc basin.

The geometry of the Nanaimo depositional area has been discussed by a number of workers. CLAPP (1914) interpreted the Nanaimo Group as having been deposited in two major and several minor depositional basins. The larger of these basins were named the Comox and Nanaimo Basins. SUTHERLAND-BROWN (1966) suggested that the outcrops of the Nanaimo Group are erosional remnants of strata deposited in a single depositional basin. MULLER (in MULLER & JELETZKY, 197O) followed sutherland-Brown in delimiting a single large depositional basin, but noted that a topographic high seems to have separated the northern and southern outcrop areas, so that all stratigraphic units thin near the town of Nanaimo.The presence of an east-west paleotopographic high is also indicated by my observations in this area. For this reason, Clapp's terms Nanaimo Basin (for the southern areas) and Comox Basin (northern areas) are utilized in this paper.

The Nanaimo Group is characterized by rapid facies changes. A wide variety of marine and non marine depositional environments are inferred to have present, including both marine and non-marine deposits, although marine environments were dominant. Turbidites, grain-flow deposits, Fluxoturbidites, and resedimented conglomerates compose the majority of strata in the southwest southeastern Nanaimo Basin; strata to the north and west are more shelf-like, and contain widespread coal deposits (Fig. 2).

Marine macrofossils are most common in non-turbiditic facies, although rich concentrations of inocerami and ammonites can be recovered from very thin, distal turbidites. The faunal assemblages discussed in this paper can be characterized as occurring in one of the following, intergradational facies:

1. Sandstones, coarse to finely-grained, sorting usually poor, with high clay fraction. Bedding is generally poorly defined; numerous burrow marks suggest bioturbation. Colors vary from light grey to olive green. Fossils, mainly thick-shelled gastropods and pelecypods, are generally abundant.

2. Siltstone, high clay fraction, sorting poor. Bedding is usually massive. Concretions may or may not be present. Color is usually dark grey to olive. Fossils common.

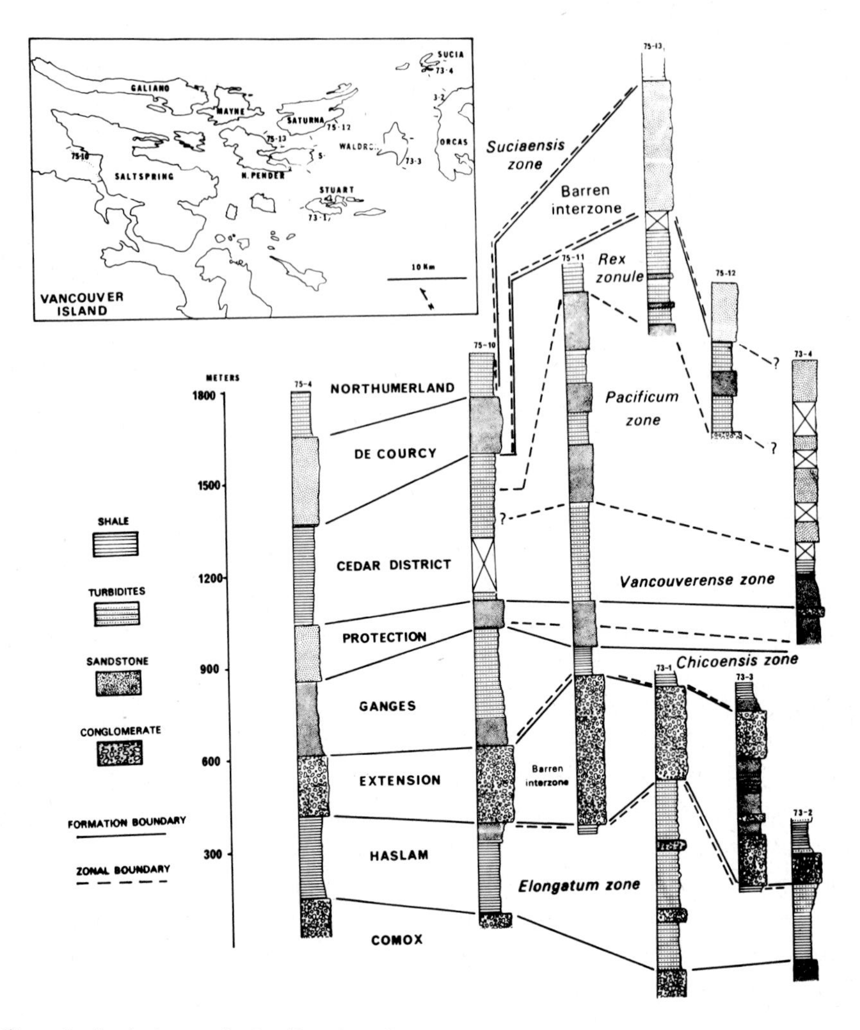

Fig. 2: Lithology of the Nanaimo Group.

3. Mudstones, slightly calcareous, with fine sand disseminated throughout. Bedding is generally massive. Sandstone interbeds and/or concretionary layers often present. Sandstone dykes extremely common. Fossils common to rare.

4. Same as 3., but without sandstone dykes.

5. Distal turbidites, thinnly bedded, composed of Bouma CD and CE divisions. Sand-shale ratios generally .3 - .6, bed thicknesses range between 2 - 20 cm. Fossils rare.

6. Mudstones, with numerous concretions packed with *Anomia* shells. Shales massive, interbedded with sandstone layers and concretionary layers.

FAUNAL ASSOCIATIONS

Elongatum Zone

The oldest known macrofauna of the Nanaimo Group is characterized by the common occurrence of two heteromorph ammonites, (*Glyptoxoceras subcompressum* (FORBES) and *Didymoceras (Bostrychoceras) elongatum* (WHITEAVES), and by *Inoceramus naumanni* YOKOYAMA and associated species such as *I. orientalis* SOKOLOV. This macrofauna first occurs in the upper beds of the Comox Formation, and ranges through varying proportions of the Haslam Formation. It is succeeded in the upper Haslam by a mostly different assemblage of ammonites and inocerami, of which the radially ribbed *Inoceramus (Sphenoceramus) schmidti* MICHAEL is the zonal index.

The *elongatum* Zone is the most areally extensive biostratigraphic unit in the Nanaimo Group, and shows the greatest macrofaunal diversity. For this reason it has been examined in greatest detail.

Assemblages of the *elongatum* Zone have been differentiated using multivariate discrimination techniques. A total of 4 194 macrofossil occurrences were tabulated during the summers of 1975 and 1976 from 34 localities at which the Comox and Haslam Formations are exposed. Even though temporal bias has been reduced by utilizing a single zone, certain ammonite and inocerami species, and possibly as yet unknown bivalve or gastropod species range only partially through the Elongatum Zone, and thus introduce temporal bias into the associative techniques. *Polyptychoceras vancouverense* (WHITEAVES) restricted to the upper Santonian beds of the *elongatum* Zone (in MULLER & JELETZKY, 1970) is the shortest ranging of the ammonites. Other short-ranging ammonites are *Eupachydiscus perplicatus* (WHITEAVES) and *E. haradai* (UHSER). According to JELETZKY (in MULLER & JELETZKY, 1970), the latter species is descendent, and completely replaces the former in the latest Santonian or early Campanian. Specimens of these two pachydiscid species may also be somewhat facies-dependent, for *E. perplicatus* is known only from the Comox Basin, while *E. haradai*, although known from several Comox Basin sections, is much more common in the Nanaimo Basin.

Each locality consisted of beds not exceeding 20 m in stratigraphic thickness and was randomly picked from measured sections. At each locality, as many fossils as possible were collected or noted, with a minimum number of 100 as goal. In many river sections the prevailing eastern dip of the Nanaimo Group coincides with the river gradient, exposing bedding-plane exposures of large areal extent. An effort was made to note every megafossil and imprint regardless of size. At two localities (GSC 77393 and 69453), counts were made on bulk material collected by other workers. Stratigraphic and geogra - phic positions of these localities are listed in WARD (1976a). The resultant data for each Elongatum Zone locality has been analyzed in two ways. First, a matrix of presence-absence data was tabulated for the sampled localities. Forty-seven taxa were identified from all collections, and for each locality these taxa were noted as either present or absent. The overall number of species within the collections is probably greater than 47, since ammonites and inocerami were differentiated at the species level, but other bivalves and gastropods were differentiated only at the generic level.

For the second method of data tabulation, each fossil was categorized as one of 15 separate variables for each locality, based on morphology (ammonites), size (inocerami) or feeding type (non-inoceramid bivalves). Gastropods, *Trigonia* and *Anomia* were also listed as separate variables. Each variable represents a percentage of the total fauna collected from the localities; the sum of all variables equal 1.00 for each locality. To insure that each variable approximated a normal distribution for each locality, a requisite of all multivariate analyses, the percentages were transformed with the expression

$$S_{ab} = \arcsin \sqrt{Y_{ab}}$$

where Y_{ab} is the proportion of the a_{th} variable from the b_{th} locality. This transformation was used to transform the binominal distributions inherent in many biological populations to normality by GOODALL (1954), and CASSIE & MICHAEL (1968), and has been discussed in detail with regard to paleoecologic applications by BUZAS (1972).

In the past, multivariate statistical techniques, such as cluster analysis, factor analysis, and ordination, have been used in differentiating faunal assemblages. Relative advantages and disadvantages of each method are discussed in SNEATH & SOKAL (1973) and VALENTINE (1973). Because cluster analysis has the advantage of a compact, two-dimensional visualization of the cladistic or ecologic relationships and of discriminating slightly dissimilar variables (ROHLF, 1972), the technique has been used here.

SOKAL & SNEATH (1963) have discussed the various types of clustering techniques available. In most ecologic and paleoecologic studies to date, either the weighted pair-group method (WPGM) or unweighted pair-group method (UPGM) agglomerative techniques has been utilized. Advantages of each are discussed by SNEATH & SOKAL (1973), HAZEL (1970)) and VALENTINE (1973). Recently, derivation of the cophenetic correlation coefficient (SOKAL & ROHLF, 1962) and its analysis with artificial data (FARRIS, 1969; ROHLF, 1970) has shown that the UPGM yields clusters with less information loss than WPGM. Consequently, the UPGM has been utilized here. The NT-SYS system of cluster

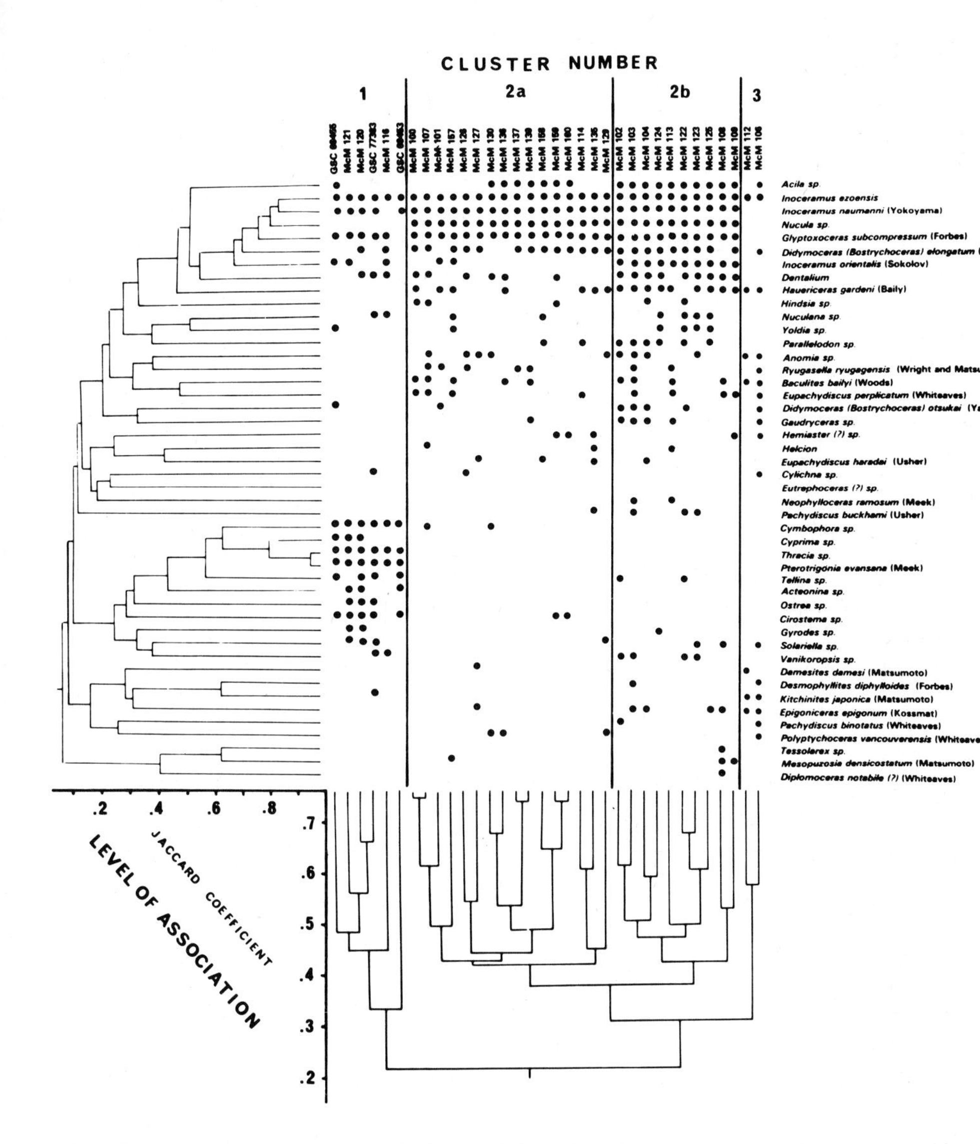

Fig. 3: Two-way phenogram of Q- and R-mode clusters using Jaccard coefficient (see text for discussion).

analysis was performed on both presence-absence data and transformed percentage data; computation was performed at the University of Rochester Computing Center. Jaccards coefficient was used for presence-absence data, and the cosine Θ coefficient for the transformed percentage data.

The resulting phenograms for presence-absence data and transformed percentage data are shown in Figures 3 and 4 respectively. The Q-mode (localities) and R-mode

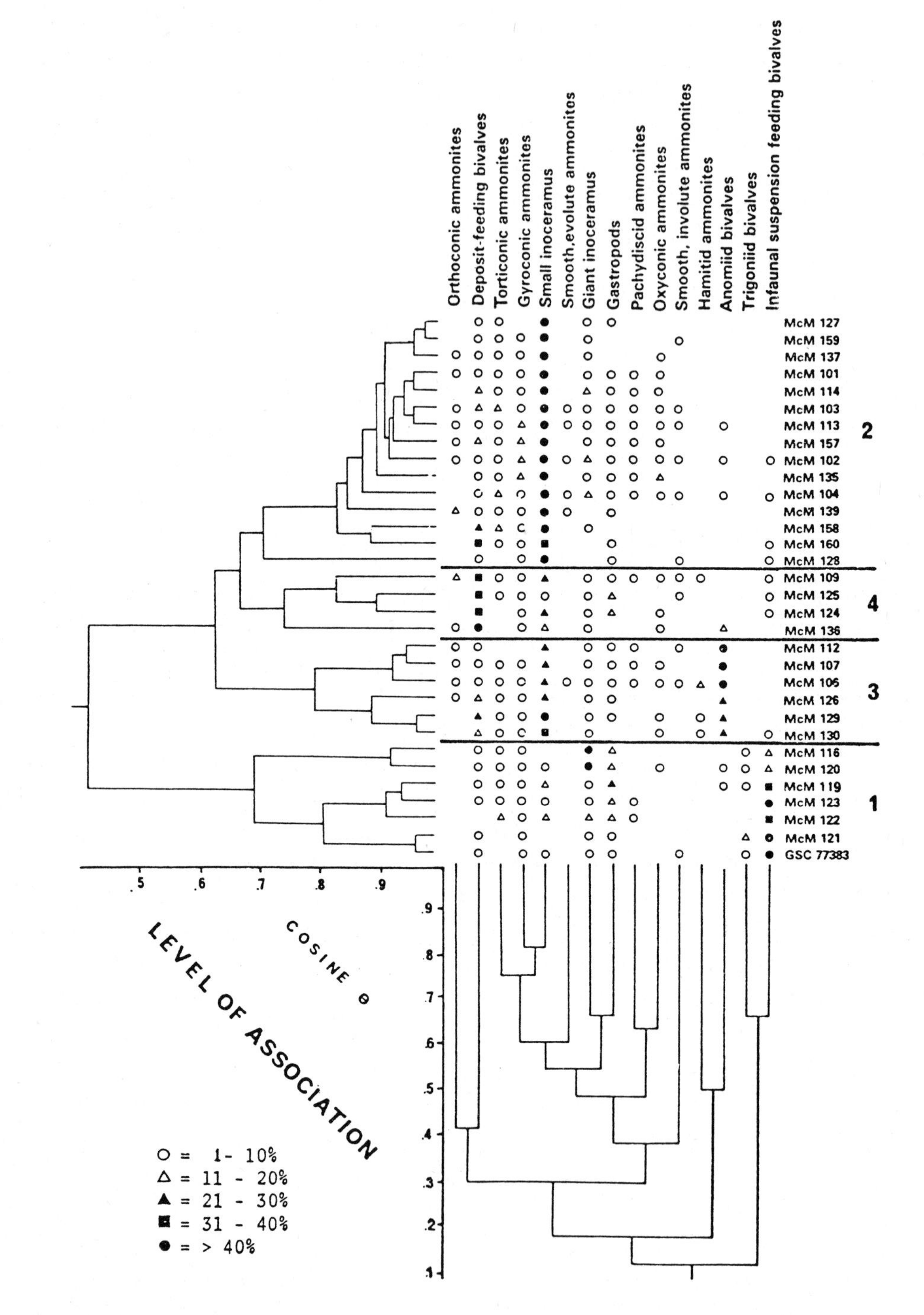

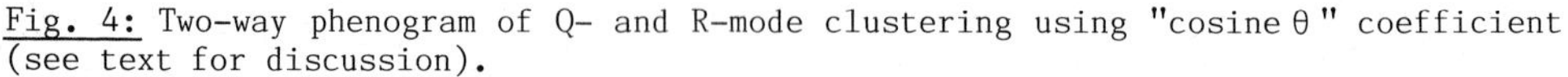
Fig. 4: Two-way phenogram of Q- and R-mode clustering using "cosine θ" coefficient (see text for discussion).

(faunal variables) phenograms for presence-absence and percentage data have been grouped together in two-way phenograms following the method of SEPKOSKI & REX (1974). This graphical method allows simultaneous presentation and intercorrelation of the Q- and R-mode clusters.

Two major clusters of localities can be differentiated in Fig. 3. These are composed, on the one hand, of sandstone localities containing a suspension-feeding bivalve and gastropod assemblage (Cluster 1), and, on the other, of mudstone, siltstone and turbidite localities with an *Inoceramus* and ammonite assemblage (Cluster 2).

Figure 4, based on the percentage at each locality of 15 faunal variables, shows better discrimination in differentiating faunal assemblages and biofacies. Four clusters of localities are discriminated, on the basis of four faunal assemblages. The bivalve-gastropod assemblage of Fig. 3 is again discriminated (also designated Cluster 1 in Fig. 4), but in this analyses the inoceramid-ammonite assemblage of Fig. 4 is divided into three distinct assemblages.

Cluster 2 of Fig. 4 is dominated by torticones *(Didymoceras (Bostrychoceras) elongatum* and *D. (B.) otsukai)*, gyrocones *(Glyptoxoceras subcompressum)*, small inocerami *(I. naumanni, I. orientalis)*, and large inocerami *(I. ezoensis)* . Also present are significant percentages of pachydiscids *(Eupachydiscus perplicatum, E. haradai)* , compressed ammonites *(Hauericeras gardeni)* involute ammonites *(Desmophyllites diphylloides, Epigoniceras epigonum, Neophylloceras ramosum, Damesties damesi)*, and evolute ammonites (*Gaudyceras* sp. indet.).

Cluster 3 is differentiated on the basis of numerical dominance by *Anomia vancouverensis* WHITEAVES) and the hamitid ammonite *Polyptychoceras vancouverensis*. Cluster 4 includes localities characterized by a faunal assemblage numerically dominated by species of infaunal deposit-feeding bivalves such as *Yoldia diminutiva* WHITEAVES, *Nucula* sps. indet. and *Nuculana* sp. indet., with lesser numbers of inoceramids and baculitids.

Based on these clusters, four assemblages can be formalized:

1. *Pterotrigonia evansana* Association.

This faunal association is comprised of infaunal suspension-feeding bivalves followed in numerical importance by gastropods, deposit-feeding bivalves and epifaunal suspension feeders. Ammonites are rare or absent. Overall fossil density is very high. The association is named after the single most common taxon, *Pterotrigonia evansana* (GABB).

The lithology present at all of the localities with this association is lithology I as defined above, and appears to be equivalent to lithologies 1 to 3 of KAUFFMAN (1967); faunal association H of Kauffman appears similar in faunal composition. This assemblage was considered to have inhabited a near-shore environment of medium to high energy. Modern analogues include communities in sandy and silty upper shelf environments at depths usually not exceeding 15 m.

RHOADS, SPEDEN & WAAGE (1972) differentiated a number of Cretaceous marine invertebrate associations based on trophic groupings; the Nanaimo Group *Pterotrigonia* assemblage appears to correspond to their *Tancredia-Ophiomorpha* assemblage, composed of infaunal suspension feeders, and in part to the Lower Timber Lake Member assemblage, composed of a mixture of deposit and suspension-feeding forms.

SAUL (196O) discussed a number of faunal associations from the Upper Cretaceous section on Chico Creek, California. The Nanaimo Group *Pterotrigonia* association contains faunal elements common to three of Saul's association: the *Gymnarus-Cymbophora* aswsociation, *Donox-semele* association, and *Calva-Glycymeris* association. All three are dominated by infaunal suspension-feeding forms.

2. *Nucula* Association.

This faunal association is characterized by infaunal deposit-feeding bivalves, including *Nucula*, *Nuculana*, *Yoldia* and *Acila* . Other common constituente include *Inoceramus ezoensis*, Nagao and Matsumoto, *Baculites bailyi* (WOODS), and *Eupachydiscus perplicatum* (WHITEAVES). Lithology 2 is characteristic.

Lithologies and fauna of this association are similar to those of lithotope and faunal assemblage P of KAUFFMAN (1967), which shows high diversity and lacks a numerically dominant organism. In addition to deposit-feeders, common constituents include large, thin-shelled inocerami, and the heteromorph ammonite *Scaphites*. The entire assemblage was considered by Kauffman as a middle shelf assemblage inhabiting silt-muds in water 6O to 1OO m deep.

Assemblage P intergrades laterally with the more shallow equivalent of the Elongatum Zone *Pterotrigionia* Association, Kauffman's assemblage H. A similar, intergradational relationship between the lithotopes and faunas of the *Pterotrigonia* and *Nucula* associations is observable in the Nanaimo Group.

The *Nucula* Association also shares faunal elements with the Trail City Member Association of RHOADS et al. (1972). In this latter association, the presence of bivalves is correlated with periods of moderate to high sedimentation, with suspension-feeding bysally-attached pterioids and inoceramids feeding well off the bottom.

A third analogous Cretaceous association is the *Nucula-Nuculana* Community of SCOTT (1974). This association from the lower Cretaceous of Texas is dominated by protobranchs and shows high diversity, but low density.

3. *Inoceramus* Assemblages.

The majority of sampled horizons contain an association dominated by epifaunal suspension-feeding bivalves which is here termed the *Inoceramus* Assemblage. Ammonites are the second most important faunal element, followed by deposit-feeding bivalves.

This assemblage shows diversity differences between the Comox and Nanaimo Basins, but appears trophically similar throughout the Nanaimo Group. The most marked diversity differences are among the ammonites; Planispiral phylloceratids and lytoceratids present in the north are rare or absent in the south.

In both the north and south, *Glyptoxeras subcompressum* is the most common ammonite, followed in numbers by *D. elongatum*. In the south the ammonites are oxycones (*Hauericeras gardeni)*, while in the north, various pachydiscid species approximately equal *H. gardeni* in absolute numbers.

Rocks containing the *Inoceramus* Assemblage are, in the Comox Basin, mainly shales with interbedded sandstone or siltstone beds and common sandstone dikes. (Lithology 3), while in the Nanaimo Basin, the assemblage is present in shales without dikes (lithology 4), and in distal turbidites (lithology 5).

The *Inoceramus* assemblage has parallels in the Cretaceous of the Western Interior. Assemblages Q, R and S of KAUFFMAN (1967) are all from dark shale facies and, like the Nanaimo *Inoceramus* association, are dominated by inoceramid and ammonite species.

4. *Anomia* Association.

The last major faunal association, named after the most common faunal element, is dominated by the epifaunal suspension-feeding′ bivalve *Anomia vancouverensis* (WHITEAVES). Wherever present, specimens of this species are the most common macrofossil.

Amonia is rarely represented as scattered valves in the shale; most commonly, it occurs as shell concentrations in pyritized concretions. The concretions are mostly spherical or ellipsoidal and are common, in the shale layers. Most concretions contain several *Inoceramus naumanni* shells; since most *Anomia* shells within these concretions bear radial ornamentation matching the curvature of the inocerami ribs, it appears that the *Anomia* were attached to the inoceramids, and to each other. Other common constituents of these concretations are the heteromorph ammonites *Ryugasella ryugasensis* (WRIGHT

& MATSUMOTO) and *Polyptychoceras vancouverense* (WHITEAVES). The latter, a heteromorph with a *Hamites*-like shape, is restricted to lithologies with the *Anomia* concretions (lithology 6), but appears free in the shale as well as in the concretions and is by far the most common ammonite at these horizons.

ELONGATUM ZONE AMMONITE ASSOCIATIONS

Cluster analyses for *elongatum* Zone macrofaunas suggest that each of the *Inoceramus*, *Nucula* and *Anomia* Assemblages contains a characteristic group of ammonites. To further test this generalization, the sample size for *elongatum* Zone Ammonites has

Table 2: *elongatum* Zone ammonites.

	Total Collected	% of total number ammonites	Occurrence: % of total No. localities
Glyptoxoceras subcompressum (Forbes)	250	27	91
Didymoceras (Bostrychoceras) elongatum (Whiteaves)	204	22	74
Polyptychoceras vancouverense (Whiteaves)	119	13	06
Baculites bailyi (Woods)	73	08	32
Eupachydiscus perplicatum (Whiteaves)	71	08	26
Hauericeras gardeni (Baily)	55	06	53
Ryugasella ryugasensis (Wright & Matsumoto)	37	04	24
Epigoniceras epigonum (Kossmat)	18	02	21
Didymoceras (Bostrychoceras) otsukai (Yabe)	17	02	21
Pachydiscus buckhami (Usher)	16	02	12
Pachydiscus haradai (Usher)	13	01	12
Gaudryceras sp.	6	01	18
Neophylloceras ramosum (Meek)	6	01	06
Desmophyllites diphylloides (Whiteaves)	6	01	06
Eutrephoceras cambelli(?) (Whiteaves)	5	01	03
Damesites damesi (Matsumoto)	4	01	06
Kichinites japonica (Matsumoto)	3	01	09
Mesopuzosia densicostatum (Matsumoto)	3	01	09
Diplomoceras sp.	2	01	03
Pachydiscus binodatus (Whiteaves)	2	01	06

been enlarged by including all known specimens of ammonites and nautilids in the collections of the Geological Survey of Canada, Ottawa, the University of Washington, and the University of British Columbia from those localities that correspond to localities used in this study. In this way the total number of cephalopods from 24 Nanaimo Group localities was increased to 910 (Table 2). As in the cluster analyses described above, each taxon was considered a variable, computed as a percentage of the total number of chambered cephalopods from that locality, and transformed to approximate a normal distribution.

The resulting phenogram (Fig. 5) differentiates three main cephalopod groups. If rare species are neglected (those species comprising 5% or less of the total cephalopod fauna), three groupings remain: an assemblage composed of *Baculites bailyi* and *Eupachydiscus perplicatum* , from localities containing the *Nucula* Assemblage; a grouping of *Didymoceras (Bostrychoceras) elongatum*, *Glyptoxoceras subcompressum*, and *Haueriiceras gardeni* , most characteristic of the Inoceramus Assemblage, and a cluster with *Polyptychoceras vancouverensis*, from the Anomia Assemblage. Although specimens of several of these ammonite species were recovered at all localities, the most common *elongatum* Zone ammonites dominate in only one characteristic facies.

Within the sampled *elongatum* Zone collections almost 75% of the ammonites are heteromorphs. Two of these species, the zonal indices *Glyptoxoceras subcompressum* and *D. (B.) elongatum*, comprise 27 and 22% respectively of the ammonites found, and are present in almost all localities. The third most common species (13%), *P. vancouve-rense*, shows a much different distribution in that it occurs at only three of 33 localities. Planispiral phyllocerastids and lytoceratids, such as *Neophylloceras ramosum*, *Desmophyllites diphylloides*, *Gaudryceras* sp. and *Damesites damesi* are rare faunal elements, and restricted almost completely to localities within the Comox Basin. Turbidite localities in the Nanaimo Basin contained mainly the zonal index and the heteromorph *Ryugasella ryugasensis*.

European workers (e.g. ZIEGLER (1969), GECZY (1971), GEYER (1971) have long recognized that the phylloceratids and lytoceratids, or "leiostracans", are most common in strata deposited in deep, offshore environments of the Tethyan region. SCOTT, (1940) in a classic work, proposed that ammonoids other than "leiostracans" could be correlated with specific sedimentary environments or habitats. During *elongatum* Zone time, Comox Basin environments appear to have been more favorable for Leiostracans than Nanaimo Basin environments.

The apparent presence of distinct ammonite biofacies within the Elongatum Zone of the Nanaimo Group raises a number of questions about mode of life and postmortal distribution of these ammonites. Restriction of *P. vancouverense* to a small number of

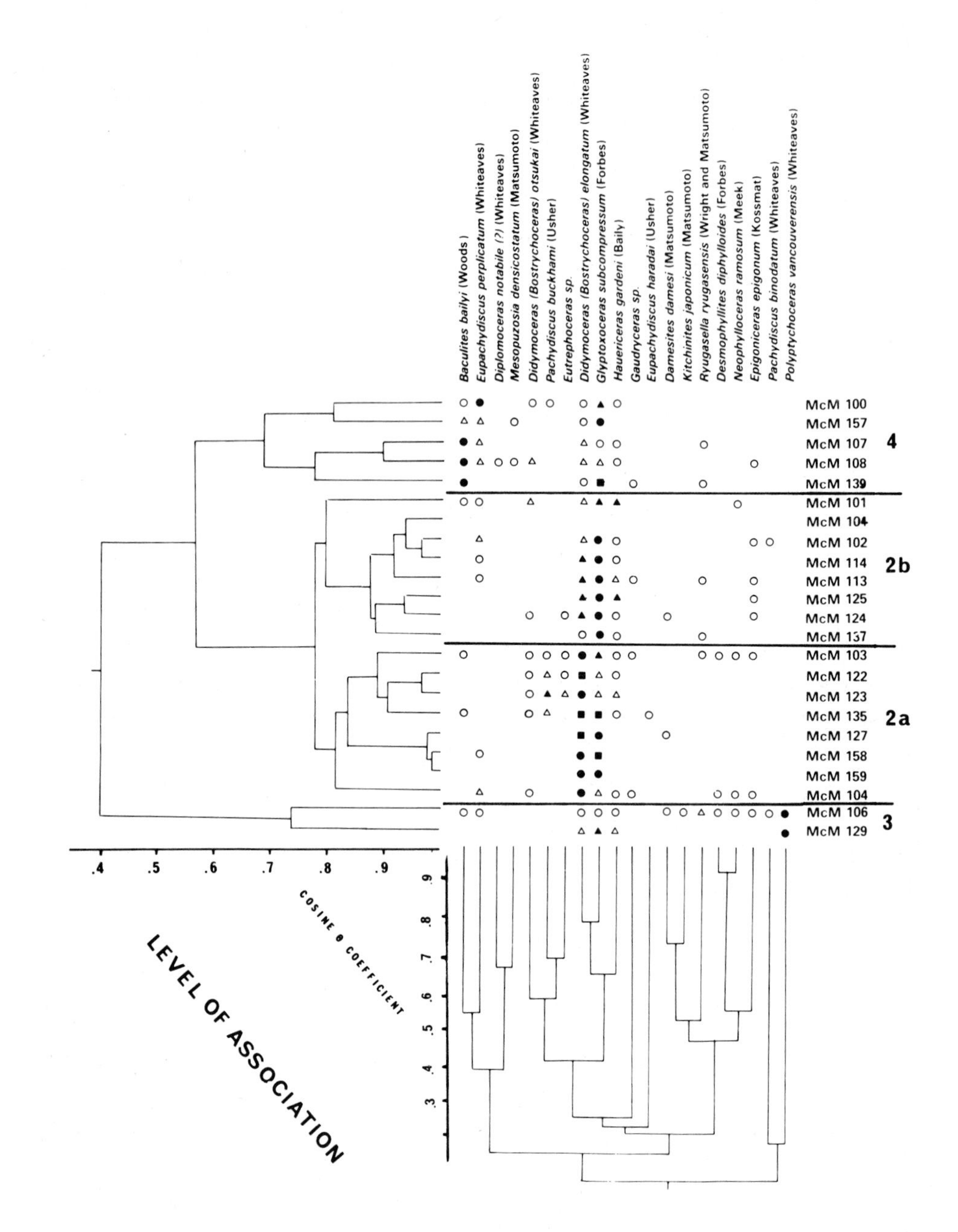

Fig. 5: Two-way phenogram using the "cosine θ" coefficient (see text for discussion).

lithologically distinct localities indicates a low degree of environmental or ecologic tolerance, and argues against extensive post-mortal drift. Similar reasoning suggests that shells of *Baculites bailyi* underwent little post-mortal drift, for it is difficult to imagine that preservation of this ammonite only could occur within one environment if extensive post-mortem shell drift was common. Conversely, the ubiquitous occurrence of the zonal index *D. elongatum* and *Glyptoxoceras subcompressum* indicates a wide environmental tolerance, extensive post-mortal drift, or both. I have postulated elsewhere that many Late Cretaceous heteromorph ammonites, especially those with U-shaped body chambers, were planktonic, and hence could be expected in a wide variety of rock types (WARD, 1976b; WARD & WESTERMANN, 1977).

POST-ELONGATUM FAUNAL ASSEMBLAGES

Post- *elongatum* Zone marine strata in the Nanaimo Group are composed of increasing proportions of non-fossiliferous rocks. By comparison with the Elongatum Zone associations, faunal changes in post-Elongatum Zone strata are most apparent among ammonites and inocerami. The majority of ammonites in younger zones are pachydiscids and baculitid species. Among inocerami, the delicately sculptured, concentrically ribbed Elongatum Zone forms such as *I. naumanni* and *I. orientalis* are replaced in the *schmidti* Zone by radially ribbed sphenoceramids, and in the *chicoensis*, *vancouverense*, and *pacificum* Zones by thick-shelled, coarsely ornamented, concentrically ribbed species such as *I. vancouverense*.

Schmidti Zone

According to JELETZKY, in MULLER & JELETZKY (1970), there is no evidence of unconformity separating the *elongatum* Zone from the *schmidti* Zone. Nevertheless, the boundary between these two zones is marked by sweeping changes in the makeup of offshore faunal associations. *Inoceramus naumanni* and *I. orientalis* are completely replaced by *I. (Sphenoceramus) schmidti* and a number of closely related species such as *I. (S.) elegans* and *I. (S.) sachalinensis*. Shells of all these species are characterized by radial ribbing. *I. schmidti* and its allies may be descendents of *I. orientalis*, since forms transitional between those with concentric and those with radially-ribbed shells have been noted by JELETZKY (in MULLER & JELETZKY, 1970) in the upper *elongatum* Zone beds on Puntledge River, Comox Basin. The *I. (S.) schmidti* fauna is found in greatest numbers in mudstones, the same rock types that yields the greatest numbers of *I. naumanni* in the *elongatum* Zone. The similarity in size and numbers of

I. schmidti relative to other macrofauna suggests that they occupied similar or identical niches as *I. naumanni*.

Other bivalve constituents of *schmidti* Zone siltstones and mudstones show little change from the preceeding zone. In the sandstone lithotopes the major change is the common appearance of the trigoniid *Yaadia* sp. in addition to *Pterotrigonia evansana*.

Among ammonites significant changes occur both in faunal composition and relative numbers of morphologic types at the *elongatum* Zone- *schmidti* Zone boundary. Perhaps the greatest change in terms of trophic relationships of the offshore faunal associations is the complete disappearance of *Glyptoxoceras subcompressum* and *Didymoceras elongatum* at the top of the *elongatum* Zone.These two species comprise 49% of all collected *elongatum* Zone ammonites which I have examined, and probably were even more prevalent relative to other species during their existence, for these shells are usually more poorly preserved than those of species with planispiral shell and thus are not as easily collected. No similarly shaped ammonites occur in the *schmidti* Zone, and the only new additions among heteromorph ammonites are *Pseudoxybeloxeras (Cyphoceras)* sp. ind. and *Neocrioceras* sp. ind., both very rare.

The most common ammonites of the *schmidti* Zone are *Canadoceras yokoyamai* and *C. multisulcatus*. These species are found in a variety of rock type, but are most common in facies intermediate between mudstones and sandstones, where they can be found vertically-imbedded at several localities. *C. yokoymai* is considered a descendent of *Eupachydiscus haradai* (JELETZKY, in MULLER & JELETZKY, 1970), and completely replaces it at the base of the *schmidti* Zone. No figures have been tabulated, but it is estimated that together these species constitute at least 50% of all ammonites collected from the *schmidti* Zone.

All other ammonites of the *elongatum* Zone also range through the *schmidti* Zone, but in greater or lesser abundance. Planispiral lytoceratids and phylloceratids seem to be slightly more common than in the *elongatum* Zone. *Baculites bailyi*, on the other hand, is much rarer in the younger than in the older zone.

Chicoensis and Vancouverense Zone

Faunal assemblages of the Chicoensis and Vancouverense Zones are characterized by mass occurrences of coarsely ribbed, concentrically ornamented *Inoceramus vancouverensis* (SHUMARD) and *I. subundatus* (MEEK) and by the presence of large, unornamented baculitids such as *Baculites chicoensis* (TRASK) and *B. inornatus* (MEEK). Both the inoceramids and baculitids are widely facies ranging. *Candoceras newberryanum*, descendent of *C. yokoyamai* of the *schmidti* Zone is also common, as is the leiostracan

Desmophyllites diphylloides. Shallow water facies are not widely exposed in strata of these two zones; where present, the infaunal suspension feeding assemblages show large number of *Pterotrigonia evansona* (GABB), *Crassatellites conradiana* GABB, *Pinna calamatoides* SHUMARD, *Volutoderma navarroensis* (SHUMARD), and *Cucullae suciaensis* (WHITEAVES). Large inoceramids such as *I. ezoensis*, important constituents of the sandy facies in *elongatum* and *schmidti* Zone strata are restricted to mudstones and the pelagic layers and of turbidites. In the latter lithologies they can be extremely common.

Post-*Vancouverense* Zone

Strata superjacent to the *vancouverense* Zone the Nanaimo Group are composed either of nonfossiliferous sandstone and conglomerates (mainly grainflows and resedimented conglomerates) or sparsely fossiliferous turbidites. The pelagic layers of the turbidites contain numerous large platyceramids such as *Inoceramus ezoensis?* or allied species, *Entocostea* sp. ind., and large baculitids such as *Baculites rex* Anderson, *Baculites anceps pacificus* Matsumoto and Obata, and *B. occidentalis* Meek. Other faunal constituents are large pachydiscids such as *Pachydiscus suciaensis* WHITEAVES, *P. ootacodensis* (STOLICZKA) and *Anapachydiscus nelchiensis* JONES. More rarely, small leiostracans such as *Gaudryceras denmanense* WHITEAVES and *Neophylloceras remosum* (MEEK) are recovered: A diverse assemblage of as yet unstudies trace fossils related to the Nereites assemblage of Seilacher (1967a) are common.

DISCUSSION

The detailed collecting procedures used in the elongatum Zone were designed to examine the degree to which ammonites within a single, refined biostratigraphic framework show characteristic facies distribution patterns. Only two clear ammonite associations could be distinguished from within the various mudstone facies:

1) the mass occurrences of the heteromorphic ammonite *Polyptychoceras vancouverense*, found in a distinctive facies characterized by mudstones with concretions packed with the bivalve *Anomia* and lesser numbers of *Inoceramus naumanni* and *I. orientalis*; and
2) mudstone facies without the *Anomia* and *Polyptychoceras* concretions, but with various associations of ammonites that were to some lesser extent related to facies type.

For example *Baculites* was most common in siltier facies, while distal turbidites were characterized by an heteromorphic ammonites of various species, to the point that planispirals were almost entirely excluded. These latter associations, however, were quite intergradational, and never as distinct as the *Polyptychoceras* assemblage.

The most distinctive suite of ammonitiferous rocks were the shales with *Anomia* concretions. These concretions always occurred large numbers of *POlyptychoceras vancouverense*. This ammonite could also be found in the surrounding mudstones, but in far fewer numbers than in the concretions. These localities with *P. vancouverense* also had relatively few *Bostrychoceras elongatum* and *G. subcompressum*, the two most common ammonites in virtually every other facies within the *elongatum* Zone. The mass occurrences of *P. vancouverense* cannot be explained in any meaningful paleobiologic or ecologic sense. This facies was rare, and only occurred at the lowest horizons of the zone. Because of the fact that this species did not seem to range very high in the zone, (but is common and characteristic of the subjacent *Baculites capensis* Zone as exposed in California), it could be said that the *Polyptychoceras* assemblage of ammonites is as much biostratigraphically as environmentally controlled. It is a curious, however, that the characteristic concretions carrying *Anomia* as well as the *Polyptychoceras* never appear higher in the section.

Baculites baileyi was the most common ammonite to be found in the siltier sections of the Haslam Formation within the Comox Basin, such as virtually the entire part of the Trent River section above the *Anomia* concretionary beds. Although also present, baculitids were much less common in the muddier shales. This trend, of commonly finding *Baculites* in mass occurrences in the sandier or siltier parts of the section is also typical of higher Upper Cretaceous Zones in the Nanaimo Group, as well as in California. On Waldron and Sucia Islands, for instance, the *B. chicoensis* Zone and superjacent *B. inornatus* beds (*H. vancouverense* Zone) are characterized by mass occurrences of *Baculites*. These beds are either siltstones or fine grained sandstones. MATSUMOTO (1959) previously noted that the baculitid occurrences in Japan appear to be more common in the siltstone and sandstone, rather than shale facies.

Other than these two facies, the remainder of the *elongatum* Zone is very questionably divided into ammonite biofacies. Ammonite species numbers in the mudstones appear to be related to rarefaction phenomena; the larger the sample of ammonites, the higher the species numbers. In some facies, especially the very fine distal turbidites of the Saanich Peninsula area, there is an impression of higher numbers of *B. elongatum* and *G. subcompressum*; these trends, however, are not very impressive. One gets the impression of very similar assemblages of the shale facies ammonites.

The Nanaimo Group dips gently to the east, resulting in lower areal exposure as one moves higher into the section. This is most unfortunate, as it limits the area of facies able to be studied within the higher zones. Multivariate analyses of the ammonite and other molluscan assemblages can really only be carried out in the *elongatum* Zone, and perhaps the *schmidti* Zone.

Of more interest is the rather stunning functional inequality of the most common ammonites of successive Nanaimo Group Zones. As shown above, the most common ammonites of the Elongatum Zone were the heteromorphs *G. subcompressum* and the zonal index, the former a gyrocone, and the latter a torticone. The most common planispiral ammonites of the zone are the coarsely ornamented pachydiscid *Eupachydiscus haradai* and the highly streamlined *Hauericeras gardeni*. The superjacaent *I. schmidti* Zone shows an abrupt change in ammonite shape commonality. In this zone, the descendent of *E. haradai* , the more compressed and less coarsely ornamented planispiral pachydiscid *Canadoceras yokoyami* is by far the most common ammonite. No torticones are present, while but a single specimen of *Glyptoxoceras* has been found in the zone (Englishman River). The most surprising aspect of this quite sweeping change in the morphological makeup of the ammonite faunas is, perhaps, the suddenness of the changeovers. The *schmidti* Zone overlies the *elongatum* Zone at a number of Nanaimo Group localities (Nanaimo River, Haslam Creek). In the areas where the zonal contact is exposed, the ammonite faunas change from heteromorphic dominated to planispiral dominated over short stratal distances.

Subsequent Nanaimo Group zones are dominated largely by baculitids, and to a lesser extent pachydiscids; it is not until the highest zone (the *suciaensis* Zone exposed on Hornby Island) that a torticonic species reappears. Even here, however, baculitids remain the most common ammonites.

There has a been much recent work on depth gradients in chambered cephalopods, using the siphuncular strength measures introduced by WESTERMANN (1971). More recent measures on a variety of Nanaimo Group ammonites (WARD & SIGNOR, 1983) have shown that a spectrum of strong to weak siphuncles were present in Nanaimo Group ammonites. Unfortunately, the facies distributions of the various ammonites of the Nanaimo Group do not really support the depth generalizations. For instance, one of the most common ammonites on Sucia Island, stratotype section of the Vancouverense Zone and one of the richest sections in numbers of ammonites collected, is the desmoceratid *Desmophyllites diphylloides*. This ammonite has one of the strongest siphuncles in terms of Westermann's measures, but is found in shallow water facies. *Gaudryceras* spp. also have "strong" siphuncles but can be found in shallow (Sucia, Denman Island) facies, as well as deeper water facies.

Ammonites were very rare in Nanaimo Group facies that appeared to be deeper water in origin. While many of the mudstones were very fine grained, they were not necessarily deepwater. Many of the siltstone and shale facies of the Comox Basin especially appeared to be shallow shelfal muds, deposited in quiet, perhaps semi-enclosed embayments of low depositional energy. In the submarine fan facies that did appear to be deepwater in origin, ammonites were very rare. Within the turbiditic formations that characterize the upper parts of the Nanaimo Group, the few ammonites collected were either gaudrycerids, phylloceratids, or rare, giant pachydiscids. Baculitids were occasionally common in some submarine fan facies in the upper Nanaimo Group, but their common orientations parallel to sole marks suggest that they were transported downslope from more shallow water areas.

REFERENCES

Blatt, H., G. Middleton and R. Murray 1972: Origin of Sedimentary Rocks. Prentice-Hall, Inc., p. 1-634.

Buzas, M. 1969: On the quantification of biofacies. Proc. North American Paleontological Convention, p. 101-116.

Buzas, M. 1972: Patterns of species diversity and their explanation. Taxon, 21: 275-286.

Cassie, R. & A. Michael 1968: Fauna and sediments of an intertidal mud flat: A multivariate analysis. J. exp. Mar. Biol. Ecol., V. 2: 1-23.

Clapp, C. 1914: Geology of the Nanaimo Map-area. Geol. Surv. Can. Mem. 51.

Danner, W. 1977: Paleozoic rocks of Northwest Washington and adjacent parts of British Columbia. In Stewart, J. and Stevens, C., eds. Paleozoic Paleogeography of the Western United States. SEPM Pacific Coast Section, Symposium 1: 481-502.

Dickinson, W. 1971: Clastic sedimentary sequences deposited in shelf, slope, and trough settings between magmatic arcs and associated trenches. Pacific Geology, V. 3, p. 15-30.

Dickinson, W. 1976: Sedimentary basins developed during evolution of Mesozoic arctrench systems in Western North America. Canadian Jour. Earth Sciences V. 13: 1268-1287.

Dzylynski, S. & K. Walton 1965: Sedimentary features of flysch and greywackes. Elsevier Pub. 274 p.

Eisbacher, G.H. 1974: Evolution of successor basins in the Canadian Cordillera. In Modern and Ancient Geosynclinal Sedimentation, Dott and Shaver (eds.), p. 274-291,

Farris, J.S. 1969a: On the cophenetic correlation coefficient. Systematic Zoology, V. 18: 279-285.

Geczy, B. 1971: Examen Quantitatif des Ammonoides Liasiques de la Montagne Bakony. Ann. Inst. Geol. Pub. Hungary, V. 54: 438-386.

Geyer, O. 1969: The ammonite genus Sutneria in the Upper Jurassic of Europe. Lethania, V. 2, No. 1: 63-72.

Geyer, O. 1971: Zur Paleobathymetrischen Zuverlässigkeit von Ammonoideen-Faunen-Spektren. Palaeo-Palaeo-Palaeo. V. 10: 265-272.

Goodall, D. 1954: Objective methods for the classification of vegetation, II: An essay on the use of factor analysis. Australian Jour. Bat. V. 2: 304-324.

Hazel, J. 1970: Binary coefficients and clustering in biostratigraphy. Geol. Soc. Amer. Bull., V. 81: 3237-3257.

Jeletzky, J. 1965: Late Upper Jurassic and early Lower Cretaceous fossil zones of the Canadian Western Cordillera, British Columbia. Canada Geol. Survey Bull. 103: 70 p.

--- 1971: Marine Cretaceous Biotic Provinces and Paleogeography of Western and Arctic Canada. Geol. Surv. Can., Paper 70-22, 92 p.

Jones, D., Silberling, N., and Hillhouse, J. 1977: Wrangellia - a displaced continental block in northwestern North America. Can. Jour. Earth Sci. V. 14: 2565-2577.

Kauffman, E.G. 1967: Coloradoan macroinvertebrate assemblages, central Western Interior, United States. In Paleoenvironments of the Cretaceous Seaway. Colorado School of Mines Symposium, p. 67-143.

--- 1973: Cretaceous bivalvia. In Hallam, A. (ed.) Atlas of Paleobiogeography. Elsevier Pub. Co. Amsterdam, 353-383.

--- 1977: Systematic, biostratigraphic, and biogeographic relationships between Middle Cretaceous Euramerican and North Pacific Inoceramidae. Paleont. Soc. Japan Special Paper 21: 169-212.

Matsumoto, T. 1959: Upper Cretaceous ammonites from California. Mem. Fac. Sci., Kyushu Univ., Ser. D., Special Vol. 1, 172 p.

Matsumoto, T. 1960: Upper Cretaceous ammonites of California, Pt. III: With Notes on Stratigraphy of the Redding area and the Santa Ana Mountains. Mem. Fac. Sci. Kyushu Univ., Ser. D. Geol., Spec. V. II, 204 p.

Muller, J. 1977: Evolution of the Pacific Margin, Vancouver Island, and adjacent regions. Can. Jour. Earth Sci. V. 14: 2062-2085.

--- & J. Jeletzky 1970: Geology of the Upper Cretaceous Nanaimo Group. Vancouver and Gulf Islands, British Columbia. Geol. Surv. Can., Paper 69-25, 77 p.

Rhoads, D., I. Speden & K. Waage 1972: Trophic Group analysis of Upper Cretaceous (Maestrichtian) Bivalve Assemblages from South Dakota. Bull. AAPG V. 56(6): 1100-1113.

Rohlf, F.J. 1970: Adaptive hierarchical clustering schemes. Systematic Zoology, 19: 58-82.

--- 1972: An empirical comparison of three ordination techniques in numerical taxonomy. Systematic Zoology, 21: 271-280.

Saul, L. 1960: Molluscan Fauna from Chico Creek, California. Unpubl. Thesis, University of California at Los Angeles, Dept. Geol., 230 p.

Scott, G. 1940: Paleoecological factors controlling the distribution and mode of life of Cretaceous ammonoids in the Texas area. Jour. Paleo., V. 14: 1164-1203.

Scott, R. 1974: Bay and shoreface benthic communities in the Lower Cretaceous. Lethaia, V. 7: 315-330.

--- 1975: Patterns of Early Cretaceous Molluscan Diversity in south-central United States. Lethaia, V. 8: 241-252.

Seilacher, A. 1967 : Bathymetry of trace fossils. Marine Geol. 5: 413-428.

Sepkoski, J. & M. Rex 1974: Distribution of freshwater mussels: coastal rivers as biogeographic islands. Systematic Zoology, V. 23, No.2: 165-188.

Sneath, P. & R. Sokal 1973: Numerical Taxonomy. W.H. Freeman & Co., San Francisco, 573 p.

Sokal, R. & P. Sneath 1963: Principles of Numerical Taxonomy. W. Freeman & Co, San Francisco, 359 p.

--- & F. Rohlf 1962: The comparison of dendograms by objective methods. Taxon, V. 11: 33-40.

Sutherland-Brown, A. 1966: Tectonic history of the insular belt of British Columbia. In Tectonic history and mineral deposits of the western Cordillera in British Columbia and neighbouring parts of the United States. Can. Inst. Mining and Metallurgy, Spec. V. 8: 83-100.

Usher, J. 1952: Ammonite faunas of the Upper Cretaceous rocks of Vancouver Island, British Columbia. Geol. Surv. Can., Bull. 21.

Valetine, J. 1973: Evolutionary Paleoecology of the Marine Biosphere. Prentice-Hall, Inc., 511 p.

Ward, P. 1976a: Stratigraphy, paleoecology, and functional morphology of hetuomorph ammonites from the Nanaimo Group. British Columbia. Unpub. theses, McMaster University, 176 p.

--- 1976b: Upper Cretaceous Ammonites (Santonian-Campanian) from Orcas Island, Washington. Jour. Paleo. 50(3): 454-461.

--- 1978: Revisions to the stratigraphy and biochronology of the Upper Cretaceous Nanaimo Group, British Columbia and Washington State. Can. Journal. Earth Sci., V. 15: 405-423.

--- & P. Signor 1983: Evolutionary tempo in Jurassic and Cretaceous ammonites. Paleobiology, V. 9: 183-198.

--- & G. Westermann 1977: First occurrence, systematics, and functional morphology of Nipponites (Cretaceous Lytoceratina) from the Americas. Jour. Paleo. 51(2): 367-372.

Westermann, G. 1971: Form, structure and function of shell and siphuncle in coiled Mesozoic ammonoids. Life Sci. Cont., Royal Ontario Museum, no. 78: 39 p.

Ziegler, B. 1967: Ammoniten-Ökologie am Beispiel des Oberjura. Geol. Rundschau, 56: 439-464.

OYSTER BEDS: MORPHOLOGIC RESPONSE TO CHANGING SUBSTRATE CONDITIONS

A. Seilacher
Tübingen

B.A. Matyja & A. Wierzbowski
Warszawa

Abstract: Morphotypes and settling behaviors of oysters in coarsening-upwards cycles of the Polish Upper Jurassic and at different localities of the South Australian Pliocene reflect temporal and spatial changes in substrate softness that the winnowed matrix fails to record. Adaptational strategies are similar in the two cases except that the lack of comissural overlap did not allow the Jurassic *Lopha* to develop gryphid mud dwellers. It remains uncertain, however, whether we deal with ecophenotypic or evolutionary phenomena.

Oysters have been the most successful group of all cemented bivalves. Part of their morphological diversity is due to the fact that during the Mesozoic and Cenozoic, they repeatedly evolved sidelines that left their original rocky substrates to become secondary soft bottom dwellers (SEILACHER, 1984). These soft bottom oysters are of particular interest to paleontologists, (1) because of their high fossilization potential and (2) because of their more regular shapes, which facilitate taxonomic classification as well as functional interpretations. In terms of constructional morphology various types of mudstickers and recliners (Fig. 1) can be distinguished and compared to similar "strategies" in other bivalves, but their adaptational significance is necessarily inferred and needs to be tested against field evidence. The present study is such an attempt. It also illustrates the problem of distinguishing ecophenotypic from genetically controlled characters in fossil forms.

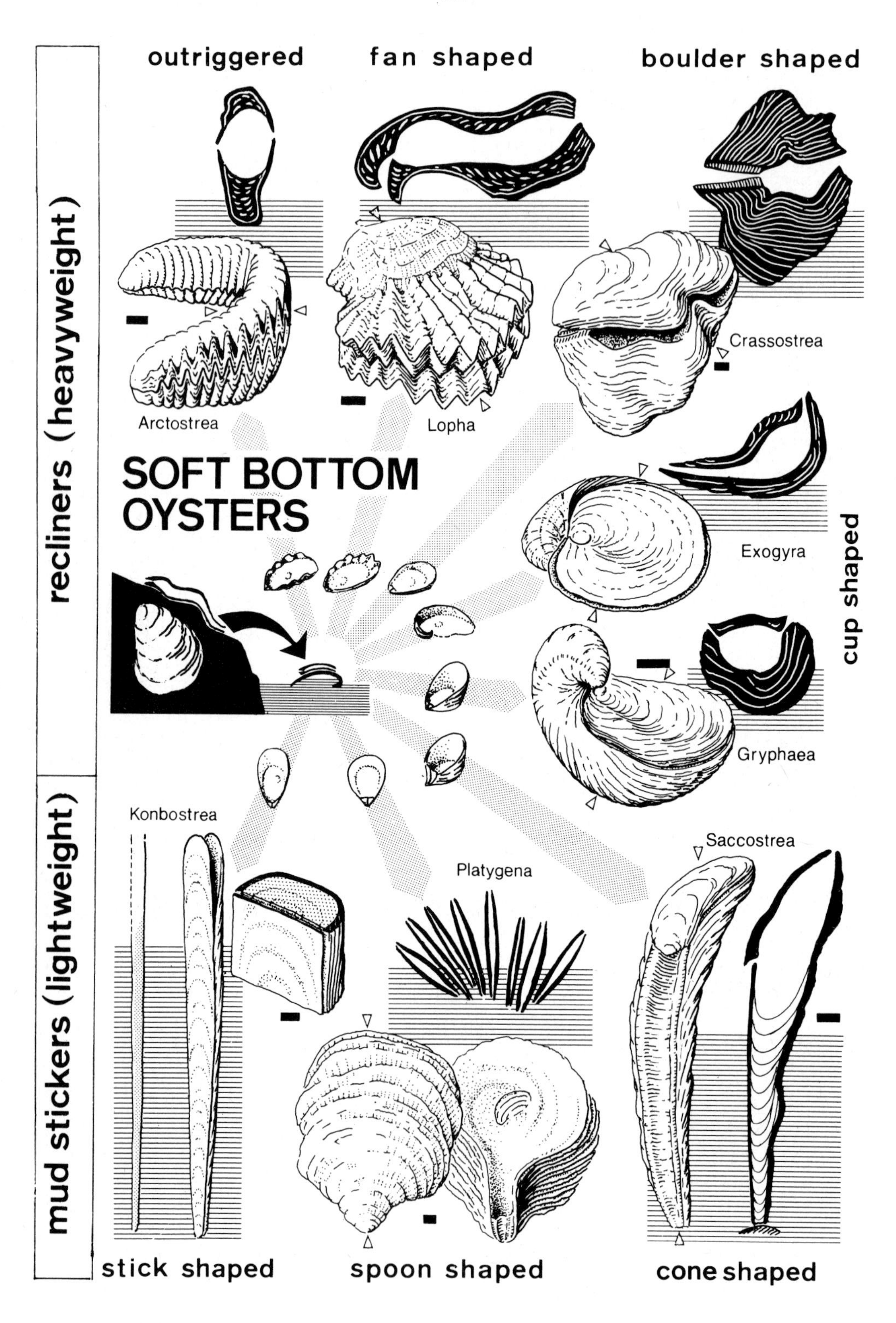
outriggered
fan shaped
boulder shaped
recliners (heavyweight)
Arctostrea
Lopha
Crassostrea
SOFT BOTTOM OYSTERS
Exogyra
cup shaped
Gryphaea
mud stickers (lightweight)
Konbostrea
Saccostrea
Platygena
stick shaped
spoon shaped
cone shaped

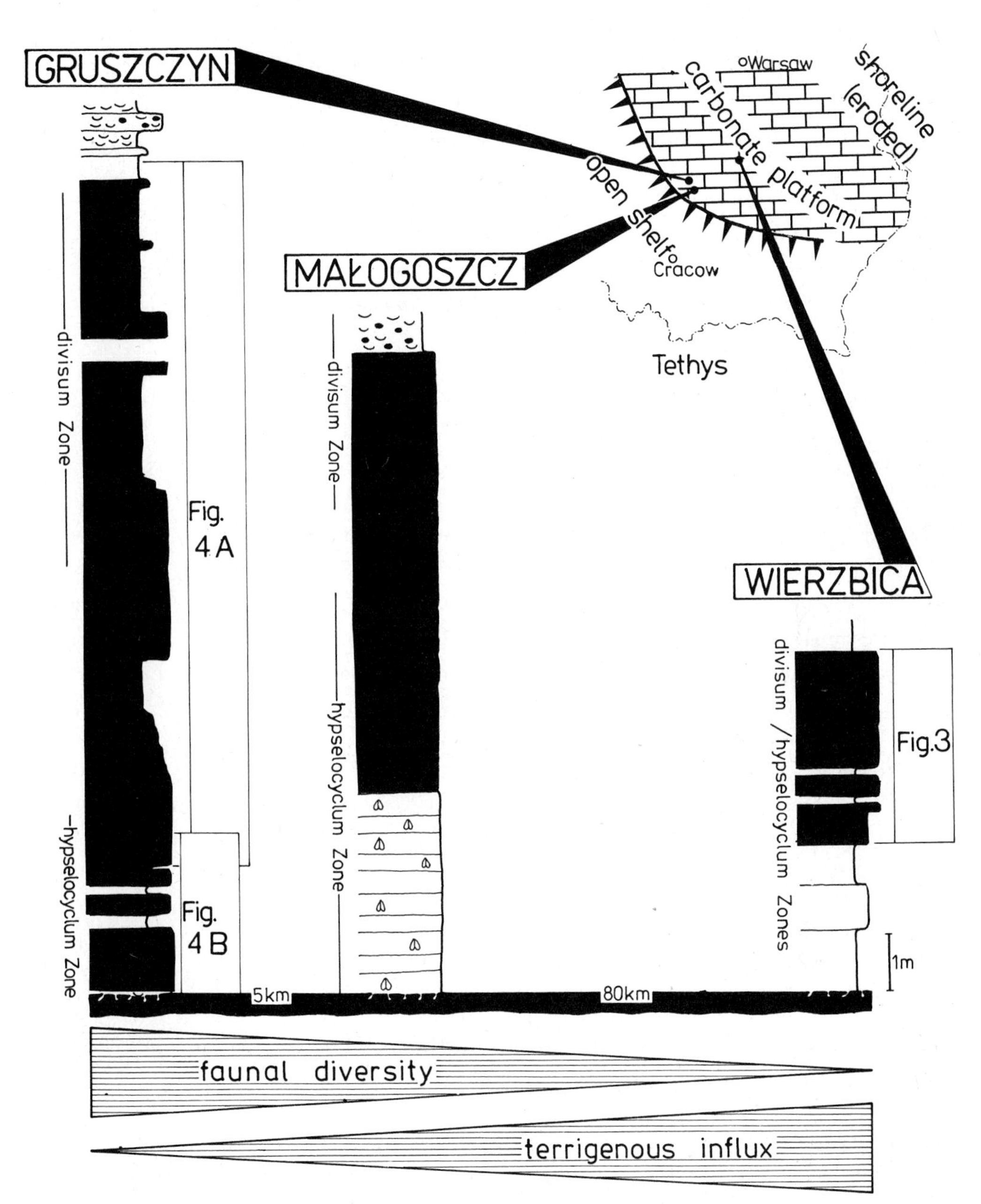

Fig. 2: The *Lopha* beds (black) of the Skorkow Lumachelle (U. Jur., Poland) increase in thickness and faunal diversity towards the edge of the carbonate platform. The basal hard ground (black bar) serves as a datum line.

Fig. 1: The unusual morphogenetic plasticity of oysters, inherited from the original rock-encrusting habit, allowed them to adapt to soft substrates by a variety of strategies. (From SEILACHER 1984).

I. UPPER JURASSIC OYSTER BEDS IN CENTRAL POLAND

A. Geologic setting

The Upper Jurassic of Poland differs from its equivalent in South Germany by that a carbonate platform facies with mainly bioclastic sedimentation follows the earlier spongy facies. Starting during the Middle Oxfordian in the north-eastern parts of the Holy Cross Mountains, this platform prograded to the southwest during the Late Oxfordian (Fig. 3) and the Kimmeridgian (KUTEK 1969). In sections the platform sediments show a vertical alternation of cross-bedded oolites or coquinas with micritic limestones and marls.

The oyster beds to be discussed are found within the Kimmeridgian part. They have a wide distribution, but become gradually replaced by oolites towards the advancing outer edge of the platform (KUTEK 1968 and 1969). Depending on whether *Lopha* or *Nanogyra* predominates, one speaks of "Alectryonia" or "Exogyra" lumachelles. In the Skorkow Lumachelles at the boundary of the hypselocyclum and divisum zones (Lower Kimmeridgian; Figs. 3-5) *Lopha* is found in the lower and *Nanogyra* in the upper part of the unit. Intercalating limestones and marly limestones consist of fine bivalve shell detritus (biomicrite and biopelmicrite) with varying amounts of onkoids and/or ooids. A hardground at the base of the unit is used as a marker horizon for regional correlation (KAZMIERCZAK & PSZCZOLKOWSKI 1968) and has been recognized in an identical stratigraphic position on the northeastern side of the Holy Cross Mountains. (Wierzbica section in Fig. 3). Since both areas have been connected before the Laramide uplift of the Paleozoic core (KUTEK & GLAZEK 1972), it can be assumed that the Skorkow Lumachelle originally covered an area of at least 1OO OO km^2.

A few additional *Nanogyra* beds are found above the Skorkow Lumachelle in the uppermost part of the Lower and the Upper Kimmeridgian. They can be individually traced over distances of several kilometers, but are insufficiently fingerprinted to assess their real extensions.

As may be expected, there is a general decrease of terrigenous material across the platform in an offshore direction. In the same direction the faunal diversity increases, with stenohaline forms such as echinoids, ammonites and brachiopods being rare or lacking in the east. One also observes an increase in thickness, but this may be largely due to the fact that the carbonate edge had to cross the SE-NW trough of the Polish-Danish aulacogene as it advanced to the west (KUTEK & GLAZEK 1972).

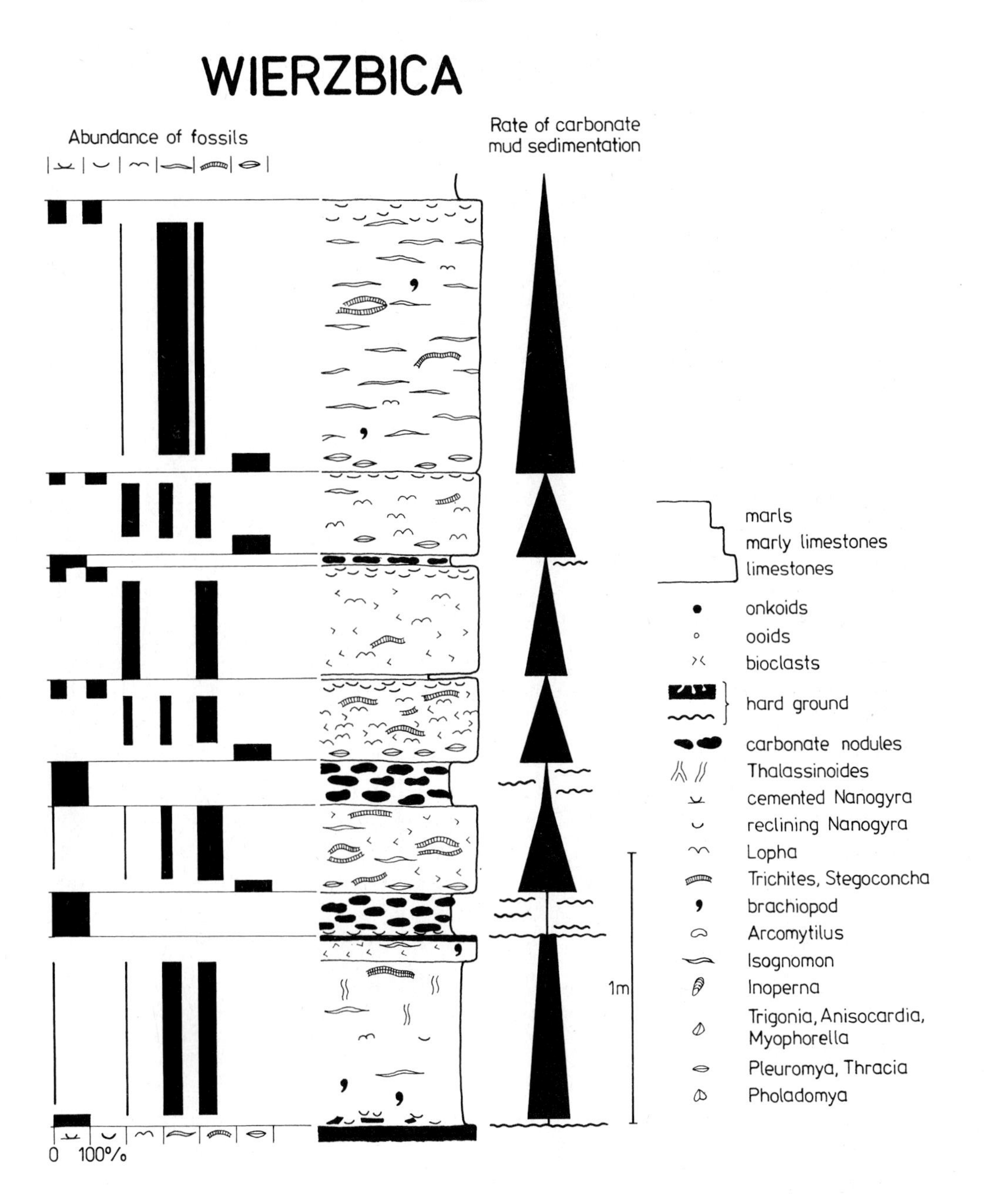

Fig. 3: Recurring faunal successions from infaunal to epibenthic to encrusting communities of bivalves in this section suggest cycles of decreasing substrate muddiness. Rather than being true ecological successions, however, the faunistic changes are probably caused by taphonomic feedback in shallowing-upwards cycles.

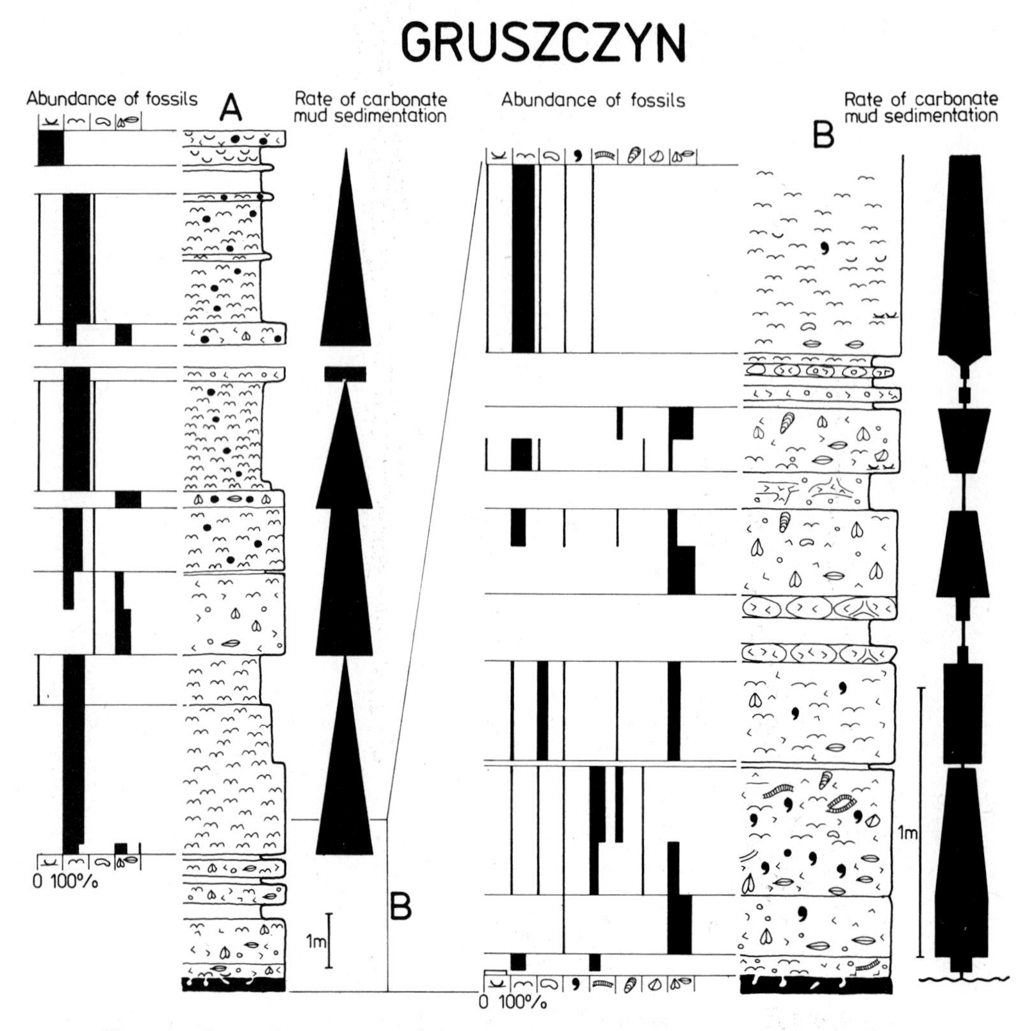

Fig. 4: At a large scale (A), the Gruszczyn section shows similar cycles as observed in Wierzbica, but cycle thicknesses are higher. At a smaller scale (B), however, the lower part of the section fails to show uniform trends, although taphonomic feedback should work at any scale.

B. Fossil associations
(Fig. 5)

In addition to the lithologic alternation of coquinas with beds of limestone, marly limestone and marl, sections of the Skorkow Lumachelle show various kinds of discontinuity surfaces (firm and hard grounds) indicating periods of reduced net-sedimentation and of erosion to already compacted or even cemented levels. The firm grounds

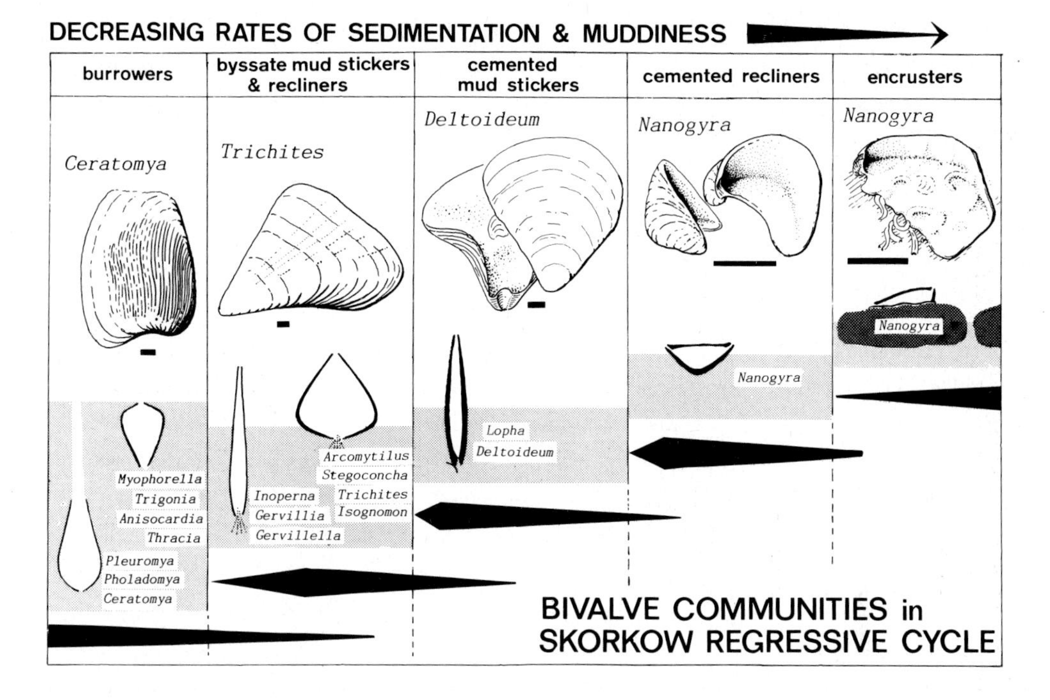

Fig. 5: Sequential bivalve associations observed in the sections (Figs. 2 - 4) reflect changing substrate conditions rather than true geologic successions. Note the similarities of *Deltoideum* to *Platygena*, and of the reclining morphotype of *Nanogyra* to *Exogyra* (Fig. 1), which has lost the alternative ability of *Nanogyra* to encrust alternative appropriate hard substrates. Compare also the better resolvable associations in the Oxfordian of western Europe (FÜRSICH 1977).

are characterized by crustacean burrows with preserved scratch marks (*Glossifungites*, *Spongeliomorpha*), the hard grounds by borings and by encrustations, mainly of *Nanogyra*. In the Skorkow Lumachelle-part of the section (Fig. 3) only the basal hardground is continuous. The others consist of nodules that are equally encrusted. This means erosion down to a level in which concretions had already formed diagenetically, followed by encrustation of the exposed pebbles and reburial under the overlying marl. It is also possible that these hardgrounds were activated repeatedly, but to corroborate this assumption it would be necessary to find hiatus concretions (VOIGT 1968) with multiple encrustations.

In any case it is clear that during the deposition of the Skorkow Lumachelle sedimentation rates and substrate conditions were subject to considerable fluctuation. In response to these fluctuations, the benthic fauna must have changed as well, but since the fossils are largely separated from their home sediments and mixed to various degrees, it is now difficult to identify the original biota. Nevertheless, some major associations of bivalves (Fig. 5) can be rather clearly distinguished. Since the constituent bivalves happen to be also mineralogically different (the burrowers have aragonitic, the oysters calcitic and the byssates composite shells), their separation might be diagenetically enhanced, but criteria of functional morphology suggest that environmental conditions had a more important control (Fig. 5).

1. Association of burrowers

The chief advantage of bivalves over other soft bottom filter feeders, such as articulate brachiopods, is their ability to burrow. Major requirements for the infaunal mode of life were the transformations of the foot into a hydraulic burrowing organ and of the fused mantle margins into siphonal tubes (STANLEY 1972). Modifications of shell geometry and shell sculpture enhance the burrowing process and the stabilization within the sediment (SEILACHER 1984) and allow to distinguish deep and shallow burrowers by morphological criteria.

In the Skorkow Lumachelle deep burrowers, such as *Pholadomya*, *Pleuromya*, *Ceratomya*, and *Thracia*, are more common than shallow burrowers (*Trigonia*, *Myophorella*, *Anisocardia* etc.). This is because the deeply burrowed shells are less commonly reworked. In many cases they are still in life position. But also their sedimentary infill becomes more easily cemented into molds that survive erosion even after the aragonitic shell around them has been dissolved.

More shelly sediments, however, make burrowing increasingly expensive, particularly for the deep burrowers, while they do favor an epibenthic mode of life (KIDWELL & AIGNER, this volume). The epifaunal niche is occupied mainly by terebratulid brachiopods and secondary soft bottom dwellers, which are phylogenetically derived from sessile rock dwellers rather than from active burrowers. The inherited inability of these bivalves to right themselves up, to crawl and to burrow was only in rare cases compensated by the evolution of new modes of locomotion (Pectinids, Limids). More characteristic is passive stabilization by particular shapes and differential weighting of the shells (SEILACHER 1984). While adaptational strategies are comparable in secondary soft bottom dwellers derived from byssate and cemented stocks, their disparate and sequential occurrence in the studied and other sections suggests that the two groups should be treated as different associations.

2. Association of byssate mudstickers and recliners

This group is represented (a) by the mytilacean genera *Arcomytilus*, *Stegoconcha* and *Trichites*, whose broad anterior resting surfaces and various degrees of shell thickening identify them as edgewise recliners, and (b) by multivincular forms such as *Iso-gnomon* and *Gervillia*, in which shell morphology alone is not as indicative for a particular mode of life, so that other criteria, such as oriented overgrowth, have to be used in addition (SEILACHER 1954). As a whole, this group characterizes soft bottoms in which sedimentation rate at the ontogenetic scale is reduced, so that smothering is not a constant menace and where the mud is shelly enough for byssal anchoring at the surface or within the sediment.

3. *Lopha* Association

Like in the previous group fixation is critical only for the early ontogenetic stages, while passive stabilization mechanisms take over as the animals grow larger. Nevertheless, byssal attachment may be maintained as an additional means of stabilization and limited mobility throughout life. Cemented recliners however, completely give up their juvenile fixation and their morphogenetic program has to change drastically conforming to the new paradigm. It is probably for this reason that encrusting recliners are less tolerant to mud sedimentation than the byssate ones.

Cemented mudstickers have an intermediate position in this respect, because they need a certain degree of mud sedimentation to become stabilized, as will be shown in the case of *Lopha*.

4. *Nanogyra* Association

The ecologic separation of the oyster *Nanogyra* from *Lopha* has probably to do with its much smaller size. In soft bottom conditions it can grow into a recliner similar to *Exogyra* (Fig. 1); but being so small it would be more sensitive to mud smothering. On the other hand, *Nanogyra* may remain an encruster throughout life, provided there are adequately large hard substrates, such as pebbles or hardgrounds to grow on. Accordingly we can distinguish between associations of reclining and of encrusting *Nanogyra* in the studied sections.

C. Ecologic response to sedimentary cycles

The distribution of the associations in the section is not random but follows a sequential pattern (Figs. 4-5). It is tempting to interpret such patterns in terms of the ecological successions observed in modern biota, particularly in plant ecology. But the time scale of true ecological successions is in the order of years, far beyond the resolution of ordinary stratigraphic sequences. More probably we deal with pseudosuccessions caused by a temporal gradient in sedimentational conditions plus an interdependence of successive associations via taphonomic feedback (KIDWELL & AIGNER, this volume).

I. Regressive cycles expressed by faunistic pseudosuccessions (Fig. 5)

The repeated sequence in the studied sections from infaunal to epibenthic and encrusting associations (Figs. 2 - 4) can be interpreted as a response to cycles of decreasing mud sedimentation, provided that this decrease was effective at the scale of individual life cycles and not simply the result of long-scale erosive events. In a way we can compare these ecological cycles to the coarsening- or shallowing-upward cycles of the sedimentologists. Nevertheless, faunistic replacement may have taken place episodically after the previous community had been locally wiped out by smothering events.

2. Regressive cycles expressed by morphotypic trends (Fig. 6)

In addition to the ecological response at the faunistic level, a sequential change of settling behavior and growth form is observed within individual oyster beds.

This phenomenon is most clearly expressed in the thick *Lopha* bed of the Malogoszcz quarry (Fig. 2). In its lower part, one finds a large number of elongated

---

Fig. 6: The reworked and winnowed nature of fossil oyster beds is commonly revealed by functional morphologies and settling behaviors that are in conflict with observed burial positions and host rock granulometries. The transition of *Lopha gregarea* from slope-oriented attachment on other mud-sticking bivalves (a-b) to self-supported mud-sticking (c) and to cup-shaped growth on dead shells (d) probably reflects decreasing rates of mud-sedimentation in accordance with the coarsening-upwards cycle of which the Skorkow Lumachelle forms the top. Convergent forms of Pliocene oysters (e-h) characterize different localities within the S-Australian basin. In this form, however, commissural overlap has also allowed the additional establishment of pseudo-commissures between adjacent mud-sticking individuals (g) and the transformation into heavy *Gryphaea*-like forms reclining on muddy bottoms (1-0) -- a mode of life that should by current opinion have become extinct by the end of the Cretaceous (LA BARBERA 1981; JABLONSKI & BOTTJER 1983). Note that actual pathways probably went from shelly to increasingly muddy substrates. The sequence of the morphotypes in the coarsening-upwards cycle is probably midleading. More likely rock-dwelling oysters have entered unstable substrates as miniaturized encrusters on shelly bottoms and only then became adapted to muddier environments. In view of the morphogenetic plasticity of oysters it also remains an open question whether these adaptations are phenotypic or evolutionary in character.

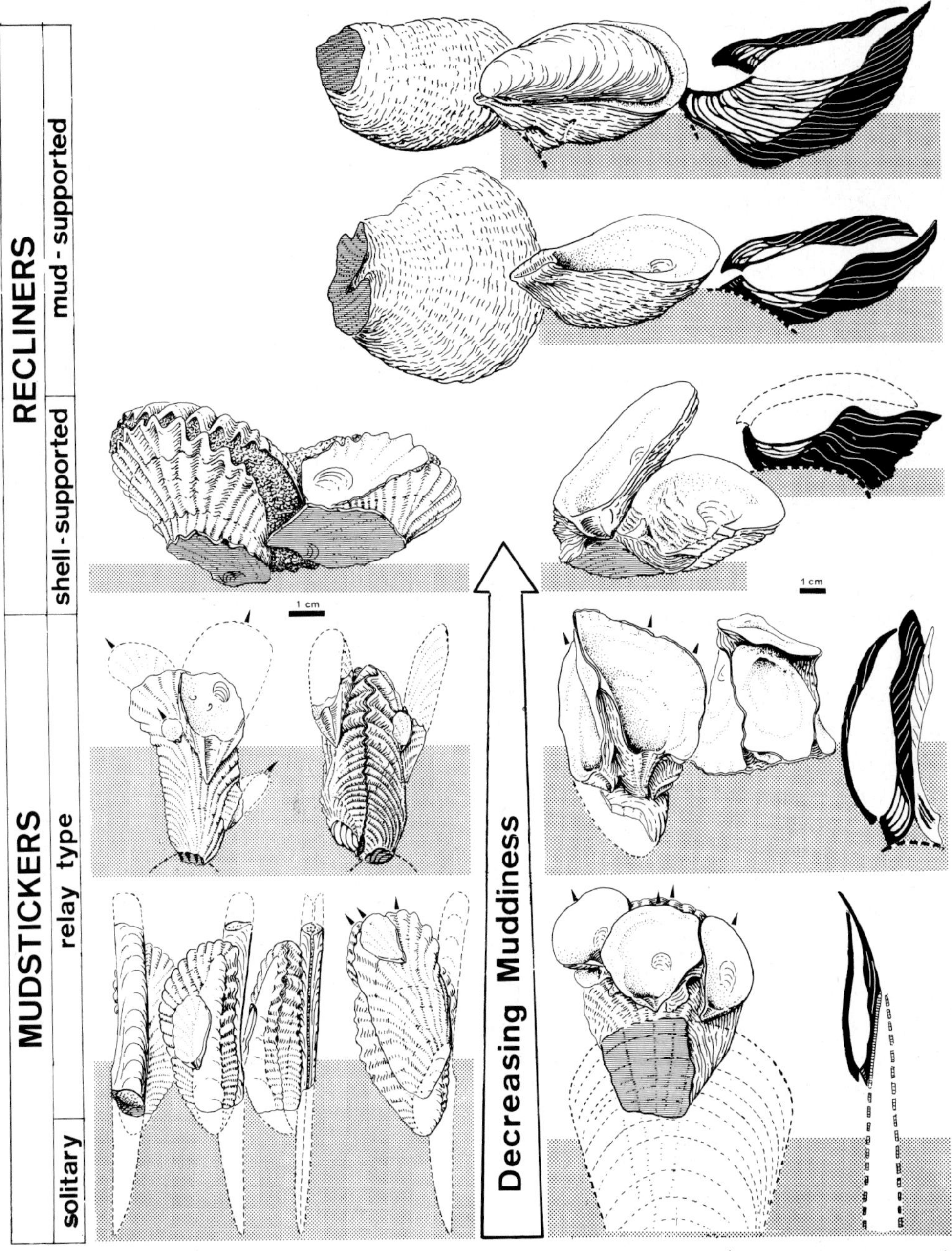

witl in-cycle (U.Jur., Poland)

within-basin (Plioc., S.Australia)

Lopha specimens attached to a very elongate species of *Gervillia*. In all studied specimens (about 20) the *Lopha* grew in the same direction as the host, as did successive generations of the same species that happened to settle on the *Gervillia*-attached founder. This and the fact that *Gervillia* is still double-valved in specimens, in which it is not only preserved as an attachment scar, proves (1) that unlike similar-shaped pendant species of the L. Toarcian (SEILACHER 1984, Fig. 7) this *Gervillia* was a mudsticker, (2) that it became encrusted by *Lopha* in a slope-oriented fashion while it was still sticking vertically in mud, and (3) that all specimens became subsequently reworked to their present horizontal position in the lumachelle. Only in one specimen the attached *Lopha* changed growth direction in a later stage, suggesting that it survived reworking and adjusted upward growth to the new orientation.

Higher up in the bed one increasingly finds *Lopha* with a similar elongate shape, but not associated with *Gervillia*. Instead we observe attachment scars (probably on dead, horizontal *Lopha* shells) that were too small to freely support the upright shell. Still it had remained in this unstable position long enough for successive *Lopha* generations to grow on it with the same slope orientation. This means that given an adequate, but not excessive, mud sedimentation, this *Lopha* species could itself become a mudsticker.

The third morphotype lacks the elongate outline and is cup-shaped instead. It could easily be considered a different species, but more probably represents simply an ecophenotypic response to an environment, in which many dead shells of previous oyster generations covered the sea floor and remained exposed for a longer time. As a consequence the encrusters could extend their cemented stage over most of the substrate shell before they grew up to form the cup with the flatter right valve as a lid.

II. PLIOCENE OYSTER BEDS IN SOUTH AUSTRALIA (Fig. 6)

It is interesting to compare the Jurassic example with oyster beds in the Pliocene of South Australia, where we meet similar, as well as additional, morphotypes not in the same bed, but at different localities within the same basin.

Counterparts to the *Gervillia*-attached forms are found in coastal exposures at Aldinga Beach, where they occur at certain levels in a very fine and well sorted sand without forming thick accumulations. They are attached to posterior ends of large *Pinna* shells and their uniform slope orientation leaves no doubt that they incrusted these byssate mudstickers while they were still in their vertical life position (Fig. 6e). Most of the *Pinna*-shells are now horizontal but double-valved (as are most of the oysters), suggesting rapid exhumation and burial during a storm event. Compared to the other occurrences these oysters are relatively flat an thinshelled (Fig. 6f).

Other morphotypes are found at Shell Hill near Black Hill to the East of Adelaide, where they form monotypic accumulations that get more than 5 m thick and are quarried for lime. Here the oysters form bunches with a characteristic relationship between the individuals (Fig. 6g-h). Not only do all members of a bunch grow in the same direction (which was upward in life). They also tend to grow along a joint front, so that the margins of adjacent left valves form a kind of commissure, which does not coincide with the true commissures because the right valves are always a little smaller than the left ones. This means that unlike in the association with *Pinna* the individuals did not overcompete each other but grew up synchronously, much like the calices in a coral head. Bunches of 2-4 oysters are the rule, while solitary specimens are conspicuously lacking. Thus we deal with gregarious mudstickers that became invariably reworked by storm, while the original host muds became winnowed away to deeper parts of the basin.

A different situation was found at a certain horizon within the oyster beds in a road cut a few hundred meters from Oyster Hill. Here the oysters have extremely large attachment scars (not the contacts resulting from communal growth), from which they grew up in a cup-like fashion, much like the similarly shaped recliners in the Jurassic example. This mode of growth (Fig. 6 i-k) allows the left valve to become excessively thickened, thus providing the differential weighting that is found in many recliners. It should be also noted that adjacent individuals on the same substrate shell never show the communal pseudo-commissures that are so typical for mud-sticking bunches.

In addition the Pliocene oysters have produced a *Gryphaea*-like variant that is lacking in the Jurassic material. In terms of morphogenesis this means that the attachment scar is relatively small and that subsequent free growth of the left valve proceeds in a spiral fashion. Functionally it is an approximation to the horn coral paradigm, which provides not only a stable position in soft substrates, but also allows the passive reburial into this position after the shells have become displaced by storms or biological activity (SEILACHER 1984, Fig. 12; see also BAYER et al., this volume).

Gryphid morphotypes are characteristic for two localities in the South-Australian Basin, but geometries are significantly different in the two. At Maslin Bay in a bed several meters above the one with the *Pinna*-attached oysters the gryphid forms have an almost equidimensional outline (Fig. 6 l-m), while they are much narrower in an oyster bed at Overland Corner (Murray Basin; Fig. 6 n-o). It should be noted that the two figured specimens are representative for samples of several dozen specimens from the first locality and several hundreds from the second.

The absence of similar gryphid types in the Jurassic *Lopha* and their substitution by *Nanogyra* (ZIEGLER 1969) is probably not coincidental. As expressed by the zigzag commissure, *Lopha* lacked the preconditioning for this type of growth, namely the slight overlap of the lower (in this case the left) valve over the other along the commissure. Radial ribs, where present in gryphid bivalves, are restricted to the more convex valve and do not reflect folding of the proper commissure.

III. CONCLUSION

In traditional descriptive paleontology, the discussed morphotypes would probably have been treated as distinct taxonomic entities at the subspecies or species level, particularly when they occur in strict stratigraphic or geographic separation. In view of the extreme morphogenetic plasticity of modern oysters, however, it is more likely that we deal only with an ecophenotypic response to temporal or regional fluctuations in rates of sedimentation and in bottom consistency. Nevertheless the modification of form was adaptational, because it followed pathways known from other, convergent lineages. This means that such modifications can be used to indicate environmental details which the lithologic record does not reveal; but they tell us little about the modes and tempoes of truly evolutionary processes.

Acknowledgements: The project originated in a joint student's excursion to the Holy Cross Mountains in summer 1983 as part of a partnership program between the universities of Warsaw and Tübingen. Additional field work was supported by the SFB 53.

REFERENCES

Carter, R.M., 1968: Functional studies on the Cretaceous oyster Arctostrea.- Palaeontology,11(3): 458-487.

Fürsich, F.T., 1977: Corallian/Upper/Jurassic marine benthic associations from England and Normandy.- Palaeontology, 20(2): 337-385.

Jablonski, D. & Bottjer, D., 1983: Soft-bottom epifaunal suspension-feeding assemblages in the Late Cretaceous. Implications for the evolution of benthic paleocommunities. In: Tevesz & McCall (eds.): Biotic interactions in Recent and fossil benthic communities.- Plenum Publishing Corp.: 747-812.

Kaźmierczak, J. & Pszczólkowski,A., 1968: Nieciagloś́ci sedymentacyjne w dolnym kimerydzie poludniowo-zachodniego obrzezenia mezozoicznego Góra Swietokrzyskich /Sedimentary discontinuities in the Lower Kimmeridgian of the Holy Cross Mts.- Acta Geol.Pol., 18/3: 587-612, Warszawa.

Kutek, J., 1968: Kimeryd i najwyzszy oksford poludniowo-zachodniego obrzezenia mezozoicznego Gór Swietokrzyskich, cz. I Stratygrafia / The Kimmeridgian and Uppermost Oxfordian in the SW margins of the Holy Cross Mts., Central Poland, Part I Stratigraphy.- Acta Geol.Pol., 18/3: 493-586, Warszawa

Kutek, J., 1969: Ibidem, cz II Paleogeografia / part II Paleogeography.- Acta Geol. Pol., 19/2: 221-321.

Kutek, J. & Glazek, J., 1972: The Holy Cross area, Central Poland, in the Alpine cycle.- Acta Geol.Pol., 22/4: 603-653, Warszawa.

La Barbera, M., 1981: The ecology of Mesozoic *Gryphaea*, *Exogyra* and *Ilymatogyra* (Bivalvia, Mollusca) in a modern ocean.- Paleobiology, 7: 510-526.

Pugaczewska, H., 1971: Jurassic Ostreidae of Poland.- Acta Palaeont.Pol., 16, no. 3: 195-311, Warszawa.

Pszczólkowski, A., 1970: Zastosowanie zdjeć lotniczych do badania utworów kimerydu poludniowo-zachodniego obrzezenia mezozoicznego Gór SwiTokrzyskich / Application of aerial photographs in the research of the Kimmeridgian deposits in the SW margin of the Holy Cross Mts.- Acta Geol.Pol., 20/2: 338-363, Warszawa.

Seilacher, A., 1954: Ökologie der triassischen Muschel Lima Lineata (Schloth) und ihrer Epöken.- Jb.Geol. u. Paläont. Mh. 1954: 163-183, Stuttgart.

Seilacher, A., 1984: Constructional Morphology of Bivalves: Evolutionary Pathways in primary versus secondary soft-bottom Dwellers.- Palaeontology, 27(2): 207-237.

Stanley, S.M., 1972: Functional morphology and evolution of bysally attached bivalve mollusks.- J.Paleont., 46(2): 165-212.

Stenzel, H.B., 1971: Oysters. In: R.C. Moore (Ed.) Treatise on Invertebrate Paleontology, Part N, 3(3) Mollusca 6. Bivalvia.

Thomson, J.M., 1954: The genera of oysters and the Australian species:-Australian Journ.Mar., Freshwater Research, 5: 132-168.

Voigt, E., 1968: Über Hiatus-Konkretion (dargestellt an Beispielen aus dem Lias).- Geol. Rdsch. 58(1).

Ziegler, B., 1969: Über Exogyra virgula/Lamellibranchiata, Oberjura.- Eclogae Geol. Helvetiae, 62(2): 685-696.

ECOLOGICAL PATTERNS IN MIDDLE JURASSIC

GRYPHAEA:

THE RELATIONSHIP BETWEEN FORM AND ENVIRONMENT

Ulf Bayer, Andrew L.A. Johnson, Joseph Brannan

Leicester & Aberdeen

Abstract: *Gryphaea* is a typical oyster, sensitive and reactive to the environment and thus variable in aspects of form, such as the size of the attachment area. Nevertheless, there are regular relationships between form and specific properties of the environment, at both the intra- and inter-specific levels and for various stratigraphic horizons throughout the Middle Jurassic of Europe.

The ecological variation in *Gryphaea* provides an example of 'correlation of growth' with a number of features varying together to produce a striking ecological trend which at least mirrors the classic 'phylogenetic' sequence and may indeed be its basis, i.e. the phylogenetic sequence may reflect ecophenotypic variation and changing environments rather than genetic evolution. However, we cannot decide at present if significant ecophenotypic variation is involved in the previously described phylogenetic trends.

Gryphaea is in addition an interesting sedimentary particle which provides information about the history of sedimentation through patterns of encrustation and reworking. Thus, it provides a source for the reconstruction of its own environment and for primary bottom conditions.

INTRODUCTION

Gryphaea was and is an interesting subject for evolutionary studies. The repeated morphological trends which occur within its phylogeny have attracted much attention and been the basis for various phylogenetic models and hypotheses about mechanisms and modes of evolution (DEECKE, 1916; DEPERET, 1909; DACQUE, 1921; CHARLES & MAUBEUGE, 1951; HALLAM,1968; HALLAM & GOULD, 1975). However, *Gryphaea* has rarely been studied as if it were a typical oyster, capable of reacting actively, ecophenotypically, to external influences, and as a member of a group which is indeed almost synonymous with high ontogenetic plasticity and intraspecific variability. *Gryphaea* has moreover rarely been studied as a sedimentary particle yet its presence in the environment may introduce a local 'hardground' for colonisation by encrusting and boring organisms, and it may also be reworked as a pebble. Although the two valves separate easily after death, they are too heavy to be transported far, thus the original substrate can be reliably inferred from the sediment in which the shells are found.

While proceeding from an ecological viewpoint we nevertheless add a further example of a morphological trend to the *Gryphaea* literature -- a trend which indeed closely resembles the well known 'phylogenetic' example (HALLAM,1968; HALLAM & GOULD, 1975) but which we document within narrow stratigraphic intervals, wherein the form of *Gryphaea* is a reflection of the environment. The phylogenetic line can thus be extended to a solid cylinder reflecting a sequence of 'ecological planes' at individual horizons. It may be that the 'phylogenetic' change is largely a reflection of changing environment eliciting different ecophenotypic responses, thus the ecological dimension is an important addition to the '*Gryphaea* story.'

Our data will be presented in strict stratigraphic order, each example involving a brief geographical, morphological and then statistical analysis, followed by a discussion of important features. In the summary the material is viewed from a more formal statistical viewpoint, and in the concluding section we extend our view to the uses of *Gryphaea* in studies in evolution, ecology and sedimentology. The statistical analysis of *Gryphaea* must be appropriate to the kind of measurements taken, and in an appendix we briefly discuss some hitherto disregarded problems.

1. EARLY MIDDLE JURASSIC *Gryphaea* -- INTERSPECIFIC TRENDS

After the Toarcian 'Black Shale Event' *Gryphaea* reappears in vast populations in the uppermost Toarcian (*levesquei* Zone). However, unlike the Lower and Middle Jurassic the uppermost Toarcian and the subsequent Aalenian occurrences of *Gryphaea* are characterized by strong provincialism within Europe. Local populations are morphologically so well separated that they constitute some of the most easily classifiable -- in a taxonomic sense -- within the entire *Gryphaea* stock (Fig. 1). In addition the morphotypes show an exceptionally close relationship to different sedimentary environments (Fig. 1; see discussion of the various 'species' in BRANNAN, 1983).

1.1 Geographic and stratigraphic distribution

In the uppermost Toarcian to lower Bajocian interval there occur three main ecologically restricted morphotypes. They can be referred to four 'species':

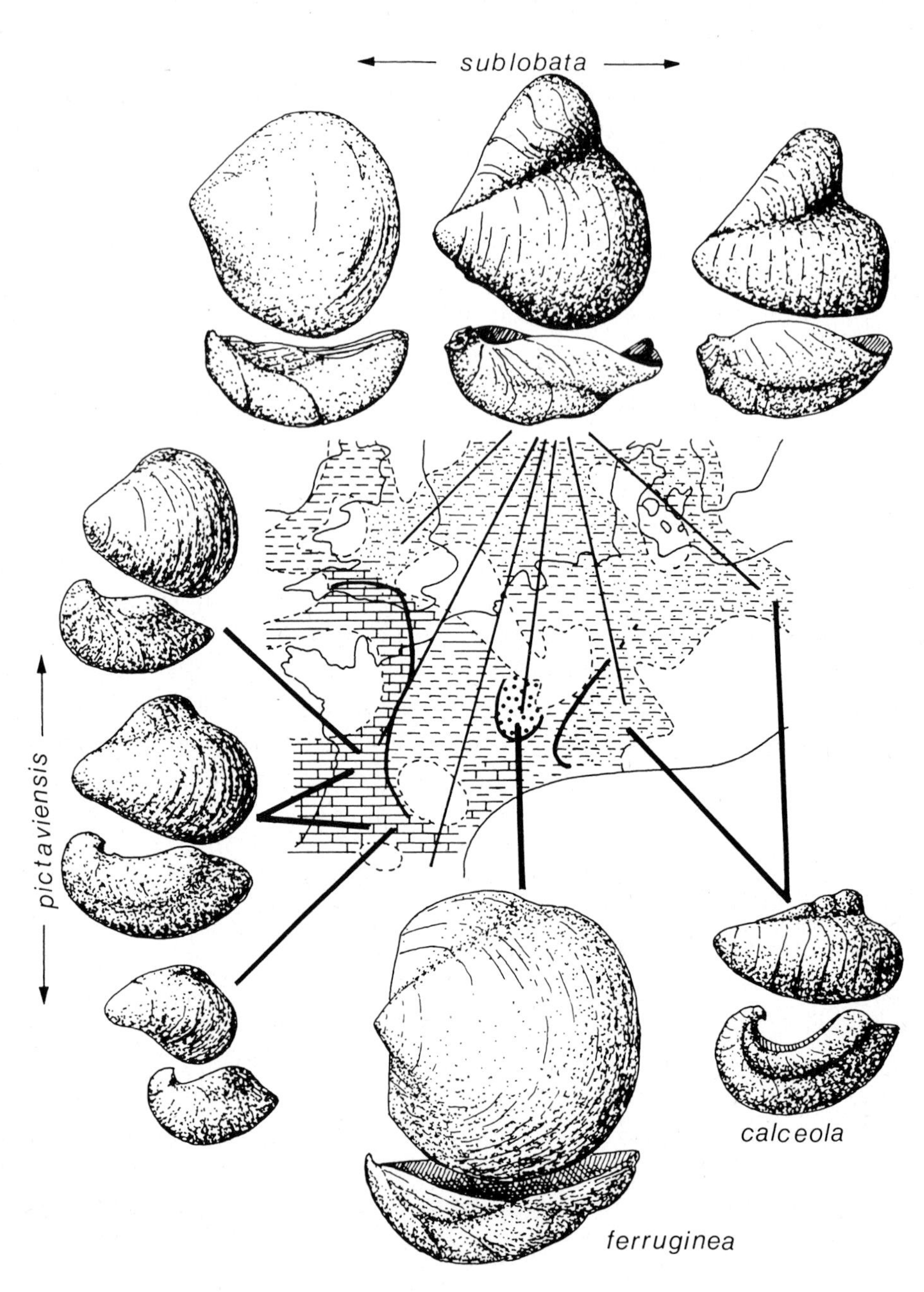

Fig. 1: Geographic distribution and morphological diversity of Upper Toarcian--- Lower Bajocian *Gryphaea* in Europe (see text for discussion). Paleogeography and generalized facies distribution adapted from BAYER & McGHEE (this volume).

Gryphaea (Bilobissa) pictaviensis HÉBERT

occurs in the *levesquei* Zone (uppermost Toarcian) of western France and northern Spain. It usually has a well-developed sulcus and closely resembles the Bajocian

Gryphaea (Bilobissa) sublobata DESHAYES

which occurs commonly in the *laeviuscula* Zone (lowermost Bajocian) and whose widespread distribution throughout Europe marks the end of the Toarcian--Aalenian period of European *Gryphaea* provincialism. *G. sublobata* differs from *G. pictaviensis* only in commonly having a larger attachment area and in usually possessing a more pronounced sulcus. The variability of *G. sublobata* is nevertheless very great, and in large populations occasional variants can be found which approach the morphology of the other species described here.

G. pictaviensis and *G. sublobata* resemble each other not only morphologically but also in their facies relationships. Both occur in marls, marly limestones and marly calcareous sands. Indeed the two can be regarded as 'stratigraphic species' with separate names having been applied to two distinct periods of mass occurrence of members of the same lineage. Throughout the Aalenian similar 'lobate' forms occur sporadically in oolitic limestones and ironstones (e.g. in the Northampton Sand Ironstone = *opalinum* Zone and in the ironstones of the Oberrhein Graben = *murchinsonae* Zone) and in sandy to silty mudstones. Such forms are nevertheless only very rare elements in Aalenian faunas (occurrences being usually restricted to single specimens) and are very much subordinate in abundance to the two following, strongly endemic, species (morphotypes):

Gryphaea (?Bilobissa) ferruginea TERQUEM

appears (like *pictaviensis*) in the *levesquei* Zone and continues into the Aalenian. It is almost entirely restricted to the 'Minette' facies (ironstones) of Lorraine. SCHÄFLE's (1929) record from Aalen, S. Germany (again from a 'Minette'-type facies) is one of the few occurrences outside Lorraine. This strongly endemic form closely resembles *G. (G.) gigantea* of the Pliensbachian and indeed HALLAM (1982, p. 357) has placed it within the latter, thus extending the range of *gigantea* into the Aalenian: 'Terquem's species *feruginea*, ... , is indistinguishable from *gigantea*'. Previously HALLAM & GOULD (1975, p. 532) have considered *ferruginea* a synonym of *sublobata*: *bilobata (= sublobata)* 'has masqueraded under the name *G. ferruginea* Terquem, almost certainly a synonym'. We prefer to regard *ferruginea* as a separate species on account of its different facies relationships and strong endemism. Occasionally occurring radial costae on the right valve (BRANNAN, 1983) indicate that it should be placed, like the other Middle Jurassic species, in the subgenus *Bilobissa* STENZEL rather than in *G. (Gryphaea)* as suggested by BRANNAN (1983).

Gryphaea (Bilobissa) calceola (QUENSTEDT)

occurs from the middle Aalenian to the lowermost Bajocian (*murchisonae* Zone to *discites* Zone) of S. Germany, its type area, and Poland (PUGACZEWSKA, 1971). It is extremely narrow and extraordinarily incurved, more so even than *G. (G.) arcuata* which it resembles in other aspects of morphology and in ecology. Although the sulcus is very much reduced the presence of radial costae on the umbo of the left valve identifies the species as a representative of *Bilobissa*. In its mass occurrences *calceola* is entirely restricted to fine grained deposits, either shales or calcareous mudstones.

The regional occurence of early Middle Jurassic gryphaeas is summarized in Fig. 1 in association with the generalized distribution of facies in Aalenian times. Central and northern Europe is divided into two major facies belts extending from north to south -- a western 'carbonate' belt and an eastern clastic belt between the North Sea and the alpine front (BAYER & McGHEE, 1984). As discussed above the late Toarcian and Aalenian *Gryphaea* species show close relationships to these facies zones. This early period of European provincialism was terminated in the early Bajocian when the endemic forms -- *pictaviensis*, *ferruginea* , *calceola* -- were replaced by the widely occuring *sublobata*, an almost cosmopolitan species (see also HALLAM & GOULD, 1975 p. 539: '*G. bilobata (= sublobata)* migrated more widely than the earlier Liassic species, perhaps as a result of the progressive Jurassic transgression'). The early Bajocian changes in the *Gryphaea* fauna, around the classical '*sowerbyi*' Zone show a close correspondence in timing to changes in the ammonite fauna, when the Graphoceratidae and the Hammatoceratidae were replaced by the Sonniniidae (BAYER & McGHEE, 1983, this vol.) These events are in turn contemporaneous with a marked global regression according to HALLAM's (1978) chart (however, a global sea-level rise in the interpretation of VAIL et al. 1984). Thus, like the change within the Lower Jurassic from *arcuata* to *mccullochi*, a sharp replacement of *Gryphaea* species can be related to a global event (HALLAM, 1982). In the case of the Lower Bajocian the eustatic event is accompanied by a widespread facies change: 'Minette'-type ironstone deposition is terminated, the S. German Aalenian shales are replaced by muddy sandstones (at least for a time) and generally throughout Europe the early *laeviuscula* Zone is characterized by stratigraphic condensation. The net result is that environments associated with 'lobate' forms of *Gryphaea* came to be dominant in Europe for approximately one ammonite Zone; the widespread occurrence of *sublobata* is thus readily explained in ecological terms.

1.2 Quantification of inter-specific diversity

The use of numerical methods has been an essential part of *Gryphaea* studies ever since TRUEMAN's (1922) famous initial paper. The measurement of *Gryphaea* and interpretation of the biometric data, however, has been a source of dispute since about 1960 when HALLAM published his statistical analysis (e.g. HALLAM, 1959 a,b, 1960, 1962; SWINNERTON, 1959, 1964; JOYSEY, 1959, 1960; PHILLIP, 1962, 1964; BURNABY, 1965). The discussion has been summarized by GOULD (1972) but there still remains the problem that measurements such as those employed by HALLAM (e.g. 1968) -- peripheral length (P), length (L), height (H) and width (B) of the shell (see Appendix A for definitions) -- do not strictly allow of a comparison with theoretical models of shell form (RAUP & MICHELSON, 1965; RAUP, 1966; BAYER, 1978).

It is for instance impossible to establish whether or not *Gryphaea* grew in a logarithmic spiral on the basis of the measurements used -- HALLAM & GOULD's (1975) statement that it did not may thus be untrue. In Appendix B is explained in some detail why a combination of the methodology employed and the ontogenetic variability of a bivalve disallow comparison with the theoretical model of a logarithmic spiral. Nevertheless, there remain no other suitable ways to quantify the morphology of *Gryphaea* so the same measurements have been used in the present study and the analysis of Appendix B has been used to generate 'rules of thumb' for the apropriate interpretation of the data.

A further problem is that samples of Middle Jurassic *Gryphaea* from a single locality and bed are often small and the specimens are commonly of similar size. Such samples are of little use in attempting to infer whether form was 'environmentally' or 'genetically' controlled unless additional information about the ontogeny of each individual can be obtained (cf. JOHNSON, 1981). The lack of such data also hampers the reconstruction of environmental and phylogenetic trends because the first stage of the analysis must always be to remove ontogenetic effects and where only 'static' data is available this introduces a further element of interpretation. This fact is well exemplified by HALLAM & GOULD's (1975) multivariate analysis. Growth lines are nevertheless very easy to discern in *Gryphaea*.

We, therefore, took measurements of shell shape for standardized values of peripheral length (as a measure of size) in order to evaluate ontogenetic changes in form. The ontogenetic trend in form can thus certainly be more accurately defined than with 'static' data -- samples can be compared to the average trend, trend lines can be compared with each other by measuring the overlap in confidence intervals and variability within each sample can be assessed from the standard deviation at each interval of peripheral length.

The 'ontogenetic trend lines' for Upper Toarcian to Lower Bajocian *Gryphaea* are summarized in Fig. 2. The lines join sample means; the vertical bars are two standard deviations in length (one on either side of the mean).

Regionally and stratigraphically homogeneous samples from single localities were available for each of *calceola (c)*, *sublobata* (s_1) and *ferruginea (f)*. These samples are well separated in terms of length (L) against periphery (P) and width (B) against periphery. Three trends are immediately apparent (Figs. 2A, 2B):

** maximum size increases in the sequence *calceola*, *sublobata*, *ferruginea*

** in the same sequence there is a decrease in curvature

** initial differences in form are exaggerated through ontogeny i.e. the trends are divergent. Plotting of B/L against P produces, however, separate sub-parallel trends (Fig. 2D, see Appendix C for discussion).

A representative sample of *pictaviensis* (p) from the Upper Toarcian of Aquitaine conforms well to the pattern described above. The sample is located between the L/P trend lines for *calceola* and *sublobata* and the standard deviations at each value of peri-

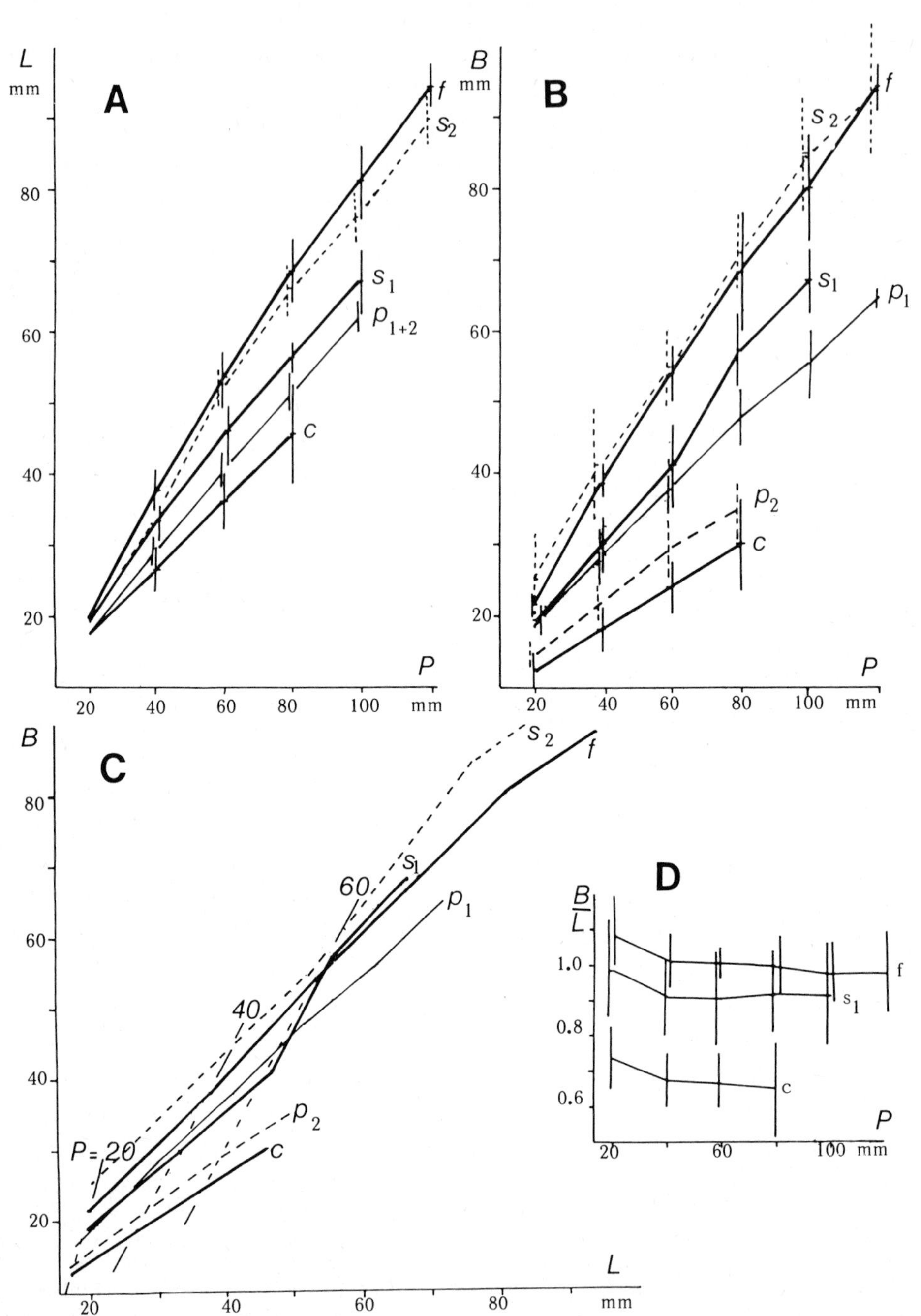

Fig. 2: Statistical analysis of U. Toarcian-- L. Bajocian *Gryphaea*: *c*: *calceola*; *p*: *pictaviensis* (p_1: Aquitaine; p_2: Mte. Noir + N. Spain); *s*: *sublobata* (s_1: Lorraine; s_2: Cotswolds); *f*: *ferruginea*. The lines represent the average ontogenetic trend for each sample; vertical bars are (twice) standard deviations for measurements at equal peripheral length. A) Shell length against peripheral length. B) Shell width against peripheral length. C) Width against length; numbered broken lines connect the mean values for equal peripheral length, i.e. they indicate the trends between samples at equal size (peripheral length). D) The width/length ratio (a 'global' measure of curvature) against peripheral length.

pheral length overlap with those of both *calceola* and *sublobata* (see also Fig. 2D). Plotted in terms of B/P (Fig. 2B) *pictaviensis* is much closer to *sublobata*; *calceola* is seen to be quite distinct. Nevertheless an inhomogeneous sample of *pictaviensis* from the Mte. Noire and from N. Spain (p_2) plots quite close to *calceola* in terms of width and there is considerable overlap in the standard deviations. This shows that there are at least some individuals within *pictaviensis* which approach the very narrow form of *calceola*.

A further unrepresentative sample, in this case of *sublobata*, was available to us in the form of a number of exceptionally large and well-preserved specimens (s_2) from the Bajocian of the Cotswolds (SYLVESTER-BRADLEY Collection; mainly material from the Geological Museum, London). This extends the range of variability in *sublobata* towards *ferruginea*; in both the L/P and B/P plots there is considerable overlap of the standard deviations for the two species. Fig. 1 (top left) shows a specimen from this sample which demonstrates that the development of a wide form in *sublobata* is associated with a reduction in the sulcus, as in *ferruginea*. Although the *sublobata* sample containing these exceptional morphologies is probably not representative, it illustrates the potential morphological variability of *sublobata* and it may point towards an ecological effect as the Gryphite Grit (calcarenite) from which these specimens are derived represents a rather unusual environment compared with other European occurrances.

The rather few samples analysed above include some which are not representative of fossil assemblages. Nevertheless the correlations observed between periphery, length and width are strong enough to suggest that the addition of further (local) samples would tend to produce a continuous distribution of sample means. This seems likely at least as far as connections between samples of *pictaviensis* and *sublobata* are concerned -- *calceola* and *ferruginea* seem to constitute relatively discrete groupings at the endpoints of the morphological space.

1.3 Ecological and Phylogenetic Trends

From the foregoing analysis it is possible to isolate two important trends:

** *there is a close correlation between environment (facies) and the occurrence of particular 'species'. This may well in fact reflect no more than an extension of patterns of 'intraspecific' variation seen in for example the samples of* <u>*sublobata*</u> *from Lorraine (s_1: from muddy, calcareous, fine-grained sandstones) and England (s_2: from a calcarenite on the evidence of the matrix).*

** *there is strong covariation of all measurements both within and between samples ('species'). A narrow shape is associated with strong incurvature and small size, while suborbicular outlines are correlated with a low degree of inflation and larger size. A sulcus is, however, most strongly developed in intermediate forms and loses its strength both in the more incurved and the relatively flat forms.*

If we interpret the different morphotypes in terms of distinct species actually or potentially able to coexist at any one time (i.e. if we lay emphasis on the observation that in the Aalenian there are several fairly distinct morphotypes) then we have species with very narrow environmental requirements. However, if we consider all (or at least a subset) of the morphotypes as variants of a single species, we have an exceptional example of polymorphism under environmental control -- the term 'ecophenotypic' as used herein may, therefore, reflect differences in gene pools. The same pattern of polymorphism and the same freedom of interpretation is found in other groups like ammonites (WESTERMANN, 1969; BAYER, 1972; BAYER & McGHEE, 1984) or scallops (JOHNSON, 1984). All these examples have the surprising 'covariation of features' in common, an observation at the base of Darwinism:

> *'Correlation of growth. I mean by this expression that the whole organisation is so tied together during its growth and development, that when slight variations in only one part occur, and are accumulated through natural selection, other parts become modified'* (Darwin, 1859).

Considering this pattern, it is not surprising that the morphological trend observed in the form of morphotypes with the environment is very similar to the phylogenetic trend in the Liassic sequence

arcuata -- mccullochi -- gigantea

(HALLAM, 1968) and in the Middle-Upper Jurassic sequence

sublobata -- dilobotes -- dilatata.

However, the question arises as to what extent these lineages are actually controlled by environmental changes. The control of stratigraphic species replacements in *Gryphaea* by global physical events was stressed by HALLAM (1982) -- with respect to the two lineages common global patterns are the Liassic and late Middle Jurassic overall transgressive trends. In order to determine exactly to what extent the ecological factors considered above can account for *Gryphaea* phylogeny, it is necessary to look for other examples: in this case the more closely studied, and better preserved *Gryphaea* of the Callovian and Oxfordian which form the major part of HALLAM & GOULD's (1975) Middle-Upper Jurassic lineage.

2. LOWER CALLOVIAN *Gryphaea* -- INTRASPECIFIC TRENDS

It is clear from the foregoing analysis of Upper Toarcian - Lower Bajocian *Gryphaea* that there is a close relationship between form and environment. The extent to which the range of forms may be subdivided is reflected in the number of specific names that are traditionally applied (*calceola*, *ferruginea*, *sublobata*, *pictaviensis*) with the interpretation of ecological and evolutionary patterns depending on the taxonomic weighting of the distinct morphotypes. Lower Callovian forms have, in contrast, usually been referred to only one species, although there are considerable differences in shape from place to place. In the following section we discuss the nature of these differences and their relationship to environment.

2.1 Morphological Trends in the Kellaways Rock

The Kellaways Sand (or Rock) is a sand unit usually restricted to the *calloviense* Zone of the Lower Callovian. The term 'Rock' is applied where the sands are more cemented. In South Yorkshire a 'Rock' unit belonging to the *enodatum* subzone overlies coarse sands containing abundant *Gryphaea*. These sands are probably equivalent in age to finer grained sands and silts in the East Midlands and to sandy clays in Dorset, both of which belong to the *calloviense* subzone (immediately preceding the *enodatum* subzone) and contain abundant *Gryphaea*. The diminution in grain size probably reflects increasing distance from a source of sand either in the North Sea or the Pennines.

Although there has been some dispute about the name which should be applied, there has been general agreement that only one *Gryphaea* species existed in England at the above-mentioned time-level:

Gryphaea (Bilobissa) dilobotes **DUFF, 1978**

= *G. bilobata* J. de C. SOWERBY
subsp. *calloviense* HALLAM & GOULD, 1975 (non J. de C. SOWERBY sp.)
= *G. calloviense* HALLAM, 1982.
The species differs from *sublobata* in having a more weakly developed sulcus.

Our three samples, which we take to be time-equivalents from the *calloviense* subzone, are from:

KT: Kettlethorpe Quarry, near South Cave, Yorkshire:
coarse sands immediately beneath the Kellaways Rock.

BT: Bletchley, Bucks (Bed 2 of CALLOMON, 1968):
silts and soft sandstones with calcareous concretions.

PL: Putton Lane, near Chickerell, Dorset (Bed 2 of ARKELL, 1947):
silty to sandy clays.

All the material is from the collection of the late P.C. Sylvester-Bradley.

The geographic locations, associated lithologies and typical morphologies of specimens in the samples are summarized in Fig. 3. The most obvious difference between the samples is in the size attained in each, which increases from Dorset to the E. Midlands to Yorkshire. This trend is paralleled by an increase in 'umbonal angle' (the angle subtended by two lines tangential to the dorsal 'shoulders' of the shell and meeting at the umbo), which is an indication of shell width. Data for this variable, derived from over 1000 specimens measured by Sylvester-Bradley, are presented in Fig. 3 in the form of cumulative frequency distributions for three classes of umbonal angle within each of the height classes by means of which Sylvester-Bradley recorded the approximate size of specimens. The clear trend towards increasing 'width' together with that towards greater size, as average grain size of the sediment increases, mirrors the inter-specific trend seen in U. Toarcian - L. Bajocian *Gryphaea*. Although the extremely wide (*ferruginea*) and the very narrow, incurved (*calceola*) forms of the latter interval are not seen in the L. Callovian, this is to be expected because the range of environments is smaller. The important point that we wish to make here is that there is the same pattern of covariation.

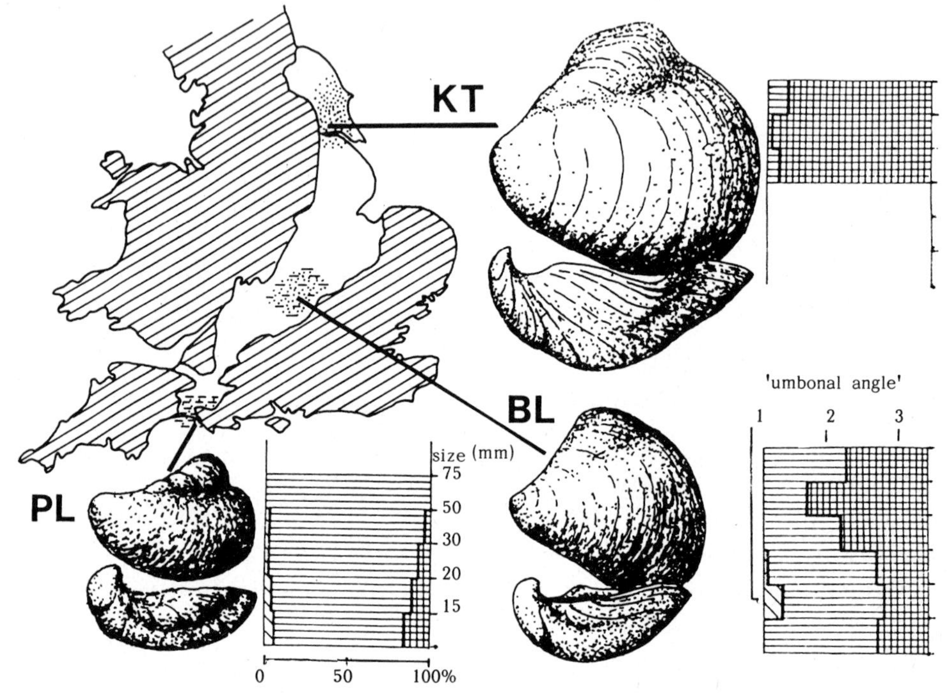

Fig. 3: The geographic locations, associated lithologies, typical morphologies and ontogenetic development of the 'umbonal angle' of Lower Callovian *G. dilobotes* in England (see text for explanation).

The similarity to the U. Toarcian - L. Bajocian *Gryphaea* becomes particularly evident in Fig. 4. Although the total variability is smaller, the trend between samples resembles Fig. 2 closely. Lower Callovian *Gryphaea* develop, like *sublobata*, a sulcus in the process of increasing width. This causes a kink in the ontogenetic trend lines for B/P and B/L which is especially well developed in the *sublobata* sample S_1 and which appears also in the *dilobotes* data. However, in the Dorset sample (PL) with a small mean size, the sulcus is reduced,as in *calceola*. If we refer to the diversity of U. Toarcian--L. Bajocian forms as interspecific variability, this variability resembles in all details the intraspecific variability of the L. Callovian *dilobotes*. However, the standard deviations in Fig. 4 overlap more than they do in the U. Toarcian--L. Bajocian example, i.e. the between sample diversity is reduced.

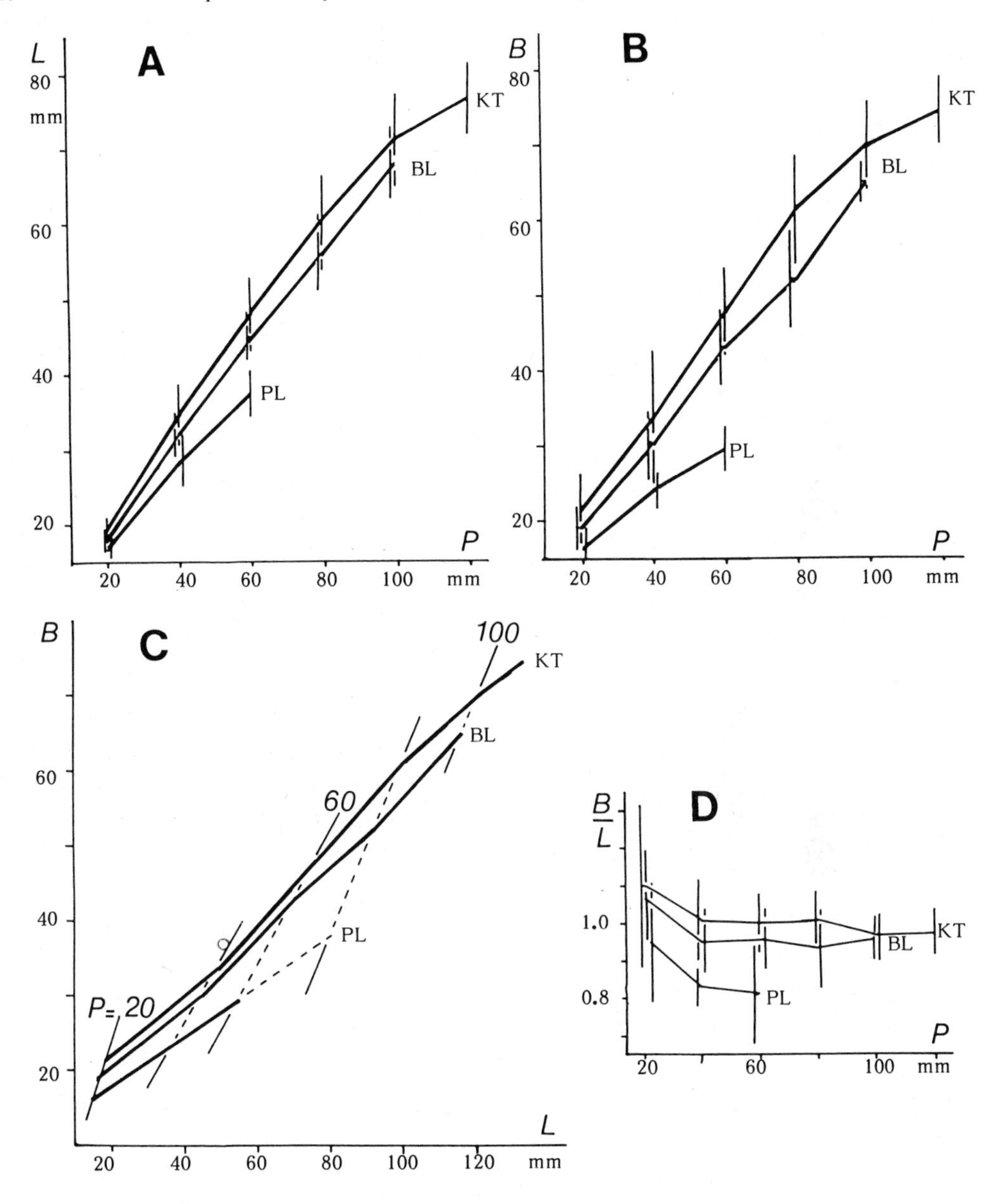

Fig. 4: Statistical analysis of L. Callovian *Gryphaea* (see text and Fig. 2 for explanation).

2.2 Form and Environment

We can combine our results from L. Callovian *Gryphaea* with those from U. Toarcian --L. Bajocian species to make the following general statements about the relationships between form and environment:

** *in shales and muddy limestones small, narrow, strongly incurved and weakly sulcate forms predominate*

** *in sands and sandy limestones we find mainly lobate forms with a switch towards width-dominated growth later in ontogeny*

** *in iron-oolites of the 'Minette' type flat, sub-orbicular forms (similar to G. gigantea of the Lias) predominate.*

At first sight it would seem that development of wider, flatter forms could be explained in terms of adaptation to higher energy environments, as such forms are more resistant to overturning by currents (HALLAM, 1968). However, more detailed analysis of the 'high energy' sediments suggests that they may in fact have been associated with quite low environmental energy at the time they were colonized by *Gryphaea*.

The sands in which *sublobata* occurs in Lorraine and S. Germany preserve rather few sedimentary structures of physical origin, but they contain a trace fossil assemblage belonging to the '*Zoophycos* facies' which suggests that for long periods of time environmental energy was low. There is a corresponding paucity of physical sedimentary structures in the sands of the L. Callovian in S. Yorkshire. Our examinations revealed that *Gryphaea* specimens are rarely broken and frequently articulated so they certainly cannot have been washed in (cf. BRASIER & BRASIER, 1978) -- they must have actually lived on the immobile sand flats which seem to constitute the environment represented in this unit. Nevertheless the two-valved specimens are not all in situ as can be seen from a specimen collected by J.D. Hudson. It contains three individuals with such different orientations that only one could possibly be in life position. Possibly the animals were disturbed by periodic erosional events which could have caused their death, either by overturning or by smothering them under a blanket of sand. The random orientation observed can easily be related to the erosional events.

While from the above it seems unreasonable to infer that the development of a relatively wide shell in coarse-grained sediments represents an adaptation to higher environmental energy, there remains a plausible alternative view that wider shells were developed to facilitate colonisation of firmer substrates such as sands, and the even firmer ironstone sediments. The fact that the very wide *Gryphaea* species of the Pliensbachian (*gigantea*) and Oxfordian (*dilatata*) occur frequently in shale sequences seems to argue against this hypothesis. However, in the next section we show that in a largely monotonous clay sequence (the Oxford Clay) there is actually considerable variation in *Gryphaea* morphology and that the horizons where the widest forms occur show indications of firm substrates.

3. OXFORD CLAY *Gryphaea* -- A 'TEST' FOR FORM/SUBSTRATE RELATIONSHIPS

The Middle and Upper Oxford Clay is mainly a monotonous sequence of grey to blue clays. The occasional limestone beds are faunistically rather similar to the intervening clays (e.g. HUDSON & PALFRAMAN, 1969). The *Gryphaea* in this facies reach considerably greater sizes than their Lower Callovian counterparts.

Two species are recognized by BRANNAN (1983), representing successive segments of one lineage:

Gryphaea (Bilobissa) lituola **LAMARCK,**

> is the earlier species, which occurs in the Callovian from the upper *jason* Zone to the *lamberti* Zone and is characterized by a high proportion of moderately-wide specimens with strong incurvature and a fairly well-developed sulcus.

Gryphaea (Bilobissa) dilatata **J. SOWERBY,**

> is the later species, from the Upper Oxford Clay (Lower Oxfordian) and is characterized by a high proportion of very wide specimens of large size and low incurvature, bearing only a weak sulcus (the shape is very similar to *G. gigantea* of the Pliensbachian).

The morphological variability of both species is considerable (Fig. 5) and individual specimens may be very difficult to assign to one or the other species if the stratigraphic horizon is not known.

3.1 Environmental Conditions

The *Gryphaea* samples for this analysis were derived from our own collection of *in situ* specimens at Bletchley, Bucks. (*athleta* Zone) and Warboys, Cambs. (*mariae* to ?*transversarium* Zone ≅ *tenuiserratum* Zone). The two *athleta* Zone samples turned out to be extremely variable (in terms of incurvature, width and sulcation; Fig. 5) but they have a similar mean morphology. We could find no evidence of any morphological or environmental differences between the samples. The data will be used here mainly for comparative purposes (B-lines in Fig. 7). Our main data set, consisting of five samples from the Upper Oxford Clay and 'Ampthill Clay' at Warboys, reveals in contrast marked differences in mean form from level to level and corresponding evidence of differences in environment.

Fig. 5: Morphological variability of Oxford Clay *Gryphaea* (see text for explanation).

The samples are as follows:

W1: from the clays of the *mariae* Zone at the base of the pit. The specimens (only 5) exhibit some encrustation by forams.

W2: from calcareous clays with interbedded marly limestones of the upper part of the *mariae* Zone. The specimens are usually encrusted with forams, sometimes with serpulids.

W3: from calcareous clays with interbedded silty limestones of the *cordatum* Zone. Most specimens are encrusted with serpulids, some heavily so.

W4: an in situ sample from a limestone bed at the base of the 'Ampthill Clay'. The specimens are heavily encrusted with serpulids, bored by *Lithophaga*, and commonly rounded by reworking. Lenses (cast infillings) of a mixture of 'fresh' and reworked shells are common.

W5: a loose sample from the top of the pit, probably from the 'Ampthill Clay'. The specimens are heavily encrusted by serpulids and bored by *Lithophaga*.

Although the sediments show few obvious differences from sample to sample and are indeed very similar throughout the Middle and Upper Oxford Clay (HUDSON & PALFRAMAN, 1969) the *Gryphaea* shells themselves provide information about depositional conditions by means of encrustation patterns, and about current activities and intensities by the frequency and intensity of reworking. These patterns indicate that there were

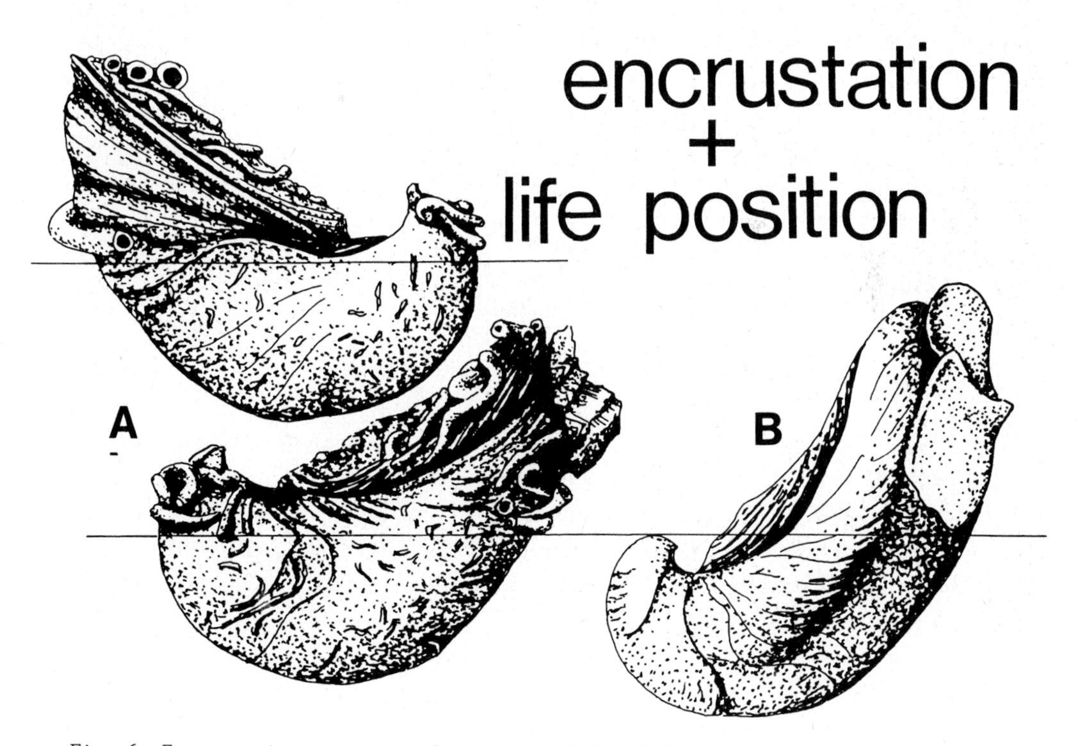

Fig. 6: Encrustation patterns of serpulids (A) and 'second generation' *Gryphaea* allow to reconstruct life positions and to infer substrate conditions (see text for discussion).

indeed environmental fluctuations during the study interval. The higher degree of encrustation and reworking of specimens at higher horizons indicate that the rate of sedimentation was probably lower and erosional events occured more frequently, and this almost certainly produced firmer substrates than in the more rapidly deposited clays at lower horizons. For these lower horizons we found additional evidence for softer substrate conditions from the pattern of encrustation because the encrusters commonly occupy a band around the edges of the shell -- sometimes with repeated bands of encrustation occurring on the shell. Fig. 6 shows double valved specimens with such a pattern of encrustation and the life position therefore infered. Certainly in the case in which a double valved shell has been used as an attachment site by other *Gryphaea* of quite large size (Fig. 6), there can be little doubt that a considerable proportion of the left valve projected above the sediment surface. In the case of the specimens in Fig. 6A, where the encrusters are fairly small, there is the possibility that not all the encrustation happened at the same time, but that the shell became successively encrusted as it grew, with the position of the commissure remaining fairly static just above the sediment/water interface and with the left valve gravitationally rotating as growth proceeded. However, the existence of specimens with multiple bands of encrustation, each with a characteristic state of preservation, is strongly suggestive of periodic colonisation of the left valve, at times when an important proportion was exposed above the sediment, and then subsequent rotation leading to burial and death of the encrusters.

The encrustation patterns found on the *Gryphaea* shells are strongly indicative of changing environmental conditions, i.e. a general trend to increasing erosional energy as reflected by the successive encrustation by forams and serpulids and followed by *Lithophaga* boring and reworking. For the environment an increasing frequency and intensity of erosional events and an increasing firmness of the substrate can be inferred.

3.2 Diversity of Oxford Clay *Gryphaea*

Just as there are differences in environment in the sequence at Warboys so there are differences in form. Fig. 7 shows that the higher samples W4 and W5 are clearly separated from samples W1 to W3 in terms of B/P and L/P (Figs. 6A, 6B). The higher samples exhibit a markedly flatter and broader shape. These morphological differences between samples appear still more pronounced in the 'subgraphs' given with Fig. 6A and 6B where the mean values of L and B have been plotted for fixed values of peripheral length for comparison between samples. Between the lower samples (W1 to W3) there are some minor differences -- statistically not significant, as indicated by the strong overlap of the standard deviations. However, between samples W2 and W3 differences in substrate conditions are indicated (see above), and the difference between the mean morphologies corresponds to the inferred environmental conditions. Sample W1 does

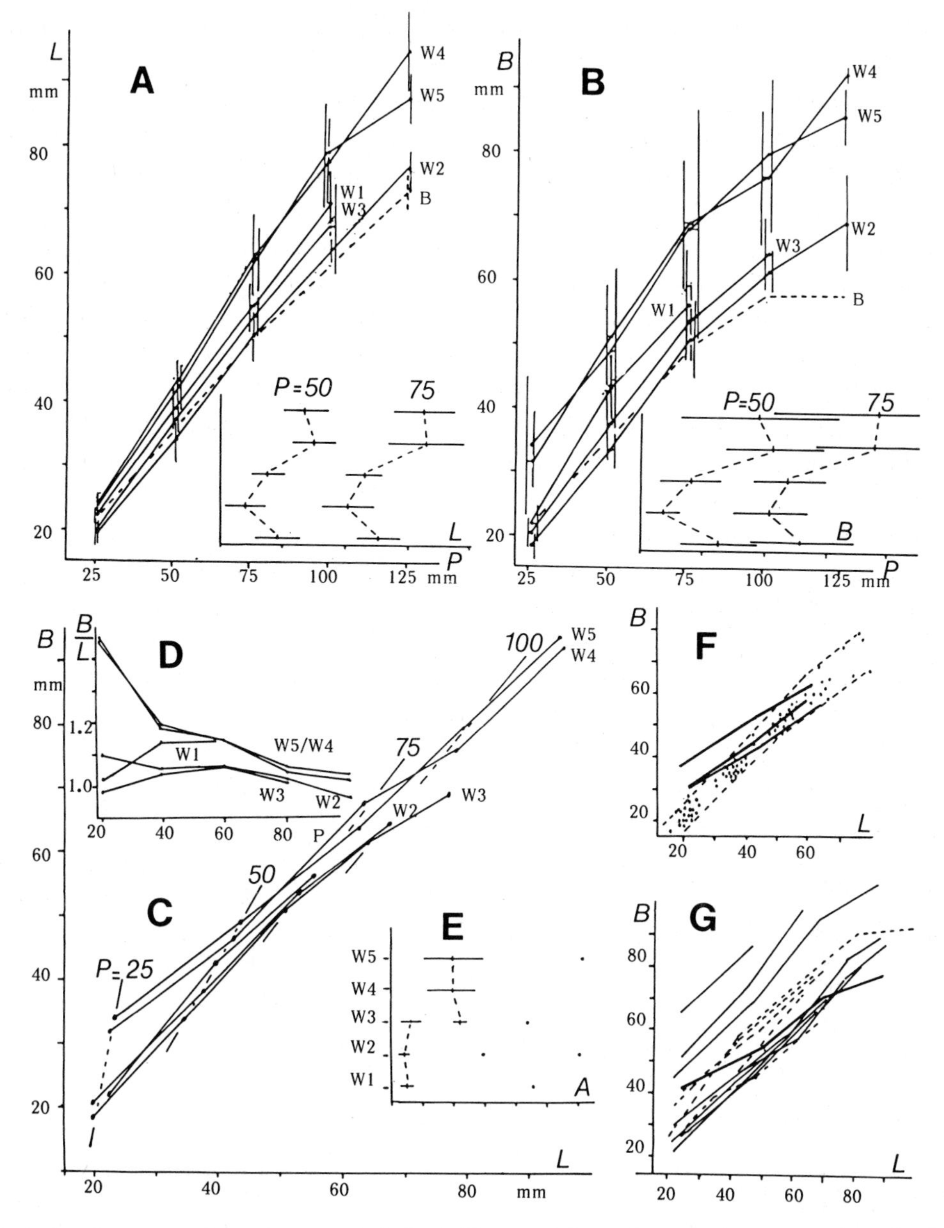

Fig. 7: Statistical analysis of Oxford Clay samples. A to D: statistical analysis as before (compare Fig. 2). The 'subgraphs' in A and B display the change of length and width values within the sample sequence for fixed values of peripheral length (sample means and standard deviations). E: Mean sizes (and standard deviations) of attachment area in the sample sequence (W1 to W5); points: extreme specimens which have not been used for the calculation of the sample statistics; the sample W3 has been split into two subsets for comparison with E and F. F: Data field of individual ontogenies in samples W2 and W3 (dots) and ontogenetic trends of specimens with large initial attachment area (heavy lines) from sample W3 (compare E). G: Individual ontogenies of samples W4 and W5 (dashed); the trend of increasing width correlates with an increase in attachment area.

not fit the general trend, but sample-size was much too small to infer the environmental and morphological conditions. The marked morphological differences between samples W1,W2,W3 and W4,W5 in morphology are paralleled by a marked difference in the size of the attachment area: it is much larger in the typically wider and flatter forms of samples W4,W5 (Fig. 7E). This morphological pattern underlines again the environmental differences -- much larger objects being available for larval settling in this case -- and it suggests an ecophenotypic control of morphology. However, it is not possible to account for the broad flat form of W4 and W5 simply in terms of a larger attachment area. Occasionally specimens with an attachment area of similar size occur in samples W1 to W3 -- and these in later growth rapidly approach the 'mean sample morphology' of the lower samples, i.e. an incurved form, as is illustrated by Fig. 7F,G in which individual ontogenetic lines are plotted. This shows that the attached growth phase and the coiled growth phase are independently controlled -- both, however, by environmental constraints.

The two samples from Bletchley are morphologically similar and plot (B line in Fig. 7) close to the ontogenetic trend lines for W2 and W3; once again the *athleta* Zone - samples show the same (low-moderate) levels of encrustation, so the substrate must have been much like that for the W2 and W3 specimens at Warboys. Morphologically the *athleta* samples have a slightly different slope in the L/P and B/P plots indicating a slightly stronger incurvature as *lituola*. However, the considerable overlap in the standard deviations between B and W2 and W3 should be noted. That the Middle Oxford Clay forms, considered as a whole, have a different mean form (more incurved) could be due to a preponderance of environments favourable to this shape (relatively soft substrate) rather than to a phyletic difference.Evidence for this is supplied by J.D. Hudson from a study of the *lamberti* Zone at Woodham, which tends to confirm our suggestions. Hudson reports that at a thin limestone horizon (the *lamberti* limestone) the incurved *lituola*-type of the under- and overlying beds is replaced by a flatter form of *dilatata*--type. The *lamberti* limestone is almost certainly a product of reduced clay input and as such probably afforded a relatively firm substrate -- the environment which we would expect to be associated with wide, flat forms of *Gryphaea*.

The sequential samples of Oxford Clay *Gryphaea* provide the same morphological trends as observed for the L. Callovian and U. Toarcian - L. Bajocian *Gryphaea* at approximately synchronous levels. Within the three sets of samples the measures of morphology L/P and B/P vary in the same manner. In the B/L plot, however, the Oxfordian *Gryphaea* samples are much less divergent than in the examples considered earlier. This is in agreement with the fact that the environmental conditions were less diverse in the Oxford Clay than in the other cases. Within the Oxford Clay the higher samples (Fig. 7G) are more variable than the lower ones (Fig. 7F) and the greater variety of patterns of encrustation and modes of accumulation suggest a corresponding diversity of environments.

3.3 Relationship between substrate and *Gryphaea* morphology

In our three case studies we have shown that the form of *Gryphaea* varies with the environment. We consider that our data from the Oxford Clay give the clearest indication of the environmental variable that determines form and that this variable is substrate firmness rather than 'current velocity'. In the Oxford Clay the association of small sediment grain size with firm substrates implies that the environment was also of rather low energy. The corollary of our results is that the probably 'higher-energy' sediments of the Aalenian ironstones in which the flat *ferruginea* occurs can be interpreted as having been quite firm. However, the question arises whether the ironstones provide the proper interpretative framework for *Gryphaea* ecology or whether the occurrence of *Gryphaea* provides, rather, information about the depositional conditions of the ironstones.

An interesting discovery in our Oxford Clay study was that although the size of the attachment area may strongly influence the growth of relatively wide shells (samples W4, W5) this does not persist into late ontogeny from the establishment of a 'growth formula' established in the early attached phase: the attached and coiled phases of growth are quite separate, and wide shells result in forms with large attachment areas only from the fact that the coiled phase of growth proceeds isometrically and from a relatively broad initial shape. The fact that a large attachment area may sometimes promote the development of a relatively wide adult shell can be viewed as an extension of the influence of substrate firmness on shell form. Possibly, growth of a relatively wide shell in the coiled phase (in forms with both large and small attachment areas) is occasioned by production of a firm substrate through the general availability of objects that also permit development of a large attachment area; namely, sizeable shell fragments. These might be concentrated through the same reduction in rate of sedimentation and increase in reworking which probably led to the non-biogenic component of the sediment being relatively firm (BAYER et al., this volume; KIDWELL & AIGNER, this volume).

A more detailed analysis of the relationship between morphology, substrate and perhaps other environmental factors requires detailed studies of the constructional morphology of *Gryphaea* as well as of the sedimentology of its environments -- both programs beyond the scope of this study.

4. SUMMARY OF ANALYSES

Three fundamental points emerge from the analyses presented above:

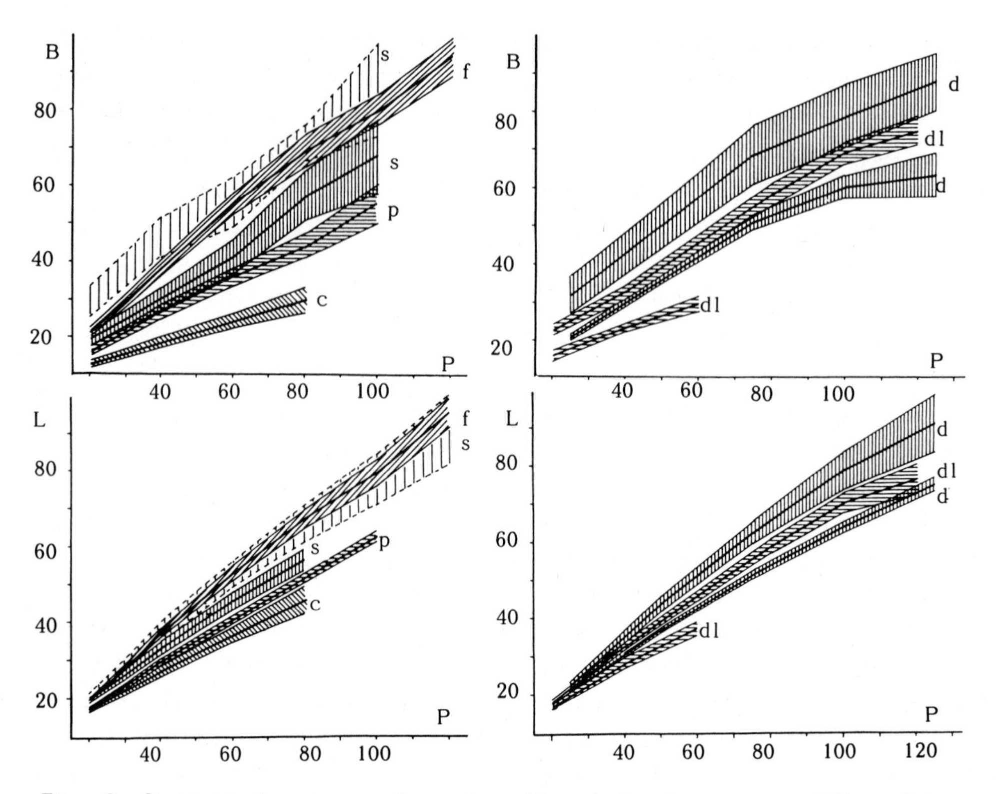

Fig. 8: Statistical summary of samples. The shadowed areas are 95% confidence intervals for the sample means (t-test). *p: pictaviensis; f: ferruginea; c: calceola; s: sublobata; dl: dilobotes; d: dilatata* (the samples W1 to W3 and the *lituola* samples have been lumped in the lower trend line because they are statistically not distinguishable; compare Figs. 2,4,7 and text for further identification).

1) *Morphological characters in Gryphaea are strongly interdependent; thus high inflation correlates with narrow shells and low inflation with wide shells. Intermediate forms (in the Middle Jurassic) are strongly sulcate.*

2) *There is strong ecological control of morphology at both intra- and inter-specific level.*

3) *Ecologically controlled geographical and stratigraphical variation produces trends which closely parallel proposed phylogenetic trends (HALLAM, 1968; HALLAM & GOULD, 1975).*

These relationships become especially evident in the data summary of Fig. 8, where the previously used standard deviations are replaced by confidence intervals (t-test) for the sample means. The confidence intervals give estimates of where the true mean value may be found with a probability of 95%, if a normal distribution of the data is assumed. It turns out that the samples within the three sample sets -- i.e. U. Toarcian - L. Bajocian; L. Callovian; U.Callovian - Oxfordian -- are surprisingly well separated. The important point, however, is that the differences between samples of similar age

i.e.

pictaviensis -- *ferruginea*
calceola -- *sublobata*
within *dilobotes*
within *(lituola) dilatata*

are clearly larger than the differences between successive stratigraphic levels. However, if the most incurved forms of the various horizons are compared the phylogenetic trend proposed by HALLAM & GOULD (1975) emerges, although they did not use such extreme forms. The present study, therefore, adds a new -- ecological -- aspect to the *Gryphaea* story; it does not disprove a phylogenetic trend. One such trend is anyway rather obvious in the Middle Jurassic *Gryphaea* in the form of a general size increase and flattening of the shell. This phylogenetic trend is indeed supported by our data (Fig. 8). However, the essentially directional aspect of HALLAM & GOULD's trend (from inflated deeply sulcate Bajocian *sublobata* to flat, non-lobed, paradigmatically stable Oxfordian *dilatata*) is not supported. We would suggest that the trend identified by HALLAM & GOULD is by no means as uniform as they suggested and that the reason for this is the great importance of adaptation to available substrates (or even ecophenotypical reaction) rather than a gradual approach to some ideal -- 'stable' -- morphology which was in fact available in the Toarcian (*ferruginea*).

Major environmental trends and overturns seem also to have affected the Liassic evolution of *Gryphaea* as HALLAM (1982) pointed out. The remaining trend in Middle Jurassic *Gryphaea* could also be related to or controlled by a large scale environmental change the overall Middle/Upper Jurassic sea-level rise which appears on both HALLAM's (1978) and VAIL's (VAIL et al., 1984) eustatic charts.

5. CONCLUSIONS

Gryphaea probably is the most celebrated case in palaeontological studies of evolution. Nevertheless, this genus is still capable of confronting us with new aspects and may well add still deeper insight into evolution through showing the importance of changes to suit local ecological conditions. *Gryphaea*, indeed, is an exceptionally good fossil for evolutionary studies, as HALLAM (1983) pointed out, because it has an extremely high fossilization potential even in diagenetically unfavourable sediments. In addition, the heavy shells are rarely transported, and *Gryphaea* is commonly found *in situ* providing the possibility of excellent paleoecological control.

The strong correlation of morphological features in both ecological and phylogenetic trends may well be an important pattern that *Gryphaea* can contribute to the evolutionary story. As recognized already by Darwin, such 'correlations of growth' have been found in other fossil groups such as ammonites (WESTERMANN, 1969; BAYER, 1972; BAYER & McGHEE, 1984) and in scallops (JOHNSON, 1984). However, their environmental control is less well understood, although an attempt has been made by BAYER & McGHEE (this volume) to relate their patterns to environmental factors and by JOHNSON (1981) to find a rule for distinguishing ecophenotypic versus genetic control. *Gryphaea*, in contrast, allows us to study phylogeny not only as a 'succession of form', but as a 's u c c e s - s i o n o f e c o l o g i e s' and evolution not simply as 'unfolding of form' but as 'u n f o l d i n g o f e c o l o g i e s' -- phylogeny as a continuous map of ecological gradients (BAYER & McGHEE, this volume) may well introduce new perspectives into the discussion of species evolution.

Gryphaea, however, is not only an evolutionary object, because its ecology is of intrinsic interest. *Gryphaea* samples contain usually few or no juveniles. The effects of winnowing and sorting of *Gryphaea* beds are still poorly understood -- at least in ecological terms. QUENSTEDT (1847) assumed a gradation of *G. calceola* into the small *'Ostrea' calceola* and HALLAM & GOULD (1975) suggested that a small *'Catinula'* of the Kellaways-Beds could simply be a small *G. dilobotes*. Are these examples of winnowed, sorted and locally accumulated small *Gryphaea* as in the *Liostrea* shell beds of the Minette (BAYER et al., this volume) -- or do they reflect i n s i t u ecological events; *Gryphaea* which just have not grown to the usual size? Or is the conventional taxonomy right which puts these small forms in even different genera? Sedimentological, ecological and evolutionary questions coincide here in the oyster *Gryphaea*.

Gryphaea is a recliner and very sensitive to substrate conditions, to which it reacts either ecophenotypically or by production of ecological variants. As the substrate controls *Gryphaea* morphology so its shells modify the substrate conditions, and *Gryphaea* may well be a useful tool in reconstruction of soft bottom conditions -- as a substrate for encrusting and boring organisms and as a sediment particle, *Gryphaea* records to some extent its own environmental and ecological history.

Acknowledgements: In the course of preparing this paper we benefited considerably from discussions with J.D. Hudson. We are grateful for valuable comments and critical remarks by S.J. Gould, A. Hallam and A. Seilacher.

We are indebted to Mrs. Sylvester-Bradley for allowing us ready access to the Collection and Data of the late P.C. Sylvester-Bradley.

The manuscript was prepared while one of us (UB) was 'Heisenberg Stipendiat' and stayed at Leicester. The study was financially supported by the DFG (UB) and the NERC (AJ).

APPENDIX

A) Measurements

For the measurements and the lettering in this study we followed HALLAM (1968) and HALLAM & GOULD (1975) to be in line with the recent *Gryphaea* literature -- the notation, however, is different from the 'standard' of STENZEL (1971). The measurements (capitals) are summarized in Fig. 9 A:

P: Periphery, the length of the curve from the attachment area to the ventral margin.

L: Length (=height of STENZEL), the maximum dorso-ventral dimension.

B: Breadth = width (=length of STENZEL), the maximum anterior--posterior dimension in the 'plane' of the commissure.

A: Attachment area, the maximum linear dimension of the attachment area.

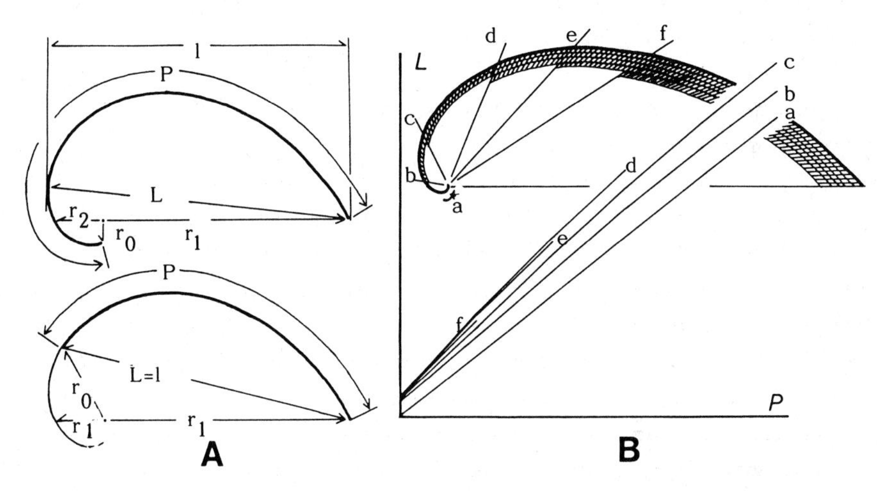

Fig. 9: A: Some measurements taken on *Gryphaea* and the theoretical parameters of the logarithmic spiral (see text for explanation). The two graphs illustrate the changes caused by the size of the attachment area.
B: Relationship between peripheral length and maximal length of the shell on an ideal logarithmic spiral and their variation due to different sizes of the attachment area (= r_0: a to f).

B) Theoretical morphology of the shell

A proper set of measurements should relate the real object to a theoretical model, because only in this way the variability of the real objects and their deviations from the model can be readily interpreted. In addition, the measured dimensions should be chosen in such a way that they are in simple mathematical relationship allowing the formulation of proper statistical hypotheses and thus permitting valid statistical testing against the expected relationship. In the present case, the most simple model -- to be used until proven inappropriate -- is the logarithmic spiral. Fig. 9A summarizes the actually measured dimensions and some theoretically useful dimensions of the logarithmic spiral:

r: radii of the spiral --

r_0: initial radius of the spiral, i. e. $r=r_0 \exp(k\phi)$;
r_1: radius at the commissure;
r_2: radius 180° opposite to r_1.

d: diameter of the shell, i.e. $d = r_1 + r_2$

l: theoretical length of the shell measured parallel to the diameter of the shell.

A logarithmic spiral

$$r = r_o e^{k\phi} \tag{1}$$

can be totally described by the parameters r_o and k (actually k is sufficient) but neither of these is measurable on a *Gryphaea* shell. Thus P or the quotient P/L is introduced as a global measurement of curvature which replaces the local (mathematical) curvature

$$\rho = \frac{\sqrt{(1 + k^2)}}{k} r \tag{2}$$

The length of the shell is given by

$$P = \frac{\sqrt{(1 + k^2)}}{k} (e^{k\phi} - r_o) \tag{3}$$

and the diameter 'd' of the shell is, according to Fig. 9

$$d = r_1 + r_2 = r_o(e^{k\phi} + e^{k(\phi - \pi)}) = r_o(1 + e^{-k\pi})\, e^{k\phi} \tag{4}$$

Eliminating $\exp(k\phi)$ from equations (3) and (4) yields a relation between P and 'd'

$$\frac{d}{r_o(1 + e^{-k\pi})} = \frac{k}{\sqrt{(1 + k^2)}} P + r_0 \tag{5}$$

i.e. a simple linear relationship

$$L \simeq d = r_o(1 + e^{-k\pi}) \left[\frac{k}{(1 + k^2)} P + r_o \right] \tag{6}$$

if our measured dimension L approximates the diameter of the spiral. However, the quality of the approximation depends on both parameters of the spiral, r_o and k. The effect of r_o is illustrated in Fig. 11. L approximates 'd' if k is small (near zero, say an ammonite) while for large k values, as in *Gryphaea*, there is a notable difference between d and l (the theoretical length) and the relationship between d and L (the measured length) becomes even a mathematically terrifying problem because by definition L is the maximum length measured from the commissure, i.e. every theoretical determination of that length requires the solution of an optimality problem. Thus by using equation (6) one cannot expect a simple relationship between L and P for *Gryphaea*, neither a linear nor an allometric one. Additional disturbance results from the initial radius determined by the attachment area of the shell (Fig. 9A) which changes the value of L considerably. Fig. 9B displays the relationships between L and P measured on a theoretical spiral for different r_0-values (attachment areas). The resulting curves are, as expected, non-linear and non-allometric and the change of r_0 clearly causes displacements of the curve similar to the ones found through the statistical analysis (e.g. Fig. 7G). Any data, using these measurements, therefore, have to be carefully checked for changes in the attachment area. In addition, these dimensions are not useful to test for logarithmic growth in *Gryphaea* (HALLAM & GOULD, 1975), and linear statistics must be used with extreme care because of the nonlinearities in the measurements. For the width measurement (and all connections between measurements) there holds a similar argument as can be analysed if an exponential growth model is introduced for width.

C) Index Numbers

The relative measurement of curvature -- the quotient B/L -- bears another problem like any index of this type. Using just a linear relationship between two parameters A,B, i.e.

$$A = a + bB \tag{7}$$

the quotient

$$A/B = (b + a/B) \tag{8}$$

turns out to be a hyperbola which approaches asymptotically a stable value -- the slope b of the straight line. The index, therefore, depends strongly on the magnitude of the measured parameters, and ontogenetic trends displayed by this index tend to be parallel.

REFERENCES

Arkell, W.J. 1947: The geology of the country around Weymouth, Swanage, Corfe and Lulworth. Mem. Geol. Surv. U.K.

Bayer, U. 1972: Zur Ontogenie und Variabilität des jurassischen Ammoniten Leioceras opalinum. N. Jb. Geol. Pal. Abh., 135: 19-41.

Bayer, U. 1978: Morphogenetic Programs, Instabilities, and Evolution - a theoretical study. N. Jb. Geol. Pal. Abh., 156: 226-261.

Bayer, U. & McGhee, G.R. 1984: Iterative evolution of middle Jurassic ammonite faunas. Lethaia, 17.

Brannan, J. 1983: Taxonomy and evolution of Triassic and Jurassic Non-lophate oysters. PhD-Thesis, Univ. of Oxford.

Brasier, M.D. & Brasier, C.J. 1978: Littoral and fluviatile facies in the 'Kellaways Beds' on the Market Weighton Swell. Proc. Yorks. Geol. Soc., 42, 1-20.

Burnaby, T.P. 1965: Reversed coiling trend in Gryphaea armata. Geol. J., 4, 257-278.

Calloman, J.H. 1968: The Kellaways Beds and the Oxford Clay. In: Sylvester-Bradley & Ford, ed. The Geology of the East Midlands. Leicester, Univ. Press.

Charles, R.P. & Maubeuge, P.L. 1951: Les Liogryphées du Jurassique inférieur de l'est du Bassin Parisien.- Bull. Soc. Géol., 60, 333-350.

Dacqué, E. 1921: Vergleichende biologische Formenkunde der fossilen niederen Tiere.

Deecke, W. 1916: Paläobiologische Studien. Sitzungsber. Heidelberger Akad. Wiss.

Dépéret, Ch. 1909: Die Umbildung der Tierwelt.

Duff, K.L. 1978: Bivalvia from the English Lower Oxford Clay (middle Jurassic). Palaeontogr. Soc. 137 pp.

Gould, S.J. 1972: Allometric fallacies and the evolution of Gryphaea: a new interpretation based on White's criterion of geometric similarity. In: Dobzhansky, T., Hecht, M.K., Steers, W.C. eds.: Evolutionary Biology, 6, 91-118.

Hallam, A. 1959a: On the supposed evolution of Gryphaea in the Lias. Geol. Mag., 96, 99-108.

Hallam, A. 1959b: The supposed evolution of Gryphaea. Geol. Mag. 96, 419-420.

Hallam, A. 1962: The evolution of Gryphaea. Geol. Mag. 99, 561-574.

Hallam, A. 1960: On Gryphaea. Geol. Mag. 97, 518-522.

Hallam, A. 1968: Morphology, palaeoecology and evolution of the genus Gryphaea in the British Lias. Phil. Transact. Roy. Soc. London, 254B, 91-128.

Hallam, A. 1978: How rare is phyletic gradualism and what is its evolutionary significance? Evidence from Jurassic bivalves. Paleobiology, 4, 16-25.

Hallam, A. 1982: Patterns of speciation in Jurassic Gryphaea. Paleobiology, 8, 354-366.

Hallam, A. 1983: Early and mid-Jurassic molluscan biogeography and the establishment of the central Atlantic seaway.Paleo[3], 43: 181-193.

Hallam, A. & Gould, S.J. 1975: The evolution of British and American Middle and Upper Jurassic Gryphaea: a biometric study.- Proc. R. Soc. Lond., B 189, 511-542.

Hudson, J.D. & Palframan, D.F.B. 1969: The ecology and preservation of the Oxford Clay fauna at Woodham, Buckinghamshire. Quart. J. Geol. Soc. London, 124, 387-418.

Johnson, A.L.A. 1981: Detection of ecophenotypic variation in fossils and its application to a Jurassic scallop. - Lethaia, 14, 277-285.

Johnson, A.L.A. 1984: The palaeobiology of the bivalve families Pectinidae and Propeamussiidae in the Jurassic of Europe.- Zitteliana, 11, 235 pp.

Joysey, K.A. 1959: The evolution of the Liassic oysters Ostrea - Gryphaea. Biol. Rev. Lambr. Phil. Soc., 34, 297-332.

Joysey, K.A. 1960: On Gryphaea.- Geol. Mag., 97, 522-524.

Philip, G.M. 1962: The evolution of Gryphaea.- Geol. Mag., 99, 327-344.

Philip, G.M. 1967: Additional observations of the evolution of Gryphaea.- Geol. Jb., 5, 329-338.

Pugaczewska,H. 1971: Jurassic Ostreidae of Poland.- Acta Pal. Polonica,16, 195-311.

Pugaczewska, H. 1971a: Aalenian Gryphaeinae from the Pieneny Klippen belt of Poland.- Acta Pal. Polonica, 16, 389-399.

Quenstedt, 1847: Der Jura.

Raup, D.M. 1966: Geometric analysis of shell coiling: General problems.- Journ. Paleontology 40: 1178-1190.

Raup, D.M. & Michelson, A. 1965: Theoretical morphology of the coiled shell.- Science, 147: 1294-1295.

Schäfle, L. 1929: Über Lias-und Doggeraustern.- Geol. Palaeont. Abh., N.F. 17, 2, 65-149.

Stenzel, H.B. 1971: Oysters. In: Moore, R.C. (ed.) Treatise on Invertebrate Palaeontology, N, Bivalvia 3, N953-1224, Univ. of Kansas Press.

Swinnerton, H. 1959: Concerning Mr. A. Hallam's article on Gryphaea. Geol. Mag., 96, 307-310.

Swinnerton, H. 1964: The early development of Gryphaea.- Geol. Mag. 101, 409-420.

Trueman, A.E. 1922: The use of Gryphaea in the correlation of the lower Lias.- Geol. Mag. 59, 256-268.

Vail, P.R., Hardenbol, J., Todd, R.G. 1984: Jurassic unconformities, chronostratigraphy and sea-level changes from seismic stratigraphy and biostratigraphy. GCPSSEPM Foundation Third Annual Res. Conference Proc.

Westermann, G.E.G. 1966: Covariation and taxonomy of the Jurassic ammonite Sonninia adicra (Waagen). N. Jb. Geol. Pal. Abh., 124: 289-312.

Addresses

T. Aigner, Institut für Geologie und Paläontologie, Universität Tübingen, Sigwartstr. 10, D-7400 Tübingen

E. Altheimer, Institut für Geologie und Paläontologie, Universität Tübingen, Sigwartstr. 10, D-7400 Tübingen

U. Bayer, Institut für Geologie und Paläontologie, Universität Tübingen, Sigwartstr. 10, D-7400 Tübingen

K. Brandt, Institut für Geologie und Paläontologie, Universität Tübingen, Sigwartstr. 10, D-7400 Tübingen

J. Brannan, BP Petroleum Development Limited, Farburn Industrial Estate, Dyce, Aberdeen AB2 OPB

W. Deutschle, Institut für Geologie und Paläontologie, Universität Tübingen, Sigwartstr. 10, D-7400 Tübingen

D.T. Donovan, Department of Geology, University College London, Gower Street London WCIE 6BT

G. Einsele, Institut für Geologie und Paläontologie, Universität Tübingen, Sigwartstr. 10, D-7400 Tübingen

A. Gorthner, Tropenmedizinisches Institut, Universität Tübingen, D-7400 Tübingen

H. Hagdorn, Konsul-Übele-Str. 14, D-7118 Künzelsau

A. Hallam, Department of Geological Sciences, The University of Birmingham, P.O. Box 363, Birmingham, B15 2TT

A.L.A. Johnson, Department of Geology, University of Leicester, University Road, Leicester LE1 7RH

M. Kabamba, Institut des Sciences de la Terre, L'Université de Dijon, 6, Boulevard Gabriel, F-21100 Dijon

S.M. Kidwell, Department of Geosciences, University of Arizona, Tucson, Arizona 85721

J. Kullmann, Institut für Geologie und Paläontologie, Universität Tübingen, Sigwartstr. 10, D-7400 Tübingen

B.A. Matyja, Universytet Warszawski, Wydziat Geologii, al. Zwirki i Wiguri 93, Warszawa

G.R. McGhee, Department of Geological Sciences, Wright Geological Laboratory, Rutgers University, New Brunswick, New Jersey 08903

C. Meier-Brook, Tropenmedizinisches Institut, Universität Tübingen, D-7400 Tübingen

R. Mundlos, Schachtstr. 6, D-7107 Bad Friedrichshall 1

W.-E. Reif, Institut für Geologie und Paläontologie, Universität Tübingen, Sigwartstr. 10, D-7400 Tübingen

A.C. Riccardi, Museo di Ciencias Naturales, La Plata, Argentina

W. Ricken, Institut für Geologie und Paläontologie, Universität Tübingen, Sigwartstr. 10, D-7400 Tübingen

A. Seilacher, Institut für Geologie und Paläontologie, Universität Tübingen, Sigwartstr. 10, D-7400 Tübingen

H. Tintant, Institut des Sciences de la Terre, L'Université de Dijon, 6, Boulevard Gabriel, F-21100 Dijon

M. Urlichs, Staatliches Museum für Naturkunde Stuttgart, Arsenalplatz 3, D-7140 Ludwigsburg

P.D. Ward, Department of Geology, University of California, Davis, California 95616

G.E.G. Westermann, Department of Geology, McMaster University, 1280 Main Street West, Hamilton, Ontario, Canada L8S 4M1

A. Wierzbowski, Universytet Warszawski, Wydziat Geologii, al. Zwirki i Wiguri 93, Warszawa

R. Willmann, Geologisch-Paläontologisches Institut der Universität Kiel, Olshausenstr. 40, D-2300 Kiel

K. O. Emery, E. Uchupi

Geology of the Atlantic Ocean

1984. 399 figures, 23 oversize charts (folded)
(In 2 volumes. Not available separately.)
XX, 1050 pages. ISBN 3-540-96032-5

Contents: Exploration. - Physiography. - Internal Igneous Structure. - The Syn-Rift Supersequence and Crustal Boundary. - Drift Supersequence. - Sediment Provenance and Properties. - Evolution of Ocean Floor. - Interfaces Between Ocean and Man. - Bibliography. - Index.

Chart set
(Available separately in mailing tube)
1984. 23 oversize charts. ISBN 3-540-96033-3

The Geology of the Atlantic Ocean is a unique account of the geologic development of a young oceanic basin and its adjacent marginal seas. Combining widely dispersed data and concepts with new investigations made especially for this book, it presents the reader with a detailed thoroughly integrated study of the topography, igneous composition, structure, and sediments of the Atlantic. The shaping and composition of both ocean floor and land areas are traced from before its beginning about 180 million years ago to the present. Effects of plate movements and changing climates upon water movements and contents of organisms and sediments are included. In addition, the potential economic value of ocean resources are carefully evaluated. A detailed bibliography complements the text.
This richly illustrated elegantly designed volume is accompanied by eleven pairs of charts summarizing data on bathymetry, physiography, seismic reflection and refraction, sediments, and tectonics. Marine geologists, geophysicists, economic geologists, and oceanographers will find this book an invaluable source of information and concepts upon which new theoretical and practical investigations will be based.

Springer-Verlag
Berlin
Heidelberg
New York
Tokyo

A. D. Miall

Principles of Sedimentary Basin Analysis

1984. 387 figures. XII, 490 pages
ISBN 3-540-90941-9

Contents: Introduction. – Collecting the data. – Stratigraphic correlation. – Facies analysis. – Basin mapping methods. – Depositional systems. – Burial history. – Regional and global stratigraphic cycles. – Sedimentation and plate tectonics. – Conclusions. – Subject Index. – Authors Index.

Principles of Sedimentary Basin Analysis provides geologists with a practical guide for the study of the geologic evolution of ancient sedimentary basins, using modern methods of facies and depositional systems analysis, seismic stratigraphy and a broad range of basin mapping techniques. A new approach to stratigraphy is demonstrated that explains the genesis of lithostratigraphic units. This book also contains a detailed treatment of the plate tectonic control of basin architecture and has a unique discussion of the widespread effects of eustatic sea level changes.

The emphasis throughout is on what geologists can actually see in outcrops, well records and cores, and what can be obtained using geophysical techniques. Profusely illustrated and containing numerous examples from around the world, this up-to-date textbook will be welcomed by the student and professional geologist alike.

Springer-Verlag
Berlin
Heidelberg
New York
Tokyo